Stat2: A Second Course
Preliminary Edition

Ann R. Cannon
Cornell College

George W. Cobb
Mount Holyoke College

Bradley A. Hartlaub
Kenyon College

Julie M. Legler
St. Olaf College

Robin H. Lock
St. Lawrence University

Thomas L. Moore
Grinnell College

Allan J. Rossman
California Polytechnic State University

Jeffrey A. Witmer
Oberlin College

ISBN-13: 978-1-4292-8989-4
ISBN-10: 1-4292-8989-9

© 2011 by W.H. Freeman & Company
All rights reserved.

Printed in 2011 in the United States of America

Custom Publishing Division
W.H. Freeman and Company
41 Madison Avenue
New York, NY 10010

www.whfreeman.com/custompub

Contents

Preface		**xvii**
0	**What Is a Statistical Model?**	**1**
0.1	Fundamental Terminology	3
0.2	Four-Step Process	7
0.3	Chapter Summary	12
0.4	Exercises	13
1	**Simple Linear Regression**	**23**
1.1	The Simple Linear Regression Model	23
	Choosing a Simple Linear Model	25
	Fitting a Simple Linear Model	26
1.2	Conditions for a Simple Linear Model	28
	Estimating the standard deviation of the error term	30
1.3	Assessing Conditions	31
	Residuals versus Fits Plots	31
	Normal plots	33
1.4	Transformations	37
1.5	Outliers & Influential Points	44
	Outliers	44
	Influential Points	47
1.6	Chapter Summary	51
1.7	Exercises	53
2	**Inference for Simple Linear Regression**	**57**
2.1	Inference for Regression Slope	57
	T-test for slope	58
	Confidence Interval for Slope	60
2.2	Partitioning Variability - ANOVA	60
	ANOVA Test for Simple Linear Regression	62
2.3	Regression and Correlation	63
	Coefficient of Determination, r^2	63

		Inference for correlation	64
	2.4	Intervals for Predictions	66
	2.5	Chapter Summary	69
	2.6	Exercises	71

3 Multiple Regression 79

	3.1	Multiple Linear Regression Model	81
		Choosing a Multiple Linear Regression Model	81
		Fitting a Multiple Linear Regression Model	82
	3.2	Assessing a Multiple Regression Model	84
		T-tests for Coefficients	84
		Confidence Intervals for Coefficients	85
		ANOVA for Multiple Regression	85
		Coefficient of Multiple Determination	87
		Confidence and Prediction Intervals	89
	3.3	Comparing Two Regression Lines	90
		Indicator Variables	92
		Comparing Intercepts	94
		Comparing Slopes	97
	3.4	New Predictors from Old	101
		Interaction	101
		Polynomial Regression	104
		Complete Second Order Model	109
	3.5	Correlated Predictors	111
		Detecting Multicollinearity	116
	3.6	Testing Subsets of Predictors	118
		Nested F-test	119
	3.7	Case Study: Predicting in Retail Clothing	122
	3.8	Chapter Summary	131
	3.9	Exercises	133

4 Additional Topics in Regression 145

	4.1	Topic: Added Variable Plots	145
	4.2	Topic: Techniques for Choosing Predictors	149
		Best subsets	151
		Backward elimination	154
		Forward selection and stepwise regression	157
		Caution about automated techniques	159
	4.3	Topic: Identifying Unusual Points in Regression	160
		Leverage	160
		Standardized and Studentized Residuals	164
		Cook's Distance	166
	4.4	Topic: Coding Categorical Predictors	169

CONTENTS

4.5	Topic: Randomization Test for a Relationship	178
4.6	Topic: Bootstrap for Regression	182
	Confidence Intervals Based on Bootstrap Distributions	185
4.7	Exercises	188

5 Analysis of Variance — 197

5.1	The One-way Model: Comparing Groups	199
	One Mean or Several?	199
	Observation = Group Mean + Error	200
	Group Mean = Grand Mean + Group Effect	201
	Conditions on the Error Terms	201
	Estimating the ANOVA model terms	202
	How Do We Use Residuals?	204
	Variance of the Error Terms = Mean Square Error	205
	The triple decomposition	205
5.2	Assessing and Using the Model	208
	Using Variation to Determine if Means Are Different?	208
	Normal Error Distribution?	210
	Equal Group Variances?	211
	Independent Errors with Mean Zero?	212
	Transformations	214
5.3	Scope of Inference	218
	Why Randomization Matters	218
	Inference about Cause	219
	Inference about Populations	219
	What You See Is All There Is	223
5.4	Fisher's Least Significant Difference	226
	Comparisons	227
	Error Rates	227
	Fisher's LSD	229
5.5	Chapter Summary	233
5.6	Exercises	236

6 Multifactor ANOVA — 249

6.1	The Two-way Additive Model (Main Effects Model)	249
	The structure of two-way ANOVA data	252
	Inference after Two-way ANOVA	261
6.2	Interaction in the Two-way Model	263
6.3	The Two-way Non-additive Model (Two-Way ANOVA with Interaction)	270
	Two-way ANOVA Calculations for Balanced Complete Factorial Data	276
6.4	Case Study	279
6.5	Chapter Summary	287
6.6	Exercises	289

7 Additional ANOVA Topics — 301
- 7.1 Topic: Levene's Test for Homogeneity of Variances — 301
- 7.2 Topic: Multiple Tests — 306
  - Why Worry About Multiple Tests? — 306
  - Family-wise Error Rate — 306
  - Fisher's LSD: A Liberal Approach — 307
  - Bonferroni Adjustment: A Conservative Approach — 307
  - Tukey's "Honest Significant Difference": A Moderate Approach — 308
  - A comparison of all three methods — 310
- 7.3 Topic: Comparisons and Contrasts — 314
  - Comparing Two Means — 314
  - The Idea of a Contrast — 315
  - Linear Combinations of Means — 316
  - The Standard Error for a Contrast — 317
  - The t-test for a Single Contrast — 318
- 7.4 Topic: ANOVA and Regression with Indicators — 324
  - Two-sample Comparison of Means as Regression — 324
  - One-way ANOVA for Means as Regression — 327
  - Two-way ANOVA for Means as Regression — 330
- 7.5 Topic: Analysis of Covariance — 338
  - Setting the Stage — 338
  - What Is the ANCOVA Model? — 340
- 7.6 Exercises — 351

8 Overview of Experimental Design — 361
- Introduction — 361
- 8.1 Comparisons and Randomization — 362
  - The need for comparisons — 362
  - The need for randomization — 363
  - Reasons to randomize — 364
  - Randomized comparative experiments; treatments and units — 368
- 8.2 Randomization F Test — 372
  - Mechanics of the Randomization F-Test — 372
  - The logic of the randomization F-test — 377
  - The randomization F-test and the ordinary F-test compared — 380
- 8.3 Design Strategy: Blocking — 383
- 8.4 Design Strategy: Factorial Crossing — 389
- 8.5 Chapter Summary — 399
- 8.6 Exercises — 401

9 Logistic Regression — 413
- 9.1 The Logistic Regression Model — 414
- 9.2 Fitting a Logistic Regression Model — 416

CONTENTS   vii

	How parameter values affect shape	421
9.3	Logistic Regression and Odds	421
	Odds and odds ratios	422
	Odds Ratios in Logistic Regression	425
9.4	Assessing the Logistic Model	427
	Conditions for Logistic Regression	427
	Assessing Model Utility: Does the predictor help explain the response?	428
9.5	Analyzing 2×2 Tables with Logistic Regression	434
	Two-sample z-test for proportions (review)	436
	Chi-square test for a 2×2 table (review)	437
9.6	Summary	440
9.7	Exercises	441

10 Multiple Logistic Regression   453
10.1	Multiple Logistic Regression Model	453
	A Nested LRT-test	462
10.2	Case Study: Bird Nests	467
10.3	Summary	471
10.4	Exercises	472

11 Logistic Regression: Additional Topics   483
11.1	Assessing Logistic Regression Models	483
	Goodness of fit: Assessing the fit of the model	483
11.2	Residual Diagnostics: Assessing the Conditions	493
	Assessing Prediction	494
	Issues	496
11.3	Randomization Tests	498
11.4	Applying Logistic Regression to Larger Tables	502
	Chi-square test for a $2 \times k$ table (review)	505
11.5	Summary	509
11.6	Exercises	510

Index   515
General Index 515
Dataset Index 518

List of Figures

0.1	Health facilities in U.S. metropolitan areas	4
0.2	Weight Loss for Control versus Incentive groups	8
0.3	Residuals from Group Weight Loss Means	10
0.4	Normality Probability Plots for Residuals of Weight Loss	10
1.1	Scatterplot of Porsche price versus mileage	26
1.2	Linear regression to predict Porsche price based on mileage	27
1.3	Minitab output for regression of Porsche *Price* on *Mileage*	28
1.4	Residuals versus fitted values plot when linearity and constant variance conditions hold	31
1.5	Residuals versus fitted values plots illustrating problems with conditions	32
1.6	Examples of normal quantile plots	33
1.7	Examples of normal probability plots	34
1.8	Plot of Porsche residuals versus fitted values	35
1.9	Histogram of Porsche residuals	35
1.10	Normal Probability Plot of residuals for Porsche data	36
1.11	Scatterplot for Doctors versus Hospitals	38
1.12	Regression for number of doctors based on number of hospitals	39
1.13	Normality Plots for Residuals of Doctors versus Hospitals	39
1.14	Least Squares Line for Sqrt(Doctors) versus Hospitals	40
1.15	Plots of Residuals for Sqrt(Doctors) versus Hospitals	41
1.16	Predicted *NumMDs* from the linear model for sqrt(*NumMDs*)	41
1.17	Number of Mammal Species versus Area for S.E. Asian Islands	42
1.18	Log transformations of *Species* versus *Area* for S.E. Asian Islands	43
1.19	Residual plot after log transform of response and predictor	43
1.20	Gold medal winning distances (m) for the men's Olympic long jump, 1900 - 2008	45
1.21	Residual plot for long jump model	45
1.22	Studentized residuals for the long jump model	46
1.23	2000 Presidential election totals in Florida counties	48
1.24	Residual plot for the butterfly ballot data	48
1.25	Regression lines with and without Palm Beach	49
1.26	Regression lines with an outlier of 3407 "moved" to different counties	50

LIST OF FIGURES

2.1 Hypothetical population with no relationship between Price and Mileage 58
2.2 Confidence intervals and prediction intervals for Porsche prices 68
2.3 Normal Probability Plot for Random Numbers 77

3.1 Linear Regressions to Predict NFL Winning Percentage 81
3.2 Minitab Output for a Multiple Regression 83
3.3 Normal probability plot of residuals for NFL model 87
3.4 Percent of potential jurors in 1998 and 2000 who report each period 92
3.5 Percent reporting jurors with regression lines for 1998 and 2000 93
3.6 Residual plots for predicting Juror Percents using *I2000* and *Period* 96
3.7 Weight versus age by sex for kids 97
3.8 Separate regressions of weight versus age for boys and girls 98
3.9 Compare regression lines by sex 98
3.10 Residual plots for the multiple regression model for weights based on age and sex . . 100
3.11 Individual predictors of perch weights 103
3.12 Residual plots for two models of perch weights 104
3.13 *TotalPrice* versus *Carat* for diamonds 105
3.14 Quadratic regression for diamond prices 107
3.15 Residual plots for simple linear and quadratic models of diamond prices 108
3.16 Residuals from the quadratic model of diamond prices 108
3.17 Length versus Width of perch 112
3.18 Scatterplots of house selling price versus two predictors 114
3.19 Regression of *Amount* on *Dollar12* 125
3.20 Regression of *Amount* on *AvgSpent12* 127
3.21 Residuals versus fits for the regression of *Amount* on *AvgSpent12* 128
3.22 Residuals plots for quadratic model to predict *Amount* based on *AvgSpent12* . . . 129
3.23 Quadratic regression fit of *Amount* on *AvgSpent12* 130
3.24 *Obama − McCain* margin in 2008 Presidential Polls 142
3.25 CO2 levels by day, April-November 2001 144

4.1 Regression of *Price* on *Lot* 146
4.2 Regression of *Size* on *Lot* 146
4.3 Added variable plot for adding *Size* to *Lot* when predicting *Price* 147
4.4 Scatterplot matrix for first year GPA data 150
4.5 *GPA* versus categorical predictors 150
4.6 Unusual observations identified in the butterfly ballot model 162
4.7 Boxplot of the number of Bush votes in Florida counties 162
4.8 High leverage cases in perch weight interaction model 164
4.9 Unusual observations in the regression for perch weights 167
4.10 *Price* versus *Mileage* for three car types 171
4.11 *Price* versus *Mileage* with equal slope fits 173
4.12 *Price* versus *Mileage* with different linear fits 175
4.13 Residual plots for *Price* model based on *Mileage*, *Porsche* and *Jaguar* 176

LIST OF FIGURES

4.14	Linear regression for GPA based on Verbal SAT	179
4.15	Randomization distribution for 1000 correlations of GPA versus Verbal SAT	180
4.16	Original regression of Porsche prices on mileage	183
4.17	Porsche regressions for bootstrap samples	184
4.18	n=5000 Bootstrap Porsche Price Slopes	184
4.19	Hypothetical Skewed Bootstrap Distribution	186
5.1	Dotplots of life spans for fruit flies	200
5.2	Histograms of error terms for two models of lifespans for fruitflies	205
5.3	Normal Probability Plot of residuals for fruit flies	211
5.4	Residuals versus fits for fruit flies	212
5.5	Plot of residuals versus fits	216
5.6	Residuals from the diamond data	216
5.7	Plots to assess the fit of the ANOVA model using log(Carats) as the response variable	217
5.8	Residual plot and normal probability plot to assess conditions	221
5.9	Residual plot and normal probability plot to assess conditions on transformed data	222
5.10	Residual plot for ANOVA	225
5.11	Normal probability plot of residuals	226
5.12	Dotplot of tail lengths by species	226
5.13	Child poverty rates in Iowa by size of county	242
5.14	Palatability scores by level of concentration of liquid component	243
5.15	Normal probability plot and Residual plot for Salary data	247
6.1	Plots to assess the fit of the ANOVA model	257
6.2	Map of the rivers in upstate New York	259
6.3	The horizontal axis shows *Site*, from source (Upstream = left) to mouth (Downstream = right). The vertical axis shows iron concentration. There are four plotting symbols, one for each river. Each set of line segments shows a downward pattern from left to right: iron concentration decreases as you go from source to mouth.	260
6.4	Plots to assess the fit of the ANOVA model	261
6.5	An interaction is a difference of differences. The vertical change of each segment equals the difference (Yes - No) due to *Vitamin B12*. Interaction is present: The two differences are different, and the segments are not parallel.	265
6.6	The other interaction plot for the PigFeed data. For this plot, the vertical change of each segment equals the difference (Yes - No) due to *Antibiotics*, with one segment for each level of the other factor, *Vitamin B12*.	266
6.7	Interaction graphs for values from the additive (no interaction) model. When there is no interaction, the difference due to *Vitamin B12* (left panel) is the same for both levels of *Antibiotics*, and the change in y is the same for both line segments, and the segments are parallel. Similarly, the difference due to *Antibiotics* (right panel) is the same for both levels of *Vitamin B12*, and the segments are parallel.	267
6.8	River iron interaction plot for hypothetical data	267
6.9	Hypothetical river iron interaction plots	269

LIST OF FIGURES

6.10 Plots to assess the fit of the ANOVA model 278
6.11 Plot of Ethanol concentration versus oxygen concentration. Solid circles: Galactose. Open circles: Glucose. 280
6.12 Interaction plot for the Sugar Metabolism data. The interaction plot in the left panel is better than the one to it's right, because that first plot makes the patterns easier to see. 281
6.13 Residual plots for the Sugar Metabolism data 281
6.14 Interaction plots in four different scales 283
6.15 Plots to assess the fit of the ANOVA model using the third root of the ethanol measurement as the response variable 286
6.16 Graph of log(standard deviations) versus log(averages) for the eight groups of the sugar metabolism study. 298
6.17 Graph of log(standard deviations) versus log(averages) for six groups of the sugar metabolism study. 298
6.18 Graph of the residuals versus the comparison values for the sugar metabolism study. 300

7.1 Residuals versus fitted values for fruit flies 303
7.2 Residuals versus fitted values for cancer survival times 304
7.3 Residual plot for walking babies ANOVA model 321
7.4 Normal probability plot of residuals 321
7.5 Life spans for 8 virgins and living alone groups 325
7.6 Life spans for fruit flies living alone and with 8 virgins 326
7.7 Plots to assess the fit of the ANOVA model 340
7.8 Plots to assess the fit of the ANCOVA model 341
7.9 Scatterplot of sales versus price 342
7.10 Plots to assess the fit of the linear regression model 342
7.11 Plots to assess the fit of the ANOVA model 345
7.12 Scatterplot of active pulse rate versus resting pulse rate 347
7.13 Plots to assess the fit of the linear regression model 347
7.14 Scatterplot of active pulse rate versus resting pulse rate for each exercise level 348
7.15 Plots to assess the fit of the ANOVA model 349
7.16 Scatterplot of active pulse rate versus resting pulse rate 350
7.17 Plots to assess the fit of the linear regression model 350
7.18 Scatterplot of active pulse rate versus resting pulse rate for each exercise level 350

8.1 Mean weight of rats in a randomly chosen control group, based on 10,000 randomizations 367
8.2 Distribution of F-statistics for 10,000 re-randomizations of the Calcium data. . . . 375
8.3 The p-value equals the area of the histogram to the right of the dashed line. 375
8.4 Distribution of F-statistics for 10,000 re-randomizations of the Milgram data. . . . 376
8.5 The p-value equals the area of the histogram to the right of the dashed line. 377
8.6 Distribution of F_{Rand} based on 10,000 re-randomizations 379
8.7 4499 of 10,000 values of F_{Rand} were at least as large as $F_{obs} = 0.731$. 380

LIST OF FIGURES   xiii

8.8   F-distributions from randomization model (histogram) and normal errors model (solid curve)for (a) Calcium study (b) Milgram study (c) Hypothetical Calcium study 383
8.9   *The four-by-four array used to tempt the bees.* Each square represents a cotton ball. Those labeled S (stung) had been previously stung; those labeled F (fresh) were pristine. 385
8.10 Two analyses for the Finger Tapping data. The analysis on the left, for the block design, has three sources of variation. The analysis on the right, for the completely randomized design, has only two sources. Notice that for both analyses, the bottom rows are the same: There are 11 df total, and the total SS is 6682. Moreover, for both analyses, there are 2 df for Drugs, and the SS for Drugs is the same, at 872. Finally, although the analysis on the right has no row for Subjects, the df and SS on the right for Error come from adding the df and SS for Subjects and Error on the left. 389
8.11 The hill. The diagonal ridge corresponds to interaction, which causes Jack's one-at-a-time method to fail. 394
8.12 Jack's one-at-a-time plan for climbing the hill. The vertical line shows Jack's one-way design. The horizontal line shows Jack's one-way follow-up design for Jill. This leads to an estimate of the top of the hill being at (250,200). The top of the hill is at (300, 300). 394
8.13 395
8.14 Percentage of words recalled 397
8.15 398
8.16 Randomization distribution of F-statistics for full Challenger data 410

9.1   Acceptance Status by GPA: jittered plot with ordinary regression line 414
9.2   Logistic regression fit for probability of medical school acceptance 418
9.3   Logistic regression for putts made: Fitted model and empirical logits versus putt length. 419
9.4   Empirical logits and proportions for medical school acceptance by GPA 420
9.5   Effects of changing β_0 or β_1 in a logistic regression model 421

10.1 Contributions by Party 455
10.2 Left-side graphs: logit scale; Right-side graphs: probability scale—Top row: Only intercepts differ; Bottom row: Both intercept and slope differ 456
10.3 Bird length and total care time versus Closed or Open nests 469

11.1 Contributions to X^2 using Delta Chi-square criterion 490
11.2 Proportion of tips within each card group 503

List of Tables

0.1	Classifying general types of models	5
0.2	Weight Loss After Four Months (pounds)	8
0.3	Weight Loss After Seven Months (pounds)	18
0.4	Winning percentage and Pythagorean predictions for baseball teams in 2009	21
1.1	Price and Mileage for Used Porsches	24
1.2	Number of MDs and Community Hospitals for sample of $n = 83$ Metropolitan Areas	37
1.3	Species and Area	42
1.4	Pages and price for textbooks	54
1.5	Major League Baseball game times	55
2.1	Gesell Aptitude and First Speak Age (in months)	73
2.2	LaDainaian Tomlinson Rushing for NFL Games in 2006	74
2.3	U.S. Infant Mortality	75
2.4	Horse Prices	76
3.1	Records and points for NFL teams in 2007 season	80
3.2	Percents of randomly selected potential jurors who appeared for jury duty	91
3.3	First few cases of fish measurements in the **Perch** datafile	102
3.4	Information on first 10 diamonds in the **Diamonds** datafile	105
3.5	First few cases of the **Clothing** data	122
3.6	Percent dry mass of eggs and age for female lake trout	135
3.7	Poll: Trade unions have too much power in Britain?	138
4.1	Unusual cases in multiple regression for perch weights	167
4.2	Several models for predicting car prices	176
4.3	Verbal SAT scores and GPA	178
4.4	Verbal SAT scores and with one set of randomized GPAs	180
5.1	Fruit fly lifetime means; overall and broken down by groups	199
5.2	Computing residuals for a subset of 5 fruit flies	204
5.3	ANOVA model applied to fruit flies data subset	206
5.4	Breakdown of Group Means in ANOVA model	206

List of Tables

6.1	Rate of finger tapping by four trained subjects	250
6.2	Rate of finger tapping by four trained subjects displayed in "case by variables" format	252
6.3	Rate of finger tapping with estimated effects	254
6.4	Partitioning observed values	254
6.5	Partitioning deviations and total variation	255
6.6	The observed river iron data transformed by the log base 10	260
6.7	Data for pig feed experiment. The left panel shows individual response values; the right panel shows diet averages.	264
6.8	Four-day alfalfa growth	294
6.9	Unpopped Popcorn by *Brand* and *Trial*	295
7.1	Fisher's LSD confidence intervals	311
7.2	Mean numbers of ants on sandwiches	335
8.1	Percentage of words identified for each of four lists	404
9.1	Putting success versus distance of the putt	418
9.2	Mother's marital status by child's sex.	439
10.1	Mean Contributions by Party (in $1,000s)	455
11.1	Putting success vs. distance of the putt	485
11.2	Goodness of Fit test: Saturated model is "full"; Linear logit model is "reduced."	485
11.3	MCAT data	488
11.4	Another putting data set to illustrate overdisperson	491
11.5	Rows are the actual admissions outcome: Deny or Accept; Columns are predicted by the model: Deny or Accept; based upon 0.5 cut-off	494
11.6	Archery improvement by sex.	498

Preface

Notes to the Instructor

Prerequisites
We assume that students using this book have successfully completed a basic introductory applied statistics course (e.g. AP Statistics or a typical Stat 101 college course). No further mathematical prerequisites are needed to continue on to a second course using this book. Some transformations and the material on logistic regression assume that students are able to work (at a calculator level) with exponential and logarithmic functions.

The "Overlap Principle"
Given the wide variability in topics covered in introductory statistics courses, we understand that students will come to a second course with different backgrounds and levels of experience. We also realize that having seen material in a first course doesn't insure that students have mastered or can readily use those ideas in a second course. To help students get up to speed and make a smooth transition from the first course to a second course we recommend repeating some material from the first course, possibly in a slightly different setting. For example, Chapter 0 uses the familiar two-sample t-test as a way to illustrate the approach of specifying, estimating, and testing a statistical model. Chapters 1 and 2 work through the specification, fit, assessment, and inference for simple linear models with a single quantitative predictor. Some topics in these chapters (for example, inference for the slope of a regression line) may be familiar to students from their introductory course, but often not in the more formal setting of a linear model. A thorough introduction of the formal linear model and other ideas in the "simple" setting makes it easier to move to situations with multiple predictors in Chapter 3. For a class with strong backgrounds, instructors may choose to move more quickly through the first chapters, treating that material mostly as review to get back up "to speed."

Organization of Chapters
Our primary goal with this text is to help students gain facility with common statistical models. After completing a second course they should be able to work with models where the response variable is either quantitative or categorical and predictors (or explanatory factors) are quantitative or categorical (or both). The chapters are grouped to consider models based on the type of response and type of predictors.

Chapter 0 Statistical Models - Basic terminology and our general approach to fitting a statistical model are illustrated in the context of a two-sample t-test. The four-step process that will be used throughout the text is explained and illustrated. In short, we want students to choose a model, fit a model, assess the fit of the model, and use the model to make appropriate inferences.

Unit A (Chapters 1-4) Linear regression - These four chapters focus on a quantitative response variable and quantitative or binary predictors.

Unit B (Chapters 5-8) ANOVA for means - These four chapters focus on a quantitative response variable with categorical explanatory factor(s) with some material on experimental design and ANCOVA.

Unit C (Chapters 9-11) Logistic regression - These three chapters focus on a binary response variable with either quantitative or categorical predictors.

The overall strategy within each unit is the same and is outlined below.

- The first chapter of a unit (or first two chapters for unit A) covers the "simple" case with a single predictor or explanatory factor. This helps students get familiar with the basic ideas for that type of model (linear regression, ANOVA for means, or logistic regression) in a relatively easy setting where graphical visualizations are most feasible.

- The next chapter (Chapter 3, 6, or 10) extends those ideas to consider models with multiple predictors/explanatory variables/factors.

- Each unit has a chapter of additional topics (Chapters 4, 7 and 11). The topics in these chapters are relatively independent of each other and extend ideas discussed in the earlier chapters. For example, Section 1.5 gives a brief, relatively informal introduction to issues with outliers and influential points in linear regression models. Topic 4.3 covers these ideas in more depth, introducing more formal methods to measure leverage, assess influence, and detect outliers. These chapters are not written sequentially so there is plenty of flexibility in choosing from the additional topics.

- Unit B also has material on experimental design (Chapter 8).

Flexibility within and between units

The chapters and units are arranged to promote flexibility in the order and depth in which topics are covered. Within a unit, some instructors may choose to "splice" in an additional topic when those ideas are first introduced. For example, Section 5.4 in the first ANOVA chapter introduces techniques for doing pairwise comparisons after doing a one-way ANOVA using Fisher's LSD. Instructors who prefer a more thorough discussion of pairwise comparison issues at this point, including alternate techniques such as the Bonferroni adjustment or Tukey's HSD, can include the material in Section 7.2. Others might want to move right away to two-way ANOVA in Chapter 6 and cover various pairwise procedures later.

Instructors can also adjust the order of topics between the units. For example, some might prefer to consider logistic regression models (Unit C) before doing ANOVA models (Unit B). Others might choose to do all three general model types in the "simple setting" (Chapters 1-2, 5 and 9), then go back and reconsider each model with multiple predictors.

Model conditions

After completing a second course students should be able to:

1. Choose the appropriate statistical model for a particular problem.

2. Know the conditions that are typically required when fitting various models.

3. Assess whether or not the conditions for a particular model are reasonably met for a specific dataset.

4. Have some strategies for dealing with data when the conditions for a standard model are not met.

5. Use the appropriate model to make appropriate inferences.

For each of the models covered in the text we describe standard conditions and discuss graphical and numerical methods for checking those conditions. Although it's not feasible in a second course to prepare students to deal with all possible contingencies when fitting models, we would like them to be able to recognize when a model has substantial faults.

In some cases conditions (such as normality of residuals) are required in order to employ classical inference techniques (such as a t-test for a regression coefficient). We offer two general approaches for dealing with data where such conditions are not met:

Transformations are considered for either the response or predictor variables. For example, using a log transformation to stabilize a variance or deal with a nonlinear relationship.

Computer intensive methods, such as bootstrapping and randomization tests, are presented. The "additional topics" chapter in each unit contains material on using resampling/bootstrap methods and randomization/permutation tests to do inference for each of the general model types when the conditions for more classical inference procedures do not hold.

Technology

We assume that students will use statistical software for fitting and assessing most of the models covered in this text. We include output from Minitab and R throughout the book, but we separate specific software commands from the main text. Our goal is to allow students to focus on the statistical concepts and interpretations while reading the text and avoid the distractions that often arise when discussing or implementing specific software commands. This decision allows instructors to use other statistical software packages (e.g., SAS, SPSS, DataDesk, JMP, etc.).

Exercises

Homework exercises appear at the end of each chapter (except for the "Additional Topic..." chapters which have exercises after each independent topic). They are grouped into one of four categories:

- **Conceptual exercises** - These questions are quick to answer and require minimal (if any) calculations. They let students practice and assess basic terminology and concepts introduced in the chapter.

- **Guided exercises** - These exercises ask students to perform various stages of an analysis process with prompts for the individual steps.

- **Open-ended exercises** - These problems tend to ask for more complete analyses, without much or any step-by-step direction.

- **Supplemental exercises** - Topics for these exercises go somewhat beyond the scope of the material covered in the chapter.

Chapter 0

What Is a Statistical Model?

The unifying theme of this book is the use of models in statistical data analysis. Statistical models are useful for answering all kinds of questions. For example:

- Can we use the number of miles that a used car has been driven to predict the price that is being asked for the car? How much less can we expect to pay for each additional 1000 miles that the car has been driven? Would it be better to base our price predictions on the age of the car in years, rather than its mileage? Is it helpful to consider both age and mileage, or do we learn about as much about price by considering only one of these? Would the impact of mileage on the predicted price be different for a Honda as opposed to a Porsche?

- Do babies begin to walk at an earlier age if they engage in a regimen of special exercises? Or does any kind of exercise suffice? Or does exercise have no connection to when a baby begins to walk?

- If we find a footprint and a handprint at the scene of a crime, are they helpful for predicting the height of the person who left them? How about for predicting whether the person is male or female?

- Can we distinguish among different species of hawks based solely on the lengths of their tails?

- Do students with a higher grade point average really have a better chance of being accepted to medical school? How much better? How well can we predict whether or not an applicant is accepted, based on his/her GPA? Is there a difference between male and female students' chances for admission? If so, does one sex retain its advantage even after GPA is accounted for?

- Can a handheld device that sends a magnetic pulse into the head reduce pain for migraine sufferers?

- When people serve ice cream to themselves, do they take more if they are using a bigger bowl? What if they are using a bigger spoon?

- Which is more strongly related to average score for professional golfers: driving distance, driving accuracy, putting performance, or iron play? Are all of these useful for predicting a golfer's average score? Which are most useful? How much of the variability in golfers' scores can be explained by knowing all of these other values?

These questions reveal several purposes of statistical modeling:

1. **Making predictions.** Examples include predicting the price of a car based on its age, mileage, and model; predicting the length of a hawk's tail based on its species; predicting the probability of acceptance to medical school based on grade point average.

2. **Understanding relationships.** For example, after taking mileage into account, how is the age of a car related to its price? How does the relationship between foot length and height differ between men and women? How are the various measures of a golfer's performance related to each other and to the golfer's scoring average?

3. **Assessing differences.** For example, is the difference in ages of first walking different enough between an exercise group and a control group to conclude that exercise really does affect age of first walking? Is the rate of headache relief for migraine sufferers who experience a magnetic pulse sufficiently higher than those in the control group to advocate for the magnetic pulse as an effective treatment?

As with all models, statistical models are simplifications of reality. George Box, a renowned statistician, famously said that "all statistical models are wrong, but some are useful." Statistical models are not deterministic, meaning that their predictions are not expected to be perfectly accurate. For example, we do not expect to predict the exact price of a used car based on its mileage. Even if we were to record every imaginable characteristic of the car and include them all in the model, we would still not be able to predict its price exactly. And we certainly do not expect to predict the exact moment that a baby first walks based on the kind of exercise she/he engaged in. Statistical models merely aim to explain as much of the variability as possible in whatever phenomenon is being modeled. In fact, because human beings are notoriously variable and unpredictable, social scientists who develop statistical models are often delighted if the model explains even a small part of the variability.

A distinguishing feature of statistical models is that we pay close attention to possible simplifications and imperfections, seeking to quantify how much the model explains and how much it does not. So, while we do not expect our model's predictions to be exactly correct, we are able to state how confident we are that our predictions fall within a certain range of the truth. And while we do not expect to determine the exact relationship between two variables, we can quantify how far off our model is likely to be. And while we do not expect to assess exactly how much two groups may differ, we can draw conclusions about how likely they are to differ and by what magnitude.

More formally, a statistical model can be written as:

$$DATA = MODEL + ERROR$$

or as

$$Y = f(X) + \epsilon$$

The Y here represents the variable being modeled, X is the variable used to do the modeling, and f is a function[1]. We start in Chapter 1 with just one quantitative, explanatory variable X and with a linear function f. Then we will consider more complicated functions for f, often by transforming Y or X or both. Later we will consider multiple explanatory variables, which can be either quantitative or categorical. In these initial models we assume that the response variable Y is quantitative. Eventually, we will allow the response variable Y to be categorical.

The ϵ term in the model above is called the "error," meaning the part of the response variable Y that remains unexplained after considering the predictor X. Our models will sometimes stipulate a probability distribution for this ϵ term, often a normal distribution. An important aspect of our modeling process will be checking whether the stipulated probability distribution for the error term seems reasonable, based on the data, and making appropriate adjustments to the model if it does not.

0.1 Fundamental Terminology

Before you begin to study statistical modeling, you will find it very helpful to review and practice applying some fundamental terminology.

The **observational units** in a study are the people, objects, or cases on which data are recorded. The **variables** are the characteristics that are measured or recorded about each observational unit.

Example 0.1: *Car Prices*

In the study about predicting the price of a used car, the observational units are the cars. The variables are the car's price, mileage, age (in years), and manufacturer (Porsche or Honda).

◇

Example 0.2: *Walking Babies*

In the study about babies' walking, the observational units are the babies. The variables are whether or not the baby was put on an exercise regimen, and the age at which the baby first walked.

◇

---

[1]The term "model" is used to refer to the entire equation or just the structural part that we have denoted by $f(X)$

	C1-T	C2	C3	C4	C5	C6
	City	NumMDs	RateMDs	NumHospitals	NumBeds	RateBeds
1	Abilene, TX	313	198	5	614	388
2	Akron, OH	1899	271	7	1825	260
3	Albany, GA	340	210	3	635	392
4	Albany-Schenectady-Troy, NY	2969	353	11	2567	305
5	Albuquerque, NM	2950	385	11	1449	189
6	Alexandria, LA	451	308	5	705	482
7	Allentown-Bethlehem-Easton, PA-NJ	2007	261	10	2206	287
8	Altoona, PA	340	267	5	469	368
9	Amarillo, TX	685	293	3	875	375
10	Ames, IA	173	216	2	292	364

Figure 0.1: Health facilities in U.S. metropolitan areas

Example 0.3: *Metropolitan Health Care*

You may find it helpful to envision the data in a spreadsheet format. The row labels are cities, which are observational units, and the columns correspond to the variables. For example, Figure 0.1 shows part of a Minitab worksheet with data compiled by the U.S. Census Bureau on health care facilities in metropolitan areas. The observational units are the metropolitan areas and the variables count the number of doctors, hospitals and beds in each city as well as rates (number of doctors or beds per 100,000 residents). The full dataset for 83 metropolitan areas is in the file **MetroHealth83**.

⋄

Variables can be classified into two types: quantitative and categorical. A **quantitative** variable records numbers about the observational units. It must be sensible to perform ordinary arithmetic operations on these numbers, so zip codes and jersey numbers are not quantitative variables. A **categorical** variable records a category designation about the observational units. If there are only two possible categories, the variable is also said to be **binary**.

Example 0.1 (continued): The price, mileage, and age of a car are all *quantitative* variables. The model of the car is a *categorical* variable.

⋄

Example 0.2 (continued): Whether or not a baby was assigned to a special exercise regimen is a *categorical* variable. The age at which the baby first walked is a *quantitative* variable.

⋄

Example 0.4: *Medical School Admission*

Whether or not an applicant is accepted for medical school is a *binary* variable, as is the gender of the applicant. The applicant's undergraduate grade point average is a *quantitative* variable.

⋄

0.1. FUNDAMENTAL TERMINOLOGY

Another important consideration is the role played by each variable in the study. The variable that measures the outcome of interest is called the **response** variable. The variables whose relationship to the response is being studied are called **explanatory** variables. (When the primary goal of the model is to make predictions, the explanatory variables are also called **predictor** variables.)

Example 0.1 (continued): The price of the car is the *response* variable. The mileage, age, and model of the car are all *explanatory* variables. ◇

Example 0.2 (continued): The age at which the baby first walked is the *response* variable. Whether or not a baby was assigned to a special exercise regimen is an *explanatory* variable. ◇

Example 0.4 (continued): Whether or not an applicant is accepted for medical school is the *response* variable. The applicant's undergraduate grade point average and sex are *explanatory* variables. ◇

One reason that these classifications are important is that the choice of the appropriate analysis procedure depends on the type of variables in the study and their roles. Regression analysis (covered in Chapters 1 - 4) is appropriate when the response variable is quantitative and the explanatory variables are also quantitative. In Chapter 3 you will also learn how to incorporate binary explanatory variables into a regression analysis. Analysis of variance (ANOVA, considered in Chapters 5-8) is appropriate when the response variable is quantitative but the explanatory variables are categorical. When the response variable is categorical, logistic regression (considered in Chapters 9-11) can be used with either quantitative or categorical explanatory variables. These various scenarios are displayed in Table 0.1.

Keep in mind that variables are not always clear-cut to measure or even classify. For example, measuring headache relief is not a straightforward proposition and could be done with a quantitative measurement (intensity of pain on a 0-10 scale), a categorical scale (much relief, some relief, no relief) or as a binary categorical variable (relief or not).

Table 0.1: Classifying general types of models

Response	Predictor/explanatory	Procedure	Chapter
Quantitative	Single quantitative	Simple linear regression	1,2
Quantitative	Single categorical	One-way analysis of variance	5
Categorical	Single quantitative	Simple logistic regression	9
Categorical	Single binary	$2x2$ table	9
Quantitative	Multiple quantitative	Multiple linear regression	3,4
Quantitative	Multiple categorical	Multi-way analysis of variance	6,7
Categorical	Multiple quantitative	Multiple logistic regression	10,11
Categorical	Multiple categories	$2xk$ table	11

We collect data and fit models in order to understand **populations**, such as all students who are applying to medical school, and **parameters**, such as the acceptance rate of all students with a grade point average of 3.5. The collected data are a **sample** and a characteristic of a sample, such as the percentage of students with grade point averages of 3.5 who were admitted to medical school, out of those who applied, is a **statistic**. Thus, sample statistics are used to estimate population parameters.

Another crucial distinction is whether a research study is a controlled experiment or an observational study. In a **controlled experiment**, the researcher manipulates the explanatory variable by assigning the explanatory group or value to the observational units. (These observational units may be called **experimental units** or **subjects** in an experiment.) In an **observational study**, the researchers do not assign the explanatory variable but rather passively observe and record its information. This distinction is important because the type of study determines the scope of conclusion that can be drawn. Controlled experiments allow for drawing *cause-and-effect* conclusions. Observational studies, on the other hand, only allow for concluding that variables are *associated*. Ideally, an observational study will anticipate alternative explanations for an association and include the additional relevant variables in the model. These additional explanatory variables are then called **covariates**.

Example 0.5: *Handwriting and SAT Essay Scores*

An article about handwriting appeared in the October 11, 2006 issue of the Washington Post. The article mentioned that among students who took the essay portion of the SAT exam in 2005-6, those who wrote in cursive style scored significantly higher on the essay, on average, than students who used printed block letters. This is an example of an observational study since there was no controlled assignment of the type of writing for each essay. While it shows an association between handwriting and essay scores, we can't tell whether better writers tend to choose to write in cursive or if graders tend to score cursive essays more generously and printed ones more harshly. We might also suspect that students with higher GPAs are more likely to use cursive writing. To examine this carefully, we could fit a model with GPA as a covariate.

The article also mentioned a different study in which the identical essay was shown to many graders, but some graders were shown a cursive version of the essay and the other graders were shown a version with printed block letters. Again, the average score assigned to the essay with the cursive style was significantly higher than the average score assigned to the essay with the printed block letters. This second study involved an experiment since the binary explanatory factor of interest (cursive versus block letters) was controlled by the researchers. In that case we can infer that using cursive writing produces better essay scores, on average, than printing block letters.

◇

0.2 Four-Step Process

We will employ a four-step process for statistical modeling throughout this book. These steps are:

- **Choose** a form for the model. This involves identifying the response and explanatory variable(s) and their types. We usually examine graphical displays to help suggest a model that might summarize relationships between these variables.

- **Fit** that model to the data. This usually entails estimating model parameters based on the sample data. We will almost always use statistical software to do the necessary number-crunching to fit models to data.

- **Assess** how well the model describes the data. One component of this involves comparing the model to other models. Are there elements of the model that are not very helpful in explaining the relationships or do we need to consider a more complicated model? Another component of the assessment step concerns analyzing residuals, which are deviations between the actual data and the model's predictions, to assess how well the model fits the data. This process of assessing model adequacy is as much art as science.

- **Use** the model to address the question that motivated collecting the data in the first place. This might be to make predictions, or explain relationships, or assess differences, bearing in mind possible limitations on the scope of inferences that can be made. For example, if the data were collected as a random sample from a population, then inference can be extended to that population; if treatments were assigned at random to subjects, then a cause-and-effect relationship can be inferred; but if the data arose in other ways, then we have little statistical basis for drawing such conclusions.

The specific details for how to carry out these steps will differ depending on the type of analysis being performed and, to some extent, on the context of the data being analyzed. But these four steps are carried out in some form in all statistical modeling endeavors. To illustrate the process we consider an example in the familiar setting of a two-sample t-procedure.

Example 0.6: *Financial Incentives for Weight Loss*

Losing weight is an important goal for many individuals. An article in the *Journal of the American Medical Association* describes a study (Volpp et. al., 2008) in which researchers investigated whether financial incentives would help people lose weight more successfully. Some participants in the study were randomly assigned to a treatment group which offered financial incentives for achieving weight loss goals, while others were assigned to a control group that did not use financial incentives. All participants were monitored over a four-month period and the net weight change ($Before - After$ in pounds) at the end of this period was recorded for each individual. Note that a positive value corresponds to a weight loss and a negative change is a weight gain. The data are given in Table 0.2 and stored in the file **WeightLossIncentive**.

Table 0.2: Weight Loss After Four Months (pounds)

Control	12.5	12.0	1.0	-5.0	3.0	-5.0	7.5	-2.5	20.0	-1.0
	2.0	4.5	-2.0	-17.0	19.0	-2.0	12.0	10.5	5.0	
Incentive	25.5	24.0	8.0	15.5	21.0	4.5	30.0	7.5	10.0	18.0
	5.0	-0.5	27.0	6.0	25.5	21.0	18.5			

Source: Volpp et. al., *JAMA* 2008

The response variable in this situation (weight change) is quantitative and the explanatory factor of interest (control versus incentive) is categorical and binary. The subjects were assigned to the groups at random so this is a statistical experiment. Thus we may investigate whether there is a statistically significant difference in the distribution of weight changes due to the use of a financial incentive.

CHOOSE

When choosing a model we generally consider the question of interest and types of variables involved, then look at graphical displays and compute summary statistics for the data. Since the weight loss incentive study has a binary explanatory factor and quantitative response, we examine dotplots of the weight losses for each of the two groups (Figure 0.2) and find the sample mean and standard deviation for each group.

```
Variable     Group        N    Mean   StDev
WeightLoss   Control      19   3.92   9.11
             Incentive    17   15.68  9.41
```

The dotplots show a pair of reasonably symmetric distributions with roughly the same variability, although the mean weight loss for the incentive group is larger than the mean for the control group. One model for these data would be for the weight losses to come from a pair of normal distributions, with different means (and perhaps different standard deviations) for the two groups. Let the parameter μ_1 denote the mean weight loss after four months without a financial incentive, μ_2 be the mean with the incentive. If σ_1 and σ_2 are the respective standard deviations and we

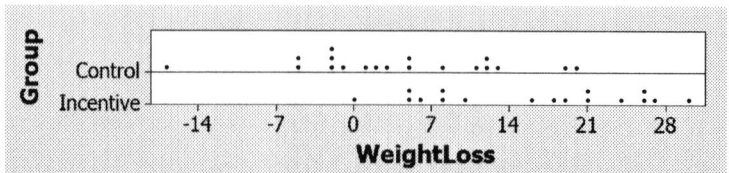

Figure 0.2: Weight Loss for Control versus Incentive groups

0.2. FOUR-STEP PROCESS

let the variable Y denote the weight losses, we can summarize the model as $Y \sim N(\mu_i, \sigma_i)$ where the subscript indicates the group membership[2] and the symbol $\sim$ signifies that the variable has a particular distribution. To see this in the $DATA = MODEL + ERROR$ format, this model could also be written as

$$Y = \mu_i + \epsilon$$

where μ_i is the population mean for the ith group and $\epsilon \sim N(0, \sigma_i)$ is the random error term. Since we only have two groups this model says that

$$Y = \mu_1 + \epsilon \sim N(\mu_1, \sigma_1) \quad \text{for individuals in the control group.}$$
$$Y = \mu_2 + \epsilon \sim N(\mu_2, \sigma_2) \quad \text{for individuals in the incentive group.}$$

FIT

To fit this model we need to estimate four parameters (the means and standard deviations for each of the two groups) using the data from the experiment. The observed means and standard deviations from the two samples provide obvious estimates. We let $\overline{y}_1 = 3.92$ estimate the mean weight loss for the control group and $\overline{y}_2 = 15.68$ estimate the mean for a population getting the incentive. Similarly $s_1 = 9.11$ and $s_2 = 9.41$ estimate the respective standard deviations. The fitted model (a prediction for the typical weight loss in either group) can then be expressed as [3]

$$\hat{y} = \overline{y}_i$$

i.e. that $\hat{y} = 3.92$ pounds for individuals without the incentive and $\hat{y} = 15.68$ pounds for those with the incentive.

Note that the error term does not appear in the fitted model since, when predicting a particular weight loss, we don't know whether the random error will be positive or negative. That does not mean that we expect there to be no error, just that the best guess for the *average* weight loss under either condition is the sample group mean, $\overline{y}_i$.

ASSESS

Our model indicates that departures from the mean in each group (the random errors) should follow a normal distribution with mean zero. To check this we examine the sample *residuals* or deviations between what is predicted by the model and the actual data weight losses.

$$residual = \text{observed} - \text{predicted} = y - \hat{y}$$

For subjects in the control group we subtract $\hat{y} = 3.92$ from each weight loss and we subtract $\hat{y} = 15.68$ for the incentive group. Dotplots of the residuals for each group are shown in Figure 0.3.

---

[2] For this example an assumption that the variances are equal, $\sigma_1^2 = \sigma_2^2$, might be reasonable, but that would lead to the less familiar pooled variance version of the t-test. We explore this situation in more detail in a later chapter.

[3] We use the ˆ symbol above a variable name to indicate predicted value, and refer to this as y-hat.

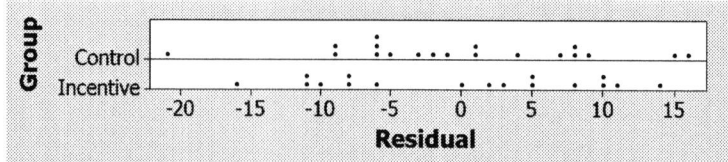

Figure 0.3: Residuals from Group Weight Loss Means

Note that the distributions of the residuals are the same as the original data, except that both are shifted to have a mean of zero. We don't see any significant departures from normality in the dotplots, but it's difficult to judge normality from dotplots with so few points. Normal probability plots (as shown in Figure 0.4) are a more informative technique for assessing normality. Departures from a linear trend in such plots indicate a lack of normality in the data. Normal probability plots will be examined in more detail in the next chapter.

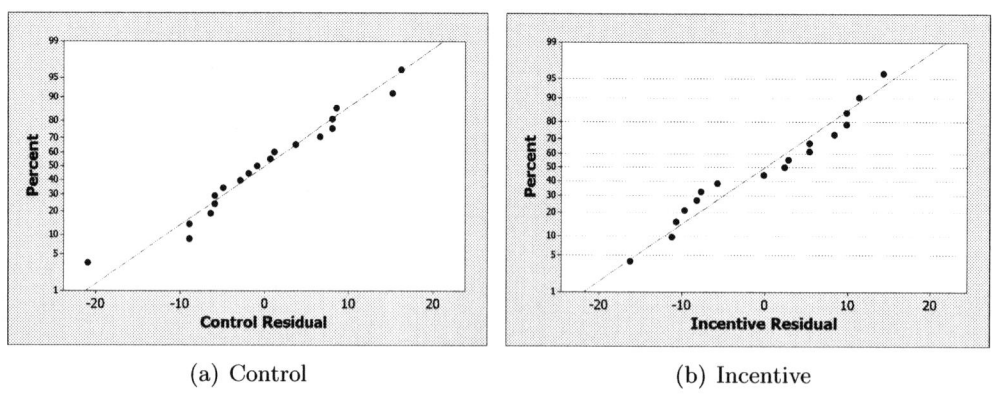

(a) Control (b) Incentive

Figure 0.4: Normality Probability Plots for Residuals of Weight Loss

As a second component of assessment, we consider whether an alternate (simpler) model might fit the data essentially as well as our model with different means for each group. This is analogous to testing the standard hypotheses for a two sample t-test:

$H_o : \mu_1 = \mu_2$
$H_a : \mu_1 \neq \mu_2$

The null hypothesis (H_o) corresponds to the simpler model $Y = \mu + \epsilon$ which uses the same mean for both the control and incentive groups. The alternative (H_a) reflects the model we have considered here that allows each group to have a different mean. Would the simpler (common mean) model suffice for the weight loss data or do the two separate groups means provide a *significantly* better explanation for the data? One way to judge this is with the results of the usual two-sample t-test (as shown in the Minitab output below).

```
Two-sample T for WeightLoss

Group      N   Mean   StDev  SE Mean
Control    19  3.92   9.11   2.1
Incentive  17  15.68  9.41   2.3

Difference = mu (Control) - mu (Incentive)
Estimate for difference:  -11.76
95% CI for difference:  (-18.05, -5.46)
T-Test of difference = 0 (vs not =): T-Value = -3.80  P-Value = 0.001  DF = 33
```

The extreme value for this test statistic ($t = -3.80$) and very small p-value (0.001) provide strong evidence that the means of the two groups are indeed significantly different. If the two group means were really the same (i.e. the common mean model was accurate and the financial incentives had no effect on weight loss) we would expect to see a difference as large as was observed in this experiment for only about 1 in 1000 replications of the experiment. Thus the model with separate means for each group does a substantially better job at explaining the results of the weight loss study.

USE

Since this was a designed experiment with random allocation of the control and incentive conditions to the subjects, we can infer that the financial incentives did produce a difference in the average weight loss over the four-month period; that is, the random allocation of conditions to subjects allows us to draw a cause-and-effect relationship. A person who is on the incentive-based treatment can be expected to lose about 11.8 pounds more ($15.68 - 3.92 = 11.76$), on average, in four months, than control subjects who are not given this treatment. Note that for most individuals, approximately 12 pounds is a substantial amount of weight to lose in four months. Moreover, if we interpret the confidence interval from the Minitab output, we can be 95% confident that the incentive treatment is worth between 5.5 and 18.1 pounds of additional weight loss, on average, over four months.

Before leaving this example we note three cautions. First, all but two of the participants in this study were adult men, so we should avoid making conclusions about the effect of financial incentives on weight loss in women. Second, if the participants in the study did not arise from taking a random sample, we would have difficulty justifying a statistical basis for generalizing the findings to other adults. Any such generalization must be justified on other grounds (such as a belief that most adults respond to financial incentives in similar ways). Third, the experimenters followed up with subjects to see if weight losses were maintained at a point seven months after the start of the study (and three months after any incentives expired). The results from the followup study appear in Exercise 0.10.

◇

0.3 Chapter Summary

In this chapter we reviewed basic terminology, introduced the 4-step approach to modeling that will be used throughout the text, and revisited a common two-sample inference problem.

After completing this chapter you should be able to distinguish between a **sample** and a **population**, describe the difference between a **parameter** and a **statistic**, and identify variables as **categorical** or **quantitative**. Prediction is a major component to modeling so identifying **explanatory** (or predictor) **variables** that can be used to develop a model for **response variable** is an important skill. Another important idea is the distinction between **observational studies** (where researchers simply observe what is happening) and **experiments** (where researchers impose "treatments").

The fundamental idea that a **statistical model** partitions data into two components, one for the model and one for error, was introduced. Even though the models will get more complex as we move through the more advanced settings, this statistical modeling idea will be a major theme throughout the text. The error term and conditions associated with this term are important features in distinguishing statistical models from mathematical models. You saw how to compute **residuals** by comparing the observed data to predictions from a model as a way to begin quantifying the errors.

The **4-step process** of choosing, fitting, assessing, and using a model is vital. Each step in the process requires careful thought and the computations will often be the easiest part of the entire process. Identifying the response and explanatory variable(s) and their types (categorical or quantitative) helps us **choose** the appropriate model(s). Statistical software will almost always be used to **fit** models and obtain estimates. Comparing models and **assessing** the adequacy of these models will require a considerable amount of practice and this is a skill that you will develop over time. Try to remember that **using** the model to make predictions, explain relationships, or assess differences is only one part of the 4-step process.

0.4 Exercises

Conceptual Exercises

0.1 *Categorical or quantitative?* Suppose that a statistics professor records the following for each student enrolled in her class:

- Gender
- Major
- Score on first exam
- Number of quizzes taken (a measure of class attendance)
- Time spent sleeping on the previous night
- Handedness (left- or right-handed)
- Political inclination (liberal, moderate, or conservative)
- Time spent on the final exam
- Score on the final exam

For the following questions, identify the response variable and the explanatory variable(s). Also classify each variable as quantitative or categorical. For categorical variables, also indicate whether the variable is binary.

a. Do the various majors differ with regard to average sleeping time?

b. Is a student's score on the first exam useful for predicting his/her score on the final exam?

c. Do male and female students differ with regard to the average time they spend on the final exam?

d. Do the proportions of left-handers differ between males and females on campus?

e. Are sleeping time, exam 1 score, and number of quizzes taken useful for predicting time spent on final exam?

f. Does knowing a student's gender help to predict his/her major?

g. Does knowing a student's political inclination and time spent sleeping help to predict his/her gender?

0.2 *Sports projects.* For each of the following sports-related projects, identify observational units and the response and explanatory variables when appropriate. Also classify the variables as quantitative or categorical.

- a. Interested in predicting how long it takes to play a Major League Baseball game, an individual recorded the following information for all 15 games played on August 26, 2008: time to complete the game, total number of runs scored, margin of victory, total number of pitchers used, ballpark attendance at the game, and which league (National or American) the game was played in.

- b. Over the course of several years, a golfer kept track of the length of all of his putts and whether or not he made the putt. He was interested in predicting whether or not he would make a putt based on how long it was.

- c. Some students recorded lots of information about all of the football games played by LaDainian Tomlinson during the 2006 season. They recorded his rushing yardage, number of rushes, rushing touchdowns, receiving yardage, number of receptions, and receiving touchdowns.

- d. A volleyball coach wants to see if a player using a jump serve is more likely to lead to winning a point than when using a standard overhand serve.

- e. To investigate whether the "home-field advantage" differs across major team sports, researchers kept track of how often the home team won a game for all games played in the 2007 and 2008 seasons in Major League Baseball, National Football League, National Basketball Association, and National Hockey League.

- f. A student compared men and women professional golfers on how far they drive a golf ball and how often their drive hits the fairway.

0.3 *Scooping ice cream.* In a study reported in the *Journal of Preventative Medicine*, 85 nutrition experts were asked to scoop themselves as much ice cream as they wanted. Some of them were randomly given a large bowl (34 ounces) as they entered the line, and the others were given a smaller bowl (17 ounces). Similarly, some were randomly given a large spoon (3 ounces) and the others were given a small spoon (2 ounces). Researchers then recorded how much ice cream each subject scooped for herself/himself. Their conjecture was that those given a larger bowl would tend to scoop more ice cream, as would those given a larger spoon.

- a. Identify the observational units in this study.

- b. Is this an observational study or a controlled experiment? Explain how you know.

- c. Identify the response variable in this study, and classify it as quantitative or categorical.

- d. Identify the explanatory variables in this study, and classify it/them as quantitative or categorical.

0.4. EXERCISES

0.4 *Wine model.* In his book *SuperCrunchers: Why Thinking by Numbers is the New Way to be Smart*, Ian Ayres writes about Orley Achenfelter, who has gained fame and generated considerable controversy by using statistical models to predict the quality of wine. Ashenfelter developed a model based on decades of data from France's Bordeaux region, which Ayers reports as:

$$WineQuality = 12.145 + .00117 WinterRain + .0614 AverageTemp - .00386 HarvestRain + \epsilon$$

where *WineQuality* is a function of the price, rainfall is measured in millimeters, and temperature is measured in °C.

a. Identify the response variable in this model. Is it quantitative or categorical?

b. Identify the explanatory variables in this model. Are they quantitative or categorical?

c. According to this model, is higher wine quality associated with more or with less winter rainfall?

d. According to this model, is higher wine quality associated with more or with less harvest rainfall?

e. According to this model, is higher wine quality associated with more or with less average growing season temperature?

f. Are the data that Ashenfelter analyzed observational or experimental? Explain.

0.5 *Measuring students.* The registrar at a small liberal arts college computes descriptive summaries for all members of the entering class on a regular basis. For example, the mean and standard deviation of the high school GPAs for all entering students in particular year were 3.16 and 0.5247, respectively. The Mathematics Department is interested in helping all students who want to take mathematics to identify the appropriate course, so they offer a placement exam. A randomly selected subset of students taking this exam during the past decade had an average score of 71.05 with a standard deviation of 8.96.

a. What is the population of interest to the registrar at this college?

b. Are the descriptive summaries computed by the registrar (3.16 and 0.5247) statistics or parameters? Explain.

c. What is the population of interest to the Mathematics Department?

d. Are the numerical summaries (71.05 and 8.96) statistics or parameters? Explain.

Guided Exercises

0.6 *Scooping ice cream.* Refer to Exercise 0.3 on self-serving ice cream. The following table reports the average amounts of ice cream scooped (in ounces) for the various treatments:

	17-ounce bowl	34-ounce bowl
2-ounce spoon	4.38	5.07
3-ounce spoon	5.81	6.58

a. Does it appear that the size of the bowl had an effect on amount scooped? Explain.

b. Does it appear that the size of the spoon had an effect on amount scooped? Explain.

c. Which appears to have more of an effect: size of bowl or size of spoon? Explain.

d. Does it appear that the effect of the bowl size is similar for both spoon sizes, or does it appear that the effect of the bowl size differs substantially for the two spoon sizes? Explain.

0.7 *Diet plans.* An article in the *Journal of the American Medical Association* (Dansinger et al., 2005) reported on a study in which 160 subjects were randomly assigned to one of four popular diet plans: Atkins, Ornish, Weight Watchers, and Zone. Among the variables measured were:

- which diet the subject was assigned to

- whether or not the subject completed the twelve-month study

- the subject's weight loss after two months, six months, and twelve months (in kilograms, with a negative value indicating weight gain)

- the degree to which the subject adhered to the assigned diet, taken as the average of 12 monthly ratings, each on a 1-10 scale (with 1 indicating complete non-adherence and 10 indicating full adherence)

a. Classify each of these variables as quantitative or categorical.

b. The primary goal of the study was to investigate whether weight loss tends to differ significantly among the four diets. Identify the explanatory and response variables for investigating this question.

c. A secondary goal of the study was to investigate whether weight loss is affected by the adherence level. Identify the explanatory and response variables for investigating this question.

d. Is this an observational study or a controlled experiment? Explain how you know.

e. If the researchers' analysis of the data leads them to conclude that there is a significant difference in weight loss among the four diets, can they legitimately conclude that the difference is because of the diet? Explain why or why not.

0.4. EXERCISES

f. If the researchers' analysis of the data analysis leads them to conclude that there is a significant association between weight loss and adherence level, can they legitimately conclude that there is a cause-and-effect association between them? Explain why or why not.

0.8 *Predicting NFL wins.* Consider the following model for predicting the number of games that a National Football League (NFL) team wins in a season:

$$wins = 4.6 + 0.5PF - 0.3PA + \epsilon$$

where PF stands for average points a team scores per game over an entire season and PA stands for points allowed per game. Currently each team plays 16 games in a season.

a. According to this model, how many more wins is a team expected to achieve in a season if they increase their scoring by an average of 3 points per game?

b. According to this model, how many more wins is a team expected to achieve in a season if they decrease their points allowed by an average of 3 points per game?

c. Based on your answers to (a) and (b), does it seem that a team should focus more on improving its offense or improving its defense?

d. Use this model to predict the number of wins for the 2009 New Orleans Saints, who scored 510 points and allowed 341 points.

e. The Saints actually won 13 games in 2009. Determine their residual from this model, and interpret what this means.

f. The largest residual value from this model belongs to the Indianapolis Colts, with a residual value of 3.21 games. The Colts actually won 14 games. Determine this model's predicted number of wins for the Colts.

g. The largest negative residual value from this model belongs to the Baltimore Ravens, with a residual value of -1.95 games. Interpret what this residual means.

0.9 *Roller coasters.* The Roller Coaster Database (*rcdb.com*) contains lots of information about roller coasters all over the world. The following statistical model for predicting the top speed (in miles per hour) of a coaster was based on more than 100 roller coasters in the United States and data displayed on the database in November of 2003:

$$TopSpeed = 54 + 7.6 TypeCode + \epsilon$$

where $TypeCode = 1$ for steel roller coasters and $TypeCode = 0$ for wooden roller coasters.

a. What top speed does this model predict for a wooden roller coaster?

b. What top speed does this model predict for a steel roller coaster?

c. Determine the difference in predicted speeds in mph for the two types of coasters. Also identify where this number appears in the model equation, and explain why that makes sense.

Some other predictor variables available at the database include: age, total length, maximum height, and maximum vertical drop. Suppose that we include all of these predictor variables in a statistical model for predicting the top speed of the coaster.

d. For each of these predictor variables, indicate whether you expect its coefficient to be positive or negative. Explain your reasoning for each variable.

e. Which of these predictor variables do you expect to be the best single variable for predicting a roller coaster's top speed? Explain why you think that.

The following statistical model was produced from these data:

$$Speed = 33.4 + 0.10 Height + 0.11 Drop + 0.0007 Length - 0.023 Age - 2.0 TypeCode + \epsilon$$

f. Comment on whether the signs of the coefficients are as you expect.

g. What top speed would this model predict for a steel roller coaster that is 10 years old, with maximum height of 150 feet, maximum vertical drop of 100 feet, and length of 4000 feet?

Open-ended Exercises

0.10 *Incentive for weight loss.* The study (Volpp et. al., 2008) on financial incentives for weight loss in Example 0.6 on page 7 used a follow-up weight check after seven months to see whether weight losses persisted after the original four months of treatment. The results are given in Table 0.3 and in the variable *Month7Loss* of the **WeightLossIncentive** data file. Note that a few participants dropped out and were not re-weighed at the seven-month point. As with the earlier example, the data are the change in weight (in pounds) from the beginning of the study and positive values correspond to weight losses. Using Example 0.6 as an outline, follow the four-step process to see whether the data provide evidence that the beneficial effects of the financial incentives still apply to the weight losses at the seven-month point.

Table 0.3: Weight Loss After Seven Months (pounds)

Control	-2.0	7.0	19.5	-0.5	-1.5	-10.0	0.5	5.0	8.5	$\bar{y}_1 = 4.64$
	18.0	16.0	-9.0	4.5	23.5	5.5	6.5	-9.5	1.5	$s_1 = 9.84$
Incentive	11.5	20.0	-22.0	2.0	7.5	16.5	19.0	18.0	-1.0	$\bar{y}_2 = 7.80$
	5.5	24.5	9.5	10.0	-8.5	4.5				$s_2 = 12.10$

Source: Volpp, John, Troxel, et. al., *JAMA* 2008

0.4. EXERCISES

0.11 *Statistics students survey.* An instructor at a small liberal arts college distributed the data collection card similar to what is shown below on the first day of class. The data for two different sections of the course are shown in the file **Day1Survey**. Note that the names have not been entered into the data set.

```
    Data Collection Card

Directions: Please answer each question and return to me.
1.  Your name (as you prefer): _____
2.  What is your current class standing? _____
3.  Sex: Male _____    Female _____
4.  How many miles (approximately) did you travel to get to campus? _____
5.  Height (estimated) in inches: _____
6.  Handedness (Left, Right, Ambidextrous): _____
7.  How much money, in coins (not bills), do you have with you? $_____
8.  Estimate the length of the white string (in inches): _____
9.  Estimate the length of the black string (in inches): _____
10. How much do you expect to read this semester (in pages/week)? _____
11. How many hours do you watch TV in a typical week? _____
12. What is your resting pulse? _____
13. How many text messages have you sent and received in the last 24 hours? _____
```

The data for this survey are stored in **Day1Survey**.

a. Apply the four-step process to the survey data to address the question: "Is there evidence that the mean resting pulse rates for women is different from the mean resting pulse rate for men?"

b. Pick another question that interests you from the survey and compare the responses of men and women.

0.12 *Statistics student survey (continued).* Refer to the survey of statistics students described in Exercise 0.11 with data in **Day1Survey**. Use the survey data to address the question: "Do women expect to do more reading than men?"

0.13 *Marathon training.* Training records for a marathon runner are provided in the file **Marathon**. The *Date*, *Miles* run, *Time* (in minutes:seconds:hundredths), and running *Pace* (in minutes:seconds:hundredths per mile) are given for a five-year period from 2002 to 2006. The time and pace have been converted to decimal minutes in *TimeMin* and *PaceMin*, respectively. The brand of the running shoe is added for 2005 and 2006. Use the four-step process to investigate if a runner has a tendency to go faster on short runs (5 or less miles) than long runs. The variable *Short* in the dataset is coded with 1 for short runs and 0 for longer runs. Assume that the data for this runner can be viewed as a sample for runners of a similar age and ability level.

0.14 *Marathon training (continued).* Refer to the data described in Exercise 0.13 that contains five years worth of daily training information for a runner. One might expect that the running patterns might change as the runner gets older. The file **Marathon** also contains a variable called *After*2004 which has the value 0 for any runs during the years 2002-2004 and 1 for runs during 2005 and 2006. Use the four-step process to see if there is evidence of a difference between these two time periods in the following aspects of the training runs.

a. the average running pace (*PaceMin*).

b. the average distance run per day (*Miles*).

Supplementary Exercises

0.15 *Pythagorean theorem of baseball.* Renowned baseball statistician Bill James devised a model for predicting a team's winning percentage. Dubbed the "Pythagorean Theorem of Baseball," this model predicts a team's winning percentage as:

$$\text{Winning Percentage} = \frac{(\text{runs scored})^2}{(\text{runs scored})^2 + (\text{runs against})^2} \times 100 + \epsilon$$

a. Use this model to predict the winning percentage for the New York Yankees, who scored 915 runs and allowed 753 runs in the 2009 season.

b. The New York Yankees actually won 103 games and lost 59 in the 2009 season. Determine the winning percentage, and also determine the residual from the Pythagorean model (by taking the observed winning percentage minus the predicted winning percentage).

c. Interpret what this residual value means for the 2009 Yankees. [Hints: Did the team do better or worse than expected, given their runs scored and runs allowed? By how much?]

d. Repeat (a)-(c) for the 2009 San Diego Padres, who scored 638 runs and allowed 769 runs.

e. Which team (Yankees or Padres) exceeded their Pythagorean expectations by more?

Table 0.4 provides data, predictions, and residuals for all 30 Major League Baseball teams in 2009.

f. Which team exceeded their Pythagorean expectations the most? Describe how this team's winning percentage compares to what is predicted by their runs scores and runs allowed.

g. Which team fell furthest below their Pythagorean expectations? Describe how this team's winning percentage compares to what is predicted by their runs scored and runs allowed.

Table 0.4: Winning percentage and Pythagorean predictions for baseball teams in 2009

TEAM	W	L	WinPct	RunScored	RunsAgainst	Predicted	Residual
Arizona Diamondbacks	70	92	43.21	720	782	45.88	-2.67
Atlanta Braves	86	76	53.09	735	641	56.80	-3.71
Baltimore Orioles	64	98	39.51	741	876	41.71	-2.20
Boston Red Sox	95	67	58.64	872	736	58.40	0.24
Chicago Cubs	83	78	51.55	707	672	52.54	-0.98
Chicago White Sox	79	83	48.77	724	732	49.45	-0.69
Cincinnati Reds	78	84	48.15	673	723	46.42	1.73
Cleveland Indians	65	97	40.12	773	865	44.40	-4.28
Colorado Rockies	92	70	56.79	804	715	55.84	0.95
Detroit Tigers	86	77	52.76	743	745	49.87	2.90
Florida Marlins	87	75	53.70	772	766	50.39	3.31
Houston Astros	74	88	45.68	643	770	41.08	4.59
Kansas City Royals	65	97	40.12	686	842	39.90	0.23
Los Angeles Angels	97	65	59.88	883	761	57.38	2.50
Los Angeles Dodgers	95	67	58.64	780	611	61.97	-3.33
Milwaukee Brewers	80	82	49.38	785	818	47.94	1.44
Minnesota Twins	87	76	53.37	817	765	53.28	0.09
New York Mets	70	92	43.21	671	757	44.00	-0.79
New York Yankees	103	59	63.58	915	753		
Oakland Athletics	75	87	46.30	759	761	49.87	-3.57
Philadelphia Phillies	93	69	57.41	820	709	57.22	0.19
Pittsburgh Pirates	62	99	38.51	636	768	40.68	-2.17
San Diego Padres	75	87	46.30	638	769		
San Francisco Giants	88	74	54.32	657	611	53.62	0.70
Seattle Mariners	85	77	52.47	640	692	46.10	6.37
St. Louis Cardinals	91	71	56.17	730	640	56.54	-0.37
Tampa Bay Rays	84	78	51.85	803	754	53.14	-1.29
Texas Rangers	87	75	53.70	784	740	52.88	0.82
Toronto Blue Jays	75	87	46.30	798	771	51.72	-5.42
Washington Nationals	59	103	36.42	710	874	39.76	-3.34

Source: *www.baseball − reference.com*

Chapter 1

Simple Linear Regression

How is the price of a used car related to the number of miles it's been driven? Is the number of doctors in a city related to the number of hospitals? How can we predict the price of a textbook from the number of pages?

In this chapter we consider a single quantitative predictor X for a quantitative response variable Y. A common model to summarize the relationship between two quantitative variables is the *simple linear regression model*. We assume that you have encountered simple linear regression as part of an introductory statistics course. Therefore we review the structure of this model, the estimation and interpretation of its parameters, the assessment of its fit, and its use in predicting values for the response. Our goal is to introduce and illustrate many of the ideas and techniques of statistical model building that will be used throughout this book in a somewhat familiar setting. In addition to recognizing when a linear model may be appropriate, we also consider methods for dealing with relationships between two quantitative variables that are not linear.

1.1 The Simple Linear Regression Model

Example 1.1: *Porsche prices*

Suppose that we are interested in purchasing a Porsche sports car. If we can't afford the high sticker price of a new Porsche, we might be interested in finding a used one. How much should we expect to pay? Obviously, the price might depend on many factors including the age, condition, and special features on the car. For this example we will focus on the relationship between $X = Mileage$ of a used Porsche and $Y = Price$. We used an internet sales site to collect data for a sample of 30 used Porsches, with price (in thousands of dollars) and mileage (in thousands of miles), as shown in Table 1.1. The data are also stored in the file named **PorschePrice**. We are interested in predicting the price of a used Porsche based on its mileage, so the explanatory variable is *Mileage*, the response is *Price* and both variables are quantitative.

⋄

Table 1.1: Price and Mileage for Used Porsches

Price ($1,000's)	Mileage (thousands)
69.4	21.5
56.9	43.0
49.9	19.9
47.4	36.0
42.9	44.0
36.9	49.8
83.0	1.3
72.9	0.7
69.9	13.4
67.9	9.7
66.5	15.3
64.9	9.5
58.9	19.1
57.9	12.9
54.9	33.9
54.7	26.0
53.7	20.4
51.9	27.5
51.9	51.7
49.9	32.4
44.9	44.1
44.8	49.8
39.9	35.0
39.7	20.5
34.9	62.0
33.9	50.4
23.9	89.6
22.9	83.4
16.0	86.0
52.9	37.4

Source: Autotrader.com, Spring 2007.

1.1. THE SIMPLE LINEAR REGRESSION MODEL

Choosing a Simple Linear Model

Recall that data can be represented by a **model** plus an **error** term:

$$Data = Model + Error$$

When the data involve a quantitative response variable Y and we have a single quantitative predictor X the model becomes

$$\begin{aligned} Y &= f(X) + \epsilon \\ &= \mu_Y + \epsilon \end{aligned}$$

where $f(X)$ is a function that gives the mean value of Y, μ_Y, at any value of X and ϵ represents the error (deviation) from that mean[1].

We generally use graphs to help visualize the nature of the relationship between the response and potential predictor variables. Scatterplots are the major tool for helping us choose a model when both the response and predictor are quantitative variables. If the scatterplot shows a consistent linear trend, then we use in our model a mean which follows a straight line relationship with the predictor. This gives a **simple linear regression** model where the function, $f(X)$, is a linear function of X. If we let β_0 and β_1 represent the intercept and slope, respectively, of that line we have

$$\mu_Y = f(X) = \beta_0 + \beta_1 X$$

and

$$Y = \beta_0 + \beta_1 X + \epsilon$$

Example 1.2: *Porsche prices (continued)*

CHOOSE

A scatterplot of price versus mileage for the sample of used Porsches is shown in Figure 1.1. The plot indicates a negative association between these two variables. It is generally understood that cars with lots of miles cost less on average than cars with only limited miles and the scatterplot supports this understanding. Since the rate of decrease in the scatterplot is relatively constant as the mileage increases, a linear model might provide a good summary of the relationship between the average prices and mileages of used Porsches for sale on this internet site. In symbols, we express the mean price as a linear function of mileage.

$$\mu_{Price} = \beta_0 + \beta_1 \cdot Mileage$$

---

[1]More formal notation for the mean value of Y at a given value of X is $\mu_{Y|X}$. To minimize distractions in most formulas we will use just μ_Y when the role of the predictor is clear.

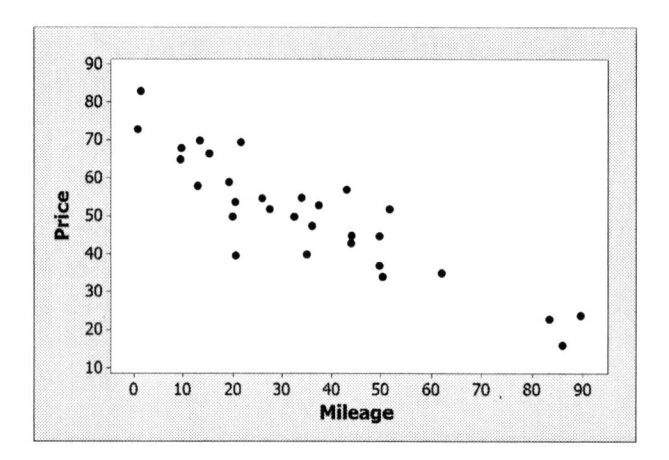

Figure 1.1: Scatterplot of Porsche price versus mileage

Thus the model for actual used Porsche prices would be

$$Price = \beta_0 + \beta_1 \cdot Mileage + \epsilon$$

This model indicates that Porsche prices should be scattered around a straight line with deviations from the line determined by the random error component, ϵ. We now turn to the question of how to choose the slope and intercept for the line that best summarizes this relationship. ◇

Fitting a Simple Linear Model

We want the best possible estimates of β_0 and β_1. Thus, we use least squares regression to fit the model to the data. This chooses coefficient estimates to minimize the sum of the squared errors and leads to the best set of predictions when we use our model to predict the data. In practice, we rely on computer technology to compute the least squares estimates for the parameters. The fitted model is represented by

$$\hat{Y} = \hat{\beta}_0 + \hat{\beta}_1 X$$

In general, we use Greek letters (β_0, β_1, etc.) to denote parameters and hats ($\hat{\beta}_0, \hat{\beta}_1$, etc.) are added to denoted estimated (fitted) values of these parameters.

A key tool for fitting a model is to compare the values it predicts for the individual data cases[2] to the actual values of the response variable in the dataset. The discrepancy in predicting each response is measured by the **residual**.

$$residual = observed\,y - predicted\,y = y - \hat{y}$$

---

[2] We generally use a lower case y when referring to the value of a variable for an individual case and an upper case Y for the variable itself.

1.1. THE SIMPLE LINEAR REGRESSION MODEL

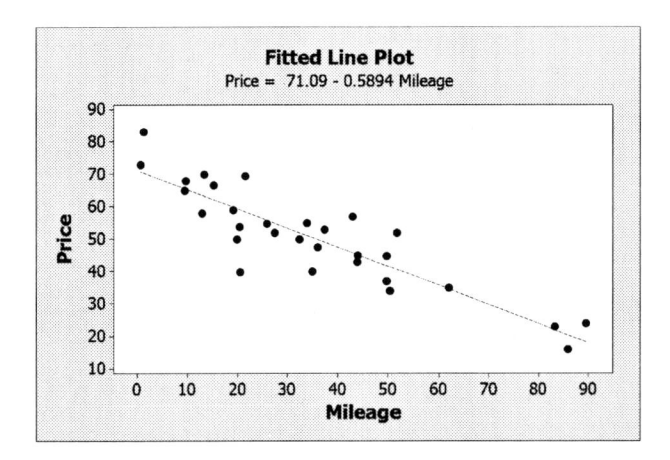

Figure 1.2: Linear regression to predict Porsche price based on mileage

The **sum of squared residuals** provides a measure of how well the line predicts the actual responses for a sample. We often denote this quantity as **SSE** for the sum of the squared errors. Statistical software calculates the fitted values of the slope and intercept so as to minimize this sum of squared residuals, hence we call this the **least squares line**.

Example 1.3: *Porsche prices (continued)*

FIT

For the ith car in the data set, with mileage x_i, the model is

$$y_i = \beta_0 + \beta_1 x_i + \epsilon_i$$

The parameters, β_0 and β_1 in the model, represent the true, population-wide intercept and slope for all Porsches for sale. The corresponding statistics, $\hat{\beta}_0$ and $\hat{\beta}_1$, are estimates derived from this particular sample of 30 Porsches. (These estimates are determined from statistical software, for example in the Minitab fitted line plot shown in Figure 1.2 or the output shown in Figure 1.3).

The least squares line is

$$\widehat{Price} = 71.09 - 0.5894 \cdot Mileage$$

Thus, for every additional 1,000 miles on a used Porsche the predicted price goes down by about $589. Also, if a (used!) Porche had zero miles on it, we would predict the price to be $71,090. In many cases the the intercept lies far from the data used to fit the model and has no practical interpretation.

```
Regression Analysis: Price versus Mileage

The regression equation is Price = 71.1 - 0.589 Mileage

Predictor        Coef   SE Coef       T      P
Constant       71.090     2.370   30.00  0.000
Mileage       -0.58940   0.05665  -10.40  0.000


S = 7.17029   R-Sq = 79.5%   R-Sq(adj) = 78.7%

Analysis of Variance

Source            DF       SS       MS        F      P
Regression         1   5565.7   5565.7   108.25  0.000
Residual Error    28   1439.6     51.4
Total             29   7005.2
```

Figure 1.3: Minitab output for regression of Porsche *Price* on *Mileage*

Note that car #1 in Table 1.1 had a mileage level of 21.5 (21,500 miles) and a price of 69.4 ($69,400), whereas the fitted line predicts a price of

$$\widehat{Price} = 71.09 - 0.5894 \cdot Mileage = 58.4$$

The residual here is $Price - \widehat{Price} = 69.4 - 58.4 = 11.0$.

If we do a similar calculation for each of the 30 cars, square each of the resulting residuals, and sum the squares we get 1439.6. If you were to choose any other straight line to make predictions for these Porsche prices based on the mileages you could never obtain an SSE less than 1439.6. ⋄

1.2 Conditions for a Simple Linear Model

We know that our model won't fit the data perfectly. The discrepancies that result from fitting the model represent what the model did not capture in each case. We want to check whether our model is reasonable and captures the main features of the dataset. Are we justified in using our model? Do the assumptions of the model appear to be reasonable? How much can we trust predictions that come from the model? Do we need to adjust or expand the model to better explain features of the data or could it be simplified without much loss of predictive power?

In specifying any model certain conditions must be satisfied for the model to make sense. We often make assumptions about the nature of the relationship between variables and the distribution of the

1.2. CONDITIONS FOR A SIMPLE LINEAR MODEL

errors. A key part of assessing any model is to check whether the conditions are reasonable for the data at hand. We hope that the residuals are small and contain no pattern that could be exploited to better explain the response variable. If our assessment shows a problem, then the model should be refined. Typically, we will rely heavily on graphs of residuals to assess the appropriateness of the model. In this section we discuss the conditions that are commonly placed on a simple linear model. The conditions we describe here for the simple linear regression model are typical of those that will be used throughout this book. In the following section we explore ways to use graphs to help us assess whether the conditions hold for a particular set of data.

Linearity – The overall relationship between the variables has a linear pattern. The average values of the response Y for each value of X fall on a common straight line.

The other conditions deal with the distribution of the errors.

Zero Mean – The error distribution is centered at zero. This means that the points are scattered at random above and below the line. (Note: By using least squares regression, we force the residual mean to be zero. Other techniques would not necessarily satisfy this condition.)

Constant Variance – The variability in the errors is the same for all values of the predictor variable. This means that the spread of points around the line remains fairly constant.

Independence – The errors are assumed to be independent from one another. Thus, one point falling above or below the line has no influence on the location of another point.

When we are interested in using the model to make formal inferences (conducting hypothesis tests or providing confidence intervals), additional assumptions are needed.

Random – The data are obtained using a random process. Most commonly this arises either from random sampling from a population of interest or from the use of randomization in a statistical experiment.

Normality – In order to use standard distributions for confidence intervals and hypothesis tests we often need to assume that the random errors follow a normal distribution.

We can summarize these conditions for a simple linear model using the following notation.

Simple Linear Regression Model

For a quantitative response variable Y and a single quantitative explanatory variable X the **simple linear regression model** is

$$Y = \beta_0 + \beta_1 X + \epsilon$$

where ϵ follows a normal distribution, that is $\epsilon \sim N(0, \sigma_\epsilon)$, and the errors are independent from one another.

Estimating the standard deviation of the error term

The simple linear regression model has three unknown parameters: the slope, β_1; the intercept, β_0; and the standard deviation σ_ϵ of the errors around the line. We have already seen that software will find the least squares estimates of the slope and intercept. Now we must consider how to estimate σ_ϵ, the standard deviation of the distribution of errors. Since the residuals estimate how much Y varies about the regression line, the sum of the squared residuals (SSE) is used to compute the estimate, $\widehat{\sigma_\epsilon}$. The value of $\widehat{\sigma_\epsilon}$ is referred to as the **regression standard error** and is interpreted as the size of a "typical" error.

Regression Standard Error

For a simple linear regression model, the estimated standard deviation of the error term based on the least squares fit to a sample of n observations is

$$\widehat{\sigma_\epsilon} = \sqrt{\frac{\sum (y - \hat{y})^2}{n - 2}} = \sqrt{\frac{SSE}{n - 2}}$$

The predicted values and resulting residuals are based on a sample slope and intercept that are calculated from the data. Therefore, we have $n - 2$ **degrees of freedom** for estimating the regression standard error[3]. In general we lose an additional degree of freedom in the denominator for each new beta parameter that is estimated in the prediction equation.

Example 1.4: *Porsche prices (continued)*

The sum of squared residuals for the Porsche data is shown in Figure 1.3 as 1439.6 (see the SS column of the Residual Error line of the Analysis of Variance table in the Minitab output). Thus, the regression standard error is

$$\widehat{\sigma_\epsilon} = \sqrt{\frac{1439.6}{30 - 2}} = 7.17$$

Using mileage to predict price of a used Porsche, the typical error will be around $7,170. So we have some feel for how far individual cases might spread above or below the regression line. Note that this value is labeled S in the Minitab output of Figure 1.3. ⋄

---

[3] If a scatterplot only has 2 points, then it's easy to fit a straight line with residuals of zero, but we have no way of estimating the variability in the distribution of the error term. This corresponds to having zero degrees of freedom.

1.3 Assessing Conditions

A variety of plots are used to assess the conditions of the simple linear model. Scatterplots, histograms, and dotplots will be helpful to begin the assessment process. However, plots of residuals versus fitted values and normal plots will provide more detailed information, and these visual displays will be used throughout the text.

Residuals versus Fits Plots

A scatterplot with the fitted line provides one visual method of checking linearity. Points will be randomly scattered above and below the line when the linear model is appropriate. Clear patterns, for example clusters of points above and below the line in a systematic fashion, indicate that the linear model is not appropriate.

A more informative way of looking at how the points vary about the regression line is a scatterplot of the residuals versus the fitted values for the prediction equation. This plot reorients the axes so that the regression line is represented as a horizontal line through zero. Positive residuals represent points that are above the regression line. The residuals versus fits plot allows us to focus on the estimated errors and look for any clear patterns without the added complexity of a sloped line.

The residual versus fits plot is especially useful for assessing the linearity and constant variance conditions of a simple linear model. The ideal pattern will be random variation above and below zero in a band of relatively constant width. Figure 1.4 shows a typical residual versus fits plot when these two conditions are satisfied.

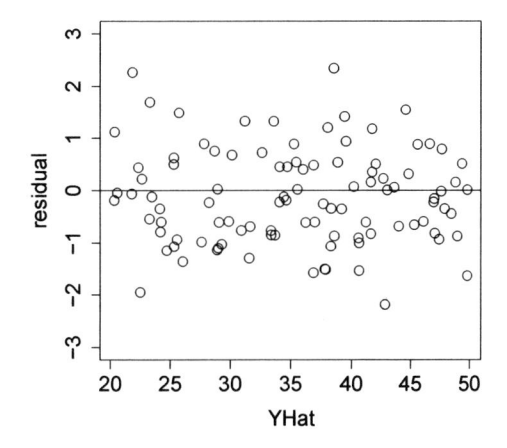

Figure 1.4: Residuals versus fitted values plot when linearity and constant variance conditions hold

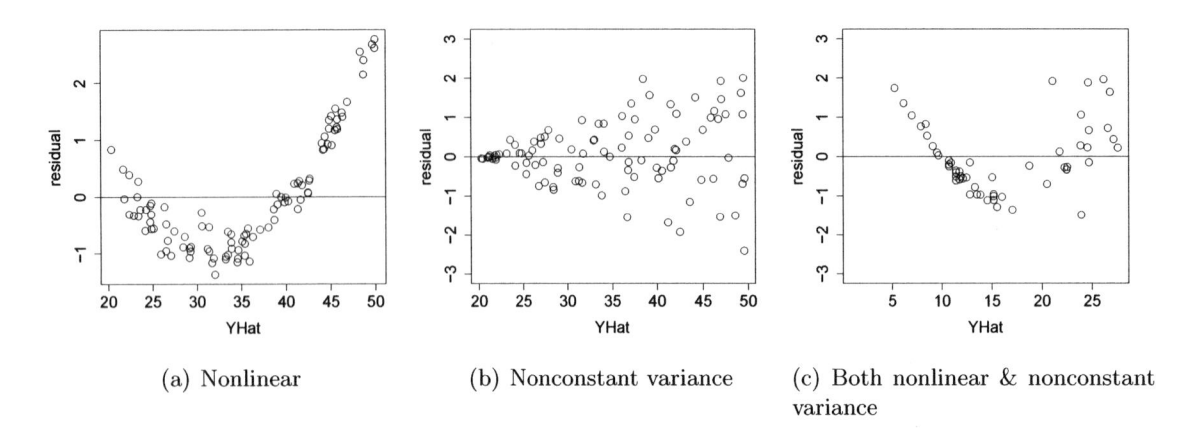

Figure 1.5: Residuals versus fitted values plots illustrating problems with conditions

(a) Nonlinear (b) Nonconstant variance (c) Both nonlinear & nonconstant variance

Figure 1.5 shows some examples of residual versus fits plots that exhibit some typical patterns indicating a problem with linearity, constant variance, or both conditions.

Figure 1.5(a) illustrates a curved pattern demonstrating a lack of linearity in the relationship. The residuals are mostly positive at either extreme of the graph and negative in the middle, indicating more of a curved relationship. Despite this pattern, the vertical width of the band of residuals is relatively constant across the graph, showing that the constant variance condition is probably reasonable for this model.

Figure 1.5(b) shows a common violation of the equal variance assumption. In many cases as the predicted response gets larger its variability also increases, producing a fan shape as in this plot. Note that a linearity assumption might still be valid in this case since the residuals are still equally dispersed above and below the zero line as we move across the graph.

Figure 1.5(c) indicates problems with both the linearity and constant variance conditions. We see a lack of linearity due to the curved pattern in the plot and, again, variance in the residuals that increases as the fitted values increase.

In practice, the assessment of a residual versus fits plot may not lead to as obvious a conclusion as in these examples. Remember that no model is "perfect" and we should not expect to always obtain the ideal plot. A certain amount of variation is natural, even for sample data that are generated from a model that meets all of the conditions. The goal is to recognize when departures from the model conditions are sufficiently evident in the data to suggest that an alternative model might be preferred or we should use some caution when the drawing conclusions from the model.

1.3. ASSESSING CONDITIONS

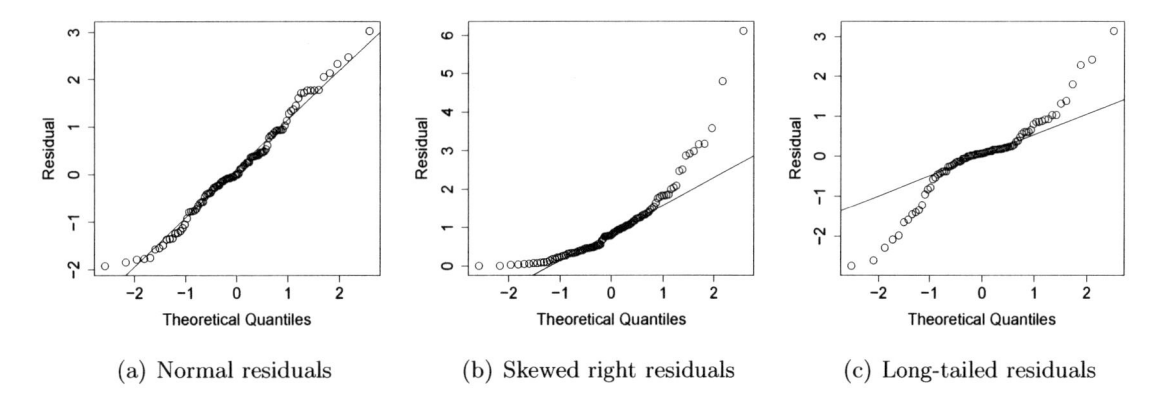

(a) Normal residuals (b) Skewed right residuals (c) Long-tailed residuals

Figure 1.6: Examples of normal quantile plots

Normal plots

Data from a normal distribution should exhibit a "bell-shaped" curve when plotted as a histogram or dotplot. However, we often need a fairly large sample to see this shape accurately and even then it may be difficult to assess whether the symmetry and curvature of the tails are consistent with a true normal curve. As an alternative, a **normal plot** shows a different view of the data where an ideal pattern for a normal sample is a straight line. Although there are a number of a variations, there are generally two common methods for constructing a normal plot.

The first, called a **normal quantile plot**, is a scatterplot of the ordered observed data versus values (the theoretical quantiles) that we would expect to see from a "perfect" normal sample of the same size. If the ordered residuals are increasing at the rate we would expect to see for a normal sample, the resulting scatterplot is a straight line. If the distribution of the residuals is skewed in one direction or has tails that are overly long due to some extreme outliers at both ends of the distribution the normal quantile plot will bend away from a straight line. Figure 1.6 shows several examples of normal quantile plots. The first (Figure 1.6(a)) was generated from residuals where the data were generated from a linear model with normal errors and the other two from models with non-normal errors.

The second common method of producing a normal plot is to use a **normal probability plot** such as those shown in Figure 1.7. Here the ordered sample data are plotted on the horizontal axis while the vertical axis is transformed to reflect the rate that normal probabilities grow. As with a normal quantile plot, the values increase as we move from left to right across the graph, but the revised scale produces a straight line when the values increase at the rate we would expect for a sample from a normal distribution. Thus, the interpretation is the same. A linear pattern (as in Figure 1.7(a)) indicates good agreement with normality and curvature or bending away from a straight line (as in Figure 1.7(b)) shows departures from normality.

(a) Normal residuals (b) Non-normal residuals

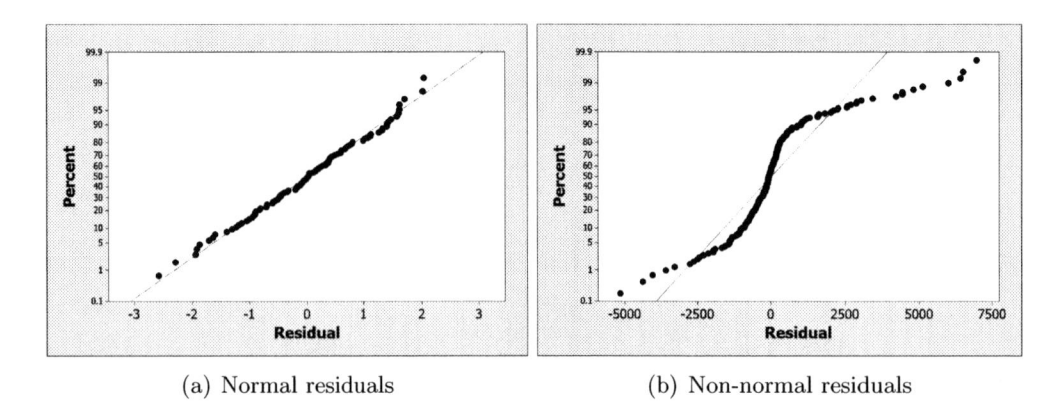

Figure 1.7: Examples of normal probability plots

Since both normal plot forms have a similar interpretation, we will use them interchangeably. The choice we make for a specific problem often depends on the options that are most readily available in the statistical software we are using.

Example 1.5: *Porsche prices (continued)*

ASSESS

We illustrate these ideas by checking the conditions for the model to predict Porsche prices based on mileage.

Linearity: Figure 1.2 shows that the linearity condition is reasonable as the scatterplot shows a consistent decline in prices with mileage and no obvious curvature. A plot of the residuals versus fitted values is shown in Figure 1.8. The horizontal band of points scattered randomly above and below the zero line illustrates that a linear model is appropriate for describing the relationship between price and mileage.

Zero mean: We used least squares regression which forces the sample mean of the residuals to be zero when estimating the intercept β_0. Also note that the residuals are scattered on either side of zero in the residual plot of Figure 1.8 and a histogram of the residuals, Figure 1.9, is centered at zero.

Constant variance: The fitted line plot in Figure 1.2 shows the data spread in roughly equal width bands on either side of the least squares line. Looking left to right in the plot of residuals versus fitted values in Figure 1.8 reinforces this finding as we see a fairly constant spread of the residuals above and below zero (where zero corresponds to actual prices that fall on the least squares regression line). This supports the constant variance condition.

1.3. ASSESSING CONDITIONS 35

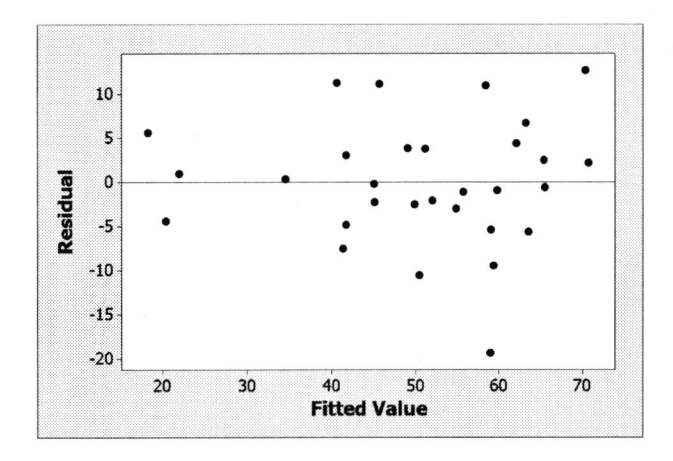

Figure 1.8: Plot of Porsche residuals versus fitted values

Independence and Random: We cannot tell from examining the data whether these conditions are satisfied. However, the context of the situation and the way the data were collected make these reasonable assumptions. There is no reason to think that one seller changing the asking price for a used car would necessarily influence the asking price of another seller. We were also told that these data were randomly selected from the Porsches for sale on the Autotrader.com website. So, at the least, we can treat it as a random sample from the population of all Porsches on that site at the particular time the sample was collected. We might want to be cautious about extending the findings to cars from a different site, an actual used car lot, or a later point in time.

Normality: In assessing normality we can refer to the histogram of residuals in Figure 1.9 where a reaqsonably bell-shaped pattern is displayed. However, a histogram based on this small sample may not be particularly informative and can change considerably depending on the bins used to

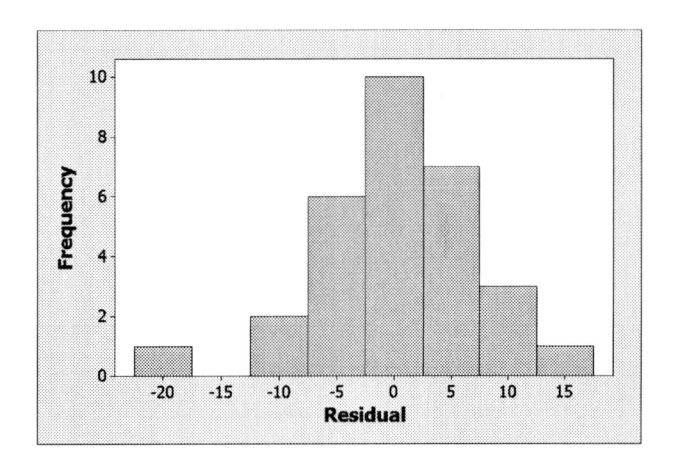

Figure 1.9: Histogram of Porsche residuals

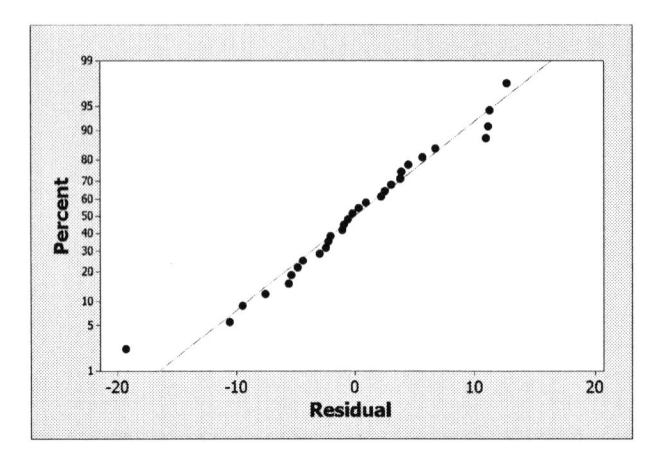

Figure 1.10: Normal Probability Plot of residuals for Porsche data

determine the bars. A more reliable plot for assessing normality is the normal probability plot of the residuals shown in Figure 1.10. This graph shows a consistent linear trend which supports the normality condition. We might have a small concern about the single point in the lower left corner of the plot, but we are looking more at the overall pattern when assessing normality.

USE

After we have decided on a reasonable model, we interpret the implications for the question of interest. For example, suppose we find a used Porsche for sale with 50 thousand miles and we believe that it is from the same population from which our sample of 30 used Porsches was drawn. What should we expect to pay for this car? Would it be an especially good deal if the owner was asking $38,000?

Based on our model, we would expect to pay

$$\widehat{Price} = 71.09 - 0.5894 \cdot 50 = 41.62$$

or $41,620. The asking price of $38,000 is below the expected price of $41,6200, but is this difference large relative to the variability in Porsche prices? We might like to know if this is a really good deal or perhaps such a low price that we should be concerned about the condition of the car. This question will be addressed in a Section 2.4 where we consider prediction intervals. For now, we can observe that the car's residual is about half of what we called a "typical error" ($\widehat{\sigma}_\epsilon = \$7.17$ thousand) below the expected price. Thus it is low, but not unusually so.   ◇

1.4 Transformations

If one or more of the conditions for a simple linear regression model are not satisfied, then we can consider transformations on one or both of the variables. In this section we provide two examples where this is the case.

Example 1.6: *Doctors and hospitals in metropolitan areas*

We expect the number of doctors in a city to be related to the number of hospitals, reflecting both the size of the city and the general level of medical care. Finding the number of hospitals in a given city is relatively easy, but counting the number of doctors is a more challenging task. Fortunately, the U.S. Census Bureau regularly collects such data for many metropolitan areas in the United States. The data in Table 1.2 show values for these two variables (and the *City* names) from the first few cases in the data file **MetroHealth83**, which has a sample of 83 metropolitan areas that have at least two community hospitals.

CHOOSE

As usual, we start the process of finding a model to predict the number of MDs ($NumMDs$) from the number of hospitals ($NumHospitals$) by examining a scatterplot of the two variables as seen in Figure 1.11. As expected this shows an increasing trend with cities having more hospitals also tending to have more doctors, suggesting that a linear model might be appropriate.

Table 1.2: Number of MDs and Community Hospitals for sample of $n = 83$ Metropolitan Areas

City	NumMDs	NumHospitals
Holland-Grand Haven, MI	349	3
Louisville, KY-IN	4042	18
Battle Creek, MI	256	3
Madison, WI	2679	7
Fort Smith, AR-OK	502	8
Sarasota-Bradenton-Venice, FL	2352	7
Anderson, IN	200	2
Honolulu, HI	3478	13
Asheville, NC	1489	5
Winston-Salem, NC	2018	6
⋮	⋮	⋮

Source: U.S. Census Bureau: 2006 State and Metropolitan Area Data Book (Table B-6)

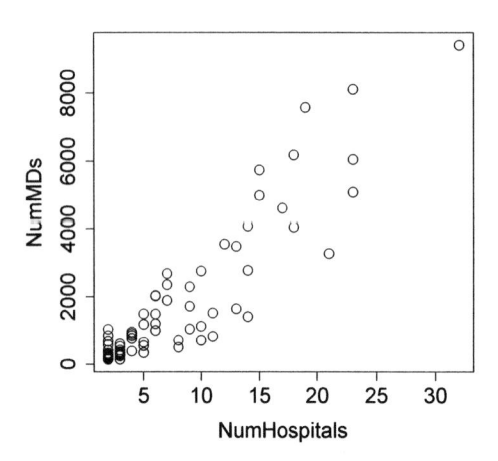

Figure 1.11: Scatterplot for Doctors versus Hospitals

FIT

Fitting a least squares line in R produces estimates for the slope and intercept as shown below, giving the prediction equation $\widehat{NumMDs} = -385.1 + 282.0 \cdot NumHospitals$. Figure 1.12(a) shows the scatterplot with this regression line as a summary of the relationship.

```
Call:
lm(formula = NumMDs ~ NumHospitals)

Coefficients:
 (Intercept)   NumHospitals
      -385.1          282.0
```

ASSESS

The line does a fairly good job of following the increasing trend in the relationship between number of doctors and number of hospitals. However a closer look at plots of the residuals shows some considerable departures from our standard regression assumptions. For example, the plot of residuals versus fitted values in Figure 1.12(b) shows a fan shape, with the variability in the residuals tending to increase as the fitted values get larger. This often occurs with count data like the number of MDs and number of hospitals where variability increases as the counts get larger. We can also observe this effect in a scatterplot with the regression line, Figure 1.12(a).

We also see from a histogram of the residuals, Figure 1.13(a), and normal quantile plot, Figure 1.13(b), that an assumption of normality would not be reasonable for the residuals in this

1.4. TRANSFORMATIONS

(a) Least squares line

(b) Residuals versus fits

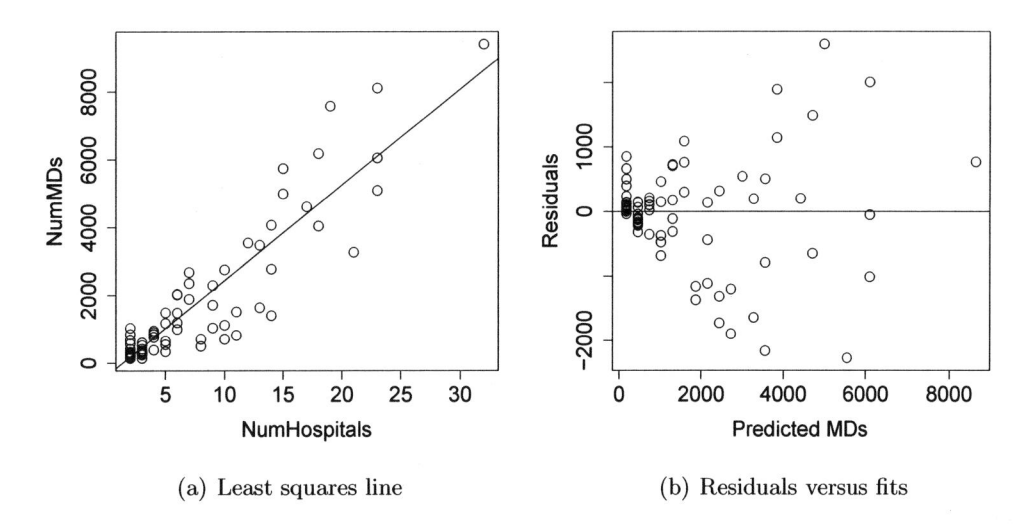

Figure 1.12: Regression for number of doctors based on number of hospitals

model. Although the histogram is relatively unimodal and symmetric, the peak is quite narrow with very long "tails." This departure from normality is seen more clearly in the normal quantile plot which has significant curvature away from a straight line at both ends.

(a) Histogram of Residuals

(b) Normal Quantile Plot

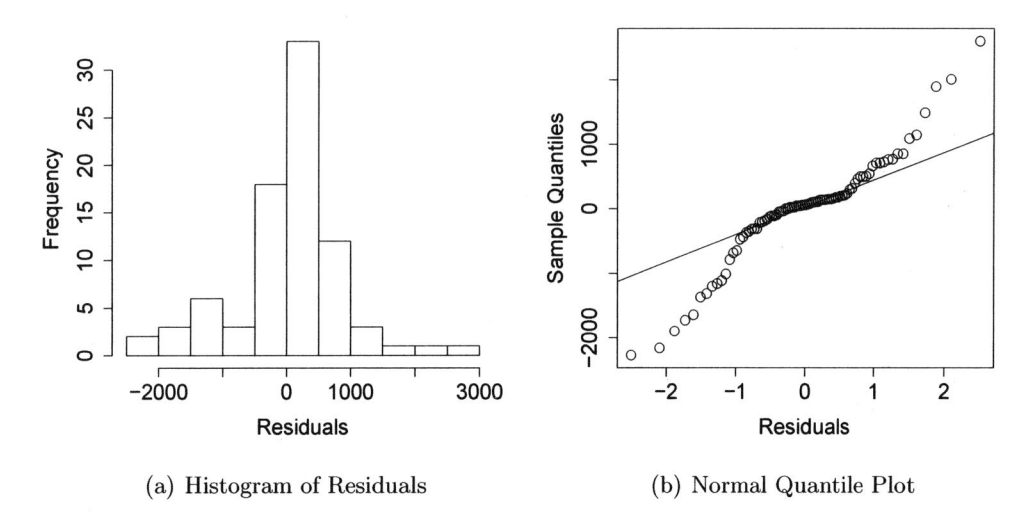

Figure 1.13: Normality Plots for Residuals of Doctors versus Hospitals

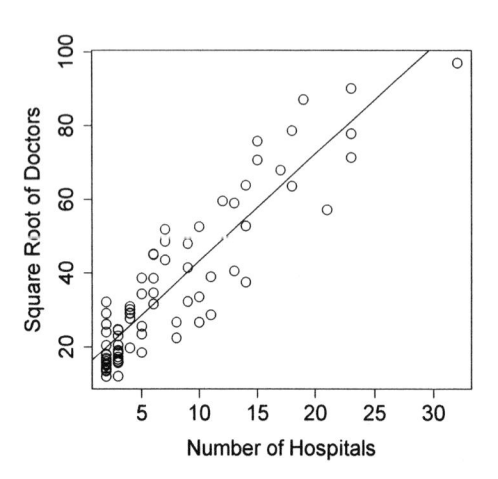

Figure 1.14: Least Squares Line for Sqrt(Doctors) versus Hospitals

CHOOSE (again)

To stabilize the variance in a response (Y) across different values of the predictor (X) we often try transformations on either Y or X. Typical options include raising a variable to a power (such as $\sqrt{Y}$, X^2, or $1/X$) or taking a logarithm (for example, using $log(Y)$ as the response). For count data, such as the number of doctors or hospitals where the variability increases along with the magnitudes of the variables, a square root transformation is often helpful. Figure 1.14 shows the least square line fit to the transformed data to predict the square root of the number of doctors based on the number of hospitals. The prediction equation is now $\widehat{\sqrt{NumMDs}} = 14.033 + 2.915 \cdot NumHospitals$.

When the equal variance assumption holds, we should see roughly parallel bands of data spread along the line. Although there might still be slightly less variability for the smallest numbers of hospitals, the situation is much better than for the data on the original scale. The residuals versus fitted values plot for the transformed data in Figure 1.15(a) and normal quantile plot of the residuals in Figure 1.15(b) also show considerable improvement at meeting the constant variance and normality conditions of our simple linear model.

USE

We must remember that our transformed linear model is predicting $\sqrt{MDs}$ so we must square its predicted values to obtain estimates for the actual number of doctors. For example, if we consider the case from the data of Louisville, Kentucky which has 18 community hospitals, the transformed model would predict

$$\widehat{\sqrt{NumMDs}} = 14.033 + 2.915 \cdot 18 = 66.50$$

1.4. TRANSFORMATIONS

(a) Residuals versus Fits

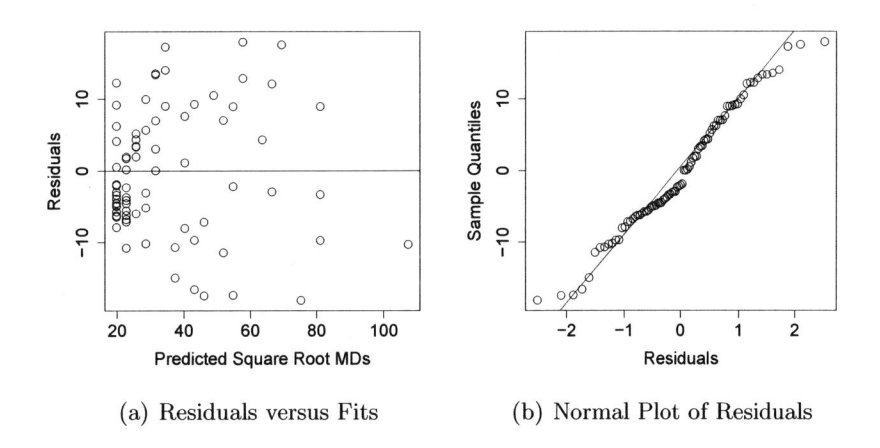

(b) Normal Plot of Residuals

Figure 1.15: Plots of Residuals for Sqrt(Doctors) versus Hospitals

so the predicted number of doctors is $66.50^2 = 4422.3$ while Louisville actually had 4042 doctors at the time of this sample. Figure 1.16 shows the scatterplot with the predicted number of doctors after transforming the linear model for the square roots of the number of doctors back to the original scale so that

$$\widehat{NumMDs} = (14.033 + 2.915 \cdot NumHospitals)^2.$$

Note that we could use this model to make predictions for other cities, but in doing so we should feel comfortable only to the extent that we believe the sample to be representative of the larger population of cities with at least two community hospitals. ◇

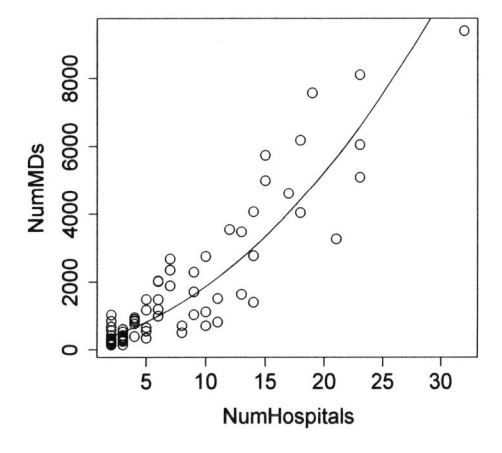

Figure 1.16: Predicted $NumMDs$ from the linear model for sqrt($NumMDs$)

Table 1.3: Species and Area

Island	Area (km^2)	Mammal Species
Borneo	743244	129
Sumatra	473607	126
Java	125628	78
Bangka	11964	38
Bunguran	1594	24
Banggi	450	18
Jemaja	194	15
Karimata Besar	130	19
Tioman	114	23
Siantan	113	16
Sirhassan	46	16
Redang	25	8
Penebangan	13	13
Perhentian Besar	8	6

Source: Heaney, Lawrence R. (1984) Mammalian species richness on islands on the Sunda Shelf, Southeast Asia, *Oecologia* .

Example 1.7: *Species by area*

The data in Table 1.3 (and the file **SpeciesArea**) show the number of mammal species and the area for 13 islands in Southeast Asia. Biologists have speculated that the number of species is related to the size of an island and would like to be able to predict the number of species given the size of an island.

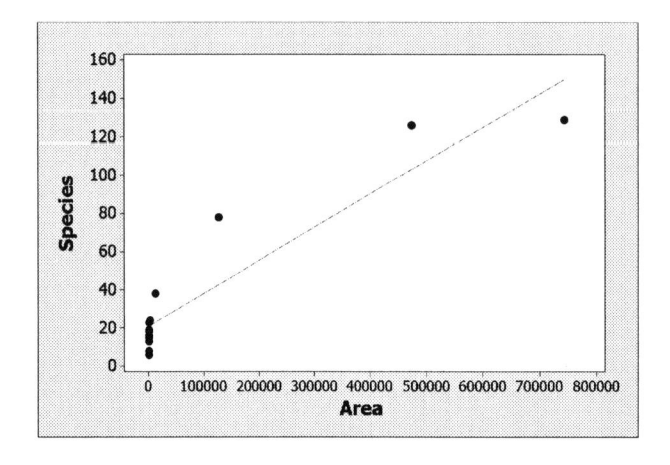

Figure 1.17: Number of Mammal Species versus Area for S.E. Asian Islands

1.4. TRANSFORMATIONS

Figure 1.17 shows a scatterplot with least squares line added. Clearly the line does not provide a good summary of this relationship, because it doesn't reflect the curved pattern shown in the plot.

In a case like this, where we see strong curvature and extreme values in a scatterplot, a logarithm transformation of either the response variable, the predictor, or possibly both, is often helpful. Applying a log transformation[4] to the species variable results in the scatterplot of Figure 1.18(a).

(a) Log *Species* versus *Area*

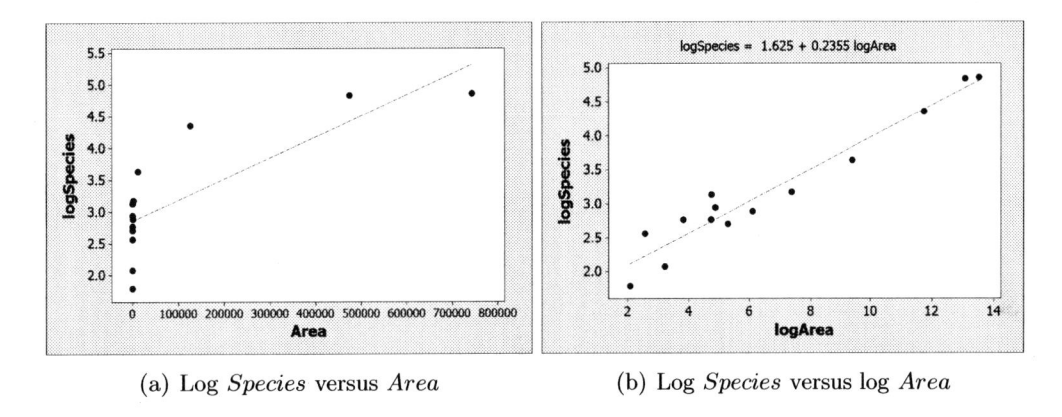

(b) Log *Species* versus log *Area*

Figure 1.18: Log transformations of *Species* versus *Area* for S.E. Asian Islands

Clearly this transformation has failed to produce a linear relationship. However if we also take a log transformation of the area we obtain the plot shown in Figure 1.18(b), which does show a linear pattern. Figure 1.19 shows a residual plot from this regression which does not show any striking patterns.

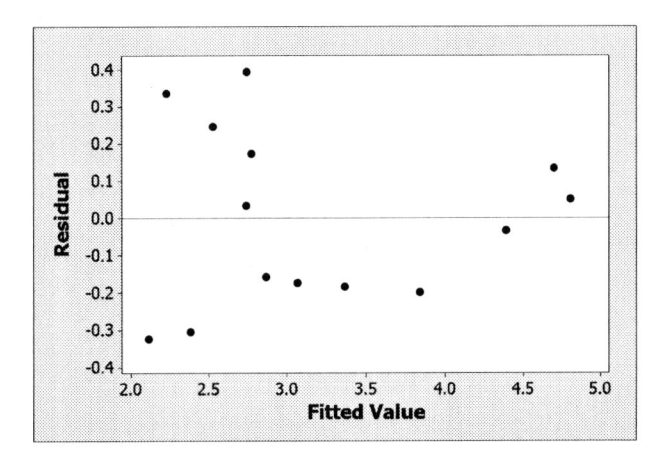

Figure 1.19: Residual plot after log transform of response and predictor

---

[4] In this text we use log to denote the natural logarithm.

Based on the fitted model we can predict $log(Species)$ based on $log(Area)$ for an island with the model

$$log(\widehat{Species}) = 1.625 + 0.2355 \cdot log(Area)$$

Suppose we wanted to use the model to find the predicted value for Java which has an area of 125,628 square kilometers. We substitute 125,628 into the equation as the *Area* and compute an estimate for the log of the number of species.

$$log(\widehat{Species}) = 1.625 + 0.2355 \cdot log(125,628) = 4.390$$

Our estimate for the number of the number of species is then

$$e^{4.390} = 80.6 \text{ species}.$$

The actual number of mammal species found on Java for this study was 78. ◇

There is no guarantee that transformations will eliminate or even reduce the problems with departures from the conditions for a simple linear regression model. Finding an appropriate transformation is as much an art as a science.

1.5 Outliers & Influential Points

Sometimes a data point just doesn't fit within a linear trend that is evident in the other points. This can be because the point doesn't fit with the other points in a scatterplot vertically — it is an *outlier* — or a point may differ from the others horizontally and vertically so that it is an *influential* point. In this section we examine methods for identifying outliers and influential points using graphs and summary statistics.

Outliers

We classify a data point as an *outlier* if it stands out away from the pattern of the rest of the data and is not well described by the model. In the simple linear model setting, an outlier is point where the magnitude of the residual is unusually large. How large must a residual be for a point to be called an outlier? That depends on the variability of all the residuals as we see in the next example.

Example 1.8: *Olympic long jump*

During the 1968 Olympics, Bob Beamon shocked the track and field world by jumping 8.9 meters (29 feet 2.5 inches), breaking the world record for the long jump by 0.65 meters (more than two feet). Figure 1.20 shows the winning men's Olympic long jump distance (labeled as *Gold*) versus *Year*, together with the least squares regression line, for the $n = 26$ Olympics held during the period 1900 - 2008. The data are stored in **LongJumpOlympics**.

1.5. OUTLIERS & INFLUENTIAL POINTS

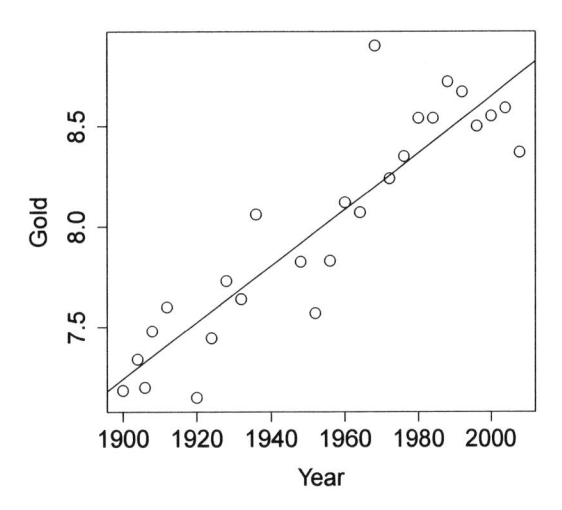

Figure 1.20: Gold medal winning distances (m) for the men's Olympic long jump, 1900 - 2008

The 1968 point clearly stands above the others and is far removed from the regression line. Because this point does not fit the general pattern in the scatterplot, it is an outlier. The unusual nature of this point is perhaps even more evident in Figure 1.21, a residual plot for the fitted least squares model.

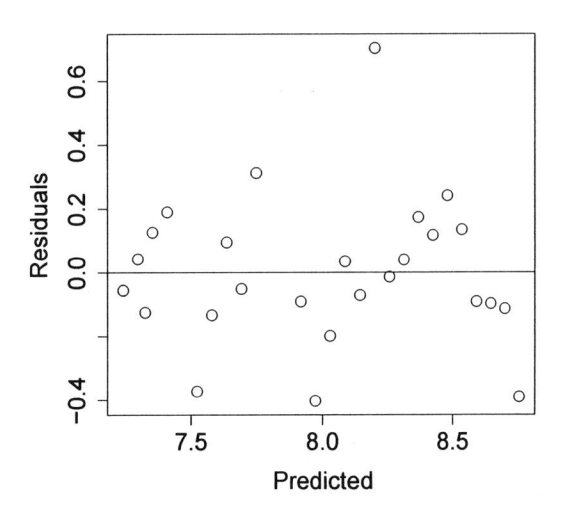

Figure 1.21: Residual plot for long jump model

The fitted regression model is

$$\widehat{Gold} = -19.48 + 0.014066 \cdot Year$$

Thus the predicted 1968 winning long jump is $\widehat{Gold} = -19.48 + 0.014066 \cdot 1968 = 8.20$ meters. The 1968 residual is $8.90 - 8.20 = 0.70$ meters.

Even when we know the context of the problem, it can be difficult to judge whether a residual of 0.70m is unusually large. One method to help decide when a residual is extreme is to put the residuals on a standard scale. For example, since the estimated standard deviation of the regression error, $\hat{\sigma}_\epsilon$, reflects the size of a "typical" error we could standardize each residual using

$$\frac{y - \hat{y}}{\hat{\sigma}_\epsilon}$$

In practice most statistical packages make some modifications to this formula when computing a **standardized** residual to account for how unusual the predicted value is for a particular case. Since an extreme outlier might have a significant effect on the estimation of σ_ϵ, another common adjustment is to estimate the standard deviation of the regression error using a model which is fit after omitting the point in question. Such residuals are often called **studentized**[5] (or **deleted-t**) residuals.

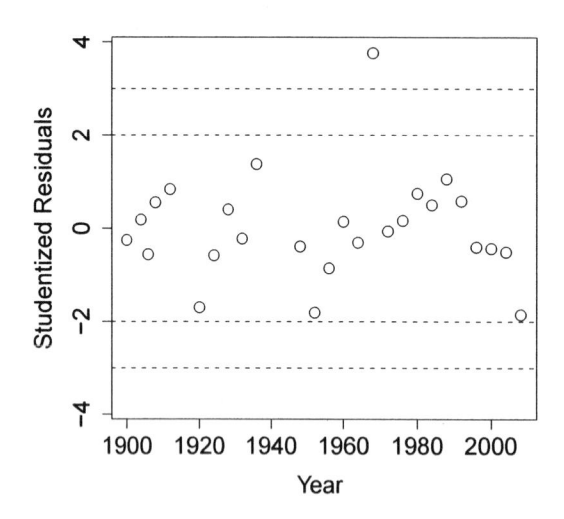

Figure 1.22: Studentized residuals for the long jump model

If the conditions of a simple linear model hold, approximately 95% of the residuals should be within 2 standard deviations of the residual mean of zero, so we would expect most standardized

---
[5] You may recall the t-distribution, which is sometimes called Student's t, from an introductory statistics course.

1.5. OUTLIERS & INFLUENTIAL POINTS

or studentized residuals to be less than 2 in absolute value. We may be slightly suspicious about points where the magnitude of the standardized or studentized residual is greater than 2 and even more wary about points beyond ±3. For example, the standardized residual for the Bob Beamon's 1968 jump is 3.03, indicating this point is an outlier. Figure 1.22 shows the studentized residuals for the long jump data plotted against the predicted values. The studentized residual for the 1968 jump is 3.77, while none of the other studentized residuals are beyond ±2, clearly pointing out the exceptional nature of that performance. ◇

Influential Points

When we fit a regression model and make a prediction we combine information across several observations, or cases, to arrive at the prediction, or fitted value, for a particular case. For example, we use the mileages and prices of many cars to arrive at a predicted price for a particular car. In doing this, we give equal weight to all of the cases in the dataset; that is, every case contributes equally to the creation of the fitted regression model and to the subsequent predictions that are based on that model.

Usually, this is a sensible and useful thing to do. Sometimes, however, this approach can be problematic, especially when the data contain one or more extreme cases that might have a significant impact on the coefficient estimates in the model.

Example 1.9: *Butterfly ballot*

The race for the presidency of the United States in the fall of 2000 was very close, with the electoral votes from Florida determining the outcome. Nationally, George W. Bush received 47.9% of the popular vote, Al Gore received 48.4%, and the rest of the popular vote was split among several other candidates. In the disputed final tally in Florida, Bush won by just 537 votes over Gore (48.847% to 48.838%) out of almost 6 million votes cast. About 2.3% of the votes cast in Florida were awarded to other candidates. One of those other candidates was Pat Buchanan, who did much better in Palm Beach County than he did anywhere else. Palm Beach County used a unique "butterfly ballot" that had candidate names on either side of the page with "chads" to be punched in the middle. This non-standard ballot seemed to confuse some voters, who punched votes for Buchanan that may have been intended for a different candidate. Figure 1.23 shows the number of votes that Buchanan received plotted against the number of votes that Bush received for each county, together with the fitted regression line ($\widehat{Buchanan} = 45.3 + 0.0049 * Bush$). The data are stored in **PalmBeach**.

The data point near the top of the scatterplot is Palm Beach County, where Buchanan picked up over 3,000 votes. Figure 1.24 is a plot of the residuals versus fitted values for this model; clearly Palm Beach County stands out from the rest of the data. Minitab computes the standardized residual for Palm Beach using this model to be 7.65 and the studentized residual to be 24.08! No

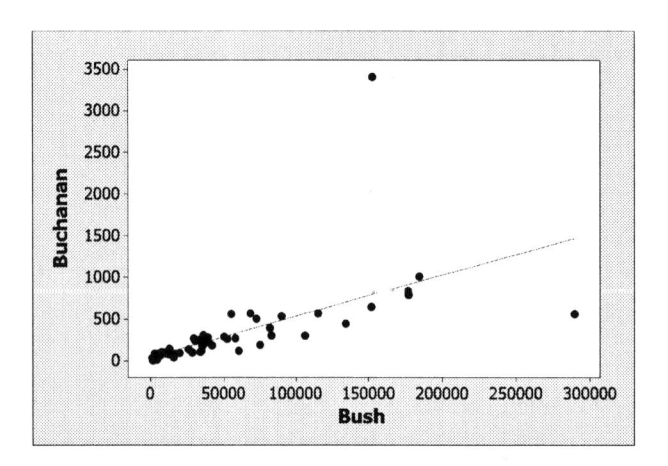

Figure 1.23: 2000 Presidential election totals in Florida counties

question that this point should be considered an outlier. Also, the data point at the far right on the plots (Dade County) has a large negative residual -907.5 which gives a standardized residual of -3.06; certainly something to consider as an outlier although not as dramatic as Palm Beach.

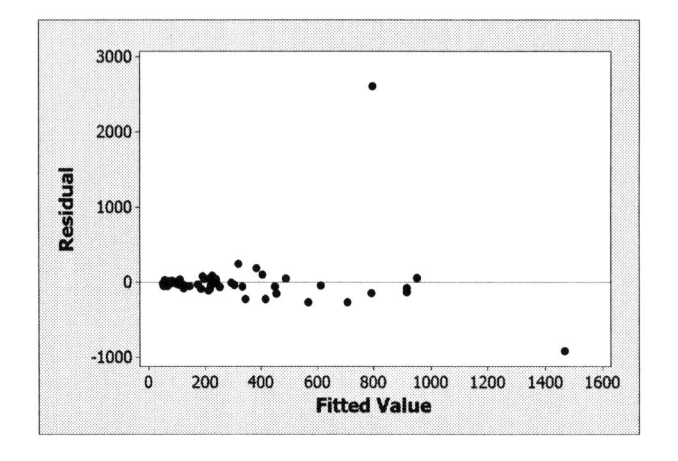

Figure 1.24: Residual plot for the butterfly ballot data

Other than recognizing that the model does a poor job of predicting Palm Beach County (and to a lesser extent Dade County), should we worry about the effect that such extreme values have on the rest of the predictions given by the model? Would removing Palm Beach County from the dataset produce much change in the regression equation? Portions of the Minitab output for fitting the simple linear model with and without the Palm Beach County data point are shown below. Figure 1.25 shows both regression lines with the steeper slope (0.0049) occurring when Palm Beach County is included and the shallower slope (0.0035) when that point is omitted. Notice that the effect of the extreme value for Palm Beach is to "pull" the regression line in its direction.

1.5. OUTLIERS & INFLUENTIAL POINTS

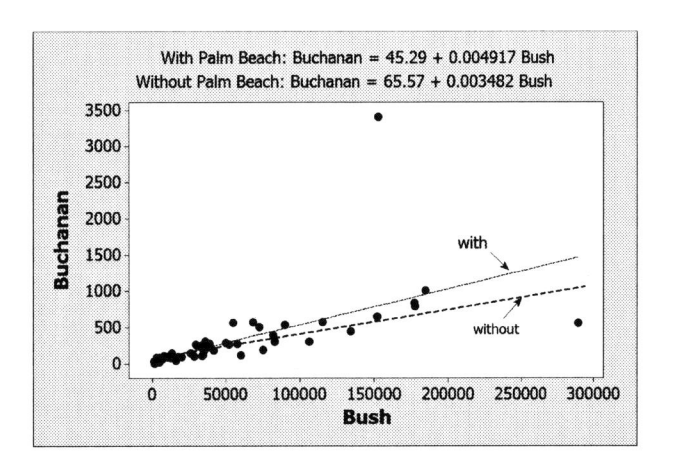

Figure 1.25: Regression lines with and without Palm Beach

Regression output with Palm Beach County:

```
The regression equation is
Buchanan = 45.3 + 0.00492 Bush

Predictor         Coef      SE Coef       T       P
Constant         45.29        54.48    0.83   0.409
Bush         0.0049168    0.0007644    6.43   0.000

S = 353.922    R-Sq = 38.9%    R-Sq(adj) = 38.0%
```

Regression output without Palm Beach County:

```
The regression equation is
Buchanan = 65.6 + 0.00348 Bush

Predictor         Coef      SE Coef       T       P
Constant         65.57        17.33    3.78   0.000
Bush         0.0034819    0.0002501   13.92   0.000

S = 112.453    R-Sq = 75.2%    R-Sq(adj) = 74.8%
```

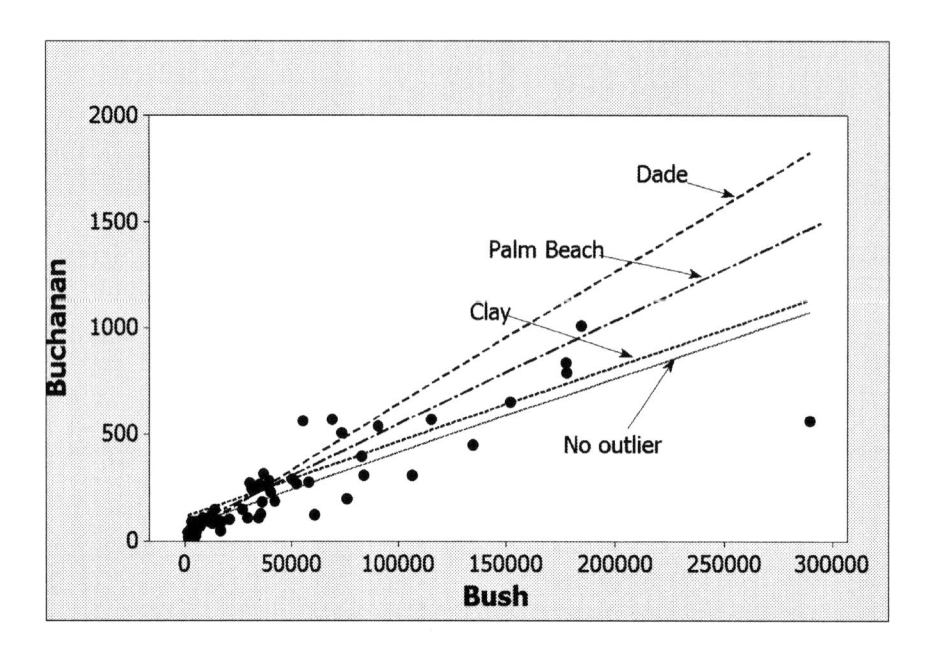

Figure 1.26: Regression lines with an outlier of 3407 "moved" to different counties

The amount of **influence** that a single point has on a regression fit depends on how well it aligns with the pattern of the rest of the points and on its value for the predictor variable. Figure 1.26 shows the regression lines we would have obtained if the extreme value (3,407 Buchanan votes) had occurred in Dade County (with 289,456 Bush votes), Palm Beach County (152,846 Bush votes), Clay County (41,745 Bush votes) or not occurred at all. Note that the more extreme values for the predictor (large Bush counts in Dade or Palm Beach) produced a bigger effect on the slope of the regression line than when the outlier was placed in a more "average" Bush county such as Clay.

Generally, points farther from the mean value of the predictor ($\bar{x}$) have greater potential to influence the slope of a fitted regression line. This concept is known as the **leverage** of a point. Points with high leverage have a greater capacity to pull the regression line in their direction than do low leverage points near the predictor mean. Although in the case of a single predictor, we could measure leverage as just the distance from the mean, we introduce a somewhat more complicated statistic in Section 4.3 that can be applied to more complicated regression settings.

1.6 Chapter Summary

In this chapter we considered a **simple linear regression model** for predicting a single quantitative response variable Y from a single quantitative predictor X.

$$Y = \beta_0 + \beta_1 \cdot X + \epsilon$$

You should be able to use statistical software to estimate and interpret the slope and intercept for this model to produce a least squares regression line.

$$\hat{Y} = \widehat{\beta_0} + \widehat{\beta_1} \cdot X$$

The coefficient estimates, $\widehat{\beta_0}$ and $\widehat{\beta_1}$, are obtained using a method of least squares that selects them to provide the smallest possible sum of squared residuals (SSE). A **residual** is the observed response (y) minus the predicted response $(\hat{y})$, or the vertical distance from a point to the line. You should be able to interpret, in context, the intercept and slope, as well as residuals.

The **scatterplot**, which shows the relationship between two quantitative variables, is an important visual tool to help choose a model. We look for the direction of association (positive, negative, or a random scatter), the strength of the relationship, and the degree of linearity. Assessing the fit of the model is a very important part of the modeling process. The conditions to check when using a simple linear regression model include **linearity**, **zero mean** (of the residuals), **constant variance** (about the regression line), **independence**, **random selection** (or random assignment), and **normality** (of the residuals). We can summarize several of these conditions by specifying the distribution of the error term, $\epsilon \sim N(0, \sigma_\epsilon)$. In addition to a scatterplot with the least squares line, various residual plots, such as a **histogram of the residuals** or a **residuals versus fits plot**, are extremely helpful in checking these conditions. Once we are satisfied with the fit of the model, we use the estimated model to make inferences or predictions.

Special plots, known as a **normal quantile plot** or **normal probability plot**, are useful in assessing the normality condition. These two plots are constructed using slightly different methods, but the interpretation is the same. A linear trend indicates that normality is reasonable, and departures from linearity indicate trouble with this condition. Be careful not to confuse linearity in normal plots with the condition of a linear relationship between the predictor and response variable.

You should be able to estimate the **standard deviation of the error** term, σ_ϵ, the third parameter in the simple linear regression model. The estimate is based on the sum of squared errors (SSE) and the associated degrees of freedom $(n-2)$,

$$\widehat{\sigma_\epsilon} = \sqrt{\frac{SSE}{n-2}}$$

and is the typical error, often referred to as the **regression standard error**.

If the conditions are not satisfied, then **transformations** on the predictor, response, or both variables should be considered. Typical transformations include the square root, reciprocal, logarithm, and raising the variable(s) to another power. Identifying a useful transformation for a particular data set (if one exists at all) is as much art as science. Trial and error is often a good approach.

You should be able to identify obvious **outliers** and **influential points**. Outliers are points that are unusually far away from the overall pattern shown by the other data. Influential points exert considerable impact on the estimated regression line. We will look at more detailed methods for identifying outliers and influential points in Section 4.3. For now, you should only worry about recognizing very extreme cases and be aware that they can affect the fitted model and analysis. One common guideline is to tag all observations with **standardized** or **studentized residuals** smaller than -2 or larger than 2 as possible outliers. To see if a point is influential, fit the model with and without that point to see if the coefficients change very much. In general, points far from the average value of the predictor variable have greater potential to influence the regression line.

1.7 Exercises

Conceptual Exercises

1.1 *Equation of a line.* Consider the fitted regression equation $\hat{Y} = 100 + 15 \cdot X$. Which of the following is *false*?

   a. The sample slope is 15.

   b. The predicted value of Y when $X = 0$ is 100.

   c. The predicted value of Y when $X = 2$ is 110.

   d. Larger values of X are associated with larger values of Y.

1.2 *Computing a residual.* Consider the fitted regression equation $\hat{Y} = 25 + 7 \cdot X$. If $x_1 = 10$ and $y_1 = 100$, what is the residual for the first data point?

1.3 *Residual plots to check conditions.* For which of the following conditions for inference in regression does a residual plot *not* aid in assessing whether the condition is satisfied?

   a. linearity

   b. constant variance

   c. independence

   d. zero mean

Guided Exercises

1.4 *Breakfast cereal.* The number of calories and number of grams of sugar per serving were measured for 36 breakfast cereals. The data are in the file **Cereal**. We are interested in trying to predict the calories using the sugar content.

   a. Make a scatterplot and comment on what you see.

   b. Find the least squares regression line for predicting calories based on sugar content.

   c. How many calories would this fitted model predict for a cereal that has 10 grams of sugar?

   d. Cheerios has 110 calories but just one gram of sugar. Find the residual for this data point.

   e. Interpret the value (not just the sign) of the slope of the fitted model in the context of this setting.

   f. Does the linear regression model appear to be a good summary of the relationship between calories and sugar content of breakfast cereals?

Table 1.4: Pages and price for textbooks

Pages	Price	Pages	Price	Pages	Price
600	95.00	150	16.95	696	130.50
91	19.95	140	9.95	294	7.00
200	51.50	194	5.95	526	41.25
400	128.50	425	58.75	1060	169.75
521	96.00	51	6.50	502	71.25
315	48.50	930	70.75	590	82.25
800	146.75	57	4.25	336	12.95
800	92.00	900	115.25	816	127.00
600	19.50	746	158.00	356	41.50
488	85.50	104	6.50	248	31.00

Source: Cal Poly student study

1.5 *Capacitor voltage.* A capacitor was charged with a nine-volt battery and then a voltmeter recorded the voltage as the capacitor was discharged. Measurements were taken every .02 seconds. The data are in the file **Volts**.

- a. Make a scatterplot with *Voltage* on the vertical axis versus *Time* on the horizontal axis. Comment on the pattern.

- b. Transform *Voltage* using a log transformation and then plot log(*Voltage*) versus *Time*. Comment on the pattern.

- c. Regress log(*Voltage*) on *Time* and write down the prediction equation.

- d. Make a plot of residuals versus fitted values from the regression from part (c). Comment on the pattern.

Open-ended Exercises

1.6 *Textbook prices.* Two undergraduate students at Cal Poly took a random sample of 30 textbooks from the campus bookstore in the fall of 2006. They recorded the price and number of pages in each book, in order to investigate the question of whether number of pages can be used to predict price. Their data are stored in the file **TextPrices** and appear in Table 1.4.

- a. Produce the relevant scatterplot to investigate the students' question. Comment on what the scatterplot reveals about the question.

- b. Determine the equation of the regression line for predicting price from number of pages.

- c. Produce and examine relevant residual plots, and comment on what they reveal about whether the conditions for inference are met with these data.

1.7. EXERCISES

Table 1.5: Major League Baseball game times

Game	League	Runs	Margin	Pitchers	Attendance	Time
CLE-DET	AL	14	6	6	38774	168
CHI-BAL	AL	11	5	5	15398	164
BOS-NYY	AL	10	4	11	55058	202
TOR-TAM	AL	8	4	10	13478	172
TEX-KC	AL	3	1	4	17004	151
OAK-LAA	AL	6	4	4	37431	133
MIN-SEA	AL	5	1	5	26292	151
CHI-PIT	NL	23	5	14	17929	239
LAD-WAS	NL	3	1	6	26110	156
FLA-ATL	NL	19	1	12	17539	211
CIN-HOU	NL	3	1	4	30395	147
MIL-STL	NL	12	12	9	41121	185
ARI-SD	NL	11	7	10	32104	164
COL-SF	NL	9	5	7	32695	180
NYM-PHI	NL	15	1	16	45204	317

(The *Time* is recorded in minutes. *Runs* and *Pitchers* are totals for both teams combined. *Margin* is the difference between the winner's and loser's scores.)

1.7 *Baseball game times.* What factors can help to predict how long a Major League Baseball game will last? The data in Table 1.5 were collected at *www.baseball-referecne.com* for the 15 games played on August 26, 2008 and stored in the file named **BaseballTimes**.

a. First analyze the distribution of the response variable (*Time* in minutes) alone. Use a graphical display (dotplot, histogram, boxplot) as well as descriptive statistics. Describe the distribution. Also identify the outlier (which game is it?), and suggest a possible explanation for it.

b. Examine scatterplots to investigate which of the quantitative predictor variables appears to be the best single predictor of time. Comment on what the scatterplots reveal.

c. Choose the one predictor variable that you consider to be the best predictor of time. Determine the regression equation for predicting time based on that predictor. Also interpret the slope coefficient of this equation.

d. Analyze appropriate residual plots, and comment on what they reveal about whether the conditions for inference appear to be met here.

1.8 *Baseball game times (continued).* Refer to the previous Exercise 1.7 on the playing time of baseball games.

   a. Which game has the largest residual (in absolute value) for the model that you selected? Is this the same game that you identified as an outlier based on your analysis of the time variable alone?

   b. Repeat the entire analysis from the previous exercise, with the outlier omitted.

   c. Comment on the extent to which omitting the outlier changed the analysis and your conclusions.

1.9 *Zero mean?* One of the neat consequences of the least squares line is that the sample means $(\bar{x}, \bar{y})$ always lie on the line, so that $\bar{y} = \hat{\beta}_0 + \hat{\beta}_1 \bar{x}$. From this we can get an easy way to calculate the intercept if we know the two means and the slope.

$$\hat{\beta}_0 = \bar{y} - \hat{\beta}_1 \cdot \bar{x}$$

We could use this formula to calculate an intercept for *any* slope, not just the one obtained by least squares estimation. See what happens if you try this. Pick any data set with two quantitative variables, find the mean for both variables, and assign one to be the predictor and the other the response. Pick any slope you like and use the formula above to compute an intercept for your line. Find predicted values and then residuals for each of the data points using this line as a prediction equation. Compute the sample mean of your residuals. What do you notice?

Chapter 2

Inference for Simple Linear Regression

Recall that in Example 1.1 we considered a simple linear regression model to predict the price of used Porches based on mileage. How can we evaluate the effectiveness of this model? Are prices significantly related to mileage? How much of the variability in Porsche prices can we explain by knowing their mileages? If we are interested in a used Porsche with about 50,000 miles, how accurately can we predict its price?

In this chapter we consider various aspects of inference based on a simple linear regression model. By inference, we mean methods such as confidence intervals and hypothesis tests that allow us to answer questions of interest about the population based on the data in our sample. Recall that the simple linear model is

$$Y = \beta_0 + \beta_1 X + \epsilon$$

where $\epsilon \sim N(0, \sigma_\epsilon)$. Many of the inference methods, such as those introduced in Section 2.1, deal with the slope parameter β_1. Note that if $\beta_1 = 0$ in the model there is no linear relationship between the predictor and the response. In Section 2.2 we examine how the variability in the response can be partitioned into one part for the linear relationship and another part for variability due to the random error. This partitioning can be used to assess how well the model explains variability in the response. The correlation coefficient r is another way to measure the strength of linear association between two quantitative variables. In Section 2.3 we connect this idea of correlation to the assessment of a simple linear model. Finally, in Section 2.4, we consider two forms of intervals that are important for quantifying the accuracy of predictions based on a regression model.

2.1 Inference for Regression Slope

We saw in Example 1.3 on page 27 that the fitted regression model for the Porsche cars dataset is

$$\widehat{Price} = 71.1 - 0.589 \cdot Mileage$$

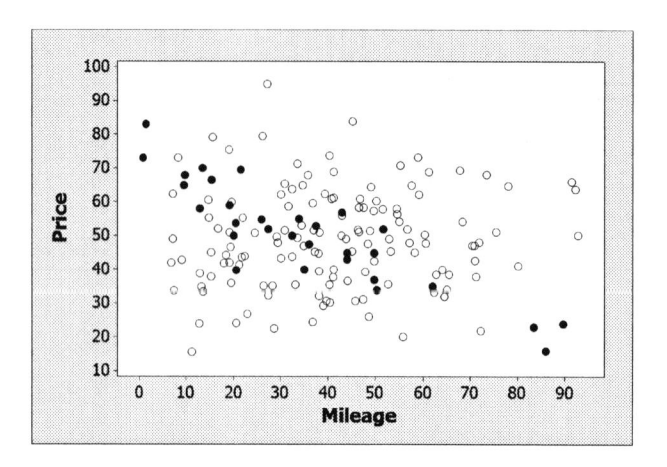

Figure 2.1: Hypothetical population with no relationship between Price and Mileage

suggesting that for every additional 1,000 miles on a used car, the price goes down by $589 on average. A skeptic might claim that price and mileage have no linear relationship, that the value of -0.589 in our fitted regression model is just a fluke, and that in the population the corresponding parameter value is really zero. The skeptic might have in mind that the population scatterplot looks something like Figure 2.1 and that our data, shown as the dark points in the plot, arose as an odd sample. Even if there is no true relationship between Y and X, any particular sample of data will show some positive or negative fitted regression slope just by chance. After all, it would be extremely unlikely that a random sample would give a slope of exactly zero between X and Y. Now we examine both a formal mechanism for assessing when the population slope is likely to be different from zero and a method to put confidence bounds around the sample slope.

T-test for slope

To assess whether the slope for sample data provides significant evidence that the slope for the population differs from zero we need to estimate its variability. The standard error of the slope, $SE_{\hat{\beta}_1}$, measures how much we expect the sample slope (i.e., the fitted slope) to vary from one random sample to another[1]. The standard errors for coefficients (slope or intercept) of a regression model are generally provided by statistical software such as the portion of the Minitab output for the Porsche prices in Figure 1.3 (page 28) that is reproduced below.

```
Predictor      Coef   SE Coef      T      P
Constant     71.090     2.370  30.00  0.000
Mileage     -0.58940   0.05665 -10.40  0.000
```

---

[1] Be careful to avoid confusing the **standard error of the slope**, $SE_{\hat{\beta}_1}$, with the **standard error of the regression**, $\hat{\sigma}_\epsilon$.

2.1. INFERENCE FOR REGRESSION SLOPE

The ratio of the slope to its standard error is one test statistic for this situation.

$$t = \frac{\hat{\beta}_1}{SE_{\hat{\beta}_1}} = \frac{-0.5894}{0.05665} = -10.4$$

Given that the test statistic is far from zero (the sample slope is more than 10 standard errors below a slope of zero), we can reject the hypothesis that the true slope, β_1, is zero.

Under the null hypotheses of no linear relationship and the regression conditions, the test statistic follows a **t-distribution** with $n-2$ degrees of freedom. Software will provide a p-value, based on this distribution, which measures the probability of getting a statistic more extreme than the observed test statistic. Our statistical inference about the linear relationship depends on the magnitude of the p-value. That is, when the p-value is below our significance level α we reject the null hypothesis and conclude that the slope differs from zero.

T-test for the Slope of a Simple Linear Model

To test whether the population slope is different from zero the hypotheses are

$H_0 : \beta_1 = 0$
$H_a : \beta_1 \neq 0$

and the test statistic is

$$t = \frac{\hat{\beta}_1}{SE_{\hat{\beta}_1}}.$$

If the conditions for the simple linear model, including normality, hold, we may compute a p-value for the test statistic using a t-distribution with $n - 2$ degrees of freedom.

Note that the $n - 2$ degrees of freedom for the t-distribution in this test is inherited from the $n - 2$ degrees of freedom in the estimate of the standard error of the regression. When the t-statistic is extreme in either tail of the t-distribution (as it is for the Porsche data), the p-value will be small (reported as 0.000 in the Minitab output), providing strong evidence that the slope in the population is different from zero. In some cases, we might be interested in testing for an association in a particular direction (for example, a positive association between number of doctors and number of hospitals in a city) in which case we would use a one-tailed alternative (such as $H_a : \beta_1 > 0$) in the t-test and compute the p-value using only the area in that tail. Software generally provides the two-sided p-value, so we need to divide by two for a one-sided p-value. We could also use the information in the computer output to perform a similar test for the intercept in the regression model, but that is rarely a question of practical importance.

Confidence Interval for Slope

The slope β_1 of the population regression line is usually the most important parameter in a regression problem. The slope is the rate of change of the mean response as the explanatory variable increases. The slope $\hat{\beta}_1$ of the least-squares line is an estimate of β_1. A confidence interval for the slope may be more useful than a test because it shows how accurate the estimate $\hat{\beta}_1$ is likely to be (and tells us more than just whether or not the slope is zero).

Confidence Interval for the Slope of a Simple Linear Model

The confidence interval for β_1 has the form

$$\hat{\beta}_1 \pm t^* \cdot SE_{\hat{\beta}_1}$$

where t^* is the critical value for the t_{n-2} density curve to obtain the desired confidence level.

In our simple linear regression example to predict Porsche prices, the sample slope, $\hat{\beta}_1$, is -0.5894. The computer output (page 58) shows us that the standard error of this estimate is 0.05665, with 28 degrees of freedom. For a 95% confidence level the t^* value is 2.05. Thus, a 95% confidence interval for the true (population) slope is $-0.589 \pm 2.05 * (0.05665)$ or -0.589 ± 0.116, which gives an interval of $(-0.705, -0.473)$. Thus, we are 95% confident that as mileage increases by 1,000 miles, the average price decreases by between $473 and $705 in the population of all used Porsches.

2.2 Partitioning Variability - ANOVA

Another way to assess the effectiveness of a model is to measure how much of the variability in the response variable is explained by the predictions based on the fitted model. This general technique is known in statistics as **analysis of variance**, abbreviated **ANOVA**. Although we will illustrate ANOVA in the context of the simple linear regression model, this approach could be applied to any situation in which a model is used to obtain predictions, as will be demonstrated in later chapters.

The basic idea is to partition the total variability in the responses into two pieces. One piece summarizes the variability explained by the model and the other piece summarizes the variability due to error and captured in the residuals. In short, we have

$$\boxed{\begin{array}{c}\text{TOTAL variation}\\\text{in Response Y}\end{array}} = \boxed{\begin{array}{c}\text{Variation explained}\\\text{by the MODEL}\end{array}} + \boxed{\begin{array}{c}\text{Unexplained variation}\\\text{in the RESIDUALS}\end{array}}$$

2.2. PARTITIONING VARIABILITY - ANOVA

In order to partition this variability, we start with deviations for individual cases. Note that we can write a deviation $y - \bar{y}$ as

$$y - \bar{y} = (\hat{y} - \bar{y}) + (y - \hat{y})$$

so that $y - \bar{y}$ is the sum of two deviations. The first deviation corresponds to the model and the second is the residual. We then sum the squares of these deviations to obtain the following relationship, known as the ANOVA sum of squares identity[2].

$$\sum (y - \bar{y})^2 = \sum (\hat{y} - \bar{y})^2 + \sum (y - \hat{y})^2$$

We summarize the partition with the following notation.

$$SSTotal = SSModel + SSE$$

The Analysis of Variance section in the Minitab output for the Porsche data in Figure 1.3 (page 28) is reproduced below.

```
Analysis of Variance

Source          DF    SS      MS      F       P
Regression       1  5565.7  5565.7  108.25  0.000
Residual Error  28  1439.6    51.4
Total           29  7005.2
```

Examining the "SS" column in the Minitab output shows the following values for this partition of variability for the Porsche regression.

$$\begin{aligned} SSmodel &= \sum (\hat{y} - \bar{y})^2 = 5565.7 \\ SSE &= \sum (y - \hat{y})^2 = 1439.6 \\ SSTotal &= \sum (y - \bar{y})^2 = 7005.2 \end{aligned}$$

and note that $7005.2 = 5565.7 + 1439.6$ (except for some round-off error).

---

[2]There is some algebra involved in deriving the ANOVA sum of squares identity that reduces the right-hand side to just the two sums of squares terms.

ANOVA Test for Simple Linear Regression

How do we tell if the model explains a significant amount of variability or if the explained variability is due to chance alone? The relevant hypotheses would be the same as those for the t-test for the slope

$H_0 : \beta_1 = 0$
$H_a : \beta_1 \neq 0$

In order to compare the *explained* (model) and *error* (residual) variabilities, we need to adjust the sums of squares by appropriate degrees of freedom. We have already seen in the computation of the regression standard error $\hat{\sigma}_\epsilon$ that the sum of squared residuals (SSE) has $n-2$ degrees of freedom. When we partition variability as in the ANOVA identity, the degrees of freedom of the components add in the same way as the sums of squares. The total sum of squares ($SSTotal$) has $n-1$ degrees of freedom when estimating the variance of the response variable (recall that we divide by $n-1$ when computing a basic variance or standard deviation). Thus, for a simple linear model with a single predictor, the sum of squares explained by the model ($SSModel$) has just 1 degree of freedom and we partition the degrees of freedom so that $n - 1 = 1 + (n - 2)$.

We divide each sum of squares by the appropriate degrees of freedom to form **mean squares**.

$$MSModel = \frac{SSModel}{1} \quad \text{and} \quad MSE = \frac{SSE}{n-2}$$

If the null hypothesis of no linear relationship is true, then both of these mean squares will estimate the variance of the error[3]. However, if the model is effective the sum of squared errors gets small and the $MSModel$ will be large relative to MSE. The test statistic is formed by dividing $MSModel$ by MSE,

$$F = \frac{MSModel}{MSE}$$

Under H_0 we expect F to be approximately equal to 1, but under H_a the F-statistic should be larger than one. If the normality condition holds, then the test statistic F follows an **F-distribution** under the null hypothesis of no relationship. The F-distribution represents a ratio of two variance estimates so we use degrees of freedom for both the numerator and the denominator. In the case of simple linear regression, the numerator degree of freedom is 1 and the denominator degrees of freedom are $n - 2$.

We summarize all of these calculations in an ANOVA table as shown below. Statistical software provides the ANOVA table as a standard part of the regression output.

---

[3]We have already seen that MSE is an estimate for σ_ϵ^2. It is not so obvious that $MSModel$ is also an estimate when the true slope is zero, but deriving this fact is beyond the scope of this book.

2.3. REGRESSION AND CORRELATION

> **ANOVA for a Simple Linear Regression Model**
>
> To test the effectiveness of the simple linear model the hypotheses are
>
> $H_0 : \beta_1 = 0$
> $H_a : \beta_1 \neq 0$
>
> and the **ANOVA table** is
>
> | Source | Degrees of freedom | Sum of Squares | Mean Square | F statistic |
> |--------|--------------------|----------------|-------------|-------------|
> | Model | 1 | SSModel | MSModel | $F = \frac{MSModel}{MSE}$ |
> | Error | $n-2$ | SSE | MSE | |
> | Total | $n-1$ | SSTotal | | |
>
> If the conditions for the simple linear model, including normality, hold, the p-value is obtained from the upper tail of an F-distribution with 1 and $n-2$ degrees of freedom.

Using the Minitab ANOVA output (page 61) for the Porsche linear regression fit, we see the $MSModel = 5565.7/1 = 5565.7$ and $MSE = 1439.6/(30-2) = 51.4$ so the test statistic is $F = 5565.7/51.4 = 108.25$. Comparing this to an F-distribution with 1 numerator and 28 denominator degrees of freedom, we find a p-value very close to 0.000 (as seen in the Minitab output) and conclude that there is a linear relationship between the price and mileage of used Porches.

2.3 Regression and Correlation

Recall that the correlation coefficient, r, is a number between -1 and +1 that measures the strength of the linear association between two quantitative variables. Thus the correlation coefficient is also useful for assessing the significance of a simple linear model. In this section we make a connection between correlation and the partitioning of variability from the previous section. We also show how the slope can be estimated from the sample correlation and how to test for a significant correlation directly.

Coefficient of Determination, r^2

Another way to assess the fit of the model using the ANOVA table is to compute the ratio of the MODEL variation to the TOTAL variation. This statistic, known as the **coefficient of determination**, tells us how much variation in the response variable Y we explain by using the explanatory variable X in the regression model. Why do we call it r^2? An interesting feature of simple linear regression is that the square of the correlation coefficient happens to be exactly the coefficient of determination. We explore a t-test for r in more detail in the next section.

> **Coefficient of Determination**
>
> The coefficient of determination for a model is
>
> $$r^2 = \frac{\text{Variability explained by the model}}{\text{Total variability in } y} = \frac{\sum(\hat{y}-\bar{y})^2}{\sum(y-\bar{y})^2} = \frac{SSModel}{SSTotal}$$
>
> Using our ANOVA partition of the variability, this formula can also be written as
>
> $$r^2 = \frac{SSTotal - SSE}{SSTotal} = 1 - \frac{SSE}{SSTotal}$$
>
> Statistical software generally provides a value for r^2 as a standard part of regression output.

If the model fit perfectly, the residuals would all be zero and r^2 would equal 1. If the model does no better than the mean $\bar{y}$ at predicting the response variable, then $SSModel$ and r^2 would both be equal to zero.

Referring once again to the Minitab output in Figure 1.3 (page 28) we see that

$$r^2 = \frac{5565.7}{7005.2} = 0.795.$$

Since r^2 is the fraction of the response variability that is explained by the model, we often convert the value to a percentage, labeled "R-sq" in the Minitab output. In this context we find that 79.5% of the variability in the prices of the Porsches in this sample can be explained by the linear model based on their mileages. The coefficient of determination is a useful tool for assessing the fit of the model and for comparing competing models.

Inference for correlation

The least-squares slope $\hat{\beta}_1$ is closely related to the correlation r between the explanatory and response variables X and Y in a sample. In fact, the sample slope can be obtained from the sample correlation together with the standard deviation of each variable as follows.

$$\hat{\beta}_1 = r \cdot \frac{s_Y}{s_X}$$

Testing the null hypothesis $H_0 : \beta_1 = 0$ is the same as testing that there is no correlation between X and Y in the population from which we sampled our data. Because correlation also makes sense when there is no explanatory-response distinction, it is handy to be able to test correlation without doing regression.

2.3. REGRESSION AND CORRELATION

> **T-test for Correlation**
>
> If we let ρ denote the population correlation, the hypotheses are
>
> $H_0 : \rho = 0$
> $H_a : \rho \neq 0$
>
> and the test statistic is
>
> $$t = \frac{r\sqrt{n-2}}{\sqrt{1-r^2}}$$
>
> If the conditions for the simple linear model, including normality, hold, we find the p-value using the t-distribution with $n-2$ degrees of freedom.

We only need the sample correlation and sample size to compute this test statistic. For the prices and mileages of the n=30 Porsches in Table 1.1 the correlation is $r = -0.891$ and the test statistic is

$$t = \frac{-0.891\sqrt{30-2}}{\sqrt{1-(-0.891)^2}} = -10.4$$

This test statistic is far in the tail of a t-distribution with 28 degrees of freedom, so we have a p-value near .0000 and conclude, once again, that there is a significant correlation between the prices and mileages of used Porsches.

Three tests for a linear relationship?

We now have three distinct ways to test for a significant linear relationship between two quantitative variables. The t-test for slope, the ANOVA for regression, and the t-test for correlation can all be used. Which test is best? If the conclusions differed which test would be more reliable? Surprisingly, these three procedures are exactly equivalent in the case of simple linear regression. In fact, the test statistics for slope and correlation are equal, and the F-statistic is the square of the t-statistic. For the Porsche cars, the t-statistic is -10.4 and the ANOVA F-statistic is $(-10.4)^2 = 108.2$.

Why do we need three different procedures for one task? While the results are equivalent in the simple linear case, we will see that these tests take on different roles when we consider multiple predictors in Chapter 3.

2.4 Intervals for Predictions

One of the most common reasons to fit a line to data is to predict the response for a particular value of the explanatory variable. In addition to a prediction, we often want a margin of error that describes how accurate the prediction is likely to be. This leads to forming an interval for the prediction, but there are two important types of intervals in practice.

To decide which interval to use, we must answer this question: Do we want to predict the *mean response* for a particular value x^* of the explanatory variable or do we want to predict the response Y for an *individual case*? Both of these predictions may be interesting, but they are two different problems. The value of the prediction is the same in both cases, but the margin of error is different. The distinction between predicting a single outcome and predicting the mean of all outcomes when $X = x^*$ determines which margin of error is used.

For example, if we would like to know the average price for all Porsches with 50 thousand miles, then we would use an interval for the mean response. On the other hand, if we want an interval to contain the price of a particular Porsche with 50 thousand miles, then we need the interval for a single prediction. To emphasize the distinction, we use different terms for the two intervals.

To estimate the *mean* response, we use a **confidence interval for** μ_Y. It has the same interpretation and properties as other confidence intervals, but it estimates the mean response of Y when X has the value x^*. This mean response

$$\mu_Y = \beta_0 + \beta_1 \cdot x^*$$

is a parameter, a fixed number whose value we don't know.

To estimate an *individual* response y, we use a **prediction interval**. A prediction interval estimates a single random response y rather than a parameter such as μ_Y. The response y is not a fixed number. If we took more observations with $X = x^*$, we would get different responses. The meaning of a prediction interval is very much like the meaning of a confidence interval. A 95% prediction interval, like a 95% confidence interval, is right 95% of the time in repeated use. "Repeated use" now means that we take an observation on Y for each of the n values of X in the original data, and then take one more observation y with $X = x^*$. We form the prediction interval, then see if it covers the observed value of y for $X = x^*$. The resulting interval will contain y in 95% of all repetitions.

The main point is that it is harder to predict one response than to predict a mean response. Both intervals have the usual form

$$\hat{y} \pm t^* \cdot SE$$

but the prediction interval is wider (has a larger SE) than the confidence interval. You will rarely need to know the details, because software automates the calculation, but here they are.

2.4. INTERVALS FOR PREDICTIONS

> **Confidence and Prediction Intervals for a Simple Linear Regression Response**
>
> A confidence interval for the mean response μ_Y when X takes the value x^* is
>
> $$\hat{y} \pm t^* \cdot SE_{\hat{\mu}}$$
>
> where the standard error is
>
> $$SE_{\hat{\mu}} = \hat{\sigma}_\epsilon \sqrt{\frac{1}{n} + \frac{(x^* - \bar{x})^2}{\sum (x - \bar{x})^2}}$$
>
> A prediction interval for a single observation of Y when X takes the value x^* is
>
> $$\hat{y} \pm t^* \cdot SE_{\hat{y}}$$
>
> where the standard error is
>
> $$SE_{\hat{y}} = \hat{\sigma}_\epsilon \sqrt{1 + \frac{1}{n} + \frac{(x^* - \bar{x})^2}{\sum (x - \bar{x})^2}}$$
>
> The value of t^* in both intervals is the critical value for the t_{n-2} density curve to obtain the desired confidence level.

There are two standard errors: $SE_{\hat{\mu}}$ for estimating the mean response μ_Y and $SE_{\hat{y}}$ for predicting an individual response y. The only difference between the two standard errors is the extra 1 under the square root sign in the standard error for prediction. The extra 1, which makes the prediction interval wider, reflects the fact that an individual response will vary from the mean response μ_Y with a standard deviation of σ_ϵ. Both standard errors are multiples of the regression standard error $\hat{\sigma}_\epsilon$. The degrees of freedom are again $n - 2$, the degrees of freedom of $\hat{\sigma}_\epsilon$.

Returning once more to the Porsche prices, suppose that we are interested in a used Porsche with about 50 thousand miles. The predicted price according to our fitted model is

$$\widehat{Price} = 71.09 - 0.5894 * 50 = 41.62$$

or an average price of about $41,620. Most software has options for computing both types of intervals using specific values of a predictor, producing output such as that shown below.

```
Predicted Values for New Observations
New
Obs     Fit  SE Fit       95% CI           95% PI
  1   41.62    1.56   (38.42, 44.83)   (26.59, 56.65)

Values of Predictors for New Observations
New
Obs  Mileage
  1     50.0
```

This confirms the predicted value of 41.62 (thousand) when $x^* = 50$. The "SE Fit" value of 1.56 is the result of the computation for $SE_{\hat{\mu}}$. The "95% CI" is the confidence interval for the mean response, so we can be 95% confident that the average price of all used Porsches with 50 thousand miles for sale on the internet is somewhere between \$38,420 and \$44,830. The "95% PI" tells us that we should expect about 95% of those Porsches with 50 thousand miles to be priced between \$26,590 and \$56,650. So if we find a Porsche with 50 thousand miles on sale for \$38,000 we know that the price is slightly better than average, but not unreasonably low.

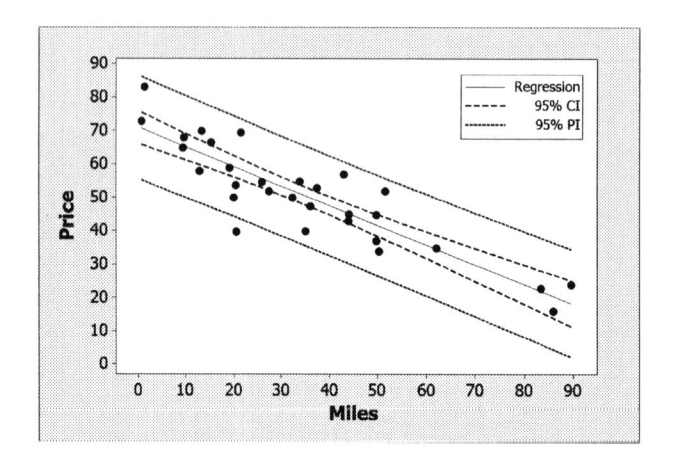

Figure 2.2: Confidence intervals and prediction intervals for Porsche prices

Figure 2.2 shows both the confidence intervals for mean Porsche prices μ_Y and prediction intervals for individual Porsche prices (y) for each mileage along with the scatterplot and regression line. Note how the intervals get wider as we move away from the mean mileage ($\bar{x} = 35$ thousand miles). Many data values lie outside of the narrower confidence bounds for μ_Y; those bounds are only trying to capture the "true" line (mean Y at each value of X). Only one of the 30 Porsches in this sample has a price which falls outside of the 95% prediction bounds at each mileage.

2.5 Chapter Summary

In this chapter we focused on statistical inference for a simple linear regression model. To test whether there is a non-zero linear association between a response variable Y and a predictor variable X, we use the estimated slope and the **standard error of the slope**, $SE_{\hat{\beta}_1}$. The standard error of the slope is different from the standard error of the regression line, so you should be able to identify and interpret both standard errors. The **slope test statistic**,

$$t = \frac{\hat{\beta}_1}{SE_{\hat{\beta}_1}}$$

is formed by dividing the estimated slope by the standard error of the slope and follows a t-distribution with $n-2$ degrees of freedom. The two-sided p-value that will be used to make our inference is provided by statistical software.

You should be able to compute and interpret **confidence intervals** for both the slope and intercept parameters. The intervals have the same general form:

$$\hat{\beta} \pm t^* \cdot SE_{\hat{\beta}}$$

where the coefficient estimate and standard error are provided by statistical software and t^* is the critical value from the t-distribution with $n-2$ degrees of freedom.

Partitioning the total variability provides an alternative way to assess the effectiveness of our linear model. The **total variation** ($SSTotal$) is partitioned into one part that is **explained by the model** ($SSModel$) and another unexplained part **due to error** (SSE).

$$SSTotal = SSModel + SSE$$

This general idea of partitioning the variability is called **ANOVA** and will be used throughout the rest of the text. Each sum of squares is divided by its corresponding degrees of freedom (1 and $n-2$, respectively) to form a mean sum of squares. An F-test statistic is formed by dividing the mean sum of squares for the model by the mean sum of squares for error

$$F = \frac{MSModel}{MSE}$$

and comparing the result to the upper tail of an F-distribution with 1 and $n-2$ degrees of freedom. For the simple linear regression model, the inferences based on this F-test and the t-test for the slope parameter will always be equivalent.

The simple linear regression model is connected with the correlation coefficient that measures the strength of linear association through a statistic called the **coefficient of determination**. There are many ways to compute the coefficient of determination r^2, but you should be able to interpret this statistic in the context of any regression setting.

$$r^2 = \frac{SSModel}{SSTotal}$$

In general, r^2 tells us how much variation in the response variable is explained by using the explanatory variable in the regression model. This is often useful for comparing competing models.

The connection between regression and correlation leads to a third equivalent test to see if our simple linear model provides an effective fit. The slope of the least squares regression line can be computed from the correlation coefficient and the standard deviations for each variable. That is,

$$\hat{\beta}_1 = r \cdot \frac{s_Y}{s_X}$$

The test statistic for the null hypothesis of no association against the alternative hypothesis of nonzero correlation is

$$t = \frac{r\sqrt{n-2}}{\sqrt{1-r^2}}$$

and the p-value is obtained from a t-distribution with $n-2$ degrees of freedom.

Prediction is an important aspect of the modeling process in most applications, and you should be aware of the important differences between a **confidence interval for a mean response** and a **prediction interval for an individual response**. The standard error for predicting an individual response will always be larger than the standard error for predicting a mean response; therefore, prediction intervals will always be wider than confidence intervals. Statistical software will provide both intervals, which have the usual form, $\hat{y} \pm t^* \cdot SE$, so you should focus on picking the correct interval and providing the appropriate interpretation. Remember that the predicted value, $\hat{y}$, and either interval both depend on a particular value for the predictor variable (x^*) which should be a part of your interpretation of a confidence or prediction interval.

2.6 Exercises

Conceptual Exercises

2.1 *Using correlation.* A regression equation was fit to a set of data for which the correlation, r, between X and Y was 0.6. Which of the following must be true?

a. The slope of the regression line is 0.6.

b. The regression model explains 60% of the variability in Y.

c. The regression model explains 36% of the variability in Y.

d. At least half of the residuals are smaller than 0.6 in absolute value.

2.2 *Interpreting the size of r^2.*

a. Does a high value of r^2, say 0.90 or 0.95, indicate that a linear relationship is the best possible model for the data? Explain.

b. Does a low value of r^2, say 0.20 or 0.30, indicate that some relationship other than linear would be the best model for the data? Explain.

2.3 *Effects on a prediction interval.* Describe the effect (if any) on the width (difference between the upper and lower endpoints) of a prediction interval if all else remains the same except:

a. the sample size is increased

b. the variability in the values of the predictor variable increases

c. the variability in the values of the response variable increases

d. the value of interest for the predictor variable moves further from the mean value of the predictor variable

Guided Exercises

2.4 *Inference for slope.* A regression model was fit to 40 data cases and the resulting sample slope, $\hat{\beta}_1$, was 15.5, with a standard error $SE_{\hat{\beta}_1}$ of 3.4. Assume that the conditions for a simple linear model, including normality, are reasonable for this situation.

a. Test the hypothesis that $\beta_1 = 0$.

b. Construct a 95% confidence interval for β_1.

2.5 *Partitioning variability.* The sum of squares for the regression model, $SSModel$, for the regression of Y on X was 110, the sum of squared errors, SSE was 40 and the total sum of squares, $SSTotal$, was 150. Calculate and interpret the value of r^2.

2.6 *Breakfast cereal.* The number of calories and number of grams of sugar per serving were measured for 36 breakfast cereals. The data are in the file **Cereal**. We are interested in trying to predict the calories using the sugar content.

  a. Test the hypothesis that sugar content has a linear relationship with calories (in an appropriate scale). Report and interpret the p-value for the test.

  b. Find and interpret a 95% confidence interval for the slope of this regression model.

2.7 *Textbook prices.* Exercise 1.6 examined data on the price and number of pages for a random sample of 30 textbooks from the Cal Poly campus bookstore. The data are stored in the file **TextPrices** and appear in Table 1.4.

  a. Perform a significance test to address the students' question of whether the number of pages is a useful predictor of a textbook's price. Report the hypotheses, test statistic, and p-value, along with your conclusion.

  b. Determine a 95% confidence interval for the population slope coefficient. Also explain what this slope coefficient means in the context of these data.

2.8 *Textbook prices (continued).* Refer to the previous Exercise 1.6 on prices and numbers of pages in textbooks.

  a. Determine a 95% confidence interval for the mean price of a 450-page textbook in the population.

  b. Determine a 95% confidence interval for the price of a particular 450-page textbook in the population.

  c. How do the midpoints of these two intervals compare? Explain why this makes sense.

  d. How do the widths of these two intervals compare? Explain why this makes sense.

  e. What value for number of pages would produce the narrowest possible prediction interval for its price? Explain.

  f. Determine a 95% prediction interval for the price of a particular 1500-page textbook in the population. Do you really have 95% confidence in this interval? Explain.

2.9 *When does a child first speak?* The data in Table 2.1 (stored in **ChildSpeaks**) are from a study about whether there is a relationship between the age at which a child first speaks (in months) and his/her score on a Gesell aptitude test taken later in childhood.

2.6. EXERCISES

Table 2.1: Gesell Aptitude and First Speak Age (in months)

Child #	Age	Gesell	Child #	Age	Gesell	Child #	Age	Gesell
1	15	95	8	11	100	15	11	102
2	26	71	9	8	104	16	10	100
3	10	83	10	20	94	17	12	105
4	9	91	11	7	113	18	42	57
5	15	102	12	9	96	19	17	121
6	20	87	13	10	83	20	11	86
7	18	93	14	11	84	21	10	100

a. Before you analyze the data, would you expect to see a positive relationship, negative relationship, or no relationship between these variables? Provide a rationale for your choice.

b. Produce a scatterplot of these data, and comment on whether age of first speaking appears to be a useful predictor of Gesell aptitude score.

c. Report the regression equation and the value of r^2. Also determine whether the relationship between these variables is statistically significant.

d. Which child has the largest (in absolute value) residual? Explain what is unusual about that child.

e. Remove the child who took 42 months to speak and produce a new scatterplot with a fitted line. Comment on how removing the one child has affected the line and the value of r^2.

f. Now remove the child who took 26 months to speak, in addition to the child who took 42 months. Repeat the analysis and comment on the effect of removing this child as in part (e).

g. Now remove a third child, the one identified as having a large residual in part (d). Repeat the analysis, and comment on how removing this child has affected the analysis.

2.10 *Tomlinson rushes in the NFL.* The data in Table 2.2 (stored in **TomlinsonRush**) are the rushing yards and number of rushing attempts for LaDainian Tomlinson in each game of the 2006 National Football League season.

a. Produce a scatterplot and determine the regression equation for predicting yardage from number of rushes.

b. Is the slope coefficient equal to Tomlinson's average number of yards per rush? Explain how you know.

c. How much of the variability in Tomlinson's yardage per game is explained by knowing how many rushes he made?

Table 2.2: LaDainaian Tomlinson Rushing for NFL Games in 2006

Game	Opponent	Attempts	Yards	Game	Opponent	Attempts	Yards
1	Raiders	31	131	9	Bengals	22	104
2	Titans	19	71	10	Broncos	20	105
3	Ravens	27	98	11	Raiders	19	109
4	Steelers	13	36	12	Bills	28	178
5	49ers	21	71	13	Broncos	28	103
6	Chiefs	15	66	14	Chiefs	25	199
7	Rams	25	183	15	Seahawks	22	123
8	Browns	18	172	16	Cardinals	16	66

d. Which game has the largest (in absolute value) residual? Explain what the value of the residual means for that game. Was this a particularly good game, or a particularly bad one, for Tomlinson?

e. Determine a 90% prediction interval for Tomlinson's rushing yards in a game that he carries the ball on 20 rushes.

f. Would you feel comfortable in using this regression model to predict rushing yardage for a different player? Explain.

Open-ended Exercises

2.11 *Baseball game times.* In Exercise 1.7 you examined factors to predict how long a Major League Baseball game will last. The data in Table 1.5 were collected for the 15 games played on August 26, 2008 and stored in the file named **BaseballTimes**.

a. Calculate the correlations of each predictor variable with the length of a game ($Time$). Identify which predictor variable is most strongly correlated with time.

b. Choose the one predictor variable that you consider to be the best predictor of time. Determine the regression equation for predicting time based on that predictor. Also interpret the slope coefficient of this equation.

c. Perform the appropriate significance test of whether this predictor is really correlated with time in the population.

d. Analyze appropriate residual plots for this model, and comment on what they reveal about whether the conditions for inference appear to be met here.

2.6. EXERCISES

Table 2.3: U.S. Infant Mortality

Mortality	Year
85.8	1920
64.6	1930
47.0	1940
29.2	1950
26.0	1960
20.0	1970
12.6	1980
9.2	1990
6.9	2000

2.12 *Infant mortality rate versus year.* Table 2.3 shows infant mortality (deaths within one year of birth per 1,000 births) in the US from 1920 - 2000. The data are stored in the file **InfantMortality**.

a. Make a scatterplot of *Mortality* versus *Year* and comment on what you see.

b. Fit a simple linear regression model and examine residual plots. Do the conditions for a simple linear model appear to hold?

c. If you found significant problems with the conditions, find a transformation to improve the linear fit, otherwise proceed to the next part.

d. Test the hypothesis that there is a linear relationship in the model you have chosen.

e. Use the final model that you have selected to make a statistical statement about infant mortality in the year 2010.

2.13 *Horses for sale.* Undergraduate students at Cal Poly collected data on prices of 50 horses advertised for sale on the internet. Predictor variables include the age and height of the horse (in hands), as well as its sex. The data appear in Table 2.4 and are stored in the file **HorsePrices**.

Analyze these data in an effort to find a useful model that predicts price from one predictor variable. Be sure to consider transformations, and you may want to consider fitting a model to males and females separately. Write a report explaining the steps in your analysis and presenting your final model.

Table 2.4: Horse Prices

Horse ID	Price	Age	Height	Sex	Horse ID	Price	Age	Height	Sex
97	38000	3	16.75	m	132	20000	14	16.50	m
156	40000	5	17.00	m	69	25000	6	17.00	m
56	10000	1	*	m	141	30000	8	16.75	m
139	12000	8	16.00	f	63	50000	6	16.75	m
65	25000	4	16.25	m	164	1100	19	16.25	f
184	35000	8	16.25	f	178	15000	0.5	14.25	f
88	35000	5	16.50	m	4	45000	14	17.00	m
182	12000	17	16.75	f	211	2000	20	16.00	f
101	22000	4	17.25	m	89	20000	3	15.75	f
135	25000	6	15.25	f	57	45000	5	16.50	m
35	40000	7	16.75	m	200	20000	12	17.00	m
39	25000	7	15.75	f	38	50000	7	17.25	m
198	4500	14	16.00	f	2	50000	8	16.50	m
107	19900	6	15.50	m	248	39000	11	17.25	m
148	45000	3	15.75	f	27	20000	11	16.75	m
102	45000	6	16.75	m	19	12000	6	16.50	f
96	48000	6	16.50	m	129	15000	2	15.00	f
71	15500	12	15.75	f	13	27500	5	16.00	f
28	8500	7	16.25	f	206	12000	2	*	f
30	22000	7	16.50	f	236	6000	0.5	*	f
31	35000	5	16.25	m	179	15000	0.5	14.50	m
60	16000	7	16.25	m	232	60000	13	16.75	m
23	16000	3	16.25	m	152	50000	4	16.50	m
115	15000	7	16.25	f	36	30000	9	16.50	m
234	33000	4	16.50	m	249	40000	7	17.25	m

Source: Cal Poly students using a horse sale website

2.6. EXERCISES

Supplemental Exercises

2.14 *Linear fit in a normal probability plot* A random number generator was used to create 100 observations for a particular variable. A normal probability plot from Minitab is shown in Figure 2.3.

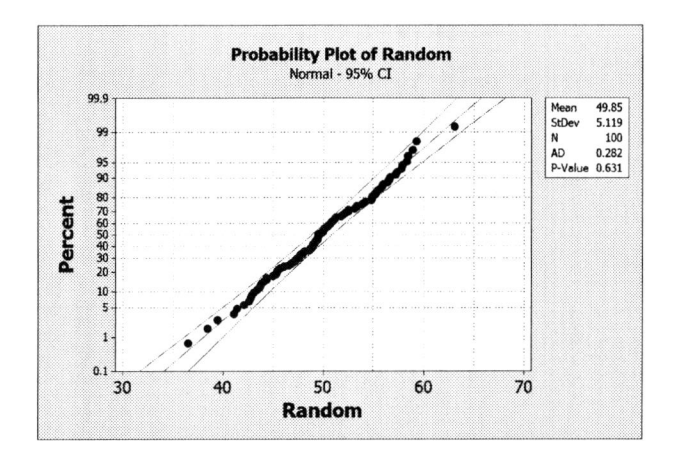

Figure 2.3: Normal Probability Plot for Random Numbers

Notice that this default output from Minitab includes a regression line and a confidence band. The band is narrow near the mean of 49.54 and gets wider as you move away from the center of the randomly generated observations.

a. Do you think Minitab used confidence intervals or prediction intervals to form the confidence band? Provide a rationale for your choice.

b. The Anderson-Darling test statistic AD = 0.282 and corresponding p-value= 0.631 are provided in the output. The null hypothesis is that the data are normal and the alternative hypothesis is that the data are not normal. Provide the conclusion for the formal Anderson-Darling procedure.

c. Do the departures from linear trend in the probability plot provide evidence for the conclusion you made in part (b)? Explain.

Chapter 3

Multiple Regression

When a scatterplot shows a linear relationship between a quantitative explanatory variable X and a quantitative response variable Y, we fit a regression line to the data in order to describe the relationship. We can also use the line to predict the value of Y for a given value of X. For example, Chapter 1 uses regression lines to describe relationships between

- The price Y of a used Porsche and its mileage X.
- The number of doctors Y in a city and the number of hospitals X.
- The abundance of mammal species Y and the area of an island X.

In all of these cases, other explanatory variables might improve our understanding of the response and help us to better predict Y.

- The price Y of a used Porsche may depend on its mileage X_1 and also its age X_2.
- The number of doctors Y in a city may depend on the number of hospitals X_1, the number of beds in those hospitals X_2, and the number of medicare recipients X_3.
- The abundance of mammal species Y may depend on the area of an island X_1, the maximum elevation X_2, and the distance to the nearest island X_3.

In Chapters 1 and 2 we studied simple linear regression with a single quantitative predictor. This chapter introduces the more general case of **multiple linear regression**, which allows several explanatory variables to combine in explaining a response variable.

Example 3.1: *NFL winning percentage*

Is offense or defense more important in winning football games? The data in Table 3.1 (stored in **NFLStanding2007**) contains the records for all NFL teams during the 2007 regular season, along with the total number of points scored (*PointsFor*) and points allowed (*PointsAgainst*).

Table 3.1: Records and points for NFL teams in 2007 season

Team	Wins	Losses	WinPct	PointsFor	PointsAgainst
New England Patriots	16	0	1.000	589	274
Dallas Cowboys	13	3	0.813	455	325
Green Bay Packers	13	3	0.813	435	291
Indianapolis Colts	13	3	0.813	450	262
Jacksonville Jaguars	11	5	0.688	411	304
San Diego Chargers	11	5	0.688	412	284
Cleveland Browns	10	6	0.625	402	382
New York Giants	10	6	0.625	373	351
Pittsburgh Steelers	10	6	0.625	393	269
Seattle Seahawks	10	6	0.625	393	291
Tennessee Titans	10	6	0.625	301	297
Tampa Bay Buccaneers	9	7	0.563	334	270
Washington Redskins	9	7	0.563	334	310
Arizona Cardinals	8	8	0.500	404	399
Houston Texans	8	8	0.500	379	384
Minnesota Vikings	8	8	0.500	365	311
Philadelphia Eagles	8	8	0.500	336	300
Buffalo Bills	7	9	0.438	252	354
Carolina Panthers	7	9	0.438	267	347
Chicago Bears	7	9	0.438	334	348
Cincinnati Bengals	7	9	0.438	380	385
Denver Broncos	7	9	0.438	320	409
Detroit Lions	7	9	0.438	346	444
New Orleans Saints	7	9	0.438	379	388
Baltimore Ravens	5	11	0.313	275	384
San Francisco 49ers	5	11	0.313	219	364
Atlanta Falcons	4	12	0.250	259	414
Kansas City Chiefs	4	12	0.250	226	335
New York Jets	4	12	0.250	268	355
Oakland Raiders	4	12	0.250	283	398
St. Louis Rams	3	13	0.188	263	438
Miami Dolphins	1	15	0.063	267	437

Source: www.nfl.com

3.1. MULTIPLE LINEAR REGRESSION MODEL 81

(a) Using Points Scored   (b) Using Points Allowed

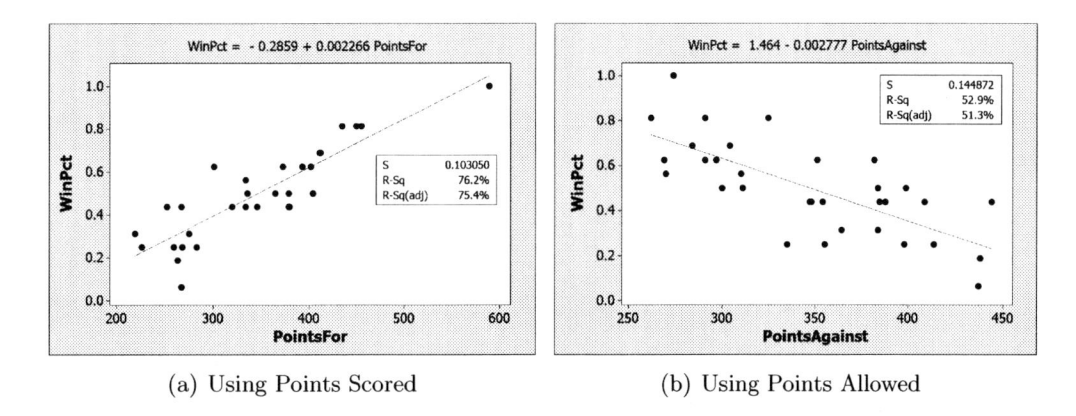

Figure 3.1: Linear Regressions to Predict NFL Winning Percentage

Winning percentage Y could be related to points scored X_1 and/or points allowed X_2. The simple linear regressions for both of these relationships are shown in Figure 3.1. Not surprisingly, scoring more points is positively associated with increased winning percentage, while points allowed has a negative relationship. By comparing the r^2 values (76.3% to 53.0%) we see that points scored is a somewhat more effective predictor of winning percentage for these data. But could we improve the prediction of winning percentage by using both variables in the same model?

$\diamond$

3.1 Multiple Linear Regression Model

Recall from Chapter 1, that the model for simple linear regression based on a single predictor X is

$$Y = \beta_0 + \beta_1 X + \epsilon \quad \text{where } \epsilon \sim N(0, \sigma_\epsilon) \text{ and the errors are independent from one another.}$$

Choosing a Multiple Linear Regression Model

Moving to the more general case of multiple linear regression we have k explanatory variables $X_1, X_2, \ldots, X_k$. The model now assumes that the mean response μ_Y for a particular set of values of the explanatory variables is a linear combination of those variables.

$$\mu_Y = \beta_0 + \beta_1 X_1 + \beta_2 X_2 + \ldots + \beta_k X_k$$

As with the simple linear regression case, the model also assumes that repeated Y responses are **independent** of each other and that Y has a **constant variance** for any combination of the predictors. When we need to do formal inference for regression parameters, we also continue to assume that the distribution of Y for any fixed set of values for the explanatory variables follows a **normal distribution**. These conditions are summarized by assuming the errors in a multiple regression model are independent values from a $N(0, \sigma_\epsilon)$ distribution.

> **The Multiple Linear Regression Model**
>
> We have n observations on k explanatory variables $X_1, X_2, \ldots, X_k$ and a response variable Y. Our goal is to study or predict the behavior of Y for the given set of the explanatory variables. The multiple linear regression model is
>
> $$Y = \beta_0 + \beta_1 X_1 + \beta_2 X_2 + \ldots + \beta_k X_k + \epsilon$$
>
> where $\epsilon \sim N(0, \sigma_\epsilon)$ and the errors are independent from one another.

This model has $k+2$ unknown parameters that we must estimate from data: the $k+1$ coefficients $\beta_0, \beta_1, \beta_2, \ldots, \beta_k$ and the standard deviation of the error σ_ϵ. Some of the $X_i's$ in the model may be interaction terms, products of two explanatory variables. Others may be squares, higher powers or other functions of quantitative explanatory variables. We can also include information from categorical predictors by coding the categories with $(0, 1)$ variables. So the model can describe quite general relationships. The main restriction is that the model is linear because each term is a constant multiple of a predictor, $\beta_i X_i$.

Fitting a Multiple Linear Regression Model

Once we have chosen a tentative set of predictors as the form for a multiple linear regression model, we need to estimate values for the coefficients based on data and then assess the fit. The estimation uses the same procedure of computing the sum of squared residuals, where the residuals are obtained as the differences between the actual Y values and the values obtained from a prediction equation of the form

$$\hat{Y} = \hat{\beta}_0 + \hat{\beta}_1 X + \hat{\beta}_2 X_2 + \ldots + \hat{\beta}_k X_k$$

As in the case of simple linear regression, statistical software chooses estimates for the coefficients, $\beta_0, \beta_1, \beta_2, \ldots, \beta_k$, that minimize the sum of the squared residuals.

Example 3.2: *NFL winning percentage (continued)*

For example, Figure 3.2 gives the Minitab output for fitting a multiple linear regression model to predict the winning percentages for NFL teams based on both the points scored and points allowed.

The fitted prediction equation in this example is

$$\widehat{WinPct} = 0.417 + 0.00177 PointsFor - 0.00153 PointsAgainst$$

If we consider the Pittsburgh Steelers who scored 393 points while allowing 260 points during the 2007 regular season, the predicted winning percentage is

$$\widehat{WinPct} = 0.417 + 0.00177 \cdot 393 - 0.00153 \cdot 269 = 0.701$$

3.1. MULTIPLE LINEAR REGRESSION MODEL

```
The regression equation is
WinPct = 0.417 + 0.00177 PointsFor - 0.00153 PointsAgainst

Predictor            Coef      SE Coef       T        P
Constant           0.4172       0.1394    2.99    0.006
PointsFor        0.0017662    0.0001870    9.45    0.000
PointsAgainst   -0.0015268    0.0002751   -5.55    0.000

S = 0.0729816    R-Sq = 88.4%    R-Sq(adj) = 87.6%

Analysis of Variance
Source            DF        SS        MS        F        P
Regression         2   1.18135   0.59068   110.90    0.000
Residual Error    29   0.15446   0.00533
Total             31   1.33581
```

Figure 3.2: Minitab Output for a Multiple Regression

Since the Steelers' 10-6 record produced an actual winning percentage of 0.625, the residual in this case is $0.625 - 0.701 = -0.076$. ◇

In addition to the estimates of the regression coefficients, the other parameter of the multiple regression model that we need to estimate is the standard deviation of the error term, σ_ϵ. Recall that for the simple linear model, we estimated the variance of the error by dividing the sum of squared residuals (SSE) by $n - 2$ degrees of freedom. For each additional predictor we add to a multiple regression model we have a new coefficient to estimate and thus lose one more degree of freedom. In general, if our model has k predictors (plus the constant term) we lose $k + 1$ degrees of freedom when estimating the error variability, leaving $n - k - 1$ degrees of freedom in the denominator. This gives the estimate for the **standard error of the multiple regression model** with k predictors as

$$\hat{\sigma}_\epsilon = \sqrt{\frac{SSE}{n - k - 1}}$$

From the multiple regression output for the NFL winning percentages in Figure 3.2 we see that the sum of squared residuals for the $n = 32$ NFL teams is $SSE = 0.15446$ and so the standard error of this two-predictor regression model is

$$\hat{\sigma}_\epsilon = \sqrt{\frac{0.15446}{32 - 2 - 1}} = \sqrt{0.0053} = 0.073$$

Note that the error degrees of freedom (29) and the estimated variance of the error term ($MSE = 0.00533$) are also given in the Analysis of Variance table of the Minitab output and the standard error is labeled as $S = 0.0729816$.

3.2 Assessing a Multiple Regression Model

T-tests for Coefficients

With multiple predictors in the model we need to ask the question of whether or not an individual predictor is helpful to include in the model. We can test this by seeing if the coefficient for the predictor is significantly different from zero.

Individual t-Tests for Coefficients in Multiple Regression

To test the coefficient for one of the predictors, X_i, in a multiple regression model the hypotheses are

$H_0 : \beta_i = 0$
$H_a : \beta_i \neq 0$

and the test statistic is

$$t = \frac{\text{parameter estimate}}{\text{standard error of estimate}} = \frac{\hat{\beta}_i}{SE_{\hat{\beta}_i}}.$$

If the conditions for the multiple linear model, including normality, hold, we compute the p-value for the test statistic using a t-distribution with $n - k - 1$ degrees of freedom.

Example 3.3: *NFL winning percentage (continued)*

The parameter estimate, standard error of the estimate, test statistic, and p-value for the t-tests for individual predictors appear in standard regression output. For example, we can use the output in Figure 3.2 to test the *PointsAgainst* predictor in the multiple regression to predict NFL winning percentages.

$H_0 : \beta_2 = 0$
$H_a : \beta_2 \neq 0$

From the line in the Minitab output for *PointsAgainst* we see that the test statistic is

$$t = \frac{-0.0015286}{0.0002751} = -5.55$$

and the p-value (based on 29 degrees of freedom) is 0.000. This provides very strong evidence that the coefficient of *PointsAgainst* is different from zero and thus the *PointsAgainst* variable has some predictive power in this model for helping explain the variability in the winning percentages of the NFL teams. Note that the other predictor, *PointsFor*, also has an individual p-value of 0.000 and can be considered an important contributor to this model. ⋄

3.2. ASSESSING A MULTIPLE REGRESSION MODEL

Confidence Intervals for Coefficients

In addition to the estimate and the hypothesis test for the difference in intercepts, we may be interested in producing a confidence interval for one or more of the regression coefficients. As usual we can find a margin of error as a multiple of the standard error of the estimator. Assuming the normality condition, we use a value from the t-distribution based on our desired level of confidence.

Confidence Interval for a Multiple Regression Coefficient

A confidence interval for the actual value of any multiple regression coefficient, β_i, has the form

$$\hat{\beta}_i \pm t^* \cdot SE_{\hat{\beta}_i}$$

where the value of t* is the critical value from the t density with degrees of freedom equal for the error df in the model ($n - k - 1$ where k is the number of predictors). The value of the standard error of the coefficient, $SE_{\hat{\beta}_1}$, is obtained from computer output.

Example 3.4: *NFL winning percentage (continued)*

For the data on NFL winning percentages, the standard error of the coefficient of the *PointsFor* is 0.000187 and the error term has 29 degrees of freedom, so a 95% confidence interval for the size of the average improvement in winning percentage for every extra point scored during a season (assuming the same number of points allowed) is

$$0.0017662 \pm 2.045 \cdot 0.000187 = 0.0017662 \pm 0.000382 = (0.001384 \text{ to } 0.002148).$$

The coefficient and confidence limits in this situation are very small since the variability in $WinPct$ is small relative to the variability in points scored over an entire season. To get a more practical interpretation, we might consider what happens if a team improves its scoring by 50 points over the entire season. Assuming no change in points allowed, we multiply the limits by 50 to find the expected improvement in winning percentage to be somewhere between 0.069 (6.9%) and 0.107 (10.7%). ⋄

ANOVA for Multiple Regression

In addition to assessing the individual predictors one-by-one, we are generally interested in testing the effectiveness of the model as a whole. To do so, we return to the idea of partitioning the variability in the data into portions explained by the model and unexplained variability in the error term.

$$SSTotal = SSModel + SSE$$

where

$$SSModel = \sum (\hat{y} - \overline{y})^2$$
$$SSE = \sum (y - \hat{y})^2$$
$$SSTotal = \sum (y - \overline{y})^2$$

Note that, although it takes a bit more effort to compute the predicted $\hat{y}$ values with multiple predictors, the formulas are exactly the same as we saw in the previous chapter for simple linear regression. To test the overall effectiveness of the model, we again construct an ANOVA table. The primary adjustment from the simple linear cases is that we are now testing k predictors simultaneously, so we have k degrees of freedom for computing the mean square for the model in the numerator and $n - k - 1$ degrees of freedom left for the mean square error in the denominator.

ANOVA for a Multiple Regression Model

To test the effectiveness of the multiple regression linear model the hypotheses are

$H_0 : \beta_1 = \beta_2 = \ldots = \beta_k = 0$
$H_a :$ At least one $\beta_i \neq 0$

and the ANOVA table is

Source	Degrees of freedom	Sum of Squares	Mean Square	F statistic
Model	k	SSModel	MSModel=$\frac{SSModel}{k}$	$F = \frac{MSModel}{MSE}$
Error	$n - k - 1$	SSE	MSE=$\frac{SSE}{n-k-1}$	
Total	$n - 1$	SSTotal		

If the conditions for the multiple linear model, including normality, hold, we compute the p-value using the upper tail of an F-distribution with k and $n - k - 1$ degrees of freedom.

The null hypothesis that the coefficients of all the predictors are zero is consistent with an ineffective model in which none of the predictors has any linear relationship with the response variable[1]. If the model does explain a statistically significant portion of the variability in the response variable, the MSModel will be large compared to the MSE and the p-value based on the ANOVA table will be small. In that case we know that one or more of the predictors is effective in the model, but the ANOVA analysis does not identify which predictors are significant. That is the role for the individual t-tests.

---

[1] Note that the constant term, β_0, is not included in the hypotheses for the ANOVA test of the overall regression model. Only the coefficients of the predictors in the model are being tested.

3.2. ASSESSING A MULTIPLE REGRESSION MODEL

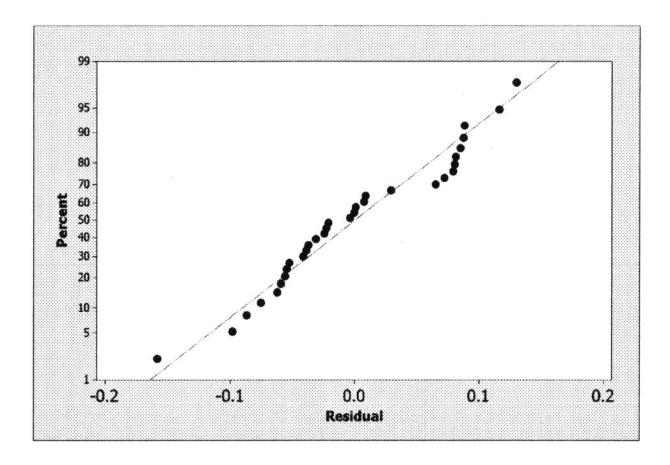

Figure 3.3: Normal probability plot of residuals for NFL model

Example 3.5: *NFL winning percentage (continued)*

The ANOVA table from the Minitab output in Figure 3.2 for the multiple regression model to predict winning percentages in the NFL is reproduced here. The normal probability plot of the residuals in Figure 3.3 indicates that the normality condition is reasonable so we proceed with interpreting the F-test.

```
Analysis of Variance
Source          DF      SS       MS        F       P
Regression       2  1.18135  0.59068   110.90   0.000
Residual Error  29  0.15446  0.00533
Total           31  1.33581
```

Using an F-distribution with 2 numerator and 29 denominator degrees of freedom produces a p-value of 0.000 for the F-statistic of 110.90. This gives strong evidence to reject a null hypothesis that $H_0 : \beta_1 = \beta_2 = 0$ and conclude that at least one of the predictors, *PointsFor* and *PointsAgainst*, is effective for explaining variability in NFL winning percentages. From the plots of Figure 3.1 and individual t-tests we know that, in fact, both of these predictors are effective in this model.   ◇

Coefficient of Multiple Determination

In the previous chapter we also encountered the use of the **coefficient of determination**, r^2, as a measure of the percentage of total variability in the response that is explained by the regression model. This concept applies equally well in the setting of multiple regression so that

$$R^2 = \frac{\text{Variability explained by the model}}{\text{Total variability in } y} = \frac{SSModel}{SSTotal} = 1 - \frac{SSE}{SSTotal}.$$

Using the information in the ANOVA table for the model to predict NFL winning percentages we see that
$$R^2 = \frac{1.18135}{1.33581} = 0.884$$

Thus we can conclude that 88.4% of the variability in winning percentage of NFL teams for the 2007 regular season can be explained by the regression model based on the points scored and points allowed.

Recall that, in the case of simple linear regression with a single predictor, we used the notation r^2 for the coefficient of determination because that value happened to be the square of the correlation between the predictor X and response Y. That interpretation does not translate directly to the multiple regression setting since we now have multiple predictors, $X_1, X_2, \ldots, X_k$, each of which has its own correlation with the response Y. So there is no longer a single correlation between predictor and response. However, we can also consider the correlation between the predictions $\hat{y}$ and the actual y values for all of the data cases. The square of this correlation is once again the coefficient of determination for the multiple regression model. For example, if we use technology to save the predicted values (fits) from the multiple regression model to predict NFL winning percentages and then compute the correlation between those fitted values and the actual winning percentages for all 32 teams we find

---

```
Pearson correlation of WinPct and FITS1 = 0.940
```

If we then compute $r^2 = (0.940)^2 = 0.8836$, we match the coefficient of determination for the multiple regression. In the case of a simple linear regression, the predicted values are a linear function of the single predictor, $\hat{Y} = \hat{\beta}_0 + \hat{\beta}_1 X$, so the correlation between X and Y must be the same as between $\hat{Y}$ and Y, up to possibly a change in $\pm$ sign. Thus, in the simple case, the coefficient of determination could be found by squaring r computed either way. Since this doesn't work with multiple predictors, we generally use a capital R^2 to denote the coefficient of determination in a multiple regression setting.

As individual predictors in separate simple linear regression models both *PointsFor* ($r^2 = 76.3\%$) and *PointsAgainst* ($r^2 = 53.0\%$) were less effective at explaining the variability in winning percentages than they are as a combination in the multiple regression model. This will always be the case. Adding a new predictor to a multiple regression model can never decrease the percentage of variability explained by that model (assuming the model is fit to the same data cases). At the very least, we could put a coefficient of zero in front of the new predictor and obtain the same level of effectiveness. In general, adding a new predictor will decrease the sum of squared errors and thus increase the variability explained by the model. But does that increase reflect important new information provided by the new predictor or just extra variability explained due to random chance? That is one of the roles for the t-tests of the individual predictors.

3.2. ASSESSING A MULTIPLE REGRESSION MODEL

Another way to account for the fact that R^2 tends to increase as new predictors are added to a model is use an **adjusted coefficient of determination** that reflects the number of predictors in the model as well as the amount of variability explained. One common way to do this adjustment is to divide the total sum of squares and sum of squared errors by their respective degrees of freedom and subtract the result from one. This subtracts the ratio of the estimated error variance, $MSE = \hat{\sigma}_\epsilon^2 = SSE/(n-k-1)$, to the ordinary sample variance of the responses $S_Y^2 = \sum(y - \bar{y})^2/(n-1)$, rather than just $SSE/SSTotal$.

Adjusted Coefficient of Determination

The **adjusted** R^2, which helps account for the number of predictors in the model, is computed with

$$R_{adj}^2 = 1 - \frac{SSE/(n-k-1)}{SStotal/(n-1)} = 1 - \frac{\hat{\sigma}_\epsilon^2}{S_Y^2}$$

Note that the denominator stays the same for all models fit to the same response variable and data cases, but the numerator in the term subtracted can actually increase when a new predictor is added to a model if the decrease in the SSE is not sufficient to offset the decrease in the error degrees of freedom. Thus the R_{adj}^2 value might go down when a weak predictor is added to a model.

Example 3.6: *NFL winning percentage (continued)*

For our two-predictor model of NFL winning percentages we can use the information in the ANOVA table to compute

$$R_{adj}^2 = 1 - \frac{0.15446/29}{1.33581/31} = 1 - \frac{0.00533}{.0431} = 1 - 0.124 = 0.876$$

and confirm the value listed as $R - Sq(adj) = 87.6\%$ for the Minitab output in Figure 3.2. While this number reveals relatively little new information on its own, it is particularly useful when comparing competing models based on different numbers of predictors. ⋄

Confidence and Prediction Intervals

Just as in Section 2.4, we can obtain interval estimates for the mean response or a future individual case given any combination of predictor values. We do not provide the details of these formulas, since computing the standard errors is more complicated with multiple predictors. However, the overall method and interpretation are the same and we rely on statistical software to manage the calculations.

Example 3.7: *NFL winning percentage (continued)*

For example, suppose an NFL team in the upcoming season scores 400 points and allows 350 points. Using the multiple regression model for *WinPct* based on *PointsFor* and *PointsAgainst*, some computer output for computing a prediction interval in this case is shown below.

```
Predicted Values for New Observations
New
Obs    Fit  SE Fit      95% CI            95% PI
  1 0.5893  0.0165  (0.5555, 0.6231)  (0.4363, 0.7424)

Values of Predictors for New Observations
New
Obs  PointsFor  PointsAgainst
 1       400            350
```

Thus we would expect this team to have a winning percentage near 59%. Since we are working with an individual team, we interpret the "95% PI" provided in the output to say with reasonable confidence that the winning percentage for this team will be between 44% and 74%. ⋄

The multiple regression model is one of the most powerful and general statistical tools for modeling the relationship between a set of predictors and a quantitative response variable. Having discussed in this section how to extend the basic techniques from simple linear models to fit and assess a multiple regression, we move in the next section(s) to consider several examples that demonstrate the versatility of this procedure. Using multiple predictors in the same model also raises some interesting challenges for choosing a set of predictors and interpreting the model, especially when the predictors might be related to each other as well as to the response variable. We consider some of these issues later in this chapter.

3.3 Comparing Two Regression Lines

In Chapter 1 we considered a simple linear regression model to summarize a linear relationship between two quantitative variables. Suppose now that we want to investigate whether such a relationship changes between groups determined by some categorical variable. For example, is the relationship between price and mileage different for Porsches offered for sale at physical car lots compared to those for sale on the internet? Does the relationship between the number of pages and price of textbooks depend on the field of study or perhaps on whether the book has a hard or soft cover? We can easily fit separate linear regression models by considering each categorical group as a different data set. However, in some circumstances we would like to formally test whether some aspect of the linear relationship (such as the slope, the intercept or possibly both) is significantly different between the two groups or use a common parameter for both groups if

3.3. COMPARING TWO REGRESSION LINES

it would work essentially as well as two different parameters. To help make these judgements we examine multiple regression models that allow us to fit and compare linear relationships for different groups determined by a categorical variable.

Example 3.8: *Potential jurors*

Tom Shields, jury commissioner for the Franklin County Municipal Court in Columbus, Ohio, is responsible for making sure that the judges have enough potential jurors to conduct jury trials. Only a small percent of the possible cases go to trial, but potential jurors must be available and ready to serve the court on short notice. Jury duty for this court is two weeks long, so Tom must bring together a new group of potential jurors twenty-six times a year. Random sampling methods are used to obtain a sample of registered voters in Franklin County every two weeks, and these individuals are sent a summons to appear for jury duty. One of the most difficult aspects of Tom's job is to get those registered voters who receive a summons to actually appear at the courthouse for jury duty. Table 3.2 shows the 1998 and 2000 data for the percentages of individuals who reported for jury duty after receiving a summons. The data are stored in **Jurors**. The reporting dates vary slightly from year to year, so they are coded sequentially from 1, the first group to report in January, to 26, the last group to report in December. A variety of methods were used after 1998 to try to increase participation rates. How successful were these methods in 2000?

A quick look at Table 3.2 shows two potentially interesting features of the relationship between these variables. First, in almost every period the percent of potential jurors who reported in 2000 is higher than in 1998, supporting the idea that participation rates did improve. Second, the percent

Table 3.2: Percents of randomly selected potential jurors who appeared for jury duty

Period	1998	2000	Period	1998	2000
1	83.3	92.6	14	65.4	94.4
2	83.6	81.1	15	65.0	88.5
3	70.5	92.5	16	62.3	95.5
4	70.7	97.0	17	62.5	65.9
5	80.5	97.0	18	65.6	87.5
6	81.6	83.3	19	63.5	80.2
7	75.3	94.6	20	75.0	94.7
8	61.3	88.1	21	67.9	76.6
9	62.7	90.9	22	62.0	75.8
10	67.8	87.1	23	71.0	80.6
11	65.0	85.4	24	62.1	80.6
12	64.1	88.3	25	58.5	71.8
13	64.7	88.3	26	50.7	63.7

Source: Franklin County Municipal Court

92   CHAPTER 3. MULTIPLE REGRESSION

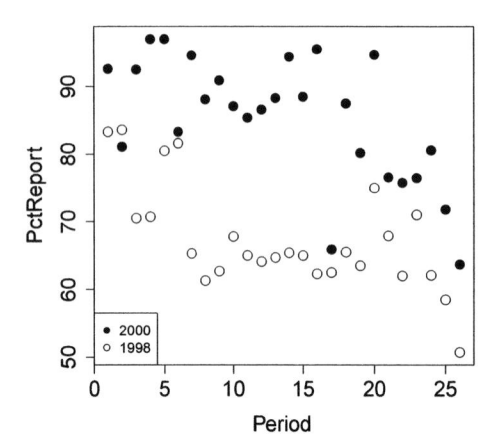

Figure 3.4: Percent of potential jurors in 1998 and 2000 who report each period

reporting appears to decrease in both years as the period number increases during the year. A scatterplot in Figure 3.4 of the percent reporting for jury duty versus biweekly period helps show these relationships by using different symbols for the years 1998 and 2000. Since the participation rates appear to be much higher in 2000 than in 1998, we should not use the same linear model to describe the decreasing trend by period during each of these years. If we consider each year as its own dataset, we can fit separate linear regression models for each year.

$$\text{For 2000:} \quad \widehat{PctReport} = 95.57 - 0.765 \cdot Period$$
$$\text{For 1998:} \quad \widehat{PctReport} = 76.43 - 0.668 \cdot Period$$

Note that the separate regression lines for the two years have similar slopes (roughly a 0.7% drop in percent reporting each period), but have very different intercepts (76.43% in 1998 versus 95.57% in 2000). Since the two lines are roughly parallel, the difference in the intercepts gives an indication of how much better the reporting percents were in 2000 after taking into account the reporting period.   ⋄

Indicator Variables

CHOOSE

Using multiple regression we can examine the *PctReturn* versus *Period* relationship for both years in a single model. The key idea is to use an **indicator variable** that distinguishes between the two groups, in this case the years 1998 and 2000. We generally use the values 0 and 1 for an indicator variable so that 1 indicates that a data case does belong to a particular group ("yes") and 0 signifies

3.3. COMPARING TWO REGRESSION LINES

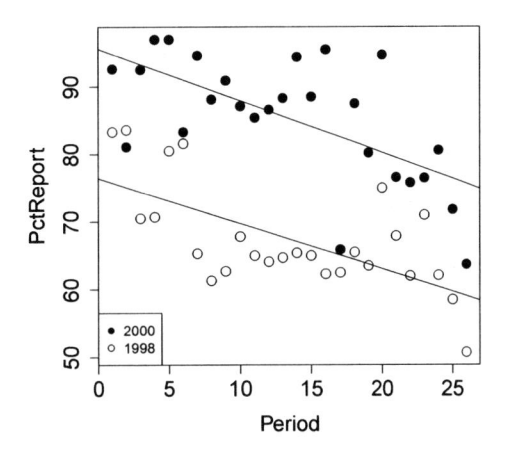

Figure 3.5: Percent reporting jurors with regression lines for 1998 and 2000

that the case is not in that group ("no"). If we have more than two categories we could define an indicator for each of the categories. For just two groups, as in this example, we can assign 1 to either of the groups, so an indicator for the year 2000 would be defined as

$$I2000 = \begin{cases} 0 & \text{if Year} = 1998 \\ 1 & \text{if Year} = 2000 \end{cases}$$

Indicator Variable

An **indicator variable** uses two values, usually 0 and 1, to indicate whether a data case does (1) or does not (0) belong to a specific category.

Using the year indicator, $I2000$, we consider the following multiple regression model

$$PctReturn = \beta_0 + \beta_1 Period + \beta_2 I2000 + \epsilon$$

For data cases from the year 1998 (where $I2000 = 0$) this model becomes

$$\begin{aligned} PctReturn &= \beta_0 + \beta_1 Period + \beta_2 \cdot 0 + \epsilon \\ &= \beta_0 + \beta_1 Period + \epsilon \end{aligned}$$

which looks like the ordinary simple linear regression, although the slope will be determined using data from both 1998 and 2000.

For data cases from the year 2000 (where $I2000 = 1$) this model becomes

$$\begin{aligned} PctReturn &= \beta_0 + \beta_1 Period + \beta_2 \cdot 1 + \epsilon \\ &= (\beta_0 + \beta_2) + \beta_1 Period + \epsilon \end{aligned}$$

which also looks similar to a simple linear regression model, although the intercept is adjusted by the amount β_2. Thus the coefficient of the $I2000$ indicator variable measures the difference in the intercepts between regression lines for 1998 and 2000 which have the same slope. This provides a convenient summary in a single model for the situation that is illustrated in Figure 3.5.

FIT

If we fit this multiple regression model to the **Jurors** data we obtain the following output

```
Coefficients:
(Intercept)      Period          I2000
   77.0816      -0.7169        17.8346
```

which gives the prediction equation

$$\widehat{PctReport} = 77.08 - 0.717 \cdot Period + 17.83 \cdot I2000$$

Comparing Intercepts

By substituting the two values of the indicator variable into the prediction equation we can obtain a least squares line for each year.

For the year 2000: $\widehat{PctReturn} = 77.08 - 0.717 \cdot Period + 17.83 = 94.91 - 0.717 \cdot Period$
For the year 1998: $\widehat{PctReturn} = 77.08 - 0.717 \cdot Period$

Note that the intercepts (94.91 in 2000 and 77.08 in 1998) are not quite the same as when we fit the two simple linear models separately (95.57 in 2000 and 76.43 in 1998). That happens since this multiple regression model forces the two prediction lines to be parallel (common slope = -0.717) rather than allowing separate slopes (-0.765 in 2000 and -0.668 in 1998). In the next example we consider a slightly more complicated multiple regression model that allows either the slope or the intercept (or both) to vary between the two groups.

ASSESS

A summary of the multiple regression model fit includes the following output:

3.3. COMPARING TWO REGRESSION LINES

```
Coefficients:
            Estimate Std. Error t value Pr(>|t|)
(Intercept) 77.0816    2.1297    36.193  < 2e-16 ***
Period      -0.7169    0.1241    -5.779  5.12e-07 ***
I2000       17.8346    1.8608     9.585  8.08e-13 ***

Residual standard error: 6.709 on 49 degrees of freedom
Multiple R-squared: 0.7188,     Adjusted R-squared: 0.7073
F-statistic: 62.63 on 2 and 49 DF,  p-value: 3.166e-14
```

The p-values for the coefficients of both *Period* and *I2000* as well as the overall ANOVA are all very small, indicating that the overall model is effective for explaining percents of jurors reporting and that each of the predictors is important in the model. The R^2 value shows that almost 72% of the variability in the reporting percents over these two years is explained by these two predictors. If we fit a model with just the *Period* predictor alone, only 19% of the variability is explained, while *I2000* alone would explain just under 53%. When we fit the simple linear regression models to the data for each year separately we obtain values of $R^2 = 42\%$ in 1998 and $R^2 = 40\%$ in 2000. Does this mean that our multiple model ($R^2 = 72\%$) does a better job at explaining the reporting percents? Unfortunately, no, since the combined model introduces extra variability in the reporting percents by including both years which the *I2000* predictor then helps to successfully explain. The regression standard error for the combined model ($\hat{\sigma}_\epsilon = 6.7$), which reflects how well the reporting percents are predicted, is close to what is found in the separate regressions for each year ($\hat{\sigma}_\epsilon = 6.2$ in 1998, $\hat{\sigma}_\epsilon = 7.3$ in 2000) and the multiple model uses fewer parameters than fitting separate regressions.

The normal quantile plot in Figure 3.6(a) is linear and the histogram in Figure 3.6(b) has a single peak and is symmetric. These indicate that a normality assumption is reasonable for the residuals in the multiple regression model to predict juror percents using the period and year. The plots of residuals versus predicted values, Figure 3.6(c), shows no clear patterns or trends. Since the periods have a sequential order across the year, Figure 3.6(d) is useful to check for independence to be sure that there are no regular trends in the residuals that occur among the periods.

USE

In this example we are especially interested in the coefficient (β_2) of the indicator variable since that reflects the magnitude of the improvement in reporting percentages in 2000 over those in 1998. From the prediction equation we can estimate the average increase to be about 17.8% and the very small p-value (8.08×10^{-13}) for this coefficient in the regression output gives strong evidence that the effect is not due to chance.

96   CHAPTER 3. MULTIPLE REGRESSION

(a) Normal Quantile Plot or Residuals

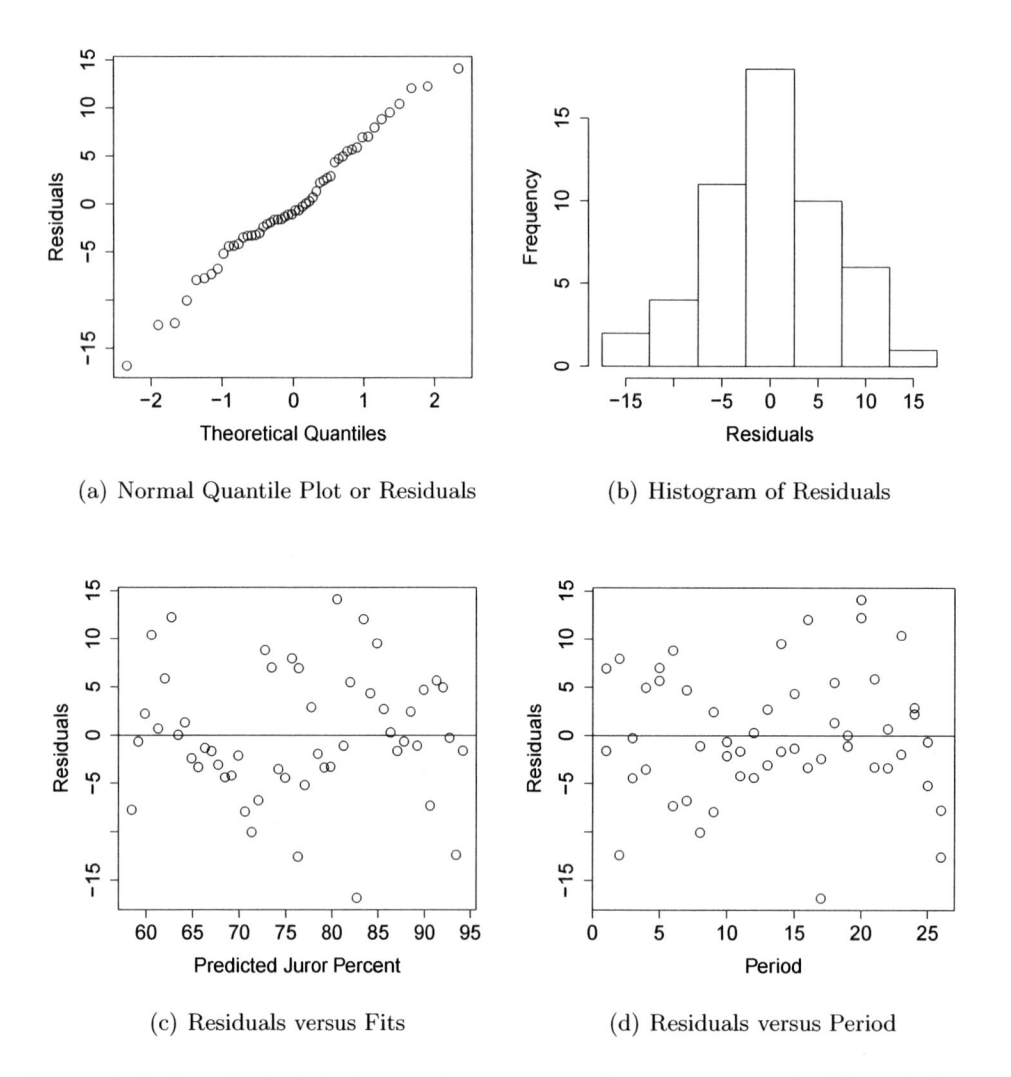

(b) Histogram of Residuals

(c) Residuals versus Fits

(d) Residuals versus Period

Figure 3.6: Residual plots for predicting Juror Percents using $I2000$ and $Period$

In addition to the estimate and the hypothesis test for the difference in intercepts, we may be interested in producing a confidence interval for the size of the average improvement. For the juror reporting model, the standard error of the coefficient of the indicator $I2000$ is 1.8608 and the error term has 49 $(52-2-1)$ degrees of freedom, so a 95% confidence interval for the size of the average improvement in percent of jurors reporting between 1998 and 2000 (after adjusting for the period) is

$$17.83 \pm 2.01 \cdot 1.8608 = 17.83 \pm 3.74 = (14.09\% \text{ to } 21.57\%).$$

Thus we may conclude with reasonable confidence that the average improvement in percent of jurors reporting is somewhere between 14% and 22%.

3.3. COMPARING TWO REGRESSION LINES

Example 3.9: *Growth rates of kids*

We all know that children tend to get bigger as they get older, but we might be interested in how growth rates compare. Do boys and girls gain weight at the same rates? The data displayed in Figure 3.7 shows the ages (in months) and weights (in pounds) for a sample of 198 kids who were part of a larger study of body measurements of children. The data are in the file **Kids198**. The plot shows a linear trend of increasing weights as the ages increase. The dots for boys and girls seem to indicate that weights are similar at younger ages, but that boys tend to weigh more at the older ages. This would suggest a larger growth rate for boys than for girls during these ages.

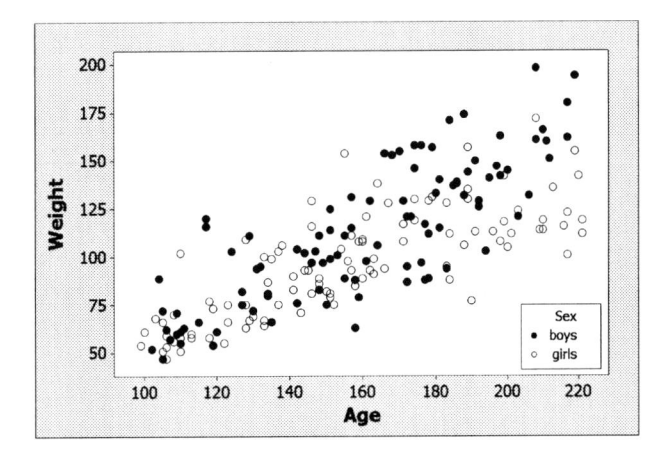

Figure 3.7: Weight versus age by sex for kids

Comparing Slopes

Figure 3.8 shows separate plots for boys and girls with the regression line drawn for each case. The line is slightly steeper for the boys (slope=0.91 pounds per month) than it is for girls (slope=0.63 pounds per month). Figure 3.9 shows both regression lines on the same plot. Does the difference in slopes for these two samples indicate that the typical growth rate is really larger for boys than it is for girls or could a difference this large reasonably occur due to random chance when selecting two samples of these sizes?

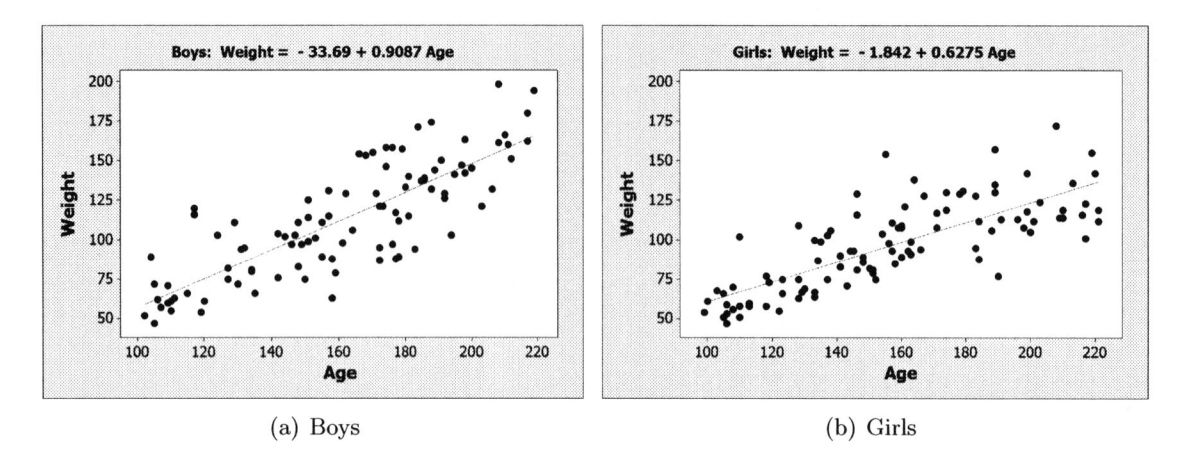

Figure 3.8: Separate regressions of weight versus age for boys and girls

(a) Boys — Boys: Weight = -33.69 + 0.9087 Age

(b) Girls — Girls: Weight = -1.842 + 0.6275 Age

CHOOSE

To compare the two slopes with a multiple regression model we add an additional term to the model considered in the previous example. Define the indicator variable *IGirl* to be 0 for the boys and 1 for the girls. You have seen that adding the *IGirl* indicator to a simple linear model of *Weight* versus *Age* allows for different intercepts for the two groups. The new predictor we add now is just the product of *IGirl* and the *Age* predictor. This gives the model

$$Weight = \beta_0 + \beta_1 Age + \beta_2 IGirl + \beta_3 Age \cdot IGirl + \epsilon$$

For the boys in the study ($IGirl = 0$) the model becomes

$$Weight = \beta_0 + \beta_1 Age + \epsilon$$

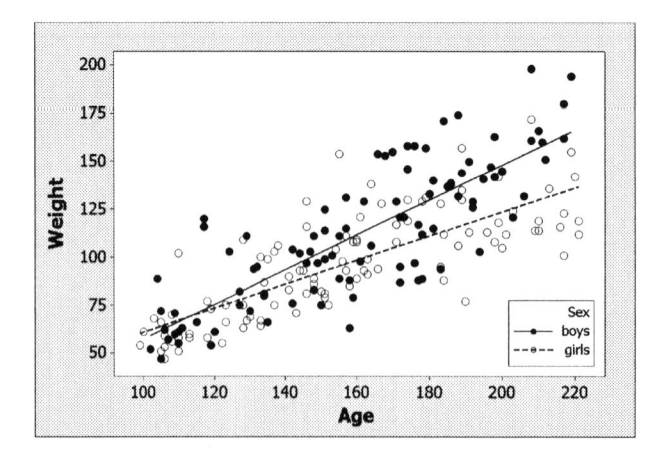

Figure 3.9: Compare regression lines by sex

3.3. COMPARING TWO REGRESSION LINES

while the model for girls ($IGirl = 1$) is

$$Weight = \beta_0 + \beta_1 Age + \beta_2 \cdot 1 + \beta_3 Age \cdot 1 + \epsilon$$
$$Weight = (\beta_0 + \beta_2) + (\beta_1 + \beta_3)Age + \epsilon$$

As in the previous model, the coefficients β_0 and β_1 give the slope and intercept for the regression line for the boys and the coefficient of the indicator variable, β_2, measures the difference in the intercepts between boys and girls. The new coefficient, β_3, shows how much the slopes change as we move from the regression line for boys to the line for girls.

FIT

Using technology to fit the multiple regression model produces the following output.

```
The regression equation is
Weight = - 33.7 + 0.909 Age + 31.9 IGirl - 0.281 AgexIGirl

Predictor        Coef   SE Coef       T       P
Constant       -33.69     10.01   -3.37   0.001
Age           0.90871   0.06106   14.88   0.000
IGirl           31.85     13.24    2.41   0.017
AgexIGirl    -0.28122   0.08164   -3.44   0.001
```

The first two terms of the prediction equation match the slope and intercept for the boys' regression line and we can obtain the least squares line for the girls by increasing the intercept by 31.9 and decreasing the slope by 0.281.

For Boys: $\widehat{Weight} = -33.7 + 0.909 \cdot Age$
For Girls: $\widehat{Weight} = -33.7 + 0.909 \cdot Age + 31.9 \cdot 1 - 0.281 \cdot 1 \cdot Age = -1.8 - 0.628 \cdot Age$

ASSESS

Figure 3.10 shows a normal probability plot of the residuals from the multiple regression model and a scatterplot of the residuals versus the predicted values. The linear pattern in the normal plot indicates that the residuals are reasonably normally distributed. The scatterplot shows no obvious patterns and a relatively consistent band of residuals on either side of zero. Neither plot raises any concerns about significant departures from conditions for the errors in a multiple regression model.

The p-values for the individual t-tests for each of the predictors are small, indicating that each of the terms have some importance in this model. Further output from the regression shown below reveals that 66.8% of the variability in the weights of these 198 kids can be explained by this model

(a) Normal Probability Plot

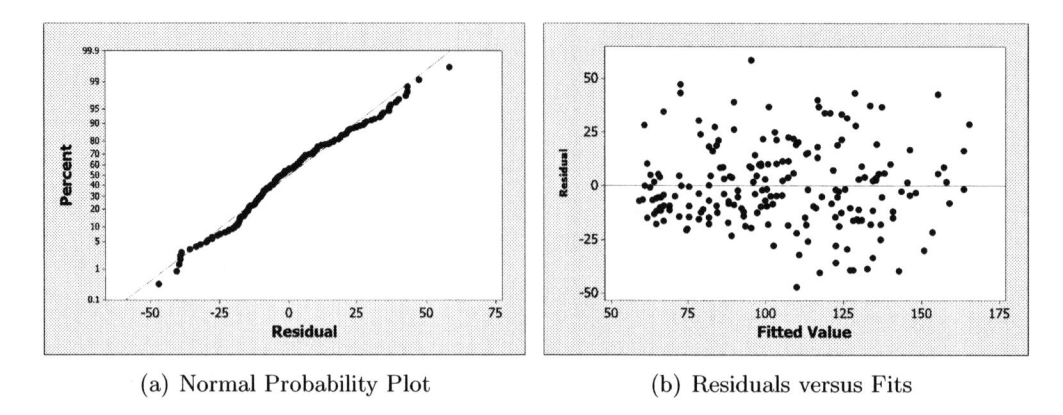

(b) Residuals versus Fits

Figure 3.10: Residual plots for the multiple regression model for weights based on age and sex

based on their age and sex. The small p-value in the analysis of variance ANOVA table indicates that this is a significant amount of variability explained and the overall model has some effectiveness for predicting kids' weights.

```
S = 19.1862    R-Sq = 66.8%    R-Sq(adj) = 66.3%

Analysis of Variance
Source           DF      SS      MS       F       P
Regression        3  143864   47955  130.27   0.000
Residual Error  194   71414     368
Total           197  215278
```

USE

One of the motivations for investigating these data was to see if there is a difference in the growth rates of boys and girls. This question could be phrased in terms of a comparison of the slopes of the regression lines for boys and girls considered separately. However, the multiple regression model allows us to fit both regressions in the same model and provides a parameter, β_3, that specifically measures the difference in the slopes between the two groups. A formal test for a difference in the slopes can now be expressed in terms of that parameter.

$H_0: \beta_3 = 0$
$H_0: \beta_3 \neq 0$

Looking at the computer output we see a p-value for testing the $Age * IGirl$ term in the model equal to 0.001. This provides strong evidence that the coefficient is different from zero and thus that the growth rates are different for boys and girls. For an estimate of the magnitude of this

3.4. NEW PREDICTORS FROM OLD

difference we can construct a confidence interval for the β_3 coefficient using the standard error given from the computer output and a t-distribution with 194 degrees of freedom. At a 95% confidence level this gives

$$-0.281 \pm 1.97 \cdot 0.08164 = -0.281 \pm .161 = (-0.442, \ -0.120)$$

Based on this interval we are 95% sure that rate of weight gain for girls in this age range is between 0.442 and 0.120 pounds per month less than the typical growth rate for boys. Note that in making this statement we are implicitly assuming that we can think of the observed measurements on the children in the sample as representative of the population of all such measurements, even if the children themselves were not a random sample. ◊

3.4 New Predictors from Old

In Section 1.4 we saw some ways to address nonlinear patterns in a simple linear regression setting by considering transformations such as a square root or logarithm. With the added flexibility of a multiple regression model, we can be more creative to include more than one function of one or more predictors in the model at the same time. This opens up a large array of possibilities, even when we start with just a few initial explanatory variables. In this section we illustrate two of the more common methods for combining predictors, using a product of two predictors in an **interaction** model and using one or more powers of a predictor in a **polynomial regression**. In the final example we combine these two methods to produce a **complete second order** model.

Interaction

In Example 3.9 we saw how adding an interaction term, a product of the quantitative predictor *Age* and categorical indicator *IGirl*, allowed the model to fit different slopes for boys and girls when predicting their weights. This idea can be extended to any two quantitative predictors. For a simple linear model, the mean response, μ_Y, follows a straight line if plotted versus the X axis. If X_1 and X_2 are two different quantitative predictors, the model $\mu_Y = \beta_0 + \beta_1 X_1 + \beta_2 X_2$ indicates that the mean response Y is a *plane* if plotted as a surface versus an (X_1, X_2) coordinate system. This implies that for any fixed value of one predictor $X_1 = c$, μ_Y follows a linear relationship with the other predictor X_2 having a slope of β_2 and intercept of $\beta_0 + \beta_1 \cdot c$. This assumes that the slope with respect to X_2 is constant for every different value of X_1. In some situations we may find that this assumption is not reasonable; the slope with respect to one predictor might change in some regular way as we move through values of another predictor. As before, we call this an **interaction** between the two predictors.

Regression Model with Interaction

For two predictors, X_1 and X_2, a multiple regression model with **interaction** has the form

$$Y = \beta_0 + \beta_1 X_1 + \beta_2 X_2 + \beta_3 X_1 X_2 + \epsilon$$

The interaction product term, $X_1 X_2$, allows the slope with respect to one predictor to change for values of the second predictor.

Example 3.10: *Perch weights*

Table 3.3 shows the first few cases from a larger sample of perch caught in a lake in Finland. The file **Perch** contains the weight (in grams), length (in centimeters) and width (in centimeters) for 56 of these fish. We would like to find a model for the weights based on the measurements of length and width.

Table 3.3: First few cases of fish measurements in the **Perch** datafile

Obs	Weight	Length	Width
104	5.9	8.8	1.4
105	32.0	14.7	2.0
106	40.0	16.0	2.4
107	51.5	17.2	2.6
108	70.0	18.5	2.9
109	100.0	19.2	3.3
110	78.0	19.4	3.1

Source: JSE Data Archive, submitted by Juha Puranen

CHOOSE

The scatterplots in Figure 3.11 show that both length and width have very similar curved relationships with the weight of these fish. While we might consider handling this curvature by adding quadratic terms (as discussed later in this section), a better explanation might be an interaction between the length and width of the fish. Fish that have a large length might show greater weight increases for every extra centimeter of width, so the slope with respect to width should be larger than it might be for shorter fish. As a very crude approximation, imagine that the fish were all two-dimensional rectangles so that the weight was essentially proportional to the area. That area would be modeled well with the *product* of the length and width. This motivates the inclusion of an interaction term to give the following model.

$$Weight = \beta_0 + \beta_1 Length + \beta_2 Width + \beta_3 Length \cdot Width + \epsilon$$

3.4. NEW PREDICTORS FROM OLD

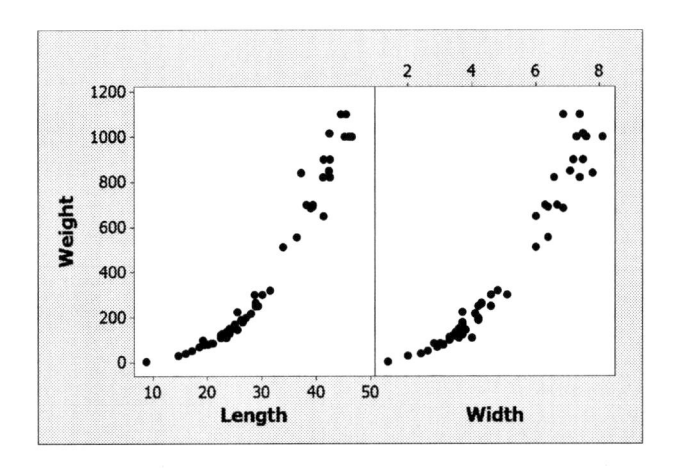

Figure 3.11: Individual predictors of perch weights

FIT

We create the new product variable (naming it *LengthxWidth*) and run the multiple regression model with the interaction term to produce the following output.

```
The regression equation is
Weight = 114 - 3.48 Length - 94.6 Width + 5.24 LengthxWidth

Predictor        Coef    SE Coef      T       P
Constant       113.93      58.78   1.94   0.058
Length         -3.483       3.152 -1.10   0.274
Width         -94.63       22.30  -4.24   0.000
LengthxWidth    5.2412      0.4131 12.69  0.000

S = 44.2381   R-Sq = 98.5%   R-Sq(adj) = 98.4%

Analysis of Variance
Source          DF       SS       MS       F       P
Regression       3  6544330  2181443  1114.68   0.000
Residual Error  52   101765     1957
Total           55  6646094
```

From this fitted model, the relationship between weight and width for perch that are 25 cm long is

$$\widehat{Weight} = 114 - 3.48(25) - 94.6 Width + 5.24(25)Width = 27.0 + 36.4 Width$$

while perch that are 30 inches long show

$$\widehat{Weight} = 114 - 3.48(30) - 94.6 Width + 5.24(30)Width = 9.6 + 62.6 Width$$

104   CHAPTER 3. MULTIPLE REGRESSION

Every extra centimeter of width for the bigger fish yields a bigger increase (on average) in the size of its weight than for the smaller fish.

(a) Without the interaction term   (b) With the interaction term

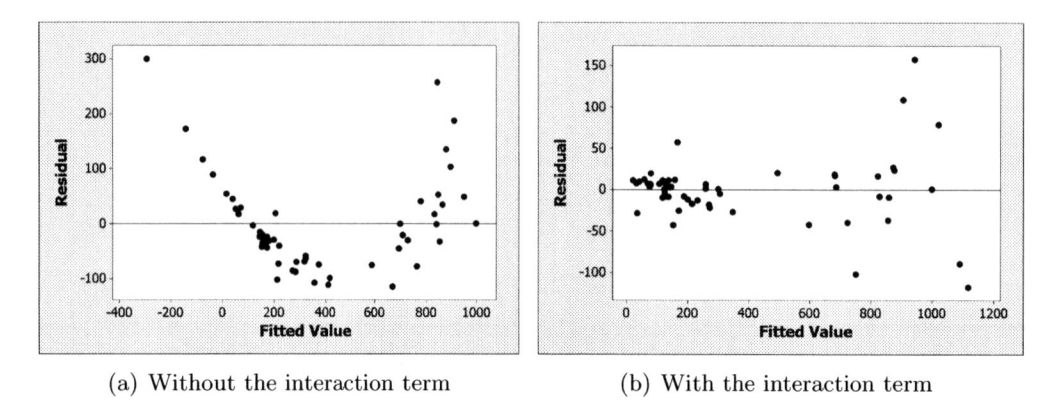

Figure 3.12: Residual plots for two models of perch weights

ASSESS

The t-tests for the individual predictors show that the interaction term, $Length \times Width$, is clearly important in this model. The $R^2 = 98.5\%$ demonstrates that this model explains most of the variability in the fish weights (compared to $R^2 = 93.7\%$ without the interaction term). The adjusted R^2 also improves by adding the interaction term, from 93.5% to 98.4%. Figure 3.12 shows the residual versus fits plots for the two predictor model with just *Length* and *Width* compared to the model with the added interaction term. While issues with nonconstant variance of the residuals, formally known as **heteroscedasticity**, remain (we tend to do better predicting weights of small fish than large ones), the interaction has successfully addressed the lack of linearity that is evident in the two-predictor model. Also, if you fit the model with just *Length* and *Width* you will find it has the unfortunate property of predicting negative weights for some of the smaller fish!   ⋄

Polynomial Regression

In Section 1.4 we saw methods for dealing with data that showed a curved rather than linear relationship by transforming one or both of the variables. Now that we have a multiple regression model, another way to deal with curvature is to add powers of one or more predictor variables to the model.

Example 3.11: *Diamond prices*

A young couple are shopping for a diamond and are interested in learning more about how these gems are priced. They have heard about the four C's: carat, color, cut, and clarity. Now they want to see if there is any relationship between these diamond characteristics and the price. Table 3.4 shows records for the first 10 diamonds from a large sample of 351 diamonds. The full data are stored in **Diamonds** and contain quantitative information on the size (*Carat*), price (*PricePerCt* and *TotalPrice*) and the *Depth* of the cut. *Color* and *Clarity* are coded as categorical variables.

3.4. NEW PREDICTORS FROM OLD

Table 3.4: Information on first 10 diamonds in the **Diamonds** datafile

Carat	Color	Clarity	Depth	PricePerCt	TotalPrice
1.08	E	VS1	68.6	6693.3	7728.8
0.31	F	VVS1	61.9	3159.0	979.3
0.31	H	VS1	62.1	1755.0	544.1
0.31	F	VVS1	60.8	3159.0	1010.9
0.33	D	IF	60.8	4758.8	1570.4
0.33	G	VVS1	61.5	2895.8	955.6
0.35	F	VS1	62.5	2457.0	860.0
0.35	F	VS1	62.3	2457.0	860.0
0.37	F	VVS1	61.4	3402.0	1258.7
0.38	D	IF	60.0	5062.5	1923.8

Source: AwesomeGems.com (July 28, 2005)

CHOOSE

Since the young couple are primarily interested in the total cost, they decide to begin by examining the relationship between *TotalPrice* and *Carat*. Figure 3.13 shows a scatterplot of this relationship for all 351 diamonds in the sample. Not surprisingly the price tends to increase as the size of a diamond increases with a bit of curvature (faster increase among larger diamonds). One way to model this sort of curvature is with a quadratic regression model.

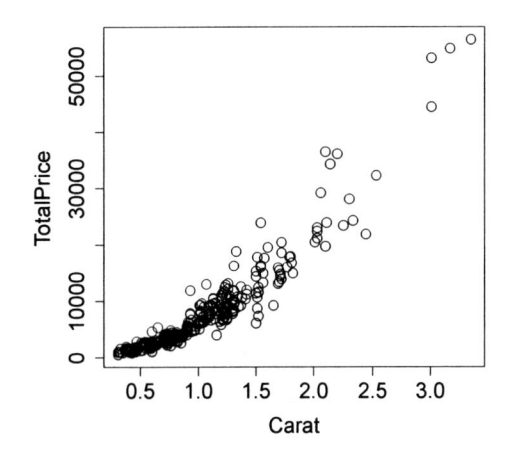

Figure 3.13: *TotalPrice* versus *Carat* for diamonds

> **Quadratic Regression Model**
>
> For a single quantitative predictor X, a **quadratic** regression model has the form
>
> $$Y = \beta_0 + \beta_1 X + \beta_2 X^2 + \epsilon$$

FIT

We add a new variable, *CaratSq*, to the dataset which is computed as the square of the *Carat* variable. Fitting a linear model to predict *TotalPrice* based on *Carat* and *CaratSq* shows

```
Coefficients:
(Intercept)        Carat        CaratSq
   -522.7         2386.0         4498.2
```

which suggests that the couple should use a fitted quadratic relationship of

$$\widehat{TotalPrice} = -522.7 + 2386 \cdot Carat + 4498.2 \cdot Carat^2$$

A scatterplot with the quadratic regression fit is shown in Figure 3.14 and illustrates how the model captures some of the curvature in the data. Note that this is still a multiple *linear* regression model since it follows the form $\mu_Y = \beta_0 + \beta_1 X_1 + \beta_2 X_2$ even though the two predictors, $X_1 = Carat$ and $X_2 = Carat^2$ are obviously related to each other. The key condition is that the model is linear in the β_i coefficients.

ASSESS

A further examination of the computer output summarizing the quadratic model shows that 92.6% of the variability in the prices of these diamonds can be explained by the quadratic relationship with the size in carats. We can compare this to the R^2 of 86.3% we would obtain if we used a simple linear model based on carats alone. Notice that the quadratic term is highly significant, as is the linear term. Sometimes a fitted quadratic model will have a significant quadratic term, but the linear term will not be statistically significant in the presence of the quadratic term. Nonetheless, it is conventional in such a situation to keep the linear term in the model.

3.4. NEW PREDICTORS FROM OLD

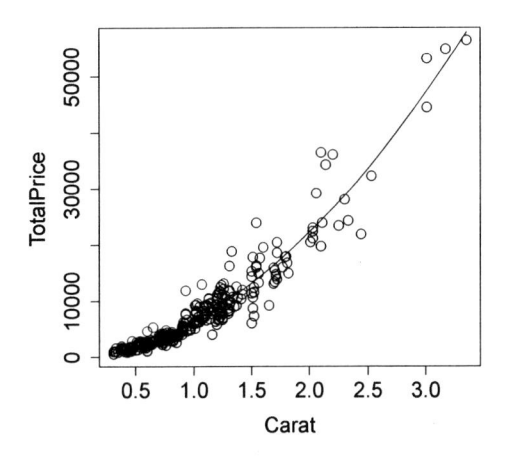

Figure 3.14: Quadratic regression for diamond prices

```
Coefficients:
             Estimate Std. Error t value Pr(>|t|)
(Intercept)    -522.7      466.3  -1.121  0.26307
Carat          2386.0      752.5   3.171  0.00166 **
CaratSq        4498.2      263.0  17.101  < 2e-16 ***
---
Signif. codes:  0 *** 0.001 ** 0.01 * 0.05 . 0.1   1

Residual standard error: 2127 on 348 degrees of freedom
Multiple R-squared: 0.9257,    Adjusted R-squared: 0.9253
F-statistic:  2168 on 2 and 348 DF,  p-value: < 2.2e-16
```

The couple should not be careless to infer from the large R^2 value that the conditions for multiple regression are automatically satisfied. Figure 3.15 shows residual versus fitted values plots for both the simple linear and quadratic regression fits. We see that the quadratic model has done a fairly good job of capturing the curvature in the relationship between carats and price of diamonds. However, the variability of the residuals still appears to increase as the fitted diamond prices increase.

Furthermore, the histogram and normal quantile plot for the quadratic model residuals in Figure 3.16 indicate that an assumption of normality would not be reasonable. Although a histogram of the residuals is relatively symmetric and centered at zero, the peak at zero is fairly sharp with lots of small magnitude residuals (typically for the smaller diamonds) and some much larger residuals (often for the larger diamonds).

108   CHAPTER 3. MULTIPLE REGRESSION

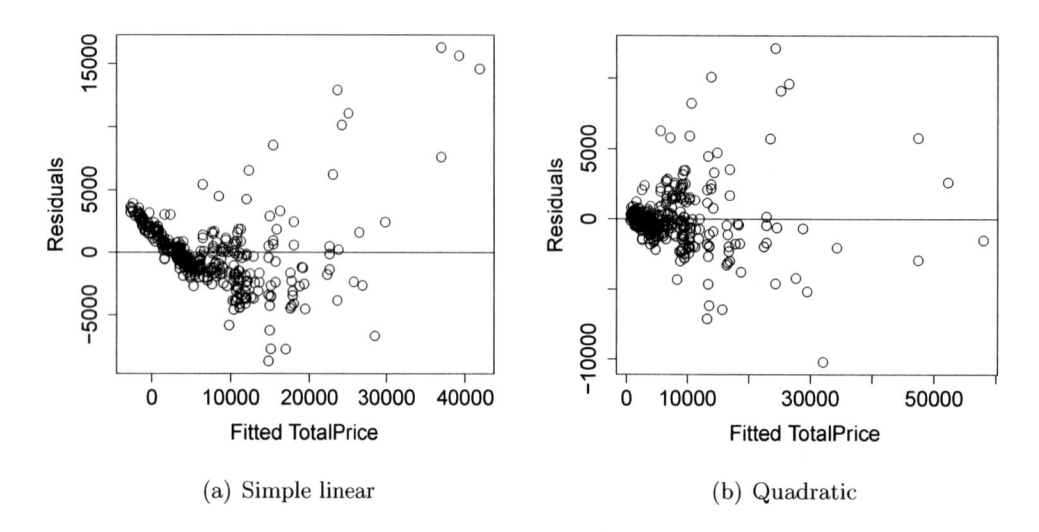

(a) Simple linear   (b) Quadratic

Figure 3.15: Residual plots for simple linear and quadratic models of diamond prices

(a) Histogram   (b) Normal quantile plot

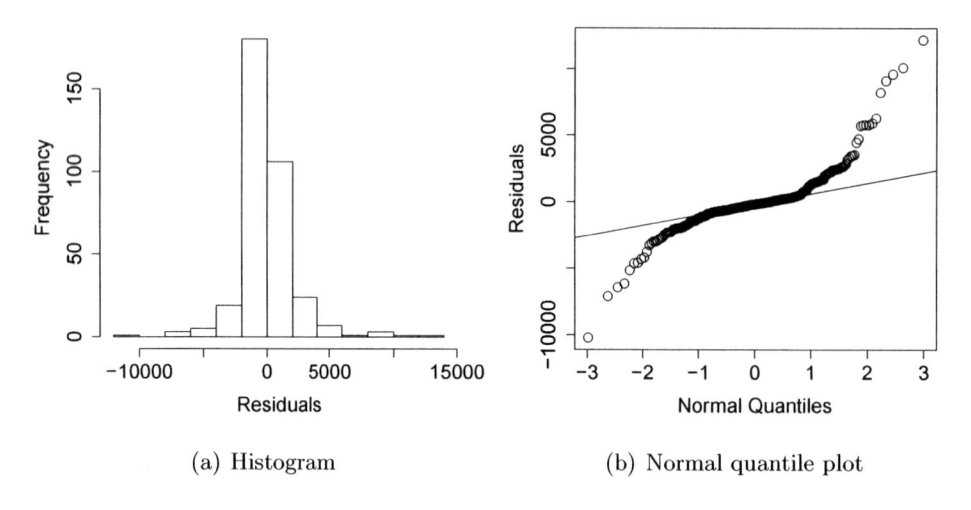

Figure 3.16: Residuals from the quadratic model of diamond prices

USE

While the heteroscedasticity and lack of normality of the residuals in the quadratic model suggests that the young couple should use some caution if they want to apply traditional inference procedures in this situation, the fitted quadratic equation can still give them a good sense of the relationship between the sizes of diamonds (in carats) and typical prices. Some exercises at the end of this chapter explore some additional regression models for these diamond prices.   ⋄

3.4. NEW PREDICTORS FROM OLD

We can easily generalize the idea of quadratic regression to include additional powers of a single quantitative predictor variable. However, note that additional polynomial terms may not improve the model much.

> **Polynomial Regression Model**
>
> For a single quantitative predictor X, a **polynomial regression** model of degree k has the form
> $$Y = \beta_0 + \beta_1 X + \beta_2 X^2 + \cdots + \beta_k X^k + \epsilon$$

For example, adding a $Carat3 = Carat^3$ predictor to the quadratic model to predict diamond prices based on 3rd degree polynomial yields the following summary statistics.

```
Coefficients:
            Estimate Std. Error t value Pr(>|t|)
(Intercept)  -723.44     875.50  -0.826  0.40919
Carat        2942.02    2185.44   1.346  0.17912
CaratSq      4077.65    1573.80   2.591  0.00997 **
Carat3         87.92     324.38   0.271  0.78652

Residual standard error: 2130 on 347 degrees of freedom
Multiple R-squared: 0.9257,    Adjusted R-squared: 0.9251
F-statistic:  1442 on 3 and 347 DF,  p-value: < 2.2e-16
```

The new cubic term in this model is clearly not significant (p-value=0.78652), the value of R^2 shows no improvement over the quadratic model (in fact, the adjusted R^2 goes down), and a plot of the fitted cubic equation with the the scatterplot is essentially indistinguishable from the quadratic fit shown in Figure 3.14.

Complete Second Order Model

In some situations we find it useful to combine the ideas from both of the previous examples. That is, we might choose to use quadratic terms to account for curvature and one or more interaction terms to handle effects that occur at particular combinations of the predictors. When doing so, we should take care that we don't overparameterize the model, making it more complicated than needed, with terms that aren't important for explaining the structure of the data. Examining t-tests for the individual terms in the model and checking how much additional variability is explained by those terms are two ways we can guard against including unnecessary complexity. In Section 3.6 we also consider a method for assessing the contribution of a set of predictors as a group.

Complete Second Order Model

For two predictors, X_1 and X_2, a **complete second order model** includes linear and quadratic terms for both predictors along with the interaction term.

$$Y = \beta_0 + \beta_1 X_1 + \beta_2 X_2 + \beta_3 X_1^2 + \beta_4 X_2^2 + \beta_5 X_1 X_2 + \epsilon$$

This extends to more than two predictors by including all linear, quadratic and pairwise interactions.

Example 3.12: *Perch weights (continued)*

We can add quadratic terms (*LengthSq* and *WidthSq*) to the interaction model of Example 3.10 to fit a complete second order model for the weights of the perch based on their lengths and widths. Here is some computer output for this model.

```
The regression equation is
Weight = 156 - 25.0 Length + 21.0 Width + 1.57 LengthSq + 34.4 WidthSq
        - 9.78 LengthxWidth

Predictor       Coef   SE Coef       T      P
Constant      156.35     61.42    2.55  0.014
Length        -25.00     14.27   -1.75  0.086
Width          20.98     82.59    0.25  0.801
LengthSq      1.5719    0.7244    2.17  0.035
WidthSq        34.41     18.75    1.84  0.072
LengthxWidth  -9.776     7.145   -1.37  0.177

S = 43.1277    R-Sq = 98.6%    R-Sq(adj) = 98.5%

Analysis of Variance
Source          DF        SS        MS       F      P
Regression       5   6553094   1310619  704.63  0.000
Residual Error  50     93000      1860
Total           55   6646094
```

3.5. CORRELATED PREDICTORS

Note that the $R^2 = 98.6\%$ for the complete second order model is only a bit higher than the $R^2 = 98.5\%$ without the two quadratic terms. Furthermore, if we use a 5% significance level, only one of the predictors (*LengthSq*) would be considered "important" in this model, although the test for overall effectiveness in the ANOVA shows the model is quite effective. We will explore this apparent contradiction in more detail in the next section. For the 56 fish in this sample, the predictions based on this more complicated complete second order model would be very similar to those obtained from the simpler model with just the linear and interaction terms. This would produce little change in the various residual plots, such as Figure 3.12(b), so we would be inclined to go with the simpler model for explaining the relationship between lengths and widths of perch and their weights. ⋄

When a higher order term, such as a quadratic term or an interaction product, is important in the model, we generally keep the lower order terms in the model, even if the coefficients for these terms are not significant.

3.5 Correlated Predictors

When fitting a multiple regression model we often encounter predictor variables that are correlated with one another. We shouldn't be surprised if predictors that are related to some response variable Y are also related to each other. This is not necessarily a bad thing, but it can lead to difficulty in fitting and interpreting the model. The next two examples illustrate some of the counterintuitive behavior that we might see when correlated predictors are added to a multiple regression model.

Example 3.13: *More perch weights*

In the previous section we examined an interaction model to predict the weights of a sample of perch caught in a Finnish lake using the length and width of the fish as predictors. The data for the 56 fish are in **Perch**. A correlation matrix for the three variables *Weight*, *Length*, and *Width* is shown below along with a p-value for testing if the sample correlations are significantly different from zero.

```
         Weight  Length
Length   0.960
         0.000

Width    0.964   0.975
         0.000   0.000
```

Not surprisingly, we see strong positive associations between both *Length* and *Width* with *Weight*, although the scatterplots in Figure 3.11 show considerable curvature in both relationships. The correlation between the predictors *Length* and *Width* is even stronger ($r = 0.975$) and Figure 3.17

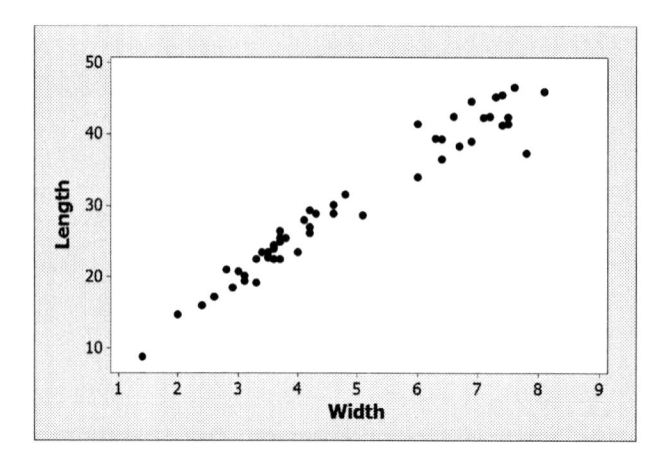

Figure 3.17: Length versus Width of perch

shows a more linear relationship between these two measurements. If we put just those two predictors in a multiple regression model to predict *Weight* we obtain the following output.

```
Weight = - 579 + 14.3 Length + 113 Width

Predictor      Coef   SE Coef       T       P
Constant    -578.76     43.67  -13.25   0.000
Length       14.307     5.659    2.53   0.014
Width        113.50     30.26    3.75   0.000

S = 88.6760    R-Sq = 93.7%    R-Sq(adj) = 93.5%


Analysis of Variance
Source           DF        SS       MS       F       P
Regression        2   6229332  3114666  396.09   0.000
Residual Error   53    416762     7863
Total            55   6646094
```

Based on the individual t-tests, both *Length* and *Width* could be considered valuable predictors in this model and, as expected, have positive coefficients. Now consider again the results when we add the *LengthxWidth* interaction term to this model. This new predictor has an even stronger correlation with *Weight* ($r = 0.989$) and, of course, is positively related to both *Length* ($r = 0.979$) and *Width* ($r = 0.988$).

3.5. CORRELATED PREDICTORS

```
The regression equation is
Weight = 114 - 3.48 Length - 94.6 Width + 5.24 LengthxWidth

Predictor        Coef    SE Coef      T       P
Constant       113.93     58.78     1.94   0.058
Length         -3.483      3.152   -1.10   0.274
Width         -94.63      22.30    -4.24   0.000
LengthxWidth    5.2412     0.4131  12.69   0.000

S = 44.2381   R-Sq = 98.5%   R-Sq(adj) = 98.4%

Analysis of Variance
Source          DF         SS         MS         F       P
Regression       3    6544330    2181443    1114.68   0.000
Residual Error  52     101765       1957
Total           55    6646094
```

What happens to the coefficients of *Length* and *Width* when we added the interaction *LengthxWidth* to the model? Both become negative, although *Length* is not significant now, *Width* would appear to have a very significant negative coefficient. Does this mean that the weight of perch goes down as the width increases? Obviously, that would not be a reasonable conclusion to draw for actual fish. The key to this apparent anomaly is that the widths are also reflected in the model through the interaction term and (to a somewhat lesser extent) through the lengths which are strongly associated with the widths. ◇

The previous example illustrates that we need to take care not to make hasty conclusions based on the individual coefficients or their t-tests when predictors in a model are related to one another. One of the challenges of dealing with multiple predictors is accounting for relationships between those predictors.

> **Multicollinearity**
>
> We say that a set of predictors exhibits **multicollinearity** when one or more of the predictors is strongly correlated with some combination of the other predictors in the set.

If one predictor has an *exact* linear relationship with one or more other predictors in a model, the least squares process to estimate the coefficients in the model does not have a unique solution. Most statistical software routines will either delete one of the perfectly correlated predictors when running the model or produce an error message. In cases where the correlation between predictors

114   CHAPTER 3. MULTIPLE REGRESSION

is high, but not one, the coefficients can be estimated, but interpretation of individual terms can be problematic. Here is another example of this phenomenon.

Example 3.14: *House prices*

The file **Houses** contains selling prices for 20 houses that were sold in 2008 in a small midwestern town. The file also contains data on the size of each house (in square feet) and the size of the lot (in square feet) that the house is on.

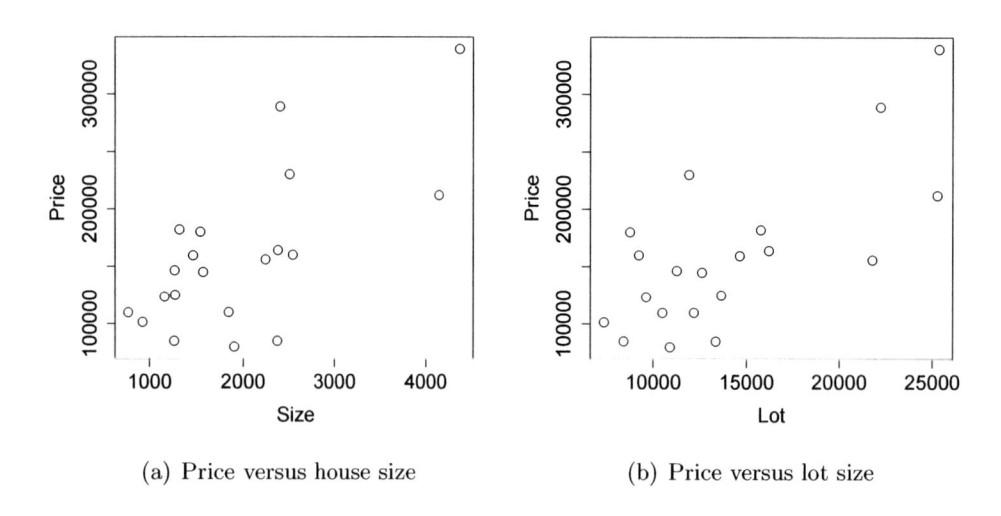

(a) Price versus house size   (b) Price versus lot size

Figure 3.18: Scatterplots of house selling price versus two predictors

CHOOSE

Figure 3.18 shows that price is positively associated with both house *Size* ($r = 0.685$) and *Lot* size ($r = 0.716$). First, we fit each predictor separately to predict *Price*.

FIT

Using technology to regress *Price* on *Size* gives the following output.

```
            Estimate Std. Error t value Pr(>|t|)
(Intercept) 64553.68   26267.76   2.458 0.024362 *
Size           48.20      12.09   3.987 0.000864 ***

Residual standard error: 50440 on 18 degrees of freedom
Multiple R-squared: 0.469,     Adjusted R-squared: 0.4395
F-statistic:  15.9 on 1 and 18 DF,  p-value: 0.0008643
```

3.5. CORRELATED PREDICTORS

Regressing *Price* on *Lot* gives similar output.

```
             Estimate Std. Error t value Pr(>|t|)
(Intercept) 36247.480  30262.135   1.198 0.246537
Lot             8.752      2.013   4.348 0.000388 ***

Residual standard error: 48340 on 18 degrees of freedom
Multiple R-squared: 0.5122,    Adjusted R-squared: 0.4851
F-statistic:  18.9 on 1 and 18 DF,  p-value: 0.0003878
```

Both of the predictors *Size* and *Lot* are highly correlated with *Price* and produce useful regression models on their own, explaining 46.9% and 51.2%, respectively, of the variability in the prices of these homes.

Let's explore what happens when we use them together. The output below shows the result of fitting the multiple regression model in which the *Price* of a house depends on *Size* and *Lot*.

```
             Estimate    Std. Error   t value   Pr(>|t|)
(Intercept) 34121.649    29716.458      1.148     0.2668
Size           23.232       17.700      1.313     0.2068
Lot             5.657        3.075      1.839     0.0834

Residual standard error: 47400 on 17 degrees of freedom
Multiple R-squared: 0.5571, Adjusted R-squared: 0.505
F-statistic: 10.69 on 2 and 17 DF,  p-value: 0.000985
```

The two-variable prediction equation is

$$\widehat{Price} = 34121.6 + 23.232 \cdot Size + 5.66 \cdot Lot$$

ASSESS

Note that the p-values for the coefficients of *Size* and of *Lot* are both below 0.25, but neither of them is statistically significant at the 0.05 level, whereas each of *Size* and *Lot* had a very small p-value and was highly significant when they were operating alone as predictors of *Price*. Moreover, the F-statistic of 10.69 and p-value of 0.000985 for the overall ANOVA show that these predictors are highly significant as a pair, even as neither individual t-test yields a small p-value. Once again, the key to this apparent contradiction is the fact that *Size* and *Lot* are themselves strongly correlated ($r = 0.767$). Remember that the individual t-tests assess how much a predictor contributes to a model *after accounting for the other predictors in the model*. Thus while house size and lot size are both strong predictors of house prices on their own, they carry similar information (bigger houses/lots tend to be more expensive). If *Size* is already in a model to predict *Price* we don't really need to add *Lot* and if *Lot* is in the model first, we don't need *Size*. This causes the individual t-tests for both predictors to be insignificant, even though at least one of them is important to have in the model to predict *Price*. ⋄

Detecting Multicollinearity

How do we know when multicollinearity might be an issue with a set of predictors? One quick check is to examine the pairwise correlations between all of the predictors. For example, a correlation matrix of the five predictors in the complete second order model for predicting perch weights based on length and width of the fish shows obvious significant correlations between each of the pairs of predictors.

```
Correlations: Length, Width, LengthxWidth, LengthSq, WidthSq

              Length      Width    LengthxWidth    LengthSq
Width         0.975
              0.000

LengthxWidth  0.979       0.988
              0.000       0.000

LengthSq      0.989       0.968       0.991
              0.000       0.000       0.000

WidthSq       0.952       0.990       0.991        0.964
              0.000       0.000       0.000        0.000
```

In some cases a dependence between predictors can be more subtle. We may have some combinations of predictors that, taken together, are strongly related to another predictor. To investigate these situations we can consider regression models for each individual variable in the model using all of the other variables as predictors. Any measure of the effectiveness of these models (for example R^2) could indicate which predictors might be strongly related to others in the model. One common form of this calculation available in many statistical packages is the **variance inflation factor** which reflects the association between a predictor and all of the other predictors.

3.5. CORRELATED PREDICTORS

> **Variance Inflation Factor**
>
> For any predictor X_i in a model, the **variance inflation factor** (VIF) is computed as
>
> $$VIF_i = \frac{1}{1 - R_i^2}$$
>
> where R_i^2 is the coefficient of multiple determination for a model to predict X_i using the other predictors in the model.
>
> As a rough rule we suspect multicollinearity with predictors for which the $VIF > 5$ which is equivalent to $R_i^2 > 80\%$.

Example 3.15: *Diamonds (continued)*

In Example 3.11 we fit a quadratic model to predict the price of diamonds based on the size (*Carat* and *CaratSq*). Suppose now that we also add the *Depth* of cut as a predictor. This produces the output shown below (after an option for the variance inflation factor has been selected).

```
The regression equation is
TotalPrice = 6343 + 2950 Carat + 4430 CaratSq - 114 Depth

Predictor     Coef   SE Coef      T      P      VIF
Constant      6343      1436   4.42  0.000
Carat       2950.0     736.1   4.01  0.000   10.942
CaratSq     4430.4     254.7  17.40  0.000   10.719
Depth      -114.08     22.66  -5.03  0.000    1.117
```

As we should expect, the VIF values for *Carat* and *CaratSq* are quite large because those two variables have an obvious relationship. The rather small VIF for *Depth* indicates that the depth of cut is not strongly related to either *Carat* or *CaratSq*, although it does appear to be an important contributor to this model (p-value=0.000). The relative independence of *Depth* as a predictor in this model allows us to make this interpretation of its individual t-test with little concern that it has been unduly influenced by the presence of the other predictors in this model. ⋄

What should we do if we detect multicollinearity in a set of predictors? First, realize that multicollinearity does not necessarily produce a poor model. In many cases, the related predictors might all be important in the model. For example, when trying to predict the perch weights we saw considerable improvement in the model when we added the *Length×Width* interaction term even though it was strongly related to both the *Length* and *Width* predictors.

Some options for dealing with correlated predictors:

a. *Drop some predictors.* If one or more predictors are strongly correlated with other predictors, try the model with those predictors left out. If those predictors are really redundant and the reduced model is essentially as effective as the bigger model, you can probably leave them out of your final model. However, if you notice a big drop in R^2 or problems appearing in the residual plots you should consider keeping one or more of those predictors in the model.

b. *Combine some predictors.* Suppose that you are working with data from a survey that has many questions on closely related topics that produce highly correlated variables. Rather than putting each of the predictors in a model individually you could create a new variable with some formula (for example, a sum) based on the group of similar predictors. This would allow you to assess the impact of that group of questions without dealing with the predictor-by-predictor variations that could occur if each was in the model individually.

c. *Discount the individual coefficients and t-tests.* As we've seen in several examples, multicollinearity can produce what initially looks like odd behavior when assessing individual predictors. A model may be highly effective overall, yet none of its individual t-tests are significant; a predictor that we expect to have a positive association with a response gets a significant negative coefficient, due to other predictors with similar information; a predictor that is clearly important in one model produces a clearly insignificant t-test in another model, when the second model includes a strongly correlated predictor. Since the model can still be quite effective with correlated predictors, we might choose to just ignore some the individual t-tests. For example, after fitting the cubic model for diamond prices the coefficients of *Carat* and $Carat^3$ were both "insignificant" (p-values=0.179 and 0.787). While that might provide good evidence that the cubic term should be dropped from the model, we shouldn't automatically throw out the *Carat* term as well.

3.6 Testing Subsets of Predictors

Individual t-tests for regression predictors ($H_0 : \beta_i = 0$) allow us to check the importance of terms in the model one at a time. The overall ANOVA for regression ($H_0 : \beta_1 = \beta_2 = \ldots = \beta_k = 0$) allows us to test the effectiveness of all of the predictors in the model as a group. Is there anything in between these two extremes? As we develop more complicated models with polynomial terms, interactions, and different kinds of predictors, we may encounter situations where we want to assess the contribution of some subset of predictors as a group. For example, do we need the second order terms (as a group) in a complete second order model for perch weights ($H_0 : \beta_3 = \beta_4 = \beta_5 = 0$)? When we are using a multiple regression model to compare regression lines for two groups (as in Example 3.9) we might want to *simultaneously* test for a significant difference in the slope and/or the intercept ($H_0 : \beta_2 = \beta_3 = 0$). Note that in both of these examples the individual predictors being tested are likely to be correlated with other predictors in the model. If there are correlated predictors within the set of terms being tested, we may avoid some multicollinearity issues by testing them as a group.

3.6. TESTING SUBSETS OF PREDICTORS

Nested F-test

The procedure we use for testing a subset of predictors is called a **nested F-test**. We say that one model is nested inside another model if its predictors are all present in the larger model. For example, an interaction model $Y = \beta_0 + \beta_1 X_1 + \beta_2 X_2 + \beta_3 X_1 X_2 + \epsilon$ is nested within a complete second order model using the same two predictors. Note that the smaller (nested) model should be entirely contained in the larger model, so models such as $Y = \beta_0 + \beta_1 X_1 + \beta_2 X_2 + \epsilon$ and $Y = \beta_0 + \beta_1 X_2 + \beta_2 X_3 + \epsilon$ have a predictor in common but neither is nested within the other.

The essence of a nested F test is to compare the full (larger) model with the reduced (nested) model that eliminates the group of predictors that we are testing. If the full model does a (statistically significant) better job of explaining the variability in the response we may conclude that at least one of those predictors is important to include in the model.

Example 3.16: *Second order model for perch weights (continued)*

In Example 3.12 we created a complete second order model using *Length* and *Width* to predict the weights of perch.

$$Weight = \beta_0 + \beta_1 Length + \beta_2 Width + \beta_3 Length^2 + \beta_4 Width^2 + \beta_5 Length \cdot Width + \epsilon$$

Some of the computer output for fitting this model is reproduced below.

```
The regression equation is
Weight = 156 - 25.0 Length + 21.0 Width + 1.57 LengthSq + 34.4 WidthSq
        - 9.78 LengthxWidth

Predictor         Coef    SE Coef       T      P
Constant        156.35      61.42    2.55  0.014
Length          -25.00      14.27   -1.75  0.086
Width            20.98      82.59    0.25  0.801
LengthSq        1.5719     0.7244    2.17  0.035
WidthSq          34.41      18.75    1.84  0.072
LengthxWidth    -9.776      7.145   -1.37  0.177

S = 43.1277    R-Sq = 98.6%    R-Sq(adj) = 98.5%

Analysis of Variance
Source           DF         SS         MS       F        P
Regression        5    6553094    1310619  704.63    0.000
Residual Error   50      93000       1860
Total            55    6646094
```

As we noted earlier only one of the individual t-tests is significant at a 5% level, but we have a lot of correlation among these five predictors so there could easily be multicollinearity issues in assessing those tests. Let's test whether adding the two quadratic terms (*LengthSq* and *WidthSq*) actually provides a substantial improvement over the interaction model with just *Length*, *Width* and *LengthxWidth*.

$H_0 : \beta_3 = \beta_4 = 0$
$H_a : \beta_3 \neq 0 \text{ or } \beta_4 \neq 0$

One quick way to compare these models is to look at their R^2 values and see how much is "lost" when the quadratic terms are dropped from the model. Here is some output for the nested interaction model.

```
The regression equation is
Weight = 114 - 3.48 Length - 94.6 Width + 5.24 LengthxWidth

Predictor        Coef    SE Coef      T      P
Constant       113.93      58.78   1.94  0.058
Length         -3.483      3.152  -1.10  0.274
Width         -94.63       22.30  -4.24  0.000
LengthxWidth   5.2412     0.4131  12.69  0.000

S = 44.2381    R-Sq = 98.5%    R-Sq(adj) = 98.4%

Analysis of Variance
Source          DF        SS       MS       F      P
Regression       3   6544330  2181443  1114.68  0.000
Residual Error  52    101765     1957
Total           55   6646094
```

The two quadratic terms only add 0.1% to the value of R^2 when they are added to the interaction model. That doesn't seem very impressive, but we would like to have a formal test of whether or not the two quadratic terms explain a significant amount of new variability. Comparing the SSModel values for the two models we see that

$$SSModel_{full} - SSModel_{reduced} = 6553094 - 6544330 = 8764$$

Is that a "significant" amount of new variability in this setting? As with the ANOVA procedure for the full model, we need to divide this amount of new variability explained by the number of predictors added to help explain it (2) to obtain a mean square. To compute an F test statistic, we then divide that mean square by the MSE for the full model.

3.6. TESTING SUBSETS OF PREDICTORS

$$F = \frac{8764/2}{93000/50} = \frac{4382}{1860} = 2.356$$

The degrees of freedom should be obvious from the way this statistic was computed; in this case we should compare this value to the upper tail of an $F_{2,50}$ distribution. Doing so produces a p-value of 0.105 which is somewhat small, but not below a 5% level. This would imply that the two quadratic terms are not especially helpful to this model and we could probably use just the interaction model for predicting perch weights. ◇

Nested F-test

To test a subset of predictors in a multiple regression model
$H_0 : \beta_i = 0$ for all predictors in the subset
$H_a : \beta_i \neq 0$ for at least one predictor in the subset

Let the **full** model denote one with all k predictors and the **reduced** model be the nested model obtained by dropping the predictors that are being tested. The test statistic is

$$F = \frac{(SSModel_{full} - SSModel_{reduced})/\#\text{ predictors tested}}{SSE_{full}/(n-k-1)}$$

The p-value is computed from an F-distribution with numerator degrees of freedom equal to the number of predictors being tested and denominator degrees of freedom equal to the error degrees of freedom for the full model.

An equivalent way to compute the amount of new variability explained by the predictors being tested is

$$SSModel_{full} - SSModel_{reduced} = SSE_{reduced} - SSE_{full}$$

Note that if we apply a nested F-test to all of the predictors at once, we get back to the original ANOVA test for an overall model. If we test just a single predictor using a nested F-test we get a test that is equivalent to the individual t-test for that predictor. The nested F-test statistic will be just the square of the test statistic from the t-test (see Exercise 3.23).

3.7 Case Study: Predicting in Retail Clothing

We will now look at a set of data with several explanatory variables to illustrate the process of arriving at a suitable multiple regression model. We first examine the data for outliers and other deviations that might unduly influence our conclusions. Next, we use descriptive statistics, especially correlations, to get an idea of which explanatory variables may be most helpful to choose for explaining the response. We fit several models using combinations of these variables, paying attention to the individual t-tests to see if any variables contribute little in a particular model, checking the regression conditions, and assessing the overall fit. Once we have settled on an appropriate model, we use it to answer the question of interest. Throughout this process we keep in mind the real-world setting of the data and use common sense to help guide our decisions.

Table 3.5: First few cases of the **Clothing** data

ID	Amount	Recency	Freq12	Dollar12	Freq24	Dollar24	Card
1	0	22	0	0	3	400	0
2	0	30	0	0	0	0	0
3	0	24	0	0	1	250	0
4	30	6	3	140	4	225	0
5	33	12	1	50	1	50	0
6	35	48	0	0	0	0	0
7	35	5	5	450	6	415	0
8	39	2	5	245	12	661	1
9	40	24	0	0	1	225	0
⋮	⋮	⋮	⋮	⋮	⋮	⋮	⋮
60	1,506,000	1	6	5000	11	8000	1

Source: Personal communication from clothing retailer David Cameron.

Example 3.17: *Predicting customer spending for a clothing retailer*

The data provided in **Clothing** represent a random sample of 60 customers from a large clothing retailer. The first few and last cases in the dataset are reproduced in Table 3.5. The manager of the store is interested in predicting how much a customer will spend on his or her next purchase based on one or more of the available explanatory variables that are described below.

3.7. CASE STUDY: PREDICTING IN RETAIL CLOTHING

Variable	Description
Amount	The net dollar amount spent by customers in their latest purchase from this retailer
Recency	The number of months since the last purchase
Freq12	The number of purchases in the last 12 months
Dollar12	The dollar amount of purchases in the last 12 months
Freq24	The number of purchases in the last 24 months
Dollar24	The dollar amount of purchases in the last 24 months
Card	1 for customers who have a private-label credit card with the retailer, 0 if not

The response variable is the *Amount* of money spent by a customer. A careful examination of Table 3.5 reveals that the first three values for *Amount* are zero because some customers purchased items and then returned them. We are not interested in modeling returns, so these observations will be removed before proceeding. The last row of the data indicates that one customer spent $1,506,000 in the store. A quick consultation with the manager reveals that this observation is a data entry error, so this customer will also be removed from our analysis. We can now proceed with the cleaned data on 56 customers.

CHOOSE

We won't go through all of the expected relationships among the variables, but we would certainly expect the amount of a purchase to be positively associated with the amount of money spent over the last 12 months (*Dollar12*) and the last 24 months (*Dollar24*). Speculating about how the frequency of purchases over the last 12 and 24 months is related to the purchase amount is not as easy. Some customers might buy small amounts on a regular basis while others might purchase large amounts of clothing at less frequent intervals because they don't like to shop. Other people like shopping and clothing so they might purchase large amounts on a regular basis.

A matrix of correlation coefficients for the 6 quantitative variables is shown below. As expected, *Amount* is strongly correlated with past spending: $r = 0.804$ with *Dollar12* and $r = 0.677$ with *Dollar24*. However, the matrix also reveals that these explanatory variables are correlated with one another. Since the variables are dollar amounts in overlapping time periods, there is a strong positive association, r = 0.827, between *Dollar12* and *Dollar24*.

```
         Amount Recency Freq12 Dollar12 Freq24 Dollar24
Amount    1.000  -0.221  0.052    0.804  0.102    0.677
Recency  -0.221   1.000 -0.584   -0.454 -0.549   -0.432
Freq12    0.052  -0.584  1.000    0.556  0.710    0.421
Dollar12  0.804  -0.454  0.556    1.000  0.485    0.827
Freq24    0.102  -0.549  0.710    0.485  1.000    0.596
Dollar24  0.677  -0.432  0.421    0.827  0.596    1.000
```

Recency (the number of months since the last purchase) is negatively associated with the purchase *Amount* and with the four explanatory variables that indicate the number of purchases or the amount of those purchases. Perhaps recent customers (low *Recency*) tend to be regular customers who visit frequently and spend more, whereas those who have not visited in some time (high *Recency*) include customers who often shop elsewhere.

To start, let's look at a simple linear regression model with the single explanatory variable most highly correlated with *Amount*. The correlations above show that this variable is *Dollar*12.

FIT & ASSESS

Some least squares regression output predicting the purchase *Amount* based on *Dollar*12 is shown below and a scatterplot with the least squares line is shown in Figure 3.19.

```
            Estimate Std. Error t value Pr(>|t|)
(Intercept) 10.0756    13.3783    0.753    0.455
Dollar12     0.3176     0.0320    9.925  8.93e-14 ***

Residual standard error: 67.37 on 54 degrees of freedom
Multiple R-squared: 0.6459,    Adjusted R-squared: 0.6393
F-statistic:  98.5 on 1 and 54 DF,  p-value: 8.93e-14
```

The simple linear model based on *Dollar*12 shows a clear increasing trend with a very small p-value for the slope. So this is clearly an effective predictor and accounts for 64.6% of the variability in purchase *Amounts* in this sample. That is a significant amount of variability, but could we do better by including more of the predictors in a multiple regression model?

Of the remaining explanatory variables, *Dollar*24 and *Recency* have the strongest associations with the purchase amounts, so let's add them to the model.

```
            Estimate  Std. Error t value Pr(>|t|)
(Intercept) -23.05236   21.59290  -1.068   0.2906
Dollar12      0.32724    0.05678   5.764 4.53e-07 ***
Dollar24      0.02151    0.04202   0.512   0.6110
Recency       2.86718    1.37573   2.084   0.0421 *

Residual standard error: 65.91 on 52 degrees of freedom
Multiple R-squared: 0.6736,    Adjusted R-squared: 0.6548
F-statistic: 35.78 on 3 and 52 DF,  p-value: 1.097e-12
```

3.7. CASE STUDY: PREDICTING IN RETAIL CLOTHING

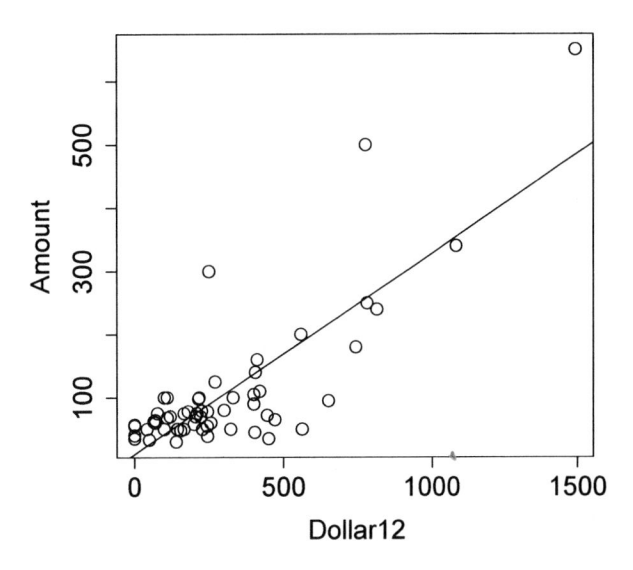

Figure 3.19: Regression of *Amount* on *Dollar*12

We see that *Dollar*12 is still a very strong predictor, but *Dollar*24 is not so helpful in this model. Although *Dollar*24 is strongly associated with *Amount* ($r = 0.677$) it is also strongly related to *Dollar*12 ($r = 0.827$) so it is not surprising that it's not particulary helpful in explaining *Amount* when *Dollar*12 is also in the model. Note that the R^2 value for this model has improved a bit, up to 67.4% and the adjusted R^2 has also increased from 63.9% to 65.5%.

What if we try a model with all six of the available predictors?

$$Amount = \beta_0 + \beta_1 Dollar12 + \beta_2 Freq12 + \beta_3 Dollar24 + \beta_4 Freq24 + \beta_5 Recency + \beta_6 Card + \epsilon$$

We omit the output for this model here, but it shows that R^2 jumps to 88.2% and adjusted R^2 is 86.8%. However only two of the six predictors (*Dollar*12 and *Freq*12) have significant individual t-tests. Again, we can expect issues of multicollinearity to be present when all six predictors are in the model.

Since adding in all six predictors helped improve the R^2 perhaps we should consider adding quadratic terms for each of the quantitative predictors (note that $Card^2$ is the same as $Card$ since the only values are 0 and 1). This 11-predictor model increases the R^2 to 89.3% but the adjusted R^2 actually drops to 86.6%. Clearly we have gone much too far in constructing an overly complicated model. Also the output for this 11-predictor model shows once again that only the coefficients of *Dollar*12 and *Freq*12 are significant at a 5% level.

Since *Dollar*12 and *Freq*12 seem to be important in most of these models, it makes sense to try a model with just those two predictors.

$$Amount = \beta_0 + \beta_1 Dollar12 + \beta_2 Freq12 + \epsilon$$

Here is some output for fitting that two-predictor model.

```
Coefficients:
            Estimate Std. Error t value Pr(>|t|)
(Intercept) 73.89763   10.46860   7.059 3.62e-09 ***
Dollar12     0.44315    0.02337  18.959  < 2e-16 ***
Freq12     -34.42587    3.56139  -9.666 2.72e-13 ***

Residual standard error: 40.91 on 53 degrees of freedom
Multiple R-squared: 0.8718,    Adjusted R-squared: 0.867
F-statistic: 180.3 on 2 and 53 DF,  p-value: < 2.2e-16
```

The R^2 drops by only 1% compared to the six-predictor model and the adjusted R^2 is essentially the same. The tests for each coefficient and the ANOVA for overall fit all have extremely small p-values so we have evidence that these are both useful predictors and together they explain a substantial portion of the variability in the purchase amounts for this sample of customers. Although we should also check the conditions on residuals, if we were restricted to models using just these six individual predictors, the two-predictor model based on *Dollar*12 and *Freq*12 would appear to be a good choice to balance simplicity and explanatory power.

CHOOSE (again)

We have used the explanatory variables that were given to us by the clothing retailer manager and come up with a reasonable two-predictor model for explaining purchase amounts. However, we should think carefully about the data and our objective. Our response variable, *Amount* measures spending of a customer on an individual visit to the store. To predict this quantity the "typical" or average purchase for that customer over a recent time period might be helpful. We have the total spending and frequency of purchases over 12 months, so we can create a new variable, *AvgSpent*12, to measure the average amount spent on each visit over the past 12 months.

$$AvgSpent12 = \frac{Dollars12}{Freq12}$$

Unfortunately, four cases had no record of any sales in the past 12 months ($Freq12 = 0$) so we need to drop those cases from the analysis as we proceed. This leaves us with $n = 52$ cases.

3.7. CASE STUDY: PREDICTING IN RETAIL CLOTHING

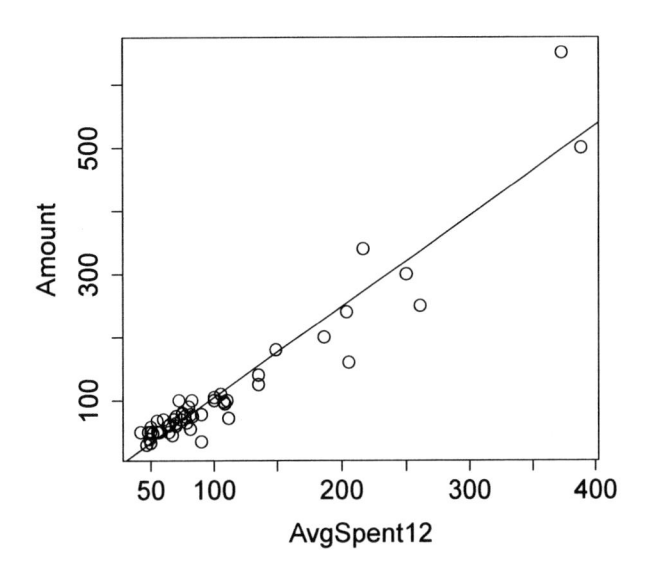

Figure 3.20: Regression of *Amount* on *AvgSpent*12

FIT

We compute values for *AvgSpent*12 for every customer in the reduced sample and try a simple linear regression model for *Amount* using this predictor. The results for this model are shown below and a scatterplot with regression line appears in Figure 3.20.

```
Coefficients:
            Estimate Std. Error t value Pr(>|t|)
(Intercept) -38.8254     8.3438  -4.653 2.43e-05 ***
AvgSpent12    1.4368     0.0642  22.380  < 2e-16 ***

Residual standard error: 35.02 on 50 degrees of freedom
Multiple R-squared: 0.9092,    Adjusted R-squared: 0.9074
F-statistic: 500.9 on 1 and 50 DF,  p-value: < 2.2e-16
```

This looks like a pretty good fit for a single predictor and the R^2 value of 90.0% is even better than $R^2 = 89.3\%$ for the 11 term quadratic model using all of the original variables!

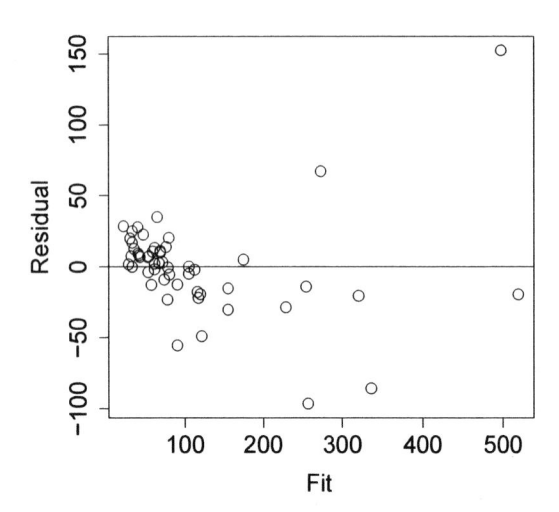

Figure 3.21: Residuals versus fits for the regression of *Amount* on *AvgSpent*12

ASSESS

The linear trend looks clear in Figure 3.20 but we should still take a look at the residuals to see if the regression conditions are reasonable. Figure 3.21 shows a plot of the residuals versus fitted values for the simple linear regression of *Amount* on *AvgSpent*12. In this plot we see a bit of curvature with positive residuals for most of the small fitted values then mostly negative residuals for the middle fitted values with a couple of large positive residuals for larger fitted values. Looking back at the scatterplots with regression line in Figure 3.20 we can also see this slight curvature, although the residual plot shows it more clearly.

This suggests adding a quadratic term $AvgSpent12Sq = (AvgSpent12)^2$ to the model. The output for fitting this model is shown below.

```
              Estimate Std. Error t value Pr(>|t|)
(Intercept)   1.402e+01  1.457e+01   0.963 0.340464
AvgSpent12    5.709e-01  2.145e-01   2.661 0.010498 *
AvgSpent12Sq  2.289e-03  5.477e-04   4.180 0.000120 ***

Residual standard error: 30.37 on 49 degrees of freedom
Multiple R-squared: 0.9331,    Adjusted R-squared: 0.9304
F-statistic: 341.7 on 2 and 49 DF,  p-value: < 2.2e-16
```

3.7. CASE STUDY: PREDICTING IN RETAIL CLOTHING

(a) Residuals versus fits

(b) Normal plot of residuals

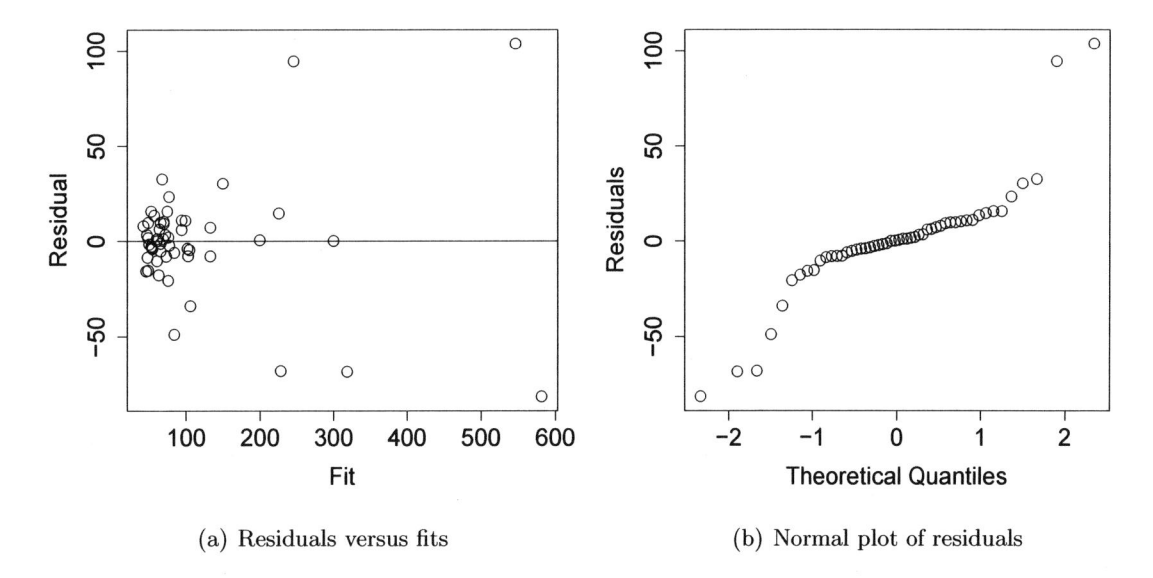

Figure 3.22: Residuals plots for quadratic model to predict *Amount* based on *AvgSpent*12

The t-test for the new quadratic term shows it is valuable to include in the model. The values for R^2 (93.3%) and adjusted R^2 (93.0%) are both improvements over any model we have considered so far. We should take care when making direct comparisons to the earlier models since the models based on *AvgSpent*12 were fit using a reduced sample that eliminated the cases for which $Freq12 = 0$. Perhaps those cases were unusual in other ways and the earlier models would also look better when fit to the smaller sample. However, a quick check of the model using *Dollar*12 and *Freq*12 using the reduced sample shows the R^2 value is virtually unchanged at 87.0%.

The plot of residuals versus fitted values in Figure 3.22(a) shows that the issue with curvature has been addressed, as the residuals appear to be more randomly scattered above and below the zero line. We may still have some problem with the equal variance condition as the residuals tend to be larger when the fitted values are larger. We might expect it to be harder to predict the purchase amounts for "big spenders" than more typical shoppers. The normal plot in Figure 3.22(b) also shows some departures from the residuals following a normal distribution. Again the issue appears to be with those few large residuals at either end of the distribution that produce somewhat longer tails than would be expected based on the rest of the residuals.

Remember that no model is perfect! While we might have some suspicions about the applicability of the equal variance and normality conditions, the tests that are based on those conditions, especially the t-test for the coefficient of the quadratic term and overall ANOVA F-test, show very small p-values in this situation; moreover we have a fairly large sample size ($n = 52$).

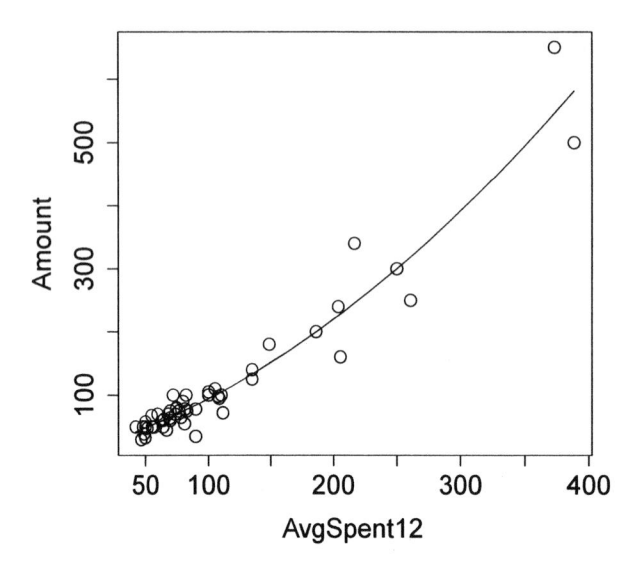

Figure 3.23: Quadratic regression fit of *Amount* on *AvgSpent*12

USE

We can be reasonably confident in recommending that the manager use the quadratic model based on average spending per visit over the past 12 months (*AvgSpent*12) to predict spending amounts for individual customers who have shopped there at least once in the past year. We might also recommend further sampling or study to deal with new or infrequent customers who haven't made any purchases in the past year (*Freq*12 = 0). Our final fitted model is given below and shown on a scatterplot of the data in Figure 3.23.

$$\widehat{Amount} = 14.02 + 0.5709 \cdot AvgSpent12 + 0.002289 \cdot (AvgSpent12)^2$$

◇

3.8 Chapter Summary

In this chapter we introduced the **multiple regression model** where two or more predictor variables are used to explain the variability in a response variable.

$$Y = \beta_0 + \beta_1 X_1 + \beta_2 X_2 + \ldots + \beta_k X_k + \epsilon$$

where $\epsilon \sim N(0, \sigma_\epsilon)$ and the errors are independent from one another.

The procedures and conditions for making inferences are very similar to those used for simple linear models in Chapters 1 and 2. The primary difference is that we have more parameters ($k + 2$) to estimate: Each of the k predictor variables has a coefficient parameter, plus we have the intercept parameter and the standard deviation of the errors. As in the case with a single predictor, the coefficients are estimated by minimizing the sum of the squared residuals. The computations are more tedious, but the main idea of using least squares to obtain the best estimates is the same. One of the major differences is that the degrees of freedom for error is now $n - k - 1$, instead of $n - 2$, so the **multiple regression standard error** is

$$\hat{\sigma}_\epsilon = \sqrt{\frac{SSE}{n - k - 1}}$$

Statistical inferences for the coefficient parameters are based on the parameter estimates, $\hat{\beta}_i$, and standard errors of the estimates, $SE_{\hat{\beta}_i}$. The test statistic for an individual parameter is

$$t = \frac{\hat{\beta}_i}{SE_{\hat{\beta}_i}}$$

which follows a t-distribution with $n - k - 1$ degrees of freedom, provided that the regression conditions, including normality of the errors, is reasonable. A confidence interval for an individual parameter is $\hat{\beta}_i \pm t^* \cdot SE_{\hat{\beta}_i}$, where t^* is the critical value for the t-distribution with $n - k - 1$ degrees of freedom. The interpretation for each interval assumes that the values of the other predictor variables are held constant and the individual t-tests assess how much a predictor contributes to a model after accounting for the other predictors in the model. In short, the individual t-tests or intervals look at the effect of each predictor variable, one-by-one in the presence of others.

Partitioning the total variability into two parts, one for the multiple regression model ($SSModel$) and another for error (SSE), allows us to test the effectiveness of all of the predictor variables as a group. The **multiple regression ANOVA** table provides the formal inference, based on an F-distribution with k numerator degrees of freedom and $n - k - 1$ denominator degrees of freedom.

The **coefficient of multiple determination**, R^2, provides a statistic that is useful for measuring the effectiveness of a model: the amount of total variability in the response ($SSTotal$) that is explained by the model ($SSModel$). In the multiple regression setting, this coefficient can also be obtained by squaring the correlation between the fitted values (based on all k predictors) and the response variable. Making a slight adjustment to this coefficient to account for the fact that adding new predictors will never decrease the variability explained by the model, we obtain an **adjusted** R^2 that is extremely useful when comparing different multiple regression models based on different numbers of predictors.

After checking the conditions for the multiple regression model, as you have done in the previous chapters, you should be able to obtain and interpret a confidence interval for a mean response and a prediction interval for an individual response.

One of the extensions of the simple linear regression model is the multiple regression model for two regression lines. This extension can be expanded to compare three or more simple linear regression lines. In fact, the point of many of the examples in this chapter is that multiple regression models are extremely flexible and cover a wide range of possibilities. Using **indicator variables**, interaction terms, squared variables, and other combinations of variables produces an incredible set of possible models for us to consider. The regression model with **interaction**, **quadratic regression**, a **complete second order model**, and **polynomial regression** were used in a variety of different examples. You should be able to choose, fit, assess, and use these multiple regression models.

When considering and comparing various multiple regression models, we must use caution, especially when the predictor variables are associated with one another. Correlated predictor variables often create **multicollinearity** problems. These multicollinearity problems are usually most noticeable when they produce counterintuitive parameter estimates or tests. An apparent contradiction between the overall F-test and the individual t-test is also a warning sign for multicollinearity problems. The easiest way to check for association in your predictor variables is to create a correlation matrix of pairwise correlation coefficients. Another way to detect multicollinearity is to compute the **variance inflation factor** (VIF), a statistic that measures the association between a predictor and all of the other predictors. As a rule of thumb, a VIF larger than 5 indicates multicollinearity. Unfortunately, there is no hard and fast rule for dealing with multicollinearlity. Dropping some predictors and combining predictors to form another variable may be helpful.

The **nested F-test** is a way to assess the importance of a subset of predictors. The general idea is to fit a full model (one with all of the predictor variables under consideration) and then fit a reduced model (one without a subset of predictors). The nested F-statistic provides a formal way to assess if the full model does a significantly better job explaining the variability in the response variable. In short, the nested F test sits between the overall ANOVA test for all of the regression coefficients and the individual t-tests for each coefficient and you should be able to apply this method for comparing models.

Finally, the case study for the clothing retailer provides a valuable modeling lesson. Think about your data and the problem at hand before blindly applying complicated models with lots of predictor variables. Keeping the model as simple and intuitive as possible is appealing for many reasons.

3.9 Exercises

Conceptual Exercises

3.1 *Predicting a statistics final exam grade.* A statistics professor assigned various grades during the semester including a midterm exam (out of 100 points) and a logistic regression project (out of 30 points). The prediction equation below was fit, using data from 24 students in the class, to predict the final exam score (out of 100 points) based on the midterm and project grades.

$$\widehat{Final} = 11.0 + 0.53 \cdot Midterm + 1.20 \cdot Project$$

a. What would this tell you about a student who got perfect scores on the midterm and project?

b. Michael got a grade of 87 on his midterm, 21 on the project and an 80 on the final. Compute his residual and write a sentence to explain what that value means in Michael's case.

3.2 *Predicting a statistics final exam grade (continued).* Does the prediction equation for final exam scores in Exercise 3.1 suggest that the project score has a stronger relationship with the final exam than the midterm exam? Explain why or why not.

3.3 *Breakfast cereals.* A regression model was fit to a sample of breakfast cereals. The response variable Y is calories per serving. The predictor variables are X_1, grams of sugar per serving, and X_2, grams of fiber per serving. The fitted regression model is

$$\hat{Y} = 109.3 + 1.0 \cdot X_1 - 3.7 \cdot X_2$$

In the context of this setting, interpret -3.7, the coefficient of X_2. That is, describe how fiber is related to calories per serving, in the presence of the sugar variable.

3.4 *Adjusting R^2.* Decide if the following statements are true or false, and explain why.

a. For a multiple regression problem the adjusted coefficient of determination, R^2_{adj}, will always be smaller than the regular, unadjusted R^2.

b. If we fit a multiple regression model and then add a new predictor to the model, the adjusted coefficient of determination, R^2_{adj}, will always increase.

3.5 *Body measurements.* Suppose that you are interested in predicting the percentage of body fat (*BodyFat*) on a man using the explanatory variables waist size (*Waist*) and *Height*.

a. Do you think that *BodyFat* and *Waist* are positively correlated? Explain.

b. For a fixed waist size (say 38 inches), would you expect *BodyFat* to be positively or negatively correlated with a man's *Height*? Explain why.

c. Suppose that *Height* does not tell you much about *BodyFat* by itself, so that the correlation between the two variables is near zero. What sort of coefficient on *Height* (positive, negative or near zero) would you expect to see in a multiple regression to predict *BodyFat* based on both *Height* and *Waist*? Explain your choice.

Guided Exercises

3.6 *Active pulse rates.* The Minitab output below comes from a study to model *Active* pulse rates (after climbing several flights of stairs) based on resting pulse rate (*Rest* in beats per minute), weight (*Wgt* in pounds), and amount of *Exercise* (in hours per week). The data were obtained from 232 students taking Stat2 courses in past semesters.

```
The regression equation is
Active = 11.8 + 1.12 Rest + 0.0342 Wgt - 1.09 Exercise

Predictor      Coef    SE Coef      T      P
Constant      11.84      11.95   0.99  0.323
Rest         1.1194     0.1192   9.39  0.000
Wgt         0.03420    0.03173   1.08  0.282
Exercise     -1.085      1.600  -0.68  0.498

S = 15.0452    R-Sq = 36.9%    R-Sq(adj) = 36.1%
```

a. Test the hypotheses that $\beta_2 = 0$ versus $\beta_2 \neq 0$ and interpret the result in the context of this problem. You may assume that the conditions for a linear model are satisfied for these data.

b. Construct and interpret a 90% confidence interval for the coefficient β_2 in this model.

c. What active pulse rate would this model predict for a 200-pound student who exercises 7 hours per week and has a resting pulse rate of 76 beats per minute?

3.7 *Major League Baseball winning percentage.* In Example 3.1 we considered a model for the winning percentages of football teams based on measures of offensive (*PointsFor*) and defensive (*PointsAgainst*) ability. The file **MLB2007Standings** contains similar data on many variables for Major League Baseball (MLB) teams from the 2007 regular season. The winning percentages are in the variable *WinPct* and scoring variables include *Runs* (scored by a team for the season) and *ERA* (essentially the average runs against a team per game).

a. Fit a multiple regression model to predict *WinPct* based on *Runs* and *ERA*. Write down the prediction equation.

3.9. EXERCISES

b. The Boston Red Sox had a winning percentage of 0.593 for the 2007 season. They scored 867 runs and had an ERA of 3.87. Use this information and the fitted model to find the residual for the Red Sox.

c. Comment on the effectiveness of each of the two predictors in this model. Would you recommend dropping one or the other (or both) from the model? Explain why or why not.

d. Does this model for team winning percentages in baseball appear to be more or less effective than the model for football teams in Example 3.1 on page 79? Give a numerical justification for your answer.

3.8 *Fish eggs.* Researchers collected samples of female lake trout from Lake Ontario in September and November of 2002 through 2004. A goal of the study was to investigate the fertility of fish that had been stocked in the lake. One measure of the viability of fish eggs is *percent dry mass(PctDM)* which reflects the energy potential stored in the eggs by recording the percentage of the total egg material that is solid. Values of the *PctDM* for a sample of 35 lake trout (14 in September and 21 in November) are given in Table 3.6 along with the age (in years) of the fish. The data are stored in three columns in a file called **FishEggs**.

Ignore the month at first and fit a simple linear regression to predict the *PctDM* based on the *Age* of the fish.

a. Write down an equation for the least squares line and comment on what it appears to indicate about the relationship between *PctDM* and *Age*.

b. What percentage of the variability in *PctDM* does *Age* explain for these fish?

Table 3.6: Percent dry mass of eggs and age for female lake trout

September							
Age	7	7	7	7	9	9	11
PctDM	34.90	37.00	37.90	38.15	33.90	36.45	35.00
Age	11	12	12	12	16	17	18
PctDM	36.15	34.05	34.65	35.40	32.45	36.55	34.00
November							
Age	7	8	8	9	9	9	9
PctDM	34.90	37.00	37.90	38.15	33.90	36.45	35.00
Age	10	10	11	11	12	12	13
PctDM	36.15	34.05	34.65	35.40	32.45	36.55	34.00
Age	13	13	14	15	16	17	18
PctDM	36.15	34.05	34.65	35.40	32.45	36.55	34.00

Source: Maternal Characteristics versus Egg Size and Energy Density, *J. Great Lakes Res.* (2008)

c. Is there evidence that the relationship in (a) is statistically significant? Explain how you know that it is or is not.

d. Produce a plot of the residuals versus the fits for the simple linear model. Does there appear to be any regular pattern?

e. Modify your plot in (d) to show the points for each *Month* (Sept/Nov) with different symbols or colors. What (if anything) do you observe about how the residuals might be related to the month? Now fit a multiple regression model, using an indicator (*Sept*) for the month and interaction product, to compare the regression lines for September and November.

f. Do you need both terms for a difference in intercepts and slopes? If not, delete any terms that aren't needed and run the new model.

g. What percentage of the variability in *PctDM* does the model in (f) explain for these fish?

h. Redo the plot in (e) showing the residuals versus fits for the model in (f) with different symbols for the months. Does this plot show an improvement over your plot in (e)? Explain why.

3.9 *More jurors.* In Example 3.8 we considered a regression model to compare the relationship between the percentage of jurors who report and the time period for two different years (1998 and 2000). The results suggested that the intercept was larger in 2000 but the slopes of the lines in the scatterplot of Figure 3.5 appeared to be relatively equal. Use a multiple regression model to formally test whether there is a statistically significant difference in the slopes of those two regression lines. The data are in a file called **Jurors**.

3.10 *More breakfast cereal.* The regression model in Exercise 3.3 was fit to a sample of 36 breakfast cereals with calories per serving as the response variable. The two predictors were grams of sugar per serving and grams of fiber per serving. The partition of the sums of squares for this model is

$$SSTotal = SSModel + SSE$$
$$17190 = 9350 + 7840$$

a. Calculate R^2 for this model and interpret the value in the context of this setting.

b. Calculate the regression standard error of this multiple regression model.

c. Calculate the F-ratio for testing the null hypothesis that neither sugar nor fiber is related to calorie content of cereals.

d. Assuming the regression conditions hold, the p-value for the F-ratio in (c) is about 0.000002. Interpret what this tells you about the variables in this situation.

3.9. EXERCISES

3.11 *Driving fatalities and speed limits.* In 1987 the federal government allowed the speed limit on interstate highways to be 65 mph in most areas. In 1995 federal restrictions were eliminated, so that states assumed control of setting speed limits on interstate highways. The datafile **Speed** contains the variables *FatalityRate*, the number of interstate fatalities per 100 million vehicle-miles of travel, *Year*, and an indicator variable *StateControl* that is 1 for years 1995 - 2007 and zero for earlier years in the period 1987 - 1994. Here are the first few rows of data:

Year	FatalityRate	StateControl
1987	2.41	0
1988	2.32	0
1989	2.17	0
1990	2.08	0
1991	1.91	0

a. Fit a regression of fatality rate on year. What is the slope of the least squares line?

b. Examine a residual plot. What is remarkable about this plot?

c. Fit the multiple regression of fatality rate on year, state control, and the interaction between year and state control. Is there a significant change in the relationship between fatality rate and year starting in 1995?

d. What are the fitted equations relating fatality rate to year before and after 1995?

3.12 *British trade unions.* The British polling company Ipsos MORI conducted several opinion polls in the UK between 1975 and 1995 in which they asked whether people agree or disagree with the statement "Trade unions have too much power in Britain today". Table 3.7 shows the dates of the polls, the agree and disagree percentages, the *NetSupport* for unions which is defined $DisagreePct - AgreePct$, the number of months after August 1975 that the poll was conducted, and a variable, *Late*, that indicates whether an observation is from the early (0) or late (1) part of the period spanned by the data. The last variable is the unemployment rate in the United Kingdom for the month of each poll. The data are also stored in **BritishUnions**.

a. Make a scatterplot of *Y=NetSupport* versus *X=Months* since August 1975 and comment on what the plot shows.

b. Fit the regression of *NetSupport* on *Months* and test whether there is a time effect (i.e., whether the coefficient of *Months* differs from zero).

c. Fit a model that produces parallel regression lines for the early and late parts of the dataset. Write down the model and fitted prediction equation.

Table 3.7: Poll: Trade unions have too much power in Britain?

Date	AgreePct	DisagreePct	NetSupport	Months	Late	Unemployment
Oct-75	75	16	-59	2	0	4.9
Aug-77	79	17	-62	23	0	5.7
Sep-78	82	16	-66	36	0	5.5
Sep-79	80	16	-64	48	0	5.4
Jul-80	72	19	-53	58	0	6.8
Nov-81	70	22	-48	74	0	10.2
Aug-82	71	21	-50	83	0	10.8
Aug-83	68	25	-43	95	0	11.5
Aug-84	68	24	-44	107	0	11.7
Aug-89	41	42	1	167	1	7.1
Jan-90	35	54	19	172	1	6.9
Aug-90	38	45	7	179	1	7.1
Feb-92	27	64	37	197	1	9.7
Dec-92	24	56	32	207	1	10.5
Aug-93	26	55	29	215	1	10.2
Aug-94	26	56	30	227	1	9.4
Aug-95	24	57	33	239	1	8.6

Source: Ipsos MORI opinion polls

d. Use the regression output from parts (b) and (c) to test the null hypothesis that a single regression line adequately describes these data against the alternative that two parallel lines are needed.

3.13 *More British trade unions.* Consider the data on opinion polls on the power of British trade unions in Exercise 3.12.

a. Create an interaction term between *Months* (since August 1975) and *Late*. Fit the regression model that produces two lines for explaining the *NetSupport*, one each for the early (*Late* = 0) and late (*Late* = 1) parts of the dataset. What is the fitted model?

b. Use a t-test to test the null hypothesis that the interaction term is not needed and parallel regression lines are adequate for describing these data.

c. Use a nested F-test to test the null hypothesis that neither of the terms involving *Late* is needed and a common regression line for both periods is adequate for describing the relationship between *NetSupport* and *Months*.

3.9. EXERCISES

3.14 *More British trade unions.* Table 3.7 in Exercise 3.12 also shows the unemployment rate in Britain for each of the months when poll data were collected.

    a. Make a scatterplot of $Y = NetSupport$ versus $X = Unemployment$ and comment on what the plot shows.

    b. Fit the regression of *NetSupport* on *Unemployment* and test whether there is a linear relationship between unemployment and net support for trade unions (i.e., whether the coefficient of *Unemployment* differs from zero) at the 0.10 level of significance.

    c. Fit the regression of *NetSupport* on *Unemployment* and *Months* since August 1975 and test whether there is a linear relationship between unemployment and net support for trade unions, at the 0.10 level of significance, when controlling for time (*Months*) in the model.

    d. How does the coefficient on *Unemployment* in the model from part (c) compare to that from part (b)? Interpret the difference in these two values.

3.15 *Combining explanatory variables.* Suppose that X_1 and X_2 are positively related with $X_1 = 2X_2 - 4$. Let $Y = 0.5X_1 + 5$ summarize a positive linear relationship between Y and X_1.

    a. Substitute the first equation into the second to show a linear relationship between Y and X_2. Comment on the direction of the association between Y and X_2 in the new equation.

    b. Now add the original two equations and rearrange terms to give an equation in the form $Y = aX_1 + bX_2 + c$. Are the coefficients of X_1 and X_2 both in the direction you would expect based on the signs in the separate equations? Combining explanatory variables that are related to each other can produce surprising results.

3.16 *Diamond prices.* In Example 3.11 we looked at quadratic and cubic polynomial models for the price of diamonds (*TotalPrice*) based on the size (*Carat*). Another variable in the **Diamonds** data file gives the *Depth* of the cut for each stone (as a percentage of the diameter). Run each of the models listed below, keeping track of the values for R^2, adjusted R^2, and which terms (according to the individual t-tests) are important in each model.

    a. A quadratic model using *Depth*.

    b. A two predictor model using *Carat* and *Depth*.

    c. A three predictor model that adds interaction for *Carat* and *Depth*.

    d. A complete second order model using *Carat* and *Depth*.

Among these four models as well as the quadratic and cubic models shown in Example 3.11, which would you recommend using for *TotalPrice* of diamonds? Explain your choice.

3.17 *More diamond prices.* One of the consistent problems with models for the *TotalPrice* of diamonds in Example 3.11 was the lack of a constant variance in the residuals. As often happens, when we try to predict the price of the larger, more expensive diamonds the variability of the residuals tends to increase.

   a. Using the model you chose in Exercise 3.16, produce one or more graphs to examine the conditions for homoscedasticity (constant variance) and normality of its residuals. Do these standard regression conditions appear to be reasonable for your model?

   b. Transform the response variable to be *logPrice* as the natural log of the *TotalPrice*. Is your "best" choice of predictors from Exercise 3.16 still a reasonable choice for predicting *logPrice*? If not, make adjustments to add or delete terms, keeping within the options offered within a complete second order model.

   c. Once you have settled on a model for *logPrice*, produce similar graphs to the ones you found in (a). Has the log transformation helped with either the constant variance or normality conditions on the residuals?

3.18 *More diamond prices.* Refer to the complete second order model you found for diamond prices in Exercise 3.16(d). Use a nested F-test to determine whether all of the terms in the model that involve the information on *Depth* could be removed as a group from the model without significantly impairing its effectiveness.

3.19 *More diamond prices.* The young couple described in in Example 3.11 has found a 0.5 carat diamond with a depth of 62% that they are interested in buying. Suppose that you decide to use the quadratic regression model for predicting the *TotalPrice* of the diamond using *Carat*. The data are stored in **Diamonds**.

   a. What average total price does the quadratic model predict for a 0.5 carat diamond?

   b. Find a 95% confidence interval for the mean total price of 0.5 carat diamonds. Write a sentence interpreting the interval in terms that will make sense to the young couple.

   c. Find a 95% prediction interval for the total price when a diamond weighs 0.5 carats. Write a sentence interpreting the interval in terms that will make sense to the young couple.

   d. Repeat the previous two intervals for the model found in part (b) of Exercise 3.17 where the response variable was *logPrice*. You should find the intervals for the log scale, but then exponentiate to give answers in terms of *TotalPrice*.

3.9. EXERCISES

3.20 *First year GPA.* The data in **FirstYearGPA** contains information from a sample of 219 first year students at a midwestern college that might be used to build a model to predict their first year GPA. Suppose that you decide to use high school GPA ($HSGPA$), verbal SAT score ($SATV$), number of humanities credits (HU), and an indicator for whether or not a student is white ($White$) as a four predictor model for GPA.

a. A white student applying to this college has a high school GPA of 3.20 and got a 600 score on his verbal SAT. If he has 10 credits in humanities, what GPA would this model predict?

b. Produce an interval that you could tell an admissions officer at this college would be 95% sure to contain the GPA of this student after his first year.

c. How much would your prediction and interval change if you added the number of credits of social science (SS) to your model? Assume this student also had 10 social science credits.

3.21 *2008 U.S. Presidential polls.* The file **Pollster08** contains data from 102 polls that were taken during the 2008 U.S. Presidential campaign. These data include all presidential polls reported on the internet site *pollster.com* that were taken between August 29th, when John McCain announced that Sarah Palin would be his running mate as the Republican nominee for vice president, and the end of September. The variable *MidDate* gives the middle date of the period when the poll was "in the field" (i.e., when the poll was being conducted). The variable *Days* measures the number of days after 28 August (the end of the Democratic convention) that the poll was conducted. The variable *Margin* shows Obama% - McCain% and is a measure of Barack Obama's lead. *Margin* is negative for those polls that showed McCain to be ahead.

The scatterplot in Figure 3.24 of *Margin* versus *Days* shows that Obama's lead dropped during the first part of September but grew during the latter part of September. A quadratic model might explain the data. However, two theories have been advanced as to what caused this pattern, which you will investigate in this exercise.

The **Pollster08** data file contains a variable *Charlie* that equals 0 if the poll was conducted before the telecast of the first ABC interview of Palin by Charlie Gibson (on 11 September) and 1 if the poll was conducted after that telecast. The variable *Meltdown* equals 0 if the poll was conducted before the bankruptcy of Lehman Brothers triggered a meltdown on Wall Street (on 15 September) and 1 if the poll was conducted after 15 September.

a. Fit a quadratic regression of *Margin* on *Days*. What is the value of R^2 for this fitted model? What is the value of SSE?

b. Fit a regression model in which *Margin* is explained by *Days* with two lines: one line before the 11 September ABC interview (i.e., $Charlie = 0$) and one line after that date ($Charlie = 1$). What is the value of R^2 for this fitted model? What is the value of SSE?

142   CHAPTER 3. MULTIPLE REGRESSION

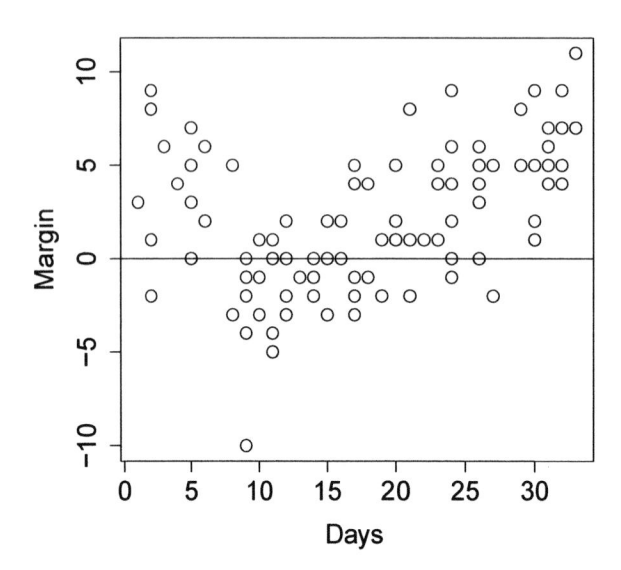

Figure 3.24: *Obama − McCain* margin in 2008 Presidential Polls

c. Fit a regression model in which *Margin* is explained by *Days* with two lines: one line before the 15 September economic meltdown (i.e., *Meltdown* = 0) and one line after 15 September (*Meltdown* = 1). What is the value of R^2 for this fitted model? What is the value of SSE?

d. Compare your answers to parts (a), (b), and (c). Which of the three models best explains the data?

3.22 *Metropolitan doctors.* In Example 1.6 we considered a simple linear model to predict the number of doctors (*NumMDs*) from the number of hospitals (*NumHospitals*) in a metropolitan area. In that example, we found that a square root transformation on the response variable, *sqrt(NumMDs)*, produced a more linear relationship. In this exercise use this transformed variable, stored as *SqrtMDs* in **MetroHealth83**, as the response variable.

a. Either the number of hospitals (*NumHospitals*) or number of beds in those hospitals (*NumBeds*) might be good predictors of the number of doctors in a city. Find the correlations between each pair of the three variables, *SqrtMDs*, *NumHospitals*, *NumBeds*. Based on these correlations, which of the two predictors would be a more effective predictor of *SqrtMDs* in a simple linear model by itself?

b. How much of the variability in the *SqrtMDs* values is explained by *NumHospitals* alone? How much by *NumBeds* alone?

3.9. EXERCISES

c. How much of the variability in the *SqrtMDs* values is explained by using a two-predictor multiple regression model with both *NumHospitals* and *NumBeds*?

d. Based on the two separate simple linear models (or the individual correlations), which of *NumHospitals* and/or *NumBeds* have significant relationship(s) with *SqrtMDs*?

e. Which of these two predictors are important in the multiple regression model? Explain what you use to make this judgement.

f. The answers to the last two parts of this exercise might appear to be inconsistent with each other. What might account for this? Hint: Look back at part (a).

3.23 *Driving fatalities and speed limits.* In Exercise 3.11 you considered a multiple regression model to compare the regression lines for highway fatalities versus year between years before and after states assumed control of setting speed limits on interstate highways. The data are in the file **Speed** with variables for *FatalityRate*, *Year*, and an indicator variable *StateControl*.

a. Use a nested F test to determine whether there is a significant difference in either the slope and/or the intercept of those two lines.

b. Use a nested F test to test only for a difference in slopes ($H_0 : \beta_3 = 0$). Compare the results of this test to the t-test for that coefficient in the original regression.

Open-ended Exercises

3.24 *Porsche versus Jaguar prices.* Recall in Example 1.1 we considered a linear relationship between the mileage and price of used Porsche sports cars based on data a student collected from internet car sites. A second student collected similar data for 30 used Jaguar sports cars that also show a decrease in price as the mileage increases. Does the nature of the price versus mileage relationship differ between the two types of sports cars? For example, does one tend to be consistently more expensive? Or does one model have faster depreciation in price as the mileage increases? Use a multiple regression model to compare these relationships and include graphical support for your conclusions. The data for all 60 cars of both models are stored in the file **PorscheJaguar** which is similar to the earlier **PorschePrice** data with the addition of an *IPorsche* variable that is 1 for the thirty Porsches and 0 for the thirty Jaguars.

3.25 *More Major League Baseball winning percentage.* The **MLB2007Standings** file records many variables for Major League Baseball (MLB) teams from the 2007 regular season. Suppose that we are interested in modeling the winning percentages (*WinPct*) for the teams. Find an example of a set of predictors to illustrate the idea that adding a predictor (or several predictors) to an existing model can cause the adjusted R^2 to *decrease*.

3.26 *Daily carbon dioxide.* Scientists at a research station in Brotjacklriegel, Germany recorded CO2 levels, in parts per million, in the atmosphere. Figure 3.25 shows the data for each day from the start of April through November in 2001. The data are stored in the file **CO2**. Find a model that captures the main trend in this scatterplot. Be sure to examine and comment on the residuals from your model. At what day (roughly) does your fitted model predict the minimum CO2 level?

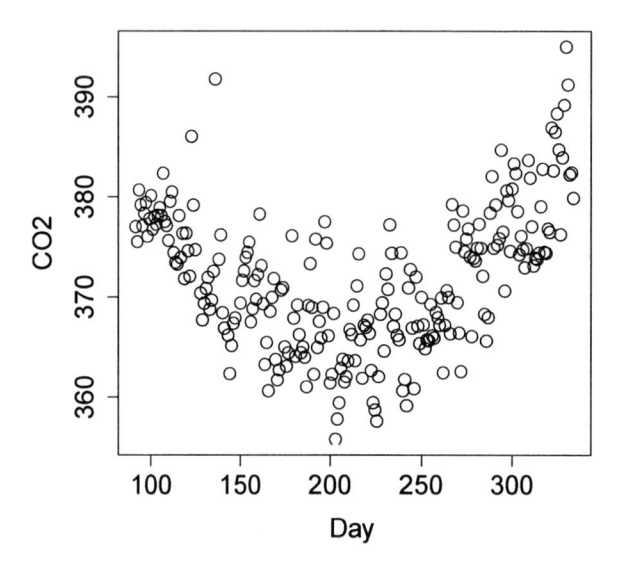

Figure 3.25: CO2 levels by day, April-November 2001

Chapter 4

Additional Topics in Regression

In Chapters 1, 2, and 3 we introduced the basic ideas of simple and multiple linear regression. In this chapter we consider some more specialized ideas and techniques involved with fitting and assessing regression models that extend some of the core ideas of the previous chapters. We prefer to think of the sections in this chapter as separate topics that stand alone and can be read in any order. For this reason, the exercises at the end of the chapter have been organized by these topics and we do not include the usual chapter summary.

4.1 Topic: Added Variable Plots

An **added variable plot** is a way to investigate the effect of one predictor in a model after accounting for the other predictors. The basic idea is to remove the predictor in question from the model and see how well the remaining predictors work to model the response. The residuals from that model represent the variability in the response that is not explained by those predictors. Next we use the same set of predictors to build a multiple regression model for the predictor we had left out. The residuals from that model represent the information in that predictor that is not related to the other predictors. Finally we construct the added variable plot by plotting one of the sets of residuals against the other. This shows what the "unique" information in that predictor can tell us about what was unexplained in the response. The next example illustrates this procedure.

Example 4.1: *House prices*

In the model to predict house prices based on the house *Size* and *Lot* size in Example 3.14 we saw that the two predictors were correlated ($r = 0.767$). Thus, to a certain extent *Size* and *Lot* are redundant: when we know one of them, we know something about the other. But they are not completely redundant, as there is information in the *Size* variable beyond what the *Lot* variable tells. We use the following steps for the data in **Houses** to construct a plot to investigate what happens when we add the variable *Size* to a model that predicts *Price* by *Lot* alone.

    a. Find the residuals when predicting *Price* without using *Size*.

        The correlation between *Price* and *Lot* is 0.716, so *Lot* tells us quite a bit about *Price*, but

146   CHAPTER 4. ADDITIONAL TOPICS IN REGRESSION

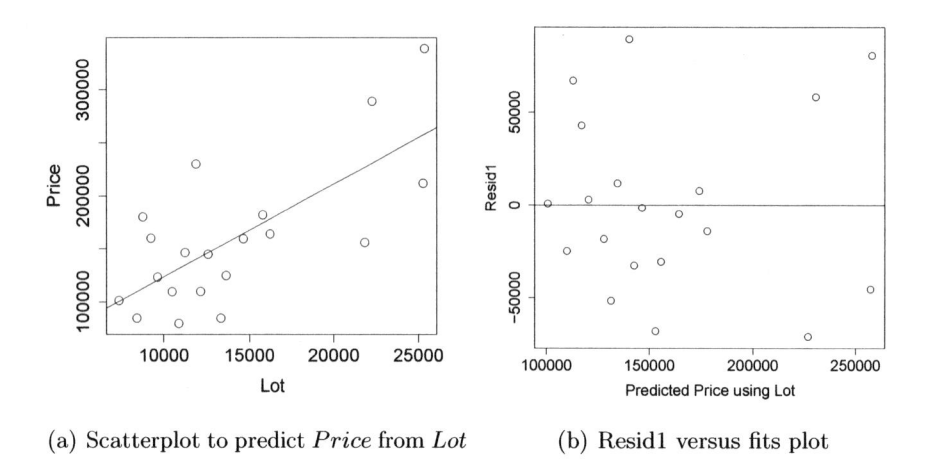

(a) Scatterplot to predict *Price* from *Lot*   (b) Resid1 versus fits plot

Figure 4.1: Regression of *Price* on *Lot*

not everything. Figure 4.1 shows the scatterplot with regression line and residual plot from the regression of *Price* on *Lot*. These residuals represent the variability in *Price* that is not explained by *Lot*. We let a new variable called *Resid*1 denote the residuals for regression of *Price* on *Lot*.

b. Find the residuals when predicting *Size* using the other predictor *Lot*.

If we regress *Size* on *Lot* then the residuals capture the information in *Size* that is not explained by *Lot*. Figure 4.2 shows the scatterplot with regression line and residual plot from the regression of *Size* on *Lot*. We store the residuals from the regression of *Size* on *Lot* in *Resid*2.

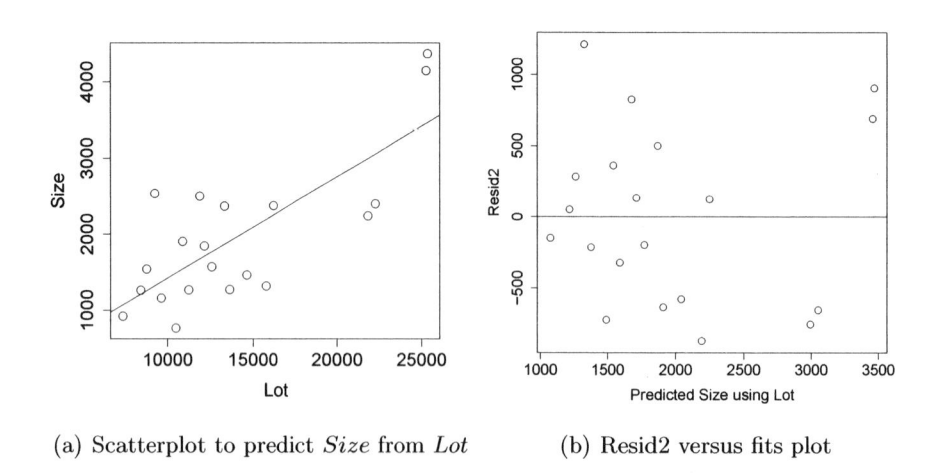

(a) Scatterplot to predict *Size* from *Lot*   (b) Resid2 versus fits plot

Figure 4.2: Regression of *Size* on *Lot*

4.1. TOPIC: ADDED VARIABLE PLOTS

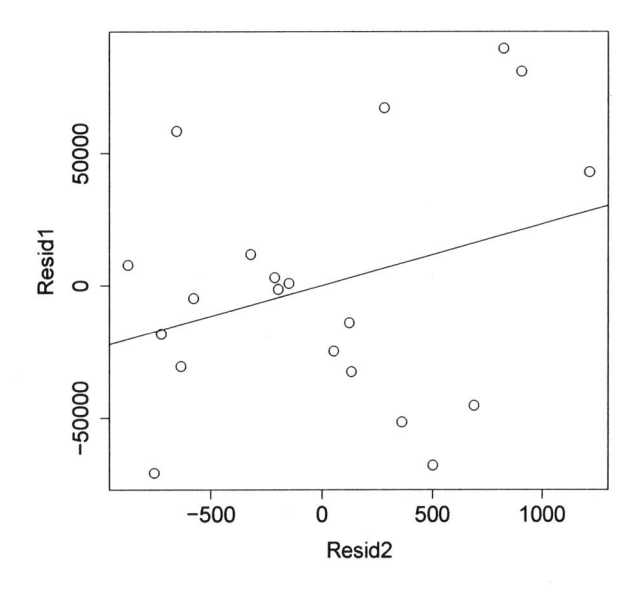

Figure 4.3: Added variable plot for adding *Size* to *Lot* when predicting *Price*

c. Plot the residuals from the first model versus the residuals from the second model to create the added variable plot.

Figure 4.3 shows a scatterplot with regression line to predict *Resid*1 (from the regression of *Price* on *Lot*) using *Resid*2 (from the regression of *Size* on *Lot*). The correlation between these two set of residuals is $r = 0.303$ which indicates that the unique information in *Size* doesn't explain much of the variation in *Price* that is not already explained by *Lot*. This helps explain why the *Size* variable is not quite significant at a 5% level when combined with *Lot* in the multiple regression model. Some summary output for the multiple regression model is shown below.

```
Coefficients:
            Estimate Std. Error t value Pr(>|t|)
(Intercept) 34121.649  29716.458   1.148   0.2668
Lot             5.657      3.075   1.839   0.0834
Size           23.232     17.700   1.313   0.2068
```

The equation of the regression line in Figure 4.3 is

$$\widehat{Resid1} = 0 + 23.232 \cdot Resid2$$

The intercept of the line must be zero since both sets of residuals have mean zero and thus the regression line must go through (0,0). Note that the slope of the line in the added variable plot is 23.232, exactly the same as the coefficient of *Size* in the multiple regression of *Price* on *Size* and *Lot*. We interpret this coefficient by saying that each additional square foot of *Size* corresponds to an additional $23.23 of *Price* while *controlling for Lot being in the model*. The added variable plot shows this graphically and the way that the added variable plot is constructed helps us understand what the words "controlling for *Lot* being in the model" mean.

◇

4.2 Topic: Techniques for Choosing Predictors

Sometimes there are only one or two variables available for predicting a response variable and it is relatively easy to fit any possible model. In other cases we often have many predictors that might be included in a model. If we include quadratic terms, interactions, and other functions of the predictors, the number of terms that can be included in a model can grow very rapidly. It is tempting to include every candidate predictor in a regression model, as the more variables we include the higher R^2 will be, but there is a lot to be said for having a model that is simple, that we can understand and explain, and that we can expect to hold up in similar situations. That is, we don't want to include in a model a predictor that is related to the response only due to chance for a particular sample.

Since we can create new "predictors" using functions of other predictors in the data, we often use the word **term** to describe any predictor, a function of a predictor (like X^2), or a quantity derived from more than one predictor (like an interaction). When choosing a model, a key question is "Which terms should be included in a final model?" If we start with k candidate terms, then there are 2^k possible models (since each term can be included or excluded from a model). This can be quite a large number. For example, in the next example we have nine predictors so we could have $2^9 = 512$ different models, even before we consider any second order terms.

Before going further, we note that often there is more than one model that does a good job of predicting a response variable. It is quite possible for different statisticians who are studying the same dataset to come up with somewhat different regression models. Thus, we are not searching for the one true, ideal model, but for a good model that helps us answer the question of interest. Some models are certainly better than others, but often there may be more than one model that is sensible to use.

Example 4.2: *First year GPA*

The file **FirstYearGPA** contains measurements on 219 college students. The response variable is GPA (grade point average after one year of college). The potential predictors are

$$
\begin{aligned}
HSGPA &= \text{high school GPA} \\
SATV &= \text{verbal/critical reading SAT score} \\
SATM &= \text{math SAT score} \\
Male &= \text{1 for male, 0 for female} \\
HU &= \text{number of credit hours earned in humanities courses in high school} \\
SS &= \text{number of credit hours earned in social science courses in high school} \\
FirstGen &= \text{1 if the student is the first in her or his family to attend college} \\
White &= \text{1 for white students, 0 for others} \\
CollegeBound &= \text{1 if attended a high school where} \geq 50\% \text{ students intended to go on to college}
\end{aligned}
$$

Figure 4.4 shows a scatterplot matrix of these data, with the top row of scatterplots showing how GPA is related to the quantitative predictors. Figure 4.5 shows boxplots of GPA by group for the four categorical predictors.

150   CHAPTER 4. ADDITIONAL TOPICS IN REGRESSION

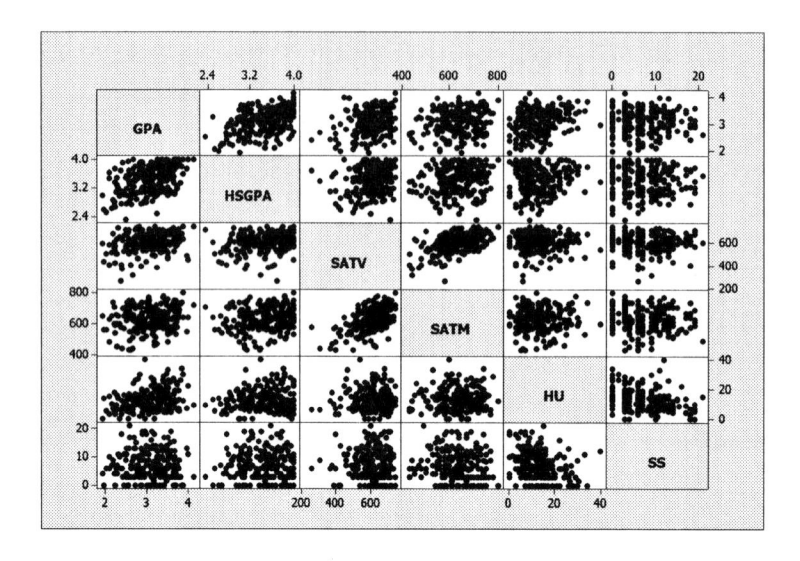

Figure 4.4: Scatterplot matrix for first year GPA data

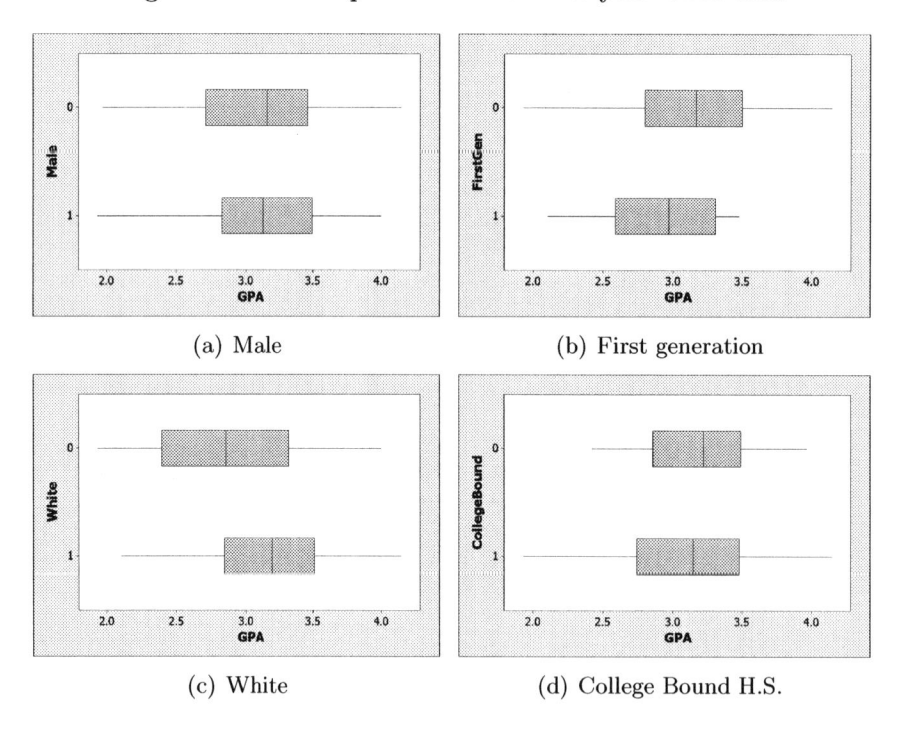

(a) Male

(b) First generation

(c) White

(d) College Bound H.S.

Figure 4.5: GPA versus categorical predictors

4.2. TOPIC: TECHNIQUES FOR CHOOSING PREDICTORS

In order to keep the explanation fairly simple, we will not consider interactions and higher order terms, but will just concentrate on the nine predictors in the dataset. As we consider which of them to include in a regression model, we need a measure of how well a model fits the data. Whenever a term is added to a regression model, R^2 goes up (or at least it doesn't go down), so if our goal were simply to maximize R^2, then we would include all nine predictors (and get $R^2 = 35.0\%$ for this example). However, some of the predictors may not add much value to the model. As we saw in Section 3.2, the adjusted R^2 statistic "penalizes" for the addition of a term, which is one way to balance the conflicting goals of predictive power and simplicity. Suppose that our goal is to maximize the adjusted R^2 using some subset of these nine predictors; how do we sort through the 512 possible models in this situation to find the optimal one? ⋄

Best subsets

Given the power and speed of modern computers it is feasible, when the number of predictors is not too large, to check all possible subsets of predictors. We don't really want to see the output for hundreds of models, so most statistical software packages (including R and Minitab) offer procedures that will display one (or several) of the best models for various numbers of predictors. For example, running a best subsets regression procedure using all nine predictors for the GPA data shows the following output.

```
                                        H              W
                                      S S S M        F h C
                                      G A A a        G i B
                            Mallows   P T T l H S    e t n
Vars  R-Sq  R-Sq(adj)   Cp        S   A V M e U S    n e d
  1   20.0     19.6   42.2   0.41737  X
  1    9.9      9.5   74.5   0.44285            X
  2   27.0     26.3   21.7   0.39962  X         X
  2   26.8     26.1   22.2   0.40007  X                   X
  3   32.3     31.4    6.5   0.38563  X         X         X
  3   30.8     29.8   11.4   0.38993  X X       X
  4   33.7     32.5    3.9   0.38239  X X       X         X
  4   33.0     31.7    6.4   0.38466  X         X    X X
  5   34.4     32.8    3.9   0.38148  X X       X X       X
  5   34.1     32.6    4.8   0.38227  X X   X X           X
  6   34.7     32.9    4.8   0.38143  X X   X X X         X
  6   34.6     32.8    5.1   0.38164  X X       X X X X
  7   34.9     32.8    6.1   0.38163  X X   X X X X X
  7   34.7     32.6    6.8   0.38226  X X       X X X   X X
  8   35.0     32.5    8.0   0.38250  X X   X X X X X X
  8   34.9     32.5    8.0   0.38251  X X X X X X X X
  9   35.0     32.2   10.0   0.38338  X X X X X X X X X
```

The default (from Minitab) shows the best two models (as measured by R^2) of each size from one to all nine predictors. For example, we see that the best individual predictor of first year GPA is $HSGPA$ and the next strongest predictor is HU. Together these predictors give the highest R^2 for a two variable model for explaining variability in GPA.

Looking at the Adjusted R^2 columns we see that value is largest at 32.9% for a six-predictor model using $HSGPA$, $SATV$, $Male$, HU, SS and $White$. But the output for fitting this model by itself (shown below) indicates that two of the predictors, $Male$ and SS, would not be significant if tested at a 5% level. Would a simpler model be as effective?

```
The regression equation is
GPA = 0.547 + 0.483 HSGPA + 0.000694 SATV + 0.0541 Male + 0.0168 HU + 0.00757 SS
      + 0.205 White

Predictor          Coef     SE Coef      T      P
Constant         0.5467      0.2835   1.93  0.055
HSGPA           0.48295     0.07147   6.76  0.000
SATV          0.0006945   0.0003449   2.01  0.045
Male            0.05410     0.05269   1.03  0.306
HU             0.016796    0.003818   4.40  0.000
SS             0.007570    0.005442   1.39  0.166
White           0.20452     0.06860   2.98  0.003

S = 0.381431   R-Sq = 34.7%   R-Sq(adj) = 32.9%

Analysis of Variance
Source            DF         SS       MS      F      P
Regression         6    16.3898   2.7316  18.78  0.000
Residual Error   212    30.8438   0.1455
Total            218    47.2336
```

Most of the criteria we have used so far to evaluate a model (for example, R^2, adjusted R^2, MSE, $\hat{\sigma}_\epsilon^2$, overall ANOVA, individual t-tests) depend only on the predictors in the model being evaluated. None of these measures takes into account what information might be available in the other potential predictors that aren't in the model. A new measure that does this was developed by statistician Colin Mallows.

4.2. TOPIC: TECHNIQUES FOR CHOOSING PREDICTORS

> **Mallow's C_p**
>
> When evaluating a regression model for a subset of m predictors from a larger set of k predictors using a sample of size n, the value of Mallow's C_p is computed by
>
> $$C_p = \frac{SSE_m}{MSE_k} + 2(m+1) - n$$
>
> where SSE_m is the sum of squared residuals from the model with just m predictors and MSE_k is the mean square error for the full model with all k predictors.
> We prefer models where C_p is small.

As we add terms to a model we drive down the first part of the C_p statistic, since SSE_m goes down while the scaling factor MSE_k remains constant. But we also increase the second part of the C_p statistic, $2(m+1)$, since $m+1$ measures the number of coefficients (including the constant term) that are being estimated in the model. This acts as a kind of penalty in calculating C_p. If the reduction in SSE_m is substantial compared to this penalty, the value of C_p will decrease, but if the new predictor explains little new variability the value of C_p will increase. A model that has a small C_p value is thought to be a good compromise between making SSE_m small and keeping the model simple. Generally, any model for which C_p is less than $m+1$ or not much greater than $m+1$ is considered to be a model worth considering.

Note that if we compute C_p for the full model with all k predictors,

$$C_p = \frac{SSE_k}{MSE_k} + 2(k+1) - n = \frac{SSE_k}{SSE_k/(n-k-1)} + 2(k+1) - n = (n-k-1) + (2k+2) - n = k+1$$

so the C_p when all predictors are included is always one more than the number of predictors (as you can verify from the last row in the best subsets output on page 151).

When we run the six-predictor model for all $n = 219$ cases using $HSGPA$, $SATV$, $Male$, HU, SS, and $White$ to predict GPA we see the $SSE_6 = 30.8438$ and for the full model with all nine predictors we have $MSE_9 = 0.14698$. This gives

$$C_p = \frac{30.8438}{0.14698} + 2(6+1) - 219 = 4.85$$

which is less than $6 + 1 = 7$ so this is a reasonable model to consider, but could we do better?

Fortunately, the values of Mallow's C_p are displayed in the best subsets regression output on page 151. The minimum $C_p = 3.9$ occurs at a four-predictor model that includes $HSGPA$, $SATV$, HU, and $White$. Interestingly, this model omits both the $Male$ and SS predictors that had insignificant t-tests in the six-predictor model with the highest adjusted R^2. When running the smaller four-predictor model all four terms have significant t-tests at a 5% level. A five-predictor model (putting SS back in) also has a C_p value of 3.9, but we would prefer the smaller model that attains essentially the same C_p.

Backward elimination

Another model selection method that is popular and rather easy to implement is called **backward elimination**.

1. Start by fitting the full model (the model that includes all terms under consideration).

2. Identify the term for which the individual t-test produces the largest p-value.

   (a) If that p-value is large (say greater than 5%) eliminate that term to produce a smaller model. Fit that model and return to the start of step 2.

   (b) If the p-value is small (less than 5%) stop since all of the predictors in the model are "significant."

Note: This process can be implemented with, for example, the goal of minimizing C_p at each step, rather than the criterion of eliminating all non-significant predictors. At each step eliminate the predictor that gives the largest drop in C_p, until we reach a point that C_p does not get smaller if any predictor left in the model is removed.

If we run the full model to predict GPA based on all nine predictors the output includes the following t-tests for individual terms.

```
Predictor          Coef      SE Coef       T       P
Constant         0.5269       0.3488    1.51   0.132
HSGPA           0.49329      0.07456    6.62   0.000
SATV          0.0005919    0.0003945    1.50   0.135
SATM          0.0000847    0.0004447    0.19   0.849
Male            0.04825      0.05703    0.85   0.398
HU             0.016187     0.003972    4.08   0.000
SS             0.007337     0.005564    1.32   0.189
FirstGen       -0.07434      0.08875   -0.84   0.403
White           0.19623      0.07002    2.80   0.006
CollegeBound     0.0215       0.1003    0.21   0.831
```

The largest p-value for a predictor is 0.849 for $SATM$ which is certainly not significant so that predictor is dropped and we fit the model with the remaining eight predictors.

```
Predictor          Coef     SE Coef      T      P
Constant         0.5552      0.3149   1.76  0.079
HSGPA           0.49502     0.07384   6.70  0.000
SATV          0.0006245   0.0003548   1.76  0.080
Male            0.05221     0.05298   0.99  0.325
HU             0.016082    0.003925   4.10  0.000
SS             0.007177    0.005487   1.31  0.192
FirstGen       -0.07559     0.08830  -0.86  0.393
White           0.19742     0.06958   2.84  0.005
CollegeBound     0.0212      0.1001   0.21  0.833
```

Now *CollegeBound* is the "weakest link" (p-value=0.833) so it is eliminated and we continue. To avoid showing excessive output, we summarize below the predictors that are identified with the largest p-value at each step.

```
9 terms:
Predictor          Coef     SE Coef      T      P
SATM         0.0000847   0.0004447   0.19  0.849

8 terms:
Predictor          Coef     SE Coef      T      P
CollegeBound    0.0212      0.1001   0.21  0.833

7 terms:
Predictor          Coef     SE Coef      T      P
FirstGen       -0.07725     0.08775  -0.88  0.380

6 terms:
Predictor          Coef     SE Coef      T      P
Male            0.05410     0.05269   1.03  0.306

5 terms:
Predictor          Coef     SE Coef      T      P
SS             0.007747    0.005440   1.42  0.156

4 terms:
Predictor          Coef     SE Coef      T      P
SATV         0.0007372   0.0003417   2.16  0.032
```

After we get a four-term model consisting of *HSGPA*, *SATV*, *HU*, and *White* we see that the weakest predictor (*SATV*, p-value=0.032) is still significant at a 5% level so we would stop and keep this model. This produces the same model that minimized the C_p in the best subsets procedure. This is a nice coincidence when it happens, but certainly not something that will occur for every set of predictors. Statistical software allows us to automate this process even further to give the output tracing the backward elimination process such as that shown below.

```
Backward elimination.  Alpha-to-Remove: 0.05

Response is GPA on 9 predictors, with N = 219

Step                  1        2        3        4        5        6
Constant         0.5269   0.5552   0.5825   0.5467   0.5685   0.6410

HSGPA             0.493    0.495    0.492    0.483    0.474    0.476
T-Value            6.62     6.70     6.81     6.76     6.68     6.70
P-Value           0.000    0.000    0.000    0.000    0.000    0.000

SATV            0.00059  0.00062  0.00063  0.00069  0.00075  0.00074
T-Value            1.50     1.76     1.79     2.01     2.19     2.16
P-Value           0.135    0.080    0.075    0.045    0.029    0.032

SATM            0.00008
T-Value            0.19
P-Value           0.849

Male              0.048    0.052    0.053    0.054
T-Value            0.85     0.99     1.00     1.03
P-Value           0.398    0.325    0.316    0.306

HU               0.0162   0.0161   0.0161   0.0168   0.0167   0.0151
T-Value            4.08     4.10     4.10     4.40     4.39     4.14
P-Value           0.000    0.000    0.000    0.000    0.000    0.000

SS               0.0073   0.0072   0.0071   0.0076   0.0077
T-Value            1.32     1.31     1.30     1.39     1.42
P-Value           0.189    0.192    0.194    0.166    0.156

FirstGen         -0.074   -0.076   -0.077
T-Value           -0.84    -0.86    -0.88
P-Value           0.403    0.393    0.380

White             0.196    0.197    0.196    0.205    0.206    0.212
T-Value            2.80     2.84     2.84     2.98     3.00     3.09
P-Value           0.006    0.005    0.005    0.003    0.003    0.002

CollegeBound       0.02     0.02
T-Value            0.21     0.21
P-Value           0.831    0.833

S                 0.383    0.383    0.382    0.381    0.381    0.382
R-Sq              34.96    34.95    34.94    34.70    34.37    33.75
R-Sq(adj)         32.16    32.47    32.78    32.85    32.83    32.51
Mallows Cp         10.0      8.0      6.1      4.8      3.9      3.9
```

4.2. TOPIC: TECHNIQUES FOR CHOOSING PREDICTORS

Backward elimination has an advantage in leaving us with a model in which all of the predictors are significant and it requires fitting relatively few models (in this example we needed to run only 6 of the 512 possible models). One disadvantage of this procedure is that the initial models tend to be the most complicated. If multicollinearity is an issue (as it often is with large sets of predictors) we know that the individual t-tests can be somewhat unreliable for correlated predictors, yet that is precisely what we use as a criteria in making decisions about which predictors to eliminate. In some situations we might eliminate a single strong predictor at an early stage of the backward elimination process if it is strongly correlated with several other predictors. Then we have no way to "get it back in" the model at a later stage when it might provide a significant benefit over the predictors that remain.

Forward selection and stepwise regression

The difficulties we see when starting with the most complicated models in backward elimination suggest that we might want to build a model from the other direction, starting with a simple model using just the best single predictor and then adding new terms. This method is known as **forward selection**.

1. Start with a model with no predictors and find the best single predictor (largest correlation with the response gives the biggest initial R^2).

2. Add the new predictor to the model, run the regression and find its individual p-value.

    (a) If that p-value is small (say less than 5%) keep that predictor in the model and try each of the remaining (unused) predictors to see which would produce the most benefit (biggest increase in R^2) when added to the existing model.

    (b) If the p-value is large (over 5%) stop and discard this predictor. At this point no (unused) predictor should be significant when added to the model and we are done.

The forward selection method generally requires fitting many more models. In our current example, we would have nine predictors to consider at the first step, where $HSGPA$ turns out to be the best with $R^2 = 20.0\%$. Next we would need to consider two predictor models combining $HSGPA$ with each of the remaining eight predictors. The forward selection in Minitab's stepwise regression procedure automates this process to give the output shown below.

Forward selection. Alpha-to-Enter: 0.05

Response is GPA on 9 predictors, with N = 219

Step	1	2	3	4
Constant	1.1799	1.0874	0.9335	0.6410
HSGPA	0.555	0.517	0.507	0.476
T-Value	7.36	7.11	7.23	6.70
P-Value	0.000	0.000	0.000	0.000
HU		0.0172	0.0153	0.0151
T-Value		4.55	4.18	4.14
P-Value		0.000	0.000	0.000
White			0.266	0.212
T-Value			4.12	3.09
P-Value			0.000	0.002
SATV				0.00074
T-Value				2.16
P-Value				0.032
S	0.417	0.400	0.386	0.382
R-Sq	19.97	26.97	32.31	33.75
R-Sq(adj)	19.60	26.30	31.36	32.51
Mallows Cp	42.2	21.7	6.5	3.9

The forward selection procedure arrives at the same four-predictor model ($HSGPA$, $SATV$, HU, and $White$) as we obtained from backward elimination and minimizing C_p with best subsets. In fact, the progression of forward steps mirrors the best one, two, three, and four term models in the best subsets output. While this may happen in many cases, it is not guaranteed to occur.

In some situations (but not this example) we may find that a predictor that was added early in a forward selection process becomes redundant when more predictors are added at later stages. Thus X_1 may be the strongest individual predictor, but X_2 and X_3 together contain much of the same information that X_1 carries about the response. We might choose X_1 to add at the first step, but later when X_2 and X_3 have been added discover that X_1 is no longer needed and has an insignificant

4.2. TOPIC: TECHNIQUES FOR CHOOSING PREDICTORS

p-value. To account for this we use **stepwise regression**, which combines features of both forward selection and backward elimination. A stepwise procedure starts with forward selection, but after any new predictor is added to the model it uses a backward elimination to delete any predictors that have become redundant in the model. In our current example, this wasn't necessary so the stepwise output would be the same as the forward selection output.

In the first year GPA example we have not considered interaction terms or power terms. However, the ideas carry over. If we are interested in, say, interactions and quadratic terms we simply create those terms and put them in the set of candidate terms before we carry out a model selection procedure such as best subsets regression or stepwise regression.

Caution about automated techniques

Model selection procedures have been implemented in computer programs such as Minitab and R, but they are not a substitute for thinking about the data and the modeling situation. At best, a model selection process should suggest to us one or more models to consider. Looking at the order in which predictors are entered in a best subsets or stepwise procedure can help us understand which predictors are relatively more important and which might be more redundant. Do not be fooled by the fact that several of these procedures gave us the same model in the first year GPA example. This will not always be the case in practice! Even when it does occur, we should not take that as evidence that we have found THE model for our data. *It is always the responsibility of the modeler to think about the possible models, to conduct diagnostic procedures (such as plotting residuals against fitted values), and to only use models that make sense.* Moreover, as we noted earlier, there may well be two or more models that are essentially the same quality. In practical situations a sightly less optimal model that uses variables that are much easier (or cheaper) to measure might be preferred over a more complicated one that squeezes an extra 0.5% of R^2 for a particular sample.

4.3 Topic: Identifying Unusual Points in Regression

In Section 1.5 we introduced the ideas of outliers and influential points in the simple linear regression setting. We return to those ideas in this section and examine more formal methods for identifying unusual points in a simple linear regression and extend to models with multiple predictors.

Leverage

Generally, in a simple linear regression situation, points farther from the mean value of the predictor ($\overline{x}$) have greater potential to influence the slope of a fitted regression line. This concept is known as **leverage**. Points with high leverage have a greater capacity to pull the regression line in their direction than do low leverage points, such as points near the predictor mean. In the case of a single predictor, we could measure leverage as just the distance from the mean, however, we will find it useful to have a somewhat more complicated statistic that can be generalized to the multiple regression setting.

Leverage for a Single Predictor

For a simple linear regression on n data points, the **leverage** of any point (x_i, y_i) is defined to be

$$h_i = \frac{1}{n} + \frac{(x_i - \overline{x})^2}{\sum (x_i - \overline{x})^2}$$

The sum of the leverages for all points in a simple linear model is $\sum h_i = 2$; thus a "typical" leverage is about $2/n$. Leverages that are more than two times the typical leverage, that is points with $h_i > 4/n$, are considered somewhat high for a single predictor model and values more than three times as large, that is points with $h_i > 6/n$, are considered to have especially high leverage.

Note that leverage depends only on the value of the predictor and not on the response. Also, a high leverage point does not necessarily exert large influence on the estimation of the least squares line. It is possible to place a new point exactly on an existing regression line at an extreme value of the predictor, thus giving it high leverage but not changing the fitted line at all. Finally, the formula for computing h_i in the simple linear case may seem familiar. If you look in Section 2.4 where we considered confidence and prediction intervals for the response Y in a simple linear model, you will see that leverage plays a prominent role in computing the standard errors for those intervals.

Example 4.3: *Butterfly ballots*

In Example 1.9 on page 47 we looked at a simple linear regression of votes for Pat Buchanan versus George Bush in Florida counties from the 2000 U.S. Presidential election. A key issue in fitting the

4.3. TOPIC: IDENTIFYING UNUSUAL POINTS IN REGRESSION

model was a significant departure from the trend of the rest of the data for Palm Beach County, which used a controversial butterfly ballot. The data for the 67 counties in that election are stored in **PalmBeach**.

Statistical software such as Minitab and R will generally have options to compute and display the leverage values for each data point when fitting a regression model. In fact, the default regression output in Minitab often includes output showing "Unusual Observations" such as the output below from the original regression using the butterfly ballot data.

Unusual Observations

Obs	Bush	Buchanan	Fit	SE Fit	Residual	St Resid
6	177279	789.0	916.9	111.1	-127.9	-0.38 X
13	289456	561.0	1468.5	193.0	-907.5	-3.06RX
29	176967	836.0	915.4	110.9	-79.4	-0.24 X
50	152846	3407.0	796.8	94.2	2610.2	7.65R
52	184312	1010.0	951.5	116.1	58.5	0.17 X

R denotes an observation with a large standardized residual.
X denotes an observation whose X value gives it large leverage.

Minitab uses the "three times typical" rule, that is $h_i > 6/n$, to identify large leverage points and flags them with an "X" in the output. For the $n = 67$ counties in this dataset, the cutoff for large leverage is $6/67 = 0.0896$ and the mean Bush vote is $\bar{x} = 43,356$. We see that there are four counties with high leverage:

- Broward (Bush=177,279 $h_6 = 0.0986$)

- Dade (Bush=289,456 $h_{13} = 0.2975$)

- Hillsborough (Bush=176,967 $h_{29} = 0.0982$)

- Pinellas (Bush=184,312 $h_{52} = 0.1076$)

By default, R uses the "two times typical" rule to identify high outliers, so with $4/67 = 0.0597$ as the cutoff R would add two more counties to the list of high leverage counties:

- Palm Beach (Bush=152,846 $h_{50} = 0.07085$)

- Dual (Bush=152,082 $h_{16} = 0.07007$)

These six "unusual" observations are highlighted in Figure 4.6. In this example the six high leverage counties by the $h_i > 4/n$ criteria are exactly the six counties that would be identified as outliers

162   CHAPTER 4. ADDITIONAL TOPICS IN REGRESSION

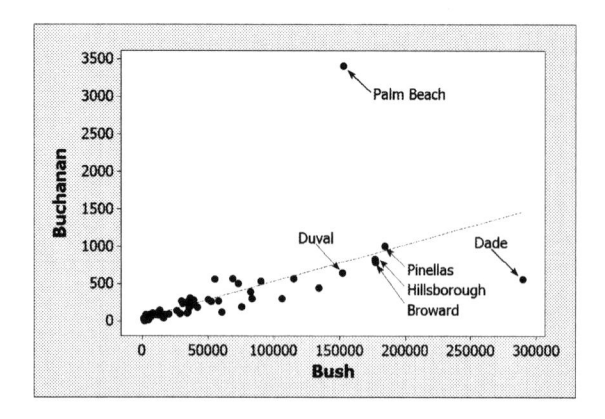

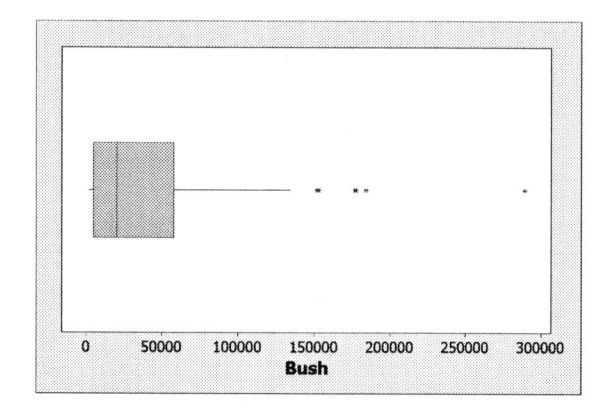

Figure 4.6: Unusual observations identified in the butterfly ballot model

Figure 4.7: Boxplot of the number of Bush votes in Florida counties

when building a boxplot (Figure 4.7) and looking for points more than $1.5 \cdot IQR$ beyond the quartiles. ⋄

In the multiple regression setting the computation of leverage is more complicated (and beyond the scope of this book) so we will depend upon statistical software. Fortunately, the interpretation is similar to the interpretation for the case of a single predictor: points with high leverage have the potential to influence the regression fit significantly. With more than one predictor, a data case may have an unusual value for any of the predictors or exhibit an unusual combination of predictor values. For example, if a model used people's *height* and *weight* as predictors, a person who is tall and thin, say 6 feet tall and 130 pounds, might have a lot of leverage on a regression fit even though neither the individual's height nor weight are particularly unusual.

For a multiple regression model with k predictors the sum of the leverages for all n data cases is $\sum h_i = (k+1)$. Thus an average or typical leverage value in the multiple case is $(k+1)/n$. As with the simple linear regression case, we identify cases with somewhat high leverage when h_i is more than twice the average, $h_i > 2(k+1)/n$, and very high leverage when $h_i > 3(k+1)/n$.

Example 4.4: *Perch weights*

In Example 3.10 we considered a multiple regression model to predict the *Weight* of perch using the *Length* and *Width* of the fish along with an interaction term, $Length \cdot Width$. Data for the $n = 56$ fish are stored in **Perch**. Some computer output for this model, including information on the unusual cases is shown below.

4.3. TOPIC: IDENTIFYING UNUSUAL POINTS IN REGRESSION

```
The regression equation is
Weight = 114 - 3.48 Length - 94.6 Width + 5.24 LengthxWidth

Predictor          Coef   SE Coef       T      P
Constant         113.93     58.78    1.94  0.058
Length           -3.483     3.152   -1.10  0.274
Width            -94.63     22.30   -4.24  0.000
LengthxWidth     5.2412    0.4131   12.69  0.000

S = 44.2381   R-Sq = 98.5%   R-Sq(adj) = 98.4%

Analysis of Variance
Source           DF        SS       MS         F      P
Regression        3   6544330  2181443   1114.68  0.000
Residual Error   52    101765     1957
Total            55   6646094

Unusual Observations
Obs   Length    Weight      Fit   SE Fit  Residual  St Resid
  1      8.8      5.90    15.38    29.05     -9.48     -0.28 X
 40     37.3    840.00   770.80    26.65     69.20      1.96 X
 50     42.4   1015.00   923.25    12.33     91.75      2.16R
 52     44.6   1100.00   918.59    15.36    181.41      4.37R
 55     46.0   1000.00  1140.11    17.18   -140.11     -3.44R
 56     46.6   1000.00  1088.68    16.75    -88.68     -2.17R

R denotes an observation with a large standardized residual.
X denotes an observation whose X value gives it large leverage.
```

Minitab identifies two high leverage cases. The first fish in the sample is just 8.8 centimeters long and 1.4 centimeters wide. Both values are considerably smaller than those for any other fish in the sample. The leverage for Fish #1 (as computed by Minitab) is $h_1 = 0.431$ while the Minitab cutoff for high leverage in this fitted model is $3(3+1)/56 = 0.214$. The case for Fish #40 is a bit more interesting. Its length (37.3 centimeters) and width (7.8 centimeters) are not the most extreme values in the dataset, but the width is unusually large for that length. The leverage for this point is $h_i = 0.363$. It is often difficult to visually spot high leverage points for models with several predictors by just looking at the individual predictor values. Figure 4.8 shows a scatterplot of the *Length* and *Width* values for this sample of fish with the two unusual points identified and a dotplot of the leverage values where those two points standout more clearly as being unusual. ◇

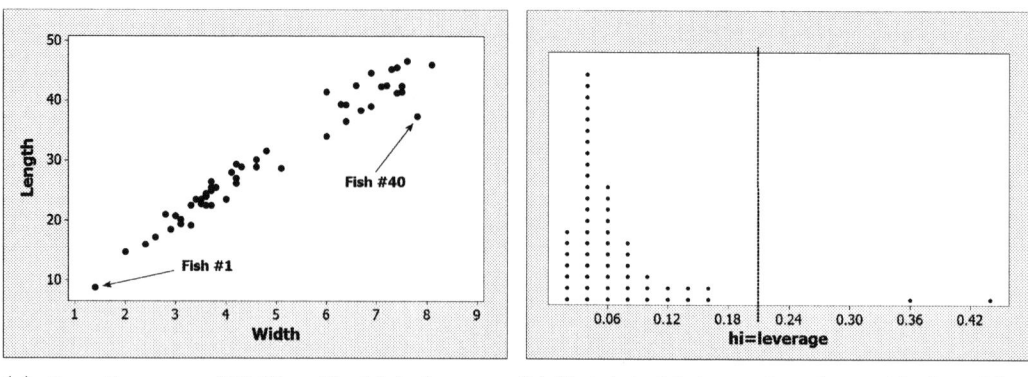

(a) *Length* versus *Width* with high leverage points identified

(b) Dotplot of leverage h_i values with dotted line showing cutoff for unusual cases

Figure 4.8: High leverage cases in perch weight interaction model

Standardized and Studentized Residuals

In Section 1.5 we introduced the idea of standardizing the residuals of a regression model so that we could more easily identify points that were poorly predicted by the fitted model. Now that we have defined the concept of leverage, we can give a more formal definition of the adjustments used to find standardized and studentized residuals.

Standardized and Studentized Residuals

The **standardized residual** for the ith data point in a regression model can be computed using

$$\text{stdres}_i = \frac{y_i - \hat{y}}{\hat{\sigma}_\epsilon \sqrt{1 - h_i}}$$

where $\hat{\sigma}_\epsilon$ is the standard deviation of the regression and h_i is the leverage for the ith point.

For a **studentized** residual (also known as a **deleted-t** residual) we replace $\hat{\sigma}_\epsilon$ with the standard deviation of the regression, $\hat{\sigma}_{(i)}$, from fitting the model with the ith point omitted.

$$\text{studres}_i = \frac{y_i - \hat{y}}{\hat{\sigma}_{(i)} \sqrt{1 - h_i}}$$

Under the usual conditions for the regression model, the standardized or studentized residuals follow a t-distribution. Thus we identify data cases with standardized or studentized residuals beyond ±2 as moderate outliers, while values beyond ±3 denote more serious outliers[1]. The adjustment in

---

[1] Minitab flags cases beyond the more liberal ±2 while R uses the ±3 bounds.

4.3. TOPIC: IDENTIFYING UNUSUAL POINTS IN REGRESSION

the standard deviation for the studentized residual helps avoid a situation where a very influential data case has a big impact on the regression fit, thus artificially making its residual smaller.

Example 4.5: *More butterfly ballots*

Here again is the "Unusual Observations" portion of the output for the simple linear model to predict the number of *Buchanan* votes from the number of *Bush* votes in the Florida counties. We see two counties flagged as having high standardized residuals, Dade (stdres=-3.06) and, of course, Palm Beach (std. res. = 7.65).

```
Unusual Observations

Obs     Bush  Buchanan     Fit   SE Fit  Residual  St Resid
  6   177279     789.0   916.9    111.1    -127.9    -0.38 X
 13   289456     561.0  1468.5    193.0    -907.5    -3.06RX
 29   176967     836.0   915.4    110.9     -79.4    -0.24 X
 50   152846    3407.0   796.8     94.2    2610.2     7.65R
 52   184312    1010.0   951.5    116.1      58.5     0.17 X

R denotes an observation with a large standardized residual.
X denotes an observation whose X value gives it large leverage.
```

Both of these points were also identified earlier as having high leverage (Dade, $h_{13} = 0.2975$ and Palm Beach, $h_{50} = 0.07085$), although Palm Beach's leverage doesn't exceed the "3 times typical" threshold to be flagged as a large leverage point by Minitab. Using the estimated standard deviation of the regression, $\hat{\sigma}_\epsilon = 353.92$, we can confirm the calculations of these standardized residuals.

$$\text{stdres}_{13} = \frac{561 - 1468.5}{353.92\sqrt{1 - 0.2975}} = -3.06 \quad \text{(Dade)}$$

$$\text{stdres}_{50} = \frac{3407 - 796.8}{353.92\sqrt{1 - 0.07085}} = 7.65 \quad \text{(Palm Beach)}$$

Although statistical software can also compute the studentized residuals, we show the explicit calculation in this situation to illustrate the effect of omitting the point and re-estimating the standard deviation of the regression. For example, if we re-fit the model without the Dade County data point the new standard deviation is $\hat{\sigma}_{(13)} = 330.00$ and without Palm Beach this goes down to $\hat{\sigma}_{(50)} = 112.45$. Thus, for the studentized residuals we have

$$\text{studres}_{13} = \frac{561 - 1468.5}{330.00\sqrt{1 - 0.2975}} = -3.38 \quad \text{(Dade)}$$

$$\text{studres}_{50} = \frac{3407 - 796.8}{112.45\sqrt{1 - 0.07085}} = 24.08 \quad \text{(Palm Beach)}$$

When comparing the two regression models, the percentage of variability explained by the model is much higher when Palm Beach County is omitted ($R^2 = 75.2\%$ compared to $R^2 = 38.9\%$) and the standard deviation of the regression is much lower ($\hat{\sigma}_\epsilon = 112.5$ compared to $\hat{\sigma}_\epsilon = 353.9$). This is not surprising since the Palm Beach County data point added a huge amount of variability to the response variable (the Buchanan vote) that was poorly explained by the model. ⋄

Cook's Distance

The amount of influence that a particular data case has on the estimated regression equation depends both on how close the case lies to the trend of the rest of the data (as measured by its standardized or studentized residual) and on its leverage (as measured by h_i). It's useful to have a statistic that reflects both of these measurements to indicate the impact of any specific case on the fitted model.

Cook's Distance

The **Cook's distance** of a data point in a regression with k predictors is given by

$$D_i = \frac{(\text{stdres}_i)^2}{k+1} \left(\frac{h_i}{1-h_i} \right)$$

A large Cook's distance indicates a point that strongly influences the regression fit. Note that this can occur with a large standardized residual, a large leverage, or some combination of the two. As a rough rule, we say that $D_i > 0.5$ indicates a moderately influential case and $D_i > 1$ shows a case is very influential. For example, in the previous linear regression both Palm Beach ($D_{50} = 2.23$) and Dade ($D_{13} = 1.98$) would be flagged as very influential counties in the least squares fit. The next biggest Cook's distance is for Orange County ($D_{48} = 0.016$), which is not very unusual at all.

Example 4.6: *More perch weights*

The computer output shown on page 163 for the model using *Length*, *Width*, and the interaction *Length · Width* to predict *Weight* for perch identified seven unusual points. Two cases (Fish #1 and #40) were flagged due to high leverage and four cases (Fish #50, #52, #55 and #56) were flagged due to high standardized residuals, although two of those cases were only slightly above the moderate boundary of ±2. Two of the high residual cases also had moderately high leverages; Fish #55 ($h_{55} = 0.1509$) and Fish #56 ($h_{56} = 0.1434$), barely beyond the $2(k+1)/n = 0.1429$ threshold.

So which of these fish (if any) might be considered influential using the criterion of Cook's distance? The results are summarized in Table 4.6. Three of these seven cases show a Cook's D value exceeding the 0.5 threshold (and none are beyond 1.0).

4.3. TOPIC: IDENTIFYING UNUSUAL POINTS IN REGRESSION

Table 4.1: Unusual cases in multiple regression for perch weights

Fish case	$Length$	$Width$	$Weight$	$stdres_i$	h_i	Cook's D_i
1	8.8	1.4	5.9	-0.28	0.431	0.015
2	14.7	2.0	32.0	0.11	0.153	0.001
40	37.3	7.8	840.0	1.96	0.363	0.547
50	42.4	7.5	1015.0	2.16	0.078	0.098
52	44.6	6.9	1100.0	4.37	0.121	0.655
55	46.0	8.1	1000.0	-3.44	0.151	0.525
56	46.6	7.6	1000.0	-2.17	0.143	0.196

- The two small fish (#1 and #2) that have high or moderately high leverage are both predicted fairly accurately so they have a small standardized residual and small Cook's D.

- Two of the larger fish (#50 and #56) have standardized residuals just beyond ±2 and #56 even has a moderately high leverage, but neither has a large Cook's D.

- Fish #40 has a standardized residual just below 2, but its very large leverage produces a large Cook's D.

- The final two cases (#52 and #55) have the most extreme standardized residuals and #55 is large enough to generate an unusual Cook's D even though it's leverage is not quite beyond the threshold.

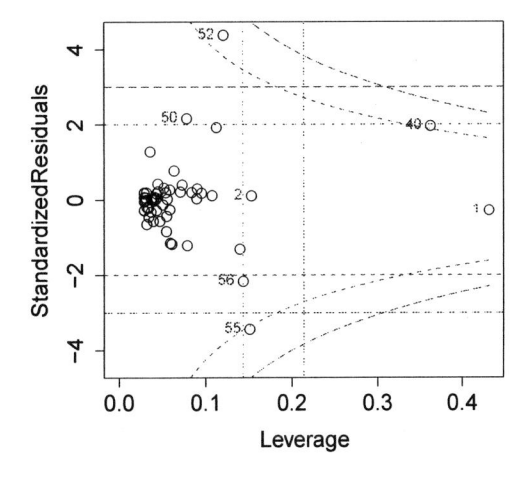

Figure 4.9: Unusual observations in the regression for perch weights

One can produce a nice plot using R to summarize these relationships, as shown in Figure 4.9. Leverage (h_i) values are shown on the horizontal axis (representing unusual combinations of the predictors) and standardized residuals are plotted on the vertical axis (showing unusual responses for a given set of predictor values). Vertical and horizontal lines show the ±2 and ±3 boundaries for each of these quantities. The curved lines show the boundaries where the combination (as measured by Cook's D) becomes unusual, beyond 0.5 or 1.0. There's a lot going on in this plot, but if you locate each of the unusual fish cases from Table 4.6 in Figure 4.9 you should get a good feel for the role of the different measures for unusual points in regression. ◊

Identifying Unusual Points in Regression - Summary

For a multiple regression model with k predictors fit with n data cases:

Statistic	Moderately unusual	Very unusual
Leverage, h_i	above $2(k+1)/n$	above $3(k+1)/n$
Standardized residual	beyond ±2	beyond ±3
Studentized residual	beyond ±2	beyond ±3
Cook's D	above 0.5	above 1.0

We conclude with a final caution about using the guidelines provided in this section to identify potential outliers or influential points while fitting a regression model. The goal of these diagnostic tools is to help us identify cases that might need further investigation. Points identified as outliers or high leverage points might be data errors that need to be fixed or special cases that need to be studied further. Doing an analysis with and without a suspicious case is often a good strategy to see how the model is affected. We should avoid blindly deleting all unusual cases until the data that remain are "nice." In many situations (like the butterfly ballot scenario) the most important features of the data would be lost if the unusual points were dropped from the analysis!

4.4 Topic: Coding Categorical Predictors

In most of our regression examples the predictors have been quantitative variables. However, we have seen how a *binary* categorical variable, such as the reporting *Period* for jurors in Example 3.8 or *Sex* in the regression model for children's growth rates in Example 3.9, can be incorporated into a model using an indicator variable. But what about a categorical variable that has more than two categories? If a {0,1} indicator worked for two categories, perhaps we should use {0,1,2} for three categories. As we show in the next example, a better method is to use multiple indicator variables for the different categories.

Example 4.7: *Car prices*

In Example 1.1 we considered a simple linear model to predict the prices (in thousands of dollars) of used Porsches offered for sale at an internet site, based on the mileages (in thousands of miles) of the cars. Suppose now that we also have similar data on the prices of two other car models, Jaguars and BMWs, as in the file **ThreeCars**. Since the car type is categorical, we might consider coding the information numerically with a variable such as

$$Car = \begin{cases} 0 & \text{if a Porsche} \\ 1 & \text{if a Jaguar} \\ 2 & \text{if a BMW} \end{cases}$$

Using the *Car* variable as a single predictor of *Price* produces the fitted regression equation

$$\widehat{Price} = 47.73 - 10.15 \cdot Car$$

Substituting the values 0, 1, and 2 into the equation for the three car types gives predicted prices of 47.73 for Porsches, 37.58 for Jaguars, and 27.42 for BMWs. If we examine the sample mean prices for each of the three cars (Porsche=50.54, Jaguar=31.96, BMW=30.23) we see one flaw in using this method for coding the car types numerically. Using the *Car* predictor in a linear model forces the fitted price for Jaguars to be exactly halfway between the predictions for Porsches and BMWs. The data indicate that Jaguar's prices are probably much closer, in general, to BMWs than to Porsches. So even though we can fit a linear model for predicting *Price* using the *Car* variable in this form and that model shows a "significant" relationship between these variables (p-value=0.0000026), we should have concerns about whether the model is an accurate representation of the true relationship.

A better way to handle the categorical information on car type is to produce separate indicator variables for each of the car models. For example,

$$Porsche = \begin{cases} 1 & \text{if a Porsche} \\ 0 & \text{if not} \end{cases} \quad Jaguar = \begin{cases} 1 & \text{if a Jaguar} \\ 0 & \text{if not} \end{cases} \quad BMW = \begin{cases} 1 & \text{if a BMW} \\ 0 & \text{if not} \end{cases}$$

If we try to fit all three indicator variables in the same multiple regression model,

$$Price = \beta_0 + \beta_1 Porsche + \beta_2 Jaguar + \beta_3 BMW + \epsilon$$

we should experience some difficulty since any one of the indicator variables is *exactly* a linear function of the other two. For example,

$$Porsche = 1 - Jaguar - BMW$$

This means there is not a unique solution when we try to estimate the coefficients by minimizing the sum of squared errors. Most software packages will either give an error message if we try to include all three predictors or automatically drop one of the indicator predictors from the model. While it might seem counterintuitive at first, including all but one of the indicator predictors is exactly the right approach.

Suppose we drop the *BMW* predictor and fit the model,

$$Price = \beta_0 + \beta_1 Porsche + \beta_2 Jaguar + \epsilon$$

to the data in **ThreeCars** and obtain a fitted prediction equation

$$\widehat{Price} = 30.233 + 20.303 Porsche + 1.723 Jaguar$$

Can we still recover useful information about BMWs as well as the other two types of cars? Notice that the constant coefficient in the fitted model, $\hat{\beta}_0 = 30.233$, matches the sample mean for the BMWs. This is no accident since the values of the other two predictors are both zero for BMWs. Thus the predicted value for BMWs from the regression is the same as the mean BMW price in the sample. The estimated coefficient of *Porsche*, $\hat{\beta}_1 = 20.303$, indicates that the prediction should increase by 20.303 when *Porsche* goes from 0 to 1. Up to a round-off difference, that is how much bigger the mean Porsche price in this sample is over the mean BMW price. Similarly, the fitted coefficient of Jaguar, $\hat{\beta}_2 = 1.723$, means we should add 1.723 to the BMW mean to get the Jaguar mean, $30.233 + 1.723 = 31.956$.

◇

Regression using Indicators for Multiple Categories

To include a categorical variable with k categories in a regression model, use indicator variables, $I_1, I_2, \ldots, I_{k-1}$, for all but one of the categories

$$Y = \beta_0 + \beta_1 I_1 + \beta_2 I_2 + \ldots + I_{k-1} + \epsilon$$

We call the category which is not included as an indicator in the model the *reference* category. The constant term represents the mean for that category and the coefficient of any indicator predictor gives the difference of that category's mean from the reference category.

The idea of leaving out one of the categories when fitting a regression model with indicators may not seem so strange if you recall that, in our previous regression examples with binary categorical

4.4. TOPIC: CODING CATEGORICAL PREDICTORS

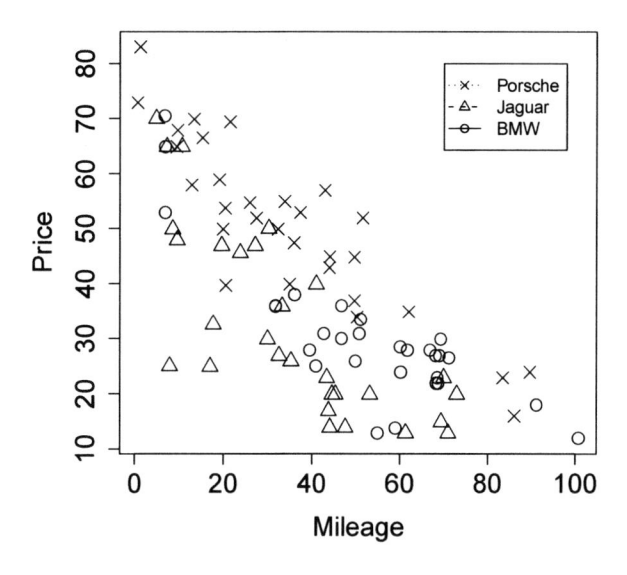

Figure 4.10: *Price* versus *Mileage* for three car types

predictors, we used just one indicator predictor. For example, to include information on gender we might use *Gender* = 0 for males and *Gender* = 1 for females, rather than using two different indictors for each gender. We can also extend the ideas of Section 3.3 to include quantitative variables in a regression model along with categorical indicators.

Example 4.8: *More car prices*

CHOOSE

Figure 4.10 shows a scatterplot of the relationship of *Price* and *Mileage* with different symbols indicating the type of car (Porsche, Jaguar, or BMW). The plot shows a decreasing trend for each of the car types with Porsches tending to be more expensive and BMWs tending to have more miles. While we could separate the data and explore separate simple linear regression fits for each of the car types, we can also combine all three in a common multiple regression model using indicator predictors along with the quantitative *Mileage* variable.

Using indicator predictors as defined in the previous example, consider the following model.

$$Price = \beta_0 + \beta_1 Mileage + \beta_2 Porsche + \beta_3 Jaguar + \epsilon$$

Once again, we omit the BMW category and use it as the reference group. This model allows for prices to change with mileage for all of the cars, but includes an adjustment for the type of car.

FIT

Here is the R output for fitting this model to the **ThreeCars** data.

```
Coefficients:
(Intercept)    Mileage     Porsche     Jaguar
    60.3007    -0.5653      9.9487    -8.0485
```

So the fitted prediction equation is

$$\widehat{Price} = 60.301 - 0.5653 Mileage + 9.949 Porsche - 8.049 Jaguar$$

For a BMW ($Porsche = 0$ and $Jaguar = 0$), this starts at a base price of 60.301 (or \$60,301 when $Mileage = 0$) and shows a decrease of about 0.5635 (or \$565.30) for every increase of one (thousand) miles driven. The coefficient of $Porsche$ indicates that the predicted price of a Porsche is 9.949 (or \$9,949) more than a BMW *which has the same number of miles.* Similarly, Jaguars with the same mileage will average about 8.049 (or \$8,049) less than the BMW. This result may seem curious when, in Example 4.7, we found that the average price for the Jaguars in the **ThreeCars** sample was slightly higher than the mean price of the BMWs. Note however that the mean mileage for the 30 Jaguars in the sample (35.9 thousand miles) is much lower than the mean mileage for the 30 BMWs (53.2 thousand miles). Thus the overall mean price of the BMWs in the sample is slightly lower than the Jaguars, even though BMWs tend to cost more than Jaguars when the mileages are similar.

This model assumes that the rate of depreciation (decrease in *Price* as more *Mileage* is driven) is the same for all three car models, about 0.5653 thousand dollars per thousand miles. Thus we can think of this model as specifying three different regression lines, each with the same slope but possibly different intercepts. Figure 4.11 shows a scatterplot of *Price* versus *Mileage* with the fitted model for each type of car.

ASSESS

The form of the model suggests two immediate questions to consider in its assessment.

(1) Are the differences between the car types statistically significant? For example, are the adjustments in the intercepts really needed or could we do just about as well by using the same regression line for each of the car types?

(2) Is the assumption of a common slope reasonable or could we gain a significant improvement in the fit by allowing for a different slope for each car type?

4.4. TOPIC: CODING CATEGORICAL PREDICTORS

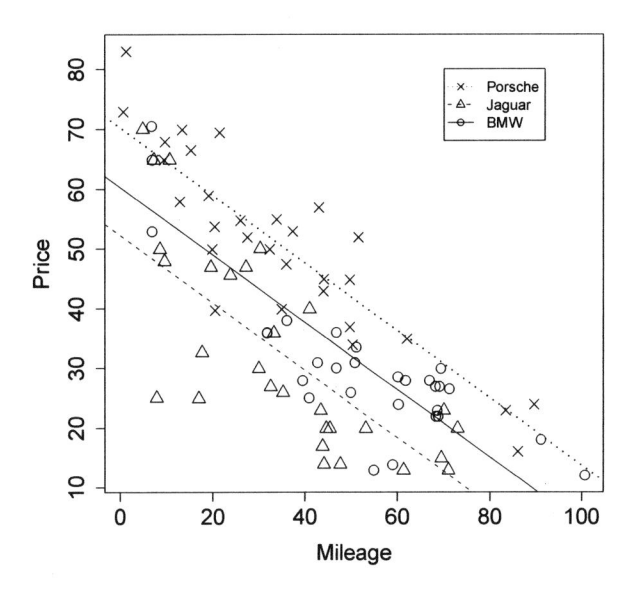

Figure 4.11: *Price* versus *Mileage* with equal slope fits

The first question asks whether the model is unnecessarily complicated and a simpler model might work just as well. The second asks whether the model is complicated enough to capture the main relationships in the data. Either of these questions are often important to ask when assessing the adequacy of any model.

We can address the first question by examining the individual t-tests for the coefficients in the model.

```
Coefficients:
            Estimate Std. Error t value Pr(>|t|)
(Intercept) 60.30070    2.71824  22.184  < 2e-16 ***
Mileage     -0.56528    0.04166 -13.570  < 2e-16 ***
Porsche      9.94868    2.35395   4.226 5.89e-05 ***
Jaguar      -8.04851    2.34038  -3.439 0.000903 ***
```

The R output shows that each of the coefficients in the model are significantly different from zero and thus all of the terms are important to keep in the model. So, not surprisingly, we find the the average price does tend to decrease as the mileage increases for all three car models. Also, the average prices at the same mileage are significantly different for the three car models, with Porsches being the most expensive, Jaguars the least, and BMWs somewhere in between.

To consider a more complicated model that would allow for different slopes we again extend the work of Section 3.3 to consider the following model.

$$Price = \beta_0 + \beta_1 Mileage + \beta_2 Porsche + \beta_3 Jaguar + \beta_4 Porsche \cdot Mileage + \beta_5 Jaguar \cdot Mileage + \epsilon$$

Before we estimate coefficients to fit the model, take a moment to anticipate the role of each of the terms. When the car is a BMW we have $Porsche = Jaguar = 0$ and the model reduces to

$$Price = \beta_0 + \beta_1 Mileage + \epsilon \quad \text{(for BMW)}$$

so β_0 and β_1 represent the slope and intercept for predicting BMW *Price* based on *Mileage*.

Going to a Porsche adds two more terms.

$$\begin{aligned} Price &= \beta_0 + \beta_1 Mileage + \beta_2 + \beta_4 Mileage + \epsilon \\ &= (\beta_0 + \beta_2) + (\beta_1 + \beta_4) Mileage + \epsilon \quad \text{(for Porsche)} \end{aligned}$$

Thus the β_2 coefficient is the *change* in intercept for a Porsche compared to a BMW and β_4 is the change in slope. Similarly, β_3 and β_5 represent the difference in intercept and slope, respectively, for a Jaguar compared to a BMW.

Here is some regression output for the more complicated model.

```
Coefficients:
                  Estimate Std. Error t value Pr(>|t|)
(Intercept)       56.29007    4.15512  13.547  < 2e-16 ***
Mileage           -0.48988    0.07227  -6.778 1.58e-09 ***
Porsche           14.80038    5.04149   2.936  0.00429 **
Jaguar            -2.06261    5.23575  -0.394  0.69462
I(Porsche*Mileage) -0.09952   0.09940  -1.001  0.31962
I(Jaguar*Mileage)  -0.13042   0.10567  -1.234  0.22057
```

We can determine separate regression lines for each of the car models by substituting into the fitted prediction equation.

$$\widehat{Price} = 56.29 - 0.49 Mileage + 14.80 Porsche - 2.06 Jaguar - 0.10 Porsche \cdot Mileage - 0.13 Jaguar \cdot Mileage$$

These lines are plotted on the scatterplot in Figure 4.12. Is there a statistically significant difference in the slopes of those lines? We see that neither of the new terms *Porsche · Mileage* and *Jaguar · Mileage* have coefficients that would be considered statistically different from zero.

4.4. TOPIC: CODING CATEGORICAL PREDICTORS

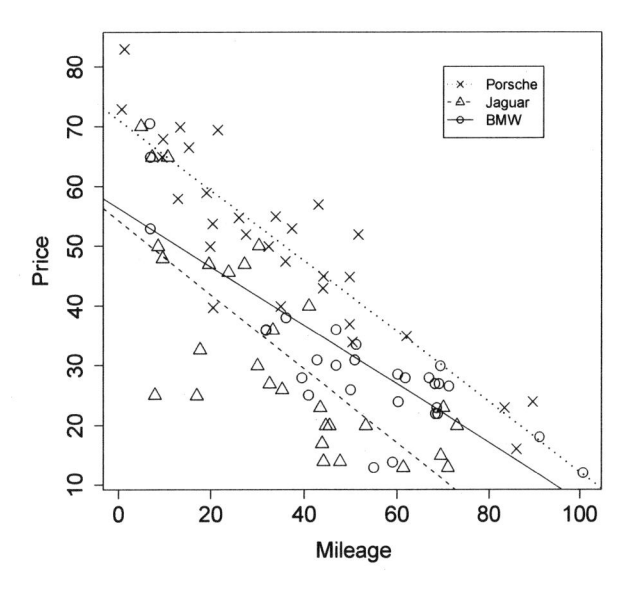

Figure 4.12: *Price* versus *Mileage* with different linear fits

However, we should take care when interpreting these individual t-tests since there are obvious relationships between the predictors (such as *Jaguar* and *Jaguar · Mileage*) in this model. For that reason we should also consider a nested F-test to assess the terms simultaneously.

To see if the different slopes are really needed we test

$H_0 : \beta_4 = \beta_5 = 0$

$H_a : \beta_4 \neq 0$ or $\beta_5 \neq 0$

For the full model with all five predictors we have $SSE_{Full} = 6268.3$ with $90 - 5 - 1 = 84$ degrees of freedom while the reduced model using just the first three terms has $SSE_{Reduced} = 6396.9$ with 86 degrees of freedom. Thus the two extra predictors account for $6396.9 - 6268.3 = 128.6$ in new variability explained. To compute the test statistic for the nested F-test in this case:

$$F = \frac{128.6/2}{6268.3/84} = 0.86$$

Comparing to an F-distribution with 2 and 84 degrees of freedom gives a p-value=0.426 which indicates that the extra terms to allow for different slopes do not produce a significantly better model.

176   CHAPTER 4. ADDITIONAL TOPICS IN REGRESSION

Table 4.2: Several models for predicting car prices

k	SSE	R^2	adj R^2	Predictors
1	6186	22.3%	21.4%	$Car = 0, 1, 2$
2	7604	27.5%	25.8%	$Porsche, Jaguar$
1	16432	59.5%	58.5%	$Mileage$
3	21301	76.9%	76.1%	$Mileage, Porsche, Jaguar$
5	21429	77.4%	76.0%	$Mileage, Porsche, Jaguar, Porsche*Mileage, Jaguar*Milege$

Table 4.2 shows a summary of several possible models for predicting *Price* based on the information in *Mileage* and *CarType*. Since the interaction terms were insignificant in the nested F-test, the model including *Mileage* and the indicators for *Porsche* and *Jaguar* would be a good choice for predicting prices of these types of cars.

Before we use the model based on *Mileage*, *Porsche*, and *Jaguar* to predict some car prices we should also check the regression conditions. Figure 4.13(a) shows a plot of the studentized residuals versus fits which produce a reasonably consistent band around zero with no studentized residuals beyond ±3. The linear pattern in the normal quantile plot of the residuals in Figure 4.13(b) indicates that the normality condition is appropriate.

(a) Studentized Residuals versus Fits   (b) Normal Plot of Residuals

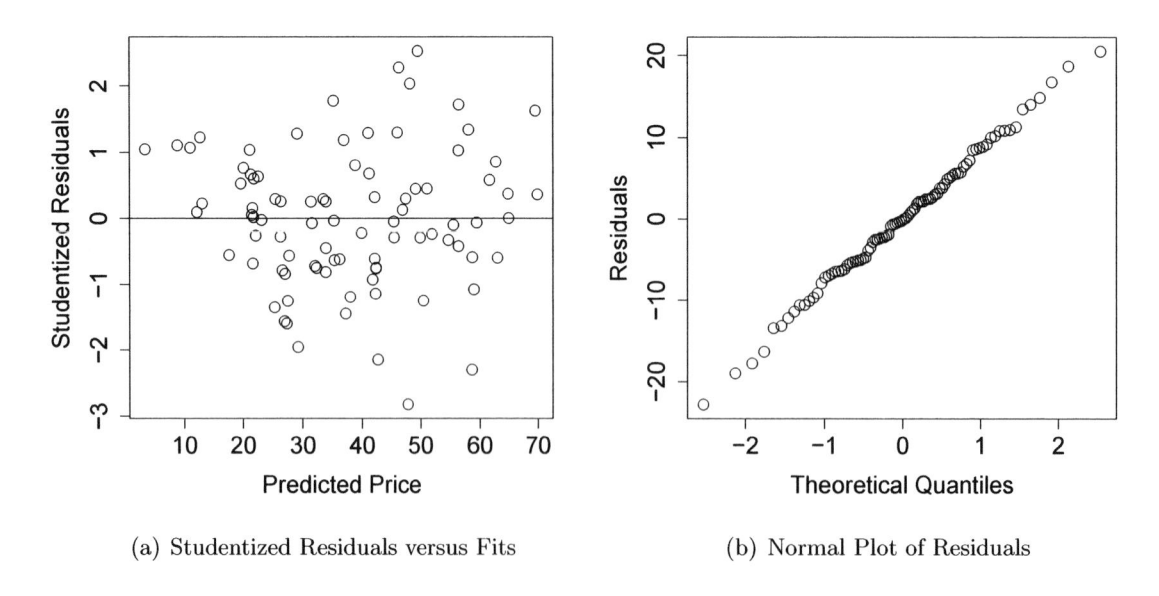

Figure 4.13: Residual plots for *Price* model based on *Mileage*, *Porsche* and *Jaguar*

4.4. TOPIC: CODING CATEGORICAL PREDICTORS

USE

Suppose that we are interested in used cars with about 50 (thousand) miles. What prices might we expect to see if we are choosing from among Porsches, Jaguars, or BMWs? Requesting 95% confidence intervals for the mean price and 95% prediction intervals for individual prices of each car model when $Mileage = 50$ yields the information below (with prices converted to dollars).

Car	Predicted Price	95% Confidence Interval	95% Prediction Interval
Porsche	$41,985	($38,614, $45,357)	($24,512, $59,459)
Jaguar	$23,988	($20.647, $27,329)	($ 6,207, $41,455)
BMW	$32,037	($28,895, $35.178)	($14,606, $49,467)

◇

4.5 Topic: Randomization Test for a Relationship

The inference procedures for linear regression depend to varying degrees on the conditions for the linear model being met. If the residuals are not normally distributed, the relationship is not linear, or the variance is not constant over values of the predictor(s), we should be wary of conclusions drawn from tests or intervals that are based on the t-distribution. In many cases we can find transformations (as described in Section 1.5) that produce new data for which the linear model conditions are more reasonable. Another alternative is to use different procedures for testing the significance of a relationship or constructing an interval that are less dependent on the conditions of a linear model.

Example 4.9: *Predicting GPAs with SAT scores*

In recent years many colleges have re-examined the traditional role the scores on the Scholastic Aptitude Tests (SAT's) play in making decisions on which students to admit. Do SAT scores really help predict success in college? To investigate this question a group of 24 introductory statistics students supplied the data in Table 4.3 showing their score on the Verbal portion of the SAT as well as their current grade point average (GPA) on a 0.0-4.0 scale (the data along with Math SAT scores are in **SATGPA**). Figure 4.14 shows a scatterplot with a least squares regression line to predict GPA using the Verbal SAT scores. The sample correlation between these variables is $r = 0.245$ which produces a p-value of 0.25 for a two-tailed t-test of $H_0: \rho = 0$ versus $H_a: \rho \neq 0$ with 22 degrees of freedom. This sample provides little evidence of a linear relationship between GPA and Verbal SAT scores.

Table 4.3: Verbal SAT scores and GPA

VerbalSAT	GPA	VerbalSAT	GPA	VerbalSAT	GPA
420	2.90	500	2.77	640	3.27
530	2.83	630	2.90	560	3.30
540	2.90	550	3.00	680	2.60
640	3.30	570	3.25	550	3.53
630	3.61	300	3.13	550	2.67
550	2.75	570	3.53	700	3.30
600	2.75	530	3.10	650	3.50
500	3.00	540	3.20	640	3.70

Source: Student survey in an introductory statistics course

One concern with fitting this model might be the potential influence of the high leverage point ($h_{13} = 0.457$) for the Verbal SAT score of 300. The correlation between GPA and Verbal SAT would increase to 0.333 if this point was omitted. However, we shouldn't ignore the fact that the student with the lowest Verbal SAT score still managed to earn a GPA slightly above the average for the whole group.

4.5. TOPIC: RANDOMIZATION TEST FOR A RELATIONSHIP

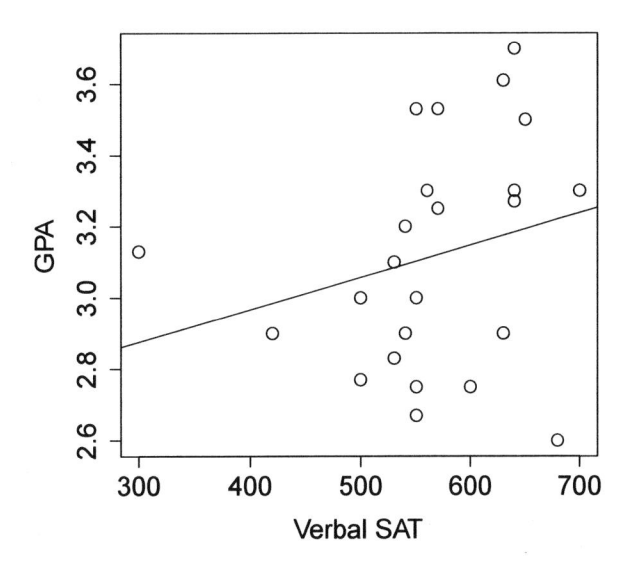

Figure 4.14: Linear regression for GPA based on Verbal SAT

If there really is no relationship between a student's Verbal SAT score and his or her GPA, we could reasonably expect to see any of the 24 GPAs in this sample associated with any of the 24 Verbal SAT scores. This key idea provides the foundation for a **randomization** test of the hypotheses

H_0: GPAs are unrelated to Verbal SAT scores ($\rho = 0$)
H_a: GPAs are related to Verbal SAT scores ($\rho \neq 0$).

The basic idea is to scramble the GPAs so they are randomly assigned to each of the 24 students in the sample (with no relationship to the Verbal SAT scores) and compute a measure of association, such as the sample correlation r, for the "new" sample. Table 4.4 shows the results of one such randomization of the GPA values from Table 4.3; for this randomization the sample correlation with Verbal SAT is $r = -0.188$. If we repeat this randomization process many, many times and record the sample correlations in each case, we can get a good picture of what the r values would look like if the GPA and Verbal SAT scores were unrelated. If the correlation from the original sample ($r = 0.244$) falls in a "typical" place in this **randomization distribution** we would conclude that the sample does not provide evidence of a relationship between GPA and Verbal SAT. On the other hand, if the original correlation falls at an extreme point in either tail of the randomization distribution, we can conclude that the sample is not consistent with the null hypothesis of "no relationship" and thus GPAs are probably related to Verbal SAT scores.

Table 4.4: Verbal SAT scores and with one set of randomized GPAs

VerbalSAT	GPA	VerbalSAT	GPA	VerbalSAT	GPA
420	3.53	500	2.90	640	2.75
530	3.50	630	2.90	560	3.61
540	3.00	550	3.70	680	3.25
640	3.27	570	2.83	550	3.00
630	2.77	300	3.30	550	2.90
550	3.30	570	3.20	700	3.13
600	3.53	530	2.60	650	2.67
500	2.75	540	3.50	640	3.30

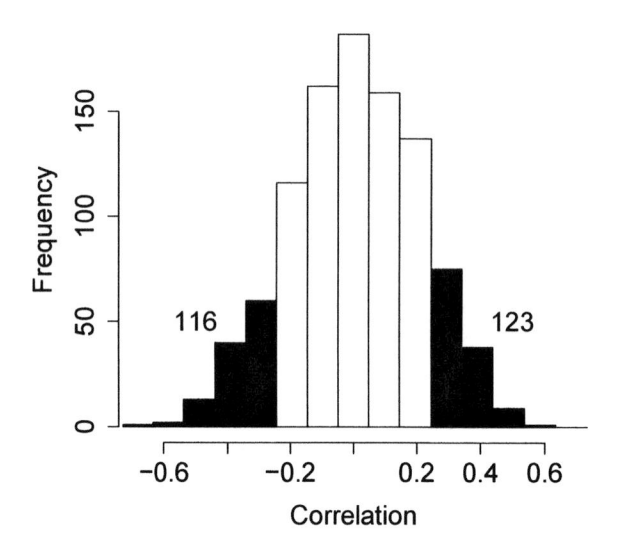

Figure 4.15: Randomization distribution for 1000 correlations of GPA versus Verbal SAT

Figure 4.15 shows a histogram of the sample correlations with Verbal SAT obtained from 1000 randomizations of the GPA data.[2] Among these permuted samples, we find 239 cases where the sample correlation was more extreme in absolute value than the $r = 0.244$ that was observed in the original sample; 116 values below -0.244 and 123 values above $+0.244$. Thus the approximate p-value from this randomization test would be $239/1000 = 0.239$ and we would conclude that the original sample does not provide evidence of a significant relationship between Verbal SAT scores

---

[2] In general we use more randomizations, 10,000 or even more, to estimate a p-value, but we chose a slightly smaller number to illustrate this first example.

4.5. TOPIC: RANDOMIZATION TEST FOR A RELATIONSHIP

and GPAs. By random chance alone, about 24% of the randomization correlations generated under the null hypothesis of no relationship were more extreme than what we observed in our sample. Note that the randomization p-value would differ slightly if a new set of 1000 randomizations was generated, but shouldn't change very much. This value is also consistent with the p-value of 0.25 that was obtained using the standard t-test for correlation with the original data. However, the t-test is only valid if the normality condition is satisfied, whereas the randomization test does not depend on normality.

◇

Modern computing power has made randomization approaches to testing hypotheses much more feasible. These procedures generally require much less stringent assumptions about the underlying structure of the data than classical tests based on the normal or t-distributions. Although we chose to use the sample correlation as the statistic to measure for each of the randomizations in the previous example, we just as easily could have used the sample slope, SSModel, the standard error of regression, or some other measure of the effectiveness of the model. Randomization techniques give us flexibility to work with different test statistics, even when a derivation of the theoretical distribution of a statistic may be unfeasible.

4.6 Topic: Bootstrap for Regression

Section 4.5 introduced the idea of a randomization test as an alternate procedure to doing a traditional t-test when the conditions for a linear model might not apply. In this section we examine another technique for doing inference on a regression model that is also less dependent on conditions such as the normality of residuals. The procedure is known an **bootstrapping**[3]. The basic idea is to use the data to generate an approximate sampling distribution for the statistic of interest, rather than relying on conditions being met to justify using some theoretical distribution.

In general, a sampling distribution shows how the values of a statistic (such as a mean, standard deviation, or regression coefficient) vary when taking many samples of the same size from the same population. In practice, we generally have just our original sample and cannot generate lots of new samples from the population. The bootstrap procedure involves creating new samples from the original sample data (not the whole population) by sampling with replacement. We are essentially assuming that the population looks roughly like many copies of the original sample, so we can simulate what additional samples might look like. For each simulated sample we calculate the desired statistic, repeating the process many times to generate a **bootstrap distribution** of possible values for the statistic. We can then use this bootstrap distribution to estimate quantities such as the standard deviation of the statistic or to find bounds on plausible values for the parameter.

Bootstrap Terminology

- A **bootstrap sample** is chosen with replacement from an existing sample, using the same sample size.

- A **bootstrap statistic** is a statistic computed for each bootstrap sample.

- A **bootstrap distribution** collects bootstrap statistics for many bootstrap samples.

Example 4.10: *Porsche prices*

In Section 2.1 we considered inference for a regression model to predict the price (in thousands of dollars) of used Porshes at an internet site based on mileage (in thousands of miles). The data are in **PorschePrice**. Some of the regression output for fitting this model is shown below; Figure 4.16 shows a scatterplot with the least squares line.

---

[3] One of the developers of this approach, Brad Efron, used the term bootstrap since the procedure allows the sample to help determine the distribution of the statistic on its own, without assistance from distributional assumptions, thus pulling itself up by its own bootstraps.

4.6. TOPIC: BOOTSTRAP FOR REGRESSION

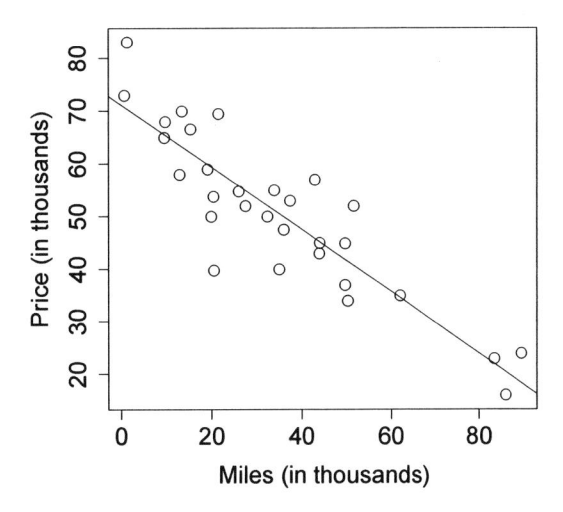

Figure 4.16: Original regression of Porsche prices on mileage

```
Coefficients:
            Estimate Std. Error t value Pr(>|t|)
(Intercept) 71.09045    2.36986   30.00  < 2e-16 ***
Mileage     -0.58940    0.05665  -10.40 3.98e-11 ***
```

Although we showed in Example 1.5 on page 34 that the residuals for the Porsche price model are fairly well-behaved, let's assume that we don't want to rely on using the t-distribution to do inference about the slope in this model and would rather construct a bootstrap distribution of sample slopes. Since the original sample size was $n = 30$ and we want to assess the accuracy for a sample slope based on this sample size, we select (with replacement) a random sample of thirty values from the original sample. One such sample is shown in Figure 4.17(a) where we add some random jitter (small displacements of the dots) to help reveal the cars at the same point. The fitted regression model for this bootstrap sample is $\widehat{Price} = 74.86 - 0.648 \cdot Mileage$. Fitted regression coefficients for this bootstrap sample, along with the estimated slopes and intercepts for nine additional bootstrap samples, are shown below, with all ten lines displayed in Figure 4.17(b).

Intercept	Slope	Intercept	Slope	Intercept	Slope
69.59	-0.570	71.36	-0.572	71.30	-0.620
73.04	-0.659	72.69	-0.628	74.67	-0.635
67.19	-0.484	71.37	-0.595	73.33	-0.638

To construct an approximate sampling distribution for the sample slope, we repeat this process many times and save all the bootstrap slope estimates. For example, Figure 4.18 shows a histogram

(a) Typical bootstrap sample (b) Regressions for ten bootstrap samples

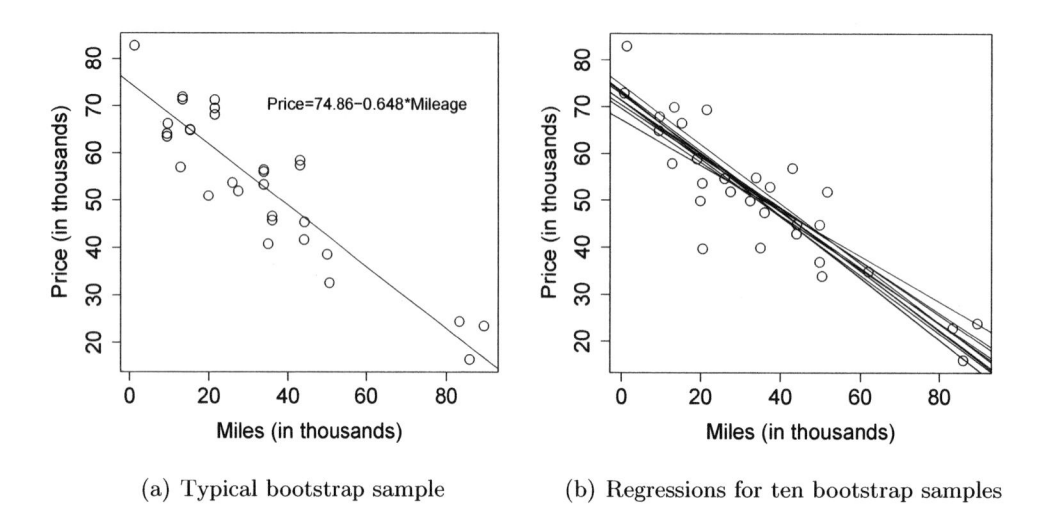

Figure 4.17: Porsche regressions for bootstrap samples

of slopes from a bootstrap distribution based on 5000 samples from the Porsche data. We can then estimate the standard deviation of the sample estimates, $SE_{\hat{\beta}_1}$, by computing the standard deviation of the slopes in the bootstrap distribution. In this case, the standard deviation of the 5000 bootstrap slopes is 0.05226 (which is similar to the standard error of the slope in the original regression output, 0.05665). In order to obtain reliable estimates of the standard error, we recommend using at least 5000 bootstrap samples. Fortunately, modern computers make this a quick task. ⋄

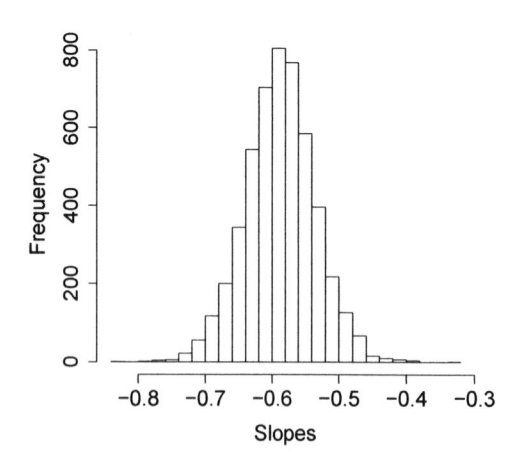

Figure 4.18: n=5000 Bootstrap Porsche Price Slopes

Confidence Intervals Based on Bootstrap Distributions

There are several methods for constructing a confidence interval for a parameter based on a bootstrap distribution of sample estimates.

Method #1: *Confidence interval based on the standard deviation of the bootstrap estimates*

When the bootstrap estimates appear to follow a normal distribution (for example, as in Figure 4.18) we can use a normal-based confidence interval with the original slope as the estimate and standard deviation (SE) based on the bootstrap distribution.

$$Estimate \pm z^* \cdot SE$$

where z^* is chosen from a standard normal distribution to reflect the desired level of confidence.

In the previous example for the Porsche prices where the original slope estimate is $\hat{\beta}_1 = -0.5894$, a 95% confidence interval for the population slope based on the bootstrap standard deviation is

$$-0.5894 \pm 1.96 \cdot 0.05226 = -0.5894 \pm 0.1024 = (-.6918, -0.4870)$$

Compare this to the traditional t-interval based on the original regression output on page 183 which is $-0.5894 \pm 0.116 = (-.7054, -0.4734)$. The lower and upper confidence limits with these two different methods are relatively close, which is not surprising since the original sample appeared to fit the regression condition of normality. A bootstrap confidence interval such as this, based on the standard deviation of the bootstrap estimates, works best when the bootstrap distribution is reasonably normally distributed, as is often the case.

Method #2: *Confidence interval based on the quantiles from the bootstrap distribution*

We can avoid any assumption of normality when constructing a confidence interval by using the quantiles directly from the bootstrap distribution itself, rather than using values from a normal table. Recall that one notion of a confidence interval is to find the middle 95% (or whatever confidence level you like) of the sampling distribution. In the classical 95% normal-based interval, we find the z^* values by locating points in each tail of the distribution that have a probability of about 2.5% beyond them. Since we already have a bootstrap distribution based on many simulated estimates, we can find these quantiles directly from the bootstrap samples. Let $q_{.025}$ represent the point in the bootstrap distribution where only 2.5% of the estimates are smaller and $q_{.975}$ be the point at the other end of the distribution where just 2.5% of the samples are larger. A 95% confidence interval can then be constructed as $(q_{.025}, q_{.975})$. For example, the quantiles using the 5000 bootstrap slopes in the previous example are $q_{.025} = -.6933$ and $q_{.975} = -.4853$.

Thus the 95% confidence interval for the slope in the Porsche price model would go from -0.6897 to -0.4845. Note that this interval is almost (but not exactly) symmetric about the estimated slope of -0.5894, and is close to the intervals based on the bootstrap standard deviation and the traditional t-methods. Using the quantiles in this way works well to give a confidence interval based on the bootstrap distribution, even when that distribution is *not* normal, provided that the distribution is at least reasonably symmetric around the estimate.

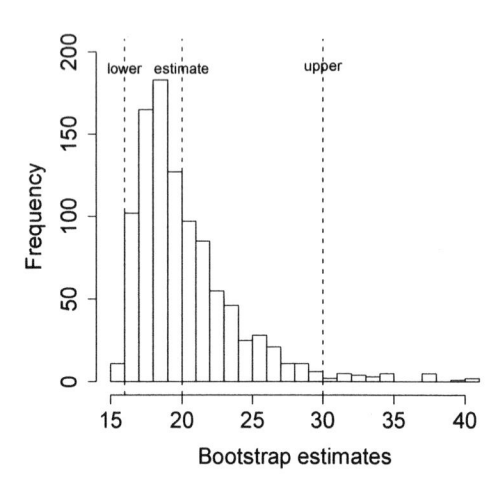

Figure 4.19: Hypothetical Skewed Bootstrap Distribution

Method #3: *Confidence interval based on reversing the quantiles from the bootstrap distribution*

The final method we discuss for generating a confidence interval from a bootstrap distribution may seem a bit counterintuitive at first, but it is useful in cases where the bootstrap distribution is not symmetric. Suppose that in a hypothetical example such as the bootstrap distribution in Figure 4.19, the original parameter estimate is 20 and the lower quantile ($q_{0.025}$) is 4 units below the original estimate, but the upper quantile ($q_{0.975}$) is 10 units above the original estimate. This would indicate that the bootstrap sampling distribution is skewed with a longer tail to the right. You might initially think that this means the confidence interval should be longer to the right than to the left (as it would be if method #2 was used). However, it turns out that exactly the reverse should be the case, i.e. the confidence interval should go from 10 units *below* the estimate to just 4 units *above* it. Thus, for the example shown in Figure 4.19 the confidence interval for the population parameter would go from 10 to 24.

Why do we need to reverse the differences? Remember that for the bootstrap samples we know (in some sense) the "population" parameter since the samples are all drawn from our original sample and we know the value of the parameter (in this case the original estimate) for that "population." While the original sample estimate may not be a perfect estimate for the parameter in the population from which the original sample was drawn, it can be regarded as the parameter for the simulated population from which the bootstrap samples are drawn. Thus if the bootstrap distri-

4.6. TOPIC: BOOTSTRAP FOR REGRESSION

bution indicates that sample estimates may reasonably go as much as ten units too high, we can infer that the original sample estimate might be as much as ten units *above* the parameter for its original population. Similarly, if the bootstrap estimates rarely go more than 4 units below the estimate, we should guess that the original estimate is probably not more than 4 units below the actual parameter.

Thus when the bootstrap distribution is skewed with lower and upper quantiles denoted by q_L and q_U respectively, we adjust the lower bound of the confidence interval to be $estimate - (q_U - estimate)$ and the new upper bound is $estimate + (estimate - q_L)$. For the 5000 bootstrapped Porsche slopes the quantiles are $q_L = -0.6897$ and $q_U = -0.4845$ so we have

$$-0.5894 - (-0.4845 - (-0.5894)) \text{ to } -0.5894 + (-0.5894 - (-0.6897))$$
$$= (-0.5894 - 0.1049, -0.5894 + 0.1003)$$
$$= (-0.6943, -0.4891)$$

Note that reversing the differences has little effect in this particular example since the bootstrap distribution is relatively symmetric to begin with. Our interval from Method #3 is very similar to the intervals produced by the other methods.

There are a number of other, more sophisticated, methods for constructing a confidence interval from a bootstrap distribution, but they are beyond the scope of this text. There are also other methods for generating bootstrap samples from a given model. For example, we could keep all of the original predictor values fixed and generate new sample responses by randomly selecting from the residuals of the original fit and adding them to the predicted values. This procedure assumes only that the errors are independent and identically distributed. The methods for constructing confidence intervals from the bootstrap distribution would remain the same. See Exercise 4.19 for an example of this approach to generating bootstrap samples.

The three methods described here for constructing bootstrap confidence intervals should help you see the reasoning behind bootstrap methods and give you tools that should work reasonably well in most situations. One of the key features of the bootstrap approach is that it can be applied relatively easily to simulate the sampling distribution of almost any estimate that is derived from the original sample. So if we want a confidence interval for a coefficient in a multiple regression model, the standard deviation of the error term, or a sample correlation the same methods can be used: Generate bootstrap samples, compute and store the statistic for each sample, then use the resulting bootstrap distribution to gain information about the sampling distribution of the statistic.

4.7 Exercises

Topic 4.1 Exercises: Added Variable Plots

4.1 *Adding variables to predict perch weights.* Consider the interaction model of Example 3.10 to predict the weights of a sample of 56 perch based on their lengths and widths.

  a. Construct an added variable plot for the interaction predictor *LengthxWidth* in this three-predictor model. Comment on how the plot shows you whether or not this is an important predictor in the model.

  b. Construct an added variable plot for the predictor *Width* in this three-predictor model. Comment on how the plot helps explain why *Width* gets a negative coefficient in the interaction model.

  c. Construct an added variable plot for the predictor *Length* in this three-predictor model. Comment on how the plot helps explain why *Length* is not a significant predictor in the interaction model.

4.2 *Adirondack high peaks.* Forty-six mountains in the Adirondacks of upstate New York are known as the High Peaks with elevations near or above 4000 feet (although modern measurements show a couple of the peaks are actually slightly under 4000 feet). A goal for hikers in the region is to become a "46er" by scaling each of these peaks. The file **HighPeaks** contains information on the elevation (in feet) of each peak along with data on typical hikes including the ascent (in feet), round trip distance (in miles), difficulty rating (on a 1 to 7 scale, with 7 being the most difficult), and expected trip time (in hours).

  a. Look at a scatterplot of $Y = Time$ of a hike versus $X = Elevation$ of the mountain and find the correlation between these two variables. Does it look like *Elevation* should be very helpful in predicting *Time*?

  b. Consider a multiple regression model using *Elevation* and *Length* to predict *Time*. Is *Elevation* important in this model? Does this two-predictor model do substantially better at explaining *Time* than either *Elevation* or *Length* alone?

  c. Construct an added variables plot to see the effect of adding *Elevation* to a model that contains just *Length* when predicting the typical trip *Time*. Does the plot show that there is information in *Elevation* that is useful for predicting *Time* after accounting for trip *Length*? Explain.

4.7. EXERCISES

Topic 4.2 Exercises: Techniques for Choosing Predictors

4.3 *Major League Baseball winning percentage.* Consider the **MLB2007Standings** file described in Exercise 3.7 on page 134. In this exercise you will consider models with four variables to predict winning percentages (*WinPct*) using any of the predictors except *Wins* and *Losses* – those would make it too easy! Don't worry if you are unfamiliar with baseball terminology and some of the acronyms for variable names have little meaning. Although knowledge of the context for data is often helpful for choosing good models, you should use software to build the models requested in this exercise rather than using specific baseball knowledge.

a. Use stepwise regression until you have a four-predictor model for *WinPct*. You may need to adjust the criteria if the procedure won't give sufficient steps initially. Write down the predictors in this model and its R^2 value.

b. Use backward elimination until you have a four-predictor model for *WinPct*. You may need to adjust the criteria if the procedure stops too soon to continue eliminating predictors. Write down the predictors in this model and its R^2 value.

c. Use a "best subsets" procedure to determine which four-predictors together would explain the most variability in *WinPct*. Write down the predictors in this model and its R^2 value.

d. Find the value of Mallow's Cp for each of the models produced in (a)-(c).

e. Assuming that these three procedures didn't all give the same four-predictor model, which model would you prefer? Explain why.

4.4 *Baseball game times.* Consider the data introduced in Exercise 1.7 on the time (in minutes) it took to play a sample of major league baseball games. The data file **BaseballTimes** contains four quantitative variables (*Runs*, *Margin*, *Pitchers* and *Attendance*) that might be useful in predicting the game times (*Time*). From among these four predictors choose a model for each of the goals below.

a. Maximize the coefficient of determination, R^2.

b. Maximize the adjusted R^2.

c. Minimize Mallow's C_p.

d. After considering the models in (a)-(c) what model would you choose to predict baseball game times? Explain your choice.

Topic 4.2 Supplemental Exercises

4.5 *Cross validation of a GPA model.* In some situations we may develop a multiple regression model that does a good job of explaining the variability in a particular sample, but does not work so well when used on other cases from the population. One way to assess this is to split the original sample into two parts: one set of cases to be used to develop and fit a model (called a *training* sample) and a second set (known as a *holdout* sample) to test the effectiveness of the model on new cases. In this exercise you will assess a model to predict first year GPA using this method.

a. Split the original **FirstYearGPA** data file to create a training sample with the first 150 cases and holdout sample with cases #151-219.

b. Use the training sample to fit a multiple regression to predict GPA using $HSGPA$, HU, and $White$. Variable selection techniques would show that this is a reasonable model to consider for the cases in the training sample. Give the prediction equation along with output to analyze the effectiveness of each predictor, estimated standard deviation of the error term, and R^2 to assess the overall contribution of the model.

c. Use the prediction equation in (b) as a formula to generate predictions of the GPA for each of the cases in the holdout sample. Also compute the residuals by subtracting the prediction from the actual GPA for each case.

d. Compute the mean and standard deviation for the residuals in (c). Is the mean reasonably close to zero? Is the standard deviation reasonably close to the standard deviation of the error term from the fit to the training sample?

e. Compute the correlation, r, between the actual and predicted GPA values for the cases in the holdout sample. This is known as the *cross validation correlation*.

f. Square the cross validation correlation to obtain an R^2 value to measure the percentage of variability in GPA for the holdout sample that is explained by the model fitted from the training sample.

g. We generally expect that the model fit for another sample will not perform as well on a new sample, but a valid model should not produce much of a drop in the effectiveness. One way to measure this, known as the *shrinkage* is to subtract the R^2 from holdout sample from the R^2 of the training sample. Compute the shrinkage for this model of first year GPA. Does it look like the training model works reasonably well for the holdout sample or has there been a considerable drop in the amount of variability explained?

Topic 4.3 Exercises: Identifying Unusual Points in Regression

4.6 *Breakfast cereals.* In Exercise 1.4 you were asked to fit a simple linear model to predict the number of calories (per serving) in breakfast cereals using the amount of sugar (gm/serving). The file **Cereal** also has a variable showing the amount of fiber (gm/serving) for each of the 36 cereals. Fit a multiple regression model to predict *Calories* based on both predictors, *Sugar*, and *Fiber*. Examine each of the measures below and identify which (if any) of the cereals you might classify as possibly "unusual" in that measure. Include specific numerical values and justification for each case.

(a) Standardized residuals (b) Studentized residuals (c) Leverage, h_i (d) Cook's D

4.7. EXERCISES

4.7 *Religiosity of countries.* Does the level of religious belief in a country predict per capita Gross Domestic Product (GDP)? The Pew Research Center's Global Attitudes Project surveyed people around the world and asked (among many other questions) whether they agreed that "belief in God is necessary for morality," whether religion is very important in their lives, and whether they pray at least once per day. The variable *Religiosity* is the sum of the percentage of positive responses on these three items, measured in each of 44 countries. This variable is part of the datafile **ReligionGDP**, which also includes the per capita GDP for each country and indicator variables that record the part of the world the country is in (East Europe, Asia, etc.).

- a. Transform the *GDP* variable by taking the log of each value, then make a scatterplot of $log(GDP)$ versus *Religiosity*.

- b. Regress $log(GDP)$ on *Religiosity*. What percentage of the variability in $log(GDP)$ is explained in this model?

- c. Interpret the coefficient of *Religiosity* in the model from part (b).

- d. Make a plot of the studentized residuals vs the predicted values. What is the magnitude of the studentized residual for Kuwait?

- e. Add the indicator variables for the regions of the world to the model, except for Africa (which will thus serve as the reference category). What percentage of the variability in $log(GDP)$ is explained by this model?

- f. Interpret the coefficient of *Religiosity* in the model from part (e).

- g. Does the inclusion of the regions variables substantially improve the model? Conduct an appropriate test at the 0.05 level.

- h. Make a plot of the studentized residuals vs the predicted values for the model with *Religiosity* and the region indicators. What is the magnitude of the studentized residual for Kuwait using the model that includes the regions variables?

4.8 *Adirondack high peaks.* Refer to the data in **HighPeaks** on 46 Adirondack high peaks that is described in Exercise 4.2 on page 188.

- a. What model would you use to predict the typical *Time* of a hike using the any combination of the other variables as predictors? Justify your choice.

- b. Examine plots using the residuals from your fitted model in (a) to assess the regression conditions of linearity, homoscedasticity, and normality in this situation. Comment on whether each of the conditions is reasonable for this model.

- c. Find the studentized residuals for the model in (a) and comment on which mountains (if any) might stand out as being unusual according to this measure.

192   CHAPTER 4. ADDITIONAL TOPICS IN REGRESSION

d. Are there any mountains that have the high leverage or may be influential on the fit? If so, identify the mountain(s) and give values for the leverage or Cook's D, as appropriate.

Topic 4.4 Exercises: Coding Categorical Predictors

4.9 *North Carolina births.* The file **NCBirths** contains data on a sample of 1450 birth records that statistician John Holcomb selected from the North Carolina State Center for Health and Environmental Statistics. One of the questions of interest is how the birth weights (in ounces) of the babies might be related to the mother's race. The variable *MomRace* codes the mother's race as white, black, Hispanic or other. We set up indicator variables for each of these categories and ran a regression model to predict birth weight using indicators for the last three categories. Here is the fitted model.

$$BirthWeightOz = 117.87 - 7.31 Black + 0.65 Hispanic - 0.73 Other$$

Explain what each of the coefficients in this fitted model is telling us about race and birth weights for babies born in North Carolina.

4.10 *More North Carolina births.* Refer to the model described in Exercise 4.9 in which the race of the mother is used to predict the birth weight of a baby. Some additional output for assessing this model is shown below.

```
Predictor     Coef   SE Coef      T      P
Constant   117.872     0.735  160.30  0.000
Black       -7.309     1.420   -5.15  0.000
Hispanic     0.646     1.878    0.34  0.731
Other       -0.726     3.278   -0.22  0.825

S = 22.1327    R-Sq = 1.9%    R-Sq(adj) = 1.7%

Analysis of Variance
Source            DF        SS      MS     F      P
Regression         3   14002.4  4667.5  9.53  0.000
Residual Error  1446  708331.7   489.9
Total           1449  722334.1
```

Assuming that the conditions for a regression model hold in this situation, interpret each of the following parts of this output. Be sure that your answers refer to the context of this problem.

a. The individual t-tests for the coefficients of the indicator variables.

b. The value of R^2.

c. The ANOVA table.

4.7. EXERCISES

4.11 *Blood Pressure.* The dataset **Blood1** contains information on the systolic blood pressure for 500 randomly chosen adults. One of the variables recorded for each subject, *Overwt*, classifies weight as 0=Normal, 1=Overweight, or 2=Obese. Fit two regression models to predict *SystolicBP*, one using *Overwt* as a single quantitative predictor and the other using indicator variables for the weight groups. Compare the results for these two models.

Topic 4.5 Exercises: Randomization Test for a Relationship

4.12 *Baseball game times.* The data in **BaseballTimes** contains information from 15 Major League Baseball games played on August 26, 2008. In Exercise 1.7 on page 55 we consider models to predict the time a game lasts (in minutes). One of the potential predictors is the number of runs scored in the game. Use a randomization procedure to test whether there is significant evidence to conclude that the correlation between *Runs* and *Time* is greater than zero.

4.13 *GPA by SAT slope.* In Example 4.9 on page 178 we look at a randomization test for the correlation between SAT scores and GPA for the data in **SATGPA**. Follow a similar procedure to obtain a randomization distribution of sample slopes of a regression model to predict GPA based on SAT score, under the null hypothesis $H_0 : \beta_1 = 0$. Use this distribution to find a p-value for the original slope if the alternative is $H_a : \beta_1 \neq 0$. Interpret the results and compare them to both the randomization test for correlation and a traditional t-test for the slope.

4.14 *More baseball game times.* Refer to the situation described in Exercise 4.14 for predicting baseball game times using the data in **BaseballTimes**. We can use the randomization procedure described in Section 4.5 to assess the effectiveness of a multiple regression model. For example, we might be interested in seeing how well number of *Runs* scored and *Attendance* can do together to predict game *Time*.

a. Fit a multiple regression model to predict *Time* based on *Runs* and *Attendance* using the original data in **BaseballTimes**.

b. Choose some value, such as R^2, SSE, $SSModel$, $ANOVA$ F-statistic, or S_ϵ, to measure the effectiveness of the original model.

c. Use technology to randomly scramble the values in the response column *Time* to create a sample in which *Time* has no consistent association with either predictor. Fit the model in (a) for this new randomization sample and record the value of the statistic you chose in (b).

d. Use technology to repeat (c) until you have values of the statistic for 10,000 randomizations (or use just 1,000 randomizations if your technology is slow). Produce a plot of the randomization distribution.

e. Explain how to use the randomization distribution in (d) to compute a p-value for your original data, under a null hypothesis of no relationship. Interpret this p-value in the context of this problem.

f. We can also test the overall effectiveness of a multiple regression using an ANOVA table. Compare the p-value of the ANOVA for the original model to the findings from your randomization test.

Topic 4.6 Exercises: Bootstrap for Regression

4.15 *Bootstrapping Adirondack hikes.* Consider a simple linear regression model to predict the *Length* (in miles) of an Adirondack hike using the typical *Time* (in hours) it takes to complete the hike. Fitting the model using the data in **HighPeaks** produces the prediction equation

$$\widehat{Length} = 1.10 + 1.077 \cdot Time$$

One rough interpretation of the slope, 1.077, is the average hiking speed (in miles per hour). In this exercise you will examine some bootstrap estimates for this slope.

a. Fit the simple linear regression model and use the estimate and standard error of the slope from the output to construct a 90% confidence interval for the slope. Give an interpretation of the interval in terms of hiking speed.

b. Construct a bootstrap distribution with slopes for 5000 bootstrap samples (each of size 46 using replacement) from high peaks data. Produce a histogram of these slopes and comment on the distribution.

c. Find the mean and standard deviation of the bootstrap slopes. How do these compare to the estimated coefficient and standard error of the slope in the original model?

d. Use the standard deviation from the bootstrap distribution to construct a 90% confidence interval for the slope.

e. Find the 5th and 95th quantiles from the bootstrap distribution of slopes (i.e. points that have 5% of the slopes more extreme) to construct a 90% percentile confidence interval for the slope of the Adirondack hike model.

f. See how far each of the endpoints for the percentile interval are from the original slope estimate. Subtract the distance to the upper bound from the original slope to get a new lower bound, then add the distance from the lower estimate to get a new upper bound.

g. Do you see much difference between the intervals of parts (a), (d), (e), and (f)?

4.16 *Bootstrap standard error of regression.* Consider the simple linear regression to predict the *Length* of Adirondack hikes using the typical hike *Time* in the **HighPeaks** data file described in Exercise 4.2. The bootstrap method can be applied to any quantity that is estimated for a regression model. Use the bootstrap procedure to find a 90% confidence interval for the standard deviation of the error term in this model, based on each of the three methods described in Section 4.6. Hint: In R, you can get the standard deviation of the error estimate for any model with `summary(model)$sigma`.

4.7. EXERCISES

4.17 *Bootstrap for a multiple regression coefficient.* Consider the multiple regression model for *Weight* of perch based on *Length*, *Width*, and an interaction term *Length · Width* that was used in Example 3.10. Use the bootstrap procedure to generate an approximate sampling distribution for the coefficient of the interaction term in this model. Construct 95% confidence intervals for the interaction coefficient using each of the three methods discussed in Section 4.6 and compare the results to an t-interval obtained from the original regression output. The data are stored in **Perch**.

4.18 *Bootstrap confidence interval for correlation* In Example 4.9 we considered a randomization test for the correlation between *VerbalSAT* scores and *GPA* for data on 24 students in **SATGPA**. Rather than doing a test, suppose that we want to construct a 95% confidence interval for the population correlation.

- a. Generate a bootstrap distribution of correlations between *VerbalSAT* and *GPA* for samples of size 24 from the data in the original sample. Produce a histogram and normality plot of the bootstrap distribution. Comment on whether assumptions of normality or symmetry appear reasonable for the bootstrap distribution.

- b. Use at least two of the methods from Section 4.6 to construct confidence intervals for the correlation between *VerbalSAT* and *GPA*. Are the results similar?

- c. Do any of your intervals for the correlation include zero? Explain what this tells you about the relationship between verbal SAT scores and GPA.

Topic 4.6 Supplemental Exercises

4.19 *Bootstrap regression based on residuals.* In Section 4.6 we generated bootstrap samples by sampling with replacement from the original sample of Porsche prices. An alternate approach is to leave the predictor values fixed and generate new values for the response variable by randomly selecting values from the residuals of the original fit and adding them to the fitted values.

- a. Run a regression to predict the *Price* of Porsche cars based on *Mileage* using the data in **PorschePrice**. Record the coefficients for the fitted model and save both the fits and residuals.

- b. Construct a new set of random errors by sampling with replacement from the original residuals. Use the same sample size as the original sample and add the errors to the original fitted values to obtain a new price for each of the cars. Produce a scatterplot of *NewPrice* versus *Mileage*.

- c. Run a regression model to predict the new prices based on mileages. Compare the slope and intercept coefficients for this bootstrap sample to the original fitted model.

- d. Repeat the process 1000 times, saving the slope from each bootstrap fit. Find the mean and standard deviation of the distribution of bootstrap slopes.

- e. Use each of the three methods discussed in Section 4.6 to construct confidence intervals for the slope in this regression model. Compare your results to each other and the intervals constructed in Section 4.6.

Chapter 5

Analysis of Variance

Do students with majors in different divisions of a college or university have different mean grade point averages? If we weigh children in elementary school, middle school, and high school, will they be equally overweight or underweight, or will they differ in how overweight or underweight they are? Do average book costs vary between college courses in the humanities, social sciences, and sciences? Do people remember a different amount of content from commercials aired during NBA, NFL, and NHL games?

As in Chapters 1 and 2, each of these examples has a response variable and an explanatory variable. And, as in Chapters 1 and 2, the response variable (grade point average, weight, book cost, amount of commercial content) is a quantitative variable. But now we are asking questions for which the explanatory variable is categorical (major, school level, course type, sport). You already saw a special case of this in your first statistics course where the explanatory variable had only two categories. For example, we might want to compare the average life span between two breeds of dogs, say Irish Setters and German Shepherds. In that case, we divide the responses into the two groups, one for each breed, and use a two-sample t-test to compare the two means. But what if we want to also compare the life span of Chihuahuas along with the life spans of Irish Setters and German Shepherds? The two-sample t-test, as it stands, is not appropriate for that case because we now have more than two groups. Fortunately there is a generalization of the two-sample t-test, called analysis of variance (abbreviated ANOVA) which handles situations that have explanatory variables with more than 2 categories. In fact, ANOVA will also work for the case in which there are exactly 2 categories, and agrees with a version of the two-sample t-test when the variances are equal.

This chapter will introduce you to the *one-way* ANOVA model for investigating the difference in several means. One-way, in this setting, means that we have one explanatory variable that breaks the sample into groups and we want to compare means of the quantitative response variable between those groups. Chapter 6 will introduce you to the *two-way* ANOVA model which allows you to have a second categorical explanatory variable. Chapter 7, like Chapter 4 in the linear regression unit, will introduce additional topics in ANOVA, and Chapter 8 will take a more in-depth look at how we collect our data and how that informs our analysis of the data.

Of course the term ANOVA should not be new to you: The ANOVA table and the idea of partitioning variability were covered in Section 2.2. So you may be asking yourself, "Why is this chapter here? Didn't we already learn about the ANOVA table in Chapter 2?" The answer is a qualified yes. In Chapter 2 we used an ANOVA table to answer the large question about whether an overall regression model was effective or not. In regression (when we were analyzing a quantitative response with a quantitative explanatory variable), however, the ANOVA table was only one of many pieces of the analysis puzzle. In this chapter we have only categorical explanatory variables and ANOVA refers not only to the table itself but also the model that we will use. Note that the general principles remain the same as in Chapter 2. We still assume that the response variable is related to the explanatory variable through a model with random errors introducing variability, and the ANOVA procedure looks at sums of squared deviations to assess whether the model is successful at explaining a significant portion of the variability in the response variable.

We conclude this introduction by describing, in some detail, an example in which we suspect, based on the design of the experiment, that an ANOVA model will be appropriate. We will follow this same example throughout the chapter as we dig deeper into the topic of ANOVA.

Example 5.1: *Fruit flies*

Does reproductive behavior reduce longevity in fruit flies? Hanley and Shapiro (1994) report on a study conducted by Partridge and Farquhar (1981) about the sexual behavior of fruit flies. It was already known that increased reproduction leads to shorter life spans for female fruit flies. But the question remained whether an increase in sexual activity would also reduce the life spans of male fruit flies. The researchers designed an experiment to answer this question. They had a total of 125 male fruit flies to use and they randomly assigned each of the 125 to one of the following five groups.

- *8 virgins*: Each male fruit fly assigned to live with 8 virgin female fruit flies.

- *1 virgin*: Each male fruit fly assigned to live with one virgin female fruit fly.

- *8 pregnant*: Each male fruit fly assigned to live with 8 pregnant female fruit flies. (The theory was that pregnant female fruit flies would not be receptive to sexual relations.)

- *1 pregnant*: Each male fruit fly assigned to live with 1 pregnant female fruit fly.

- *none*: Each male fruit fly was subjected to a lonely existence without females at all.

⋄

The ANOVA model seems appropriate for the data from this experiment because there are five groups that constitute the values of the categorical variable (the number and type of females living with a male fruit fly), and a quantitative response (the life span of the male fruit flies). Of course there is more to this model than just the general description that we have given so far. Our first task in this chapter is to describe the model in detail and how to determine if it is appropriate for a particular data set.

Once we have determined that the ANOVA model is appropriate, then we can turn to the questions that are the main interest of the researchers. Typical questions include: Are there statistically significant differences between the means of at least two groups? And if there are statistically significant differences, can we determine which *specific* differences seem to exist? Are there specific comparisons between a subset of groups that it makes sense to focus on? These are the questions that we will be attempting to answer in this chapter (and will be expanded on in Chapter 7). And we will use the fruit flies as our guide both through our discussion of the model in detail and our attempt to answer these questions.

5.1 The One-way Model: Comparing Groups

The most basic question that the ANOVA model is used to address is whether the population mean of the quantitative response variable is the same for all groups, or whether the mean response is different for at least one group. That is, we want to know if we are better able to predict the response value if we know which group the observation came from, or if all of the groups are essentially the same.

One Mean or Several?

As with any data analysis, our first instinct should be to explore the data at hand, particularly visually. So how can we visualize the results in this kind of setting? Figure 5.1(a) shows a dotplot of the life spans of the male fruit flies discussed in the introduction, regardless of which treatment they received. Figure 5.1(b) shows a dotplot of those same life spans, this time broken down by the experimental group that the individual fruit flies were in. This data can also be found in the file **FruitFlies** and the means for each group, as well as the overall mean, can be found in Table 5.1. A quick look at these dotplots suggests that there might be at least one difference between the groups. But is that difference between the sample means enough to be statistically significant? This is the question that we will ultimately want to answer.

Group	Treatment	n	$\bar{y}$
1	None	25	63.56
2	1 Pregnant	25	64.80
3	8 Pregnant	25	63.36
4	1 Virgin	25	56.76
5	8 Virgin	25	38.72
	All	125	57.44

Table 5.1: Fruit fly lifetime means; overall and broken down by groups

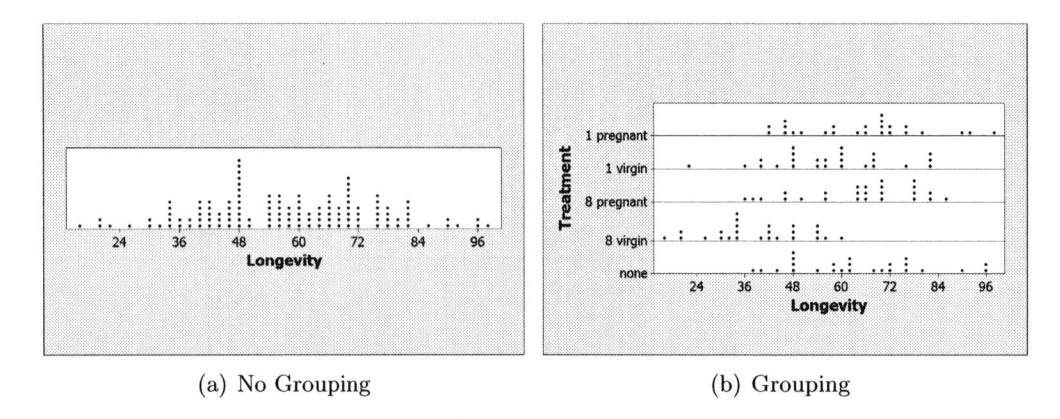

Figure 5.1: Dotplots of life spans for fruit flies

To start, we notice that we essentially have two models that we want to compare. The first model says that, in general (ignoring random variation), the value of the response variable does not differ between the groups. The second model says that it does. It says that at least one group has values of the response variable that are different, even after accounting for random variation, from the values of the response variable of another group. A good starting place for comparing these models is to look at graphs such as the dotplots in Figure 5.1. The dotplot in Figure 5.1(a) represents the data using the simpler model, while Figure 5.1(b) gives us a way of visualizing the data subject to the second model.

If the simpler model is a good fit, then there will be a great amount of overlap in all subgroups of the second dotplot. But that is not the case here. In particular, the dotplot in Figure 5.1(b) suggests that male fruit flies living with 8 virgin female fruit flies do have shorter life spans. The question, of course, is whether this difference is large enough to be statistically significant or if it could be due to chance. The rest of this chapter gives the details of this type of analysis.

Observation = Group Mean + Error

We introduced you to the idea of the ANOVA procedure by saying that it is a way of generalizing the idea of a two-sample t-test to those cases in which we have (potentially) more than two groups. In the two-sample t-test case we test the null hypothesis that the two groups have the same mean. In the ANOVA case we will be testing the null hypothesis that all groups have the same population mean value. In other words we will be testing

$$H_0 : \mu_1 = \mu_2 = \mu_3 = ... = \mu_K$$

where K is the total number of categories in the explanatory variable and μ_k is the mean response for the kth group. If the null hypothesis is true, then all groups have the same population mean value for the response variable; in other words, all potential observations start with the same value

5.1. THE ONE-WAY MODEL: COMPARING GROUPS

and any differences observed are due only to random variation. Therefore we start our modeling process by identifying a common value that we call the *grand mean* and denote it simply by μ with no subscript. All observations are based on this grand mean and, if the null hypothesis is true, will only differ from it by some amount of random variation. The null model, therefore, is

$$Y = \mu + \epsilon$$

where ϵ represents the random variation from one observation to the next.

Group Mean = Grand Mean + Group Effect

The alternative hypothesis is that there is at least one group that differs from the other groups, in terms of the mean response. Under this hypothesis we need to model each group mean separately rather than by simply using the overall mean for each predicted value. We still use the grand mean as our basis for the group means, but we recognize that any group mean may differ by some constant from that grand mean. We will let α_k denote this constant for the kth group and call it the *group effect*. Symbolically, this translates to $\mu_k = \mu + \alpha_k$ for the kth group. This leads us to the following model.

One-way Analysis of Variance Model

The **ANOVA model** for a quantitative response variable and one categorical explanatory variable with K values is

$$\begin{array}{ccccccc} \text{Response} & = & \text{Grand Mean} & + & \text{Group Effect} & + & \text{Error Term} \\ Y & = & \mu & + & \alpha_k & + & \epsilon \end{array}$$

where k refers to the specific category of the explanatory variable and $k = 1, \ldots, K$.

Notice that, if we use the fact that $\mu_k = \mu + \alpha_k$, we can rewrite the model in the box as $Y = \mu_k + \epsilon$. Also, note that using the model shown in the definition, we can rewrite the hypotheses about the group means as follows:

$H_0 : \alpha_1 = \alpha_2 = \alpha_3 = \ldots = \alpha_K = 0$
$H_a :$ At least one $\alpha_k \neq 0$

Conditions on the Error Terms

Of course with any model we need to think about when we can use it. As with a regression model, there are four conditions that must be met by the error terms for the ANOVA model to be appropriate. These conditions are precisely the same as the conditions we had on the error terms for regression models in the previous chapters.

> **ANOVA model conditions**
>
> The error terms must meet the following conditions for the ANOVA model to be applicable:
>
> - have mean zero
> - have the same standard deviation for each group (commonly referred to as the equal variances condition)
> - follow a normal distribution
> - be independent.
>
> These conditions can be summarized by stating that $\epsilon \sim N(0, \sigma_\epsilon)$ and are independent.

You might also recognize these conditions from an introductory statistics course as the conditions required by the two-sample t-test, although the equal variances condition is commonly omitted in that setting. Since ANOVA is the generalization of the two-sample t-test, it should not be surprising that the conditions are similar. The extra condition of equal variances is consistent with the condition needed in a regression model and important for the ANOVA procedure. In essence, this is saying that all populations (groups as defined by the categories of the explanatory variable) must have the same amount of variability for ANOVA to be an appropriate analysis tool.[1]

Estimating the ANOVA model terms

The next step is to actually find an estimate of the model. This means we need to estimate the overall grand mean, each of the group effects, and all of the error terms. Given the name we have chosen for the first part of the model, grand mean, it should not surprise you that our estimate for this parameter is the mean of all observations regardless of which group they belong to. The grand mean is $\hat{\mu} = \bar{y}$. Notice that under the null hypothesis, the model is $Y = \mu + \epsilon$. If the null hypothesis is true, our predicted value for any observation will be $\hat{y} = \bar{y}$.

Under the alternative hypothesis we need the second part of the model, the set of group effect terms (or α_k), which account for the differences between the individual groups and the common (grand) mean. On the surface, we might be tempted to estimate these terms using the mean values from each of the individual groups since that would be analogous to how we estimated the grand mean. But we need to keep in mind that we have already estimated part of the values of these terms by using the term for the grand mean. What we really need to compute now is how much each group *differs from the grand mean*. So each of the estimates of the α_k terms will be a difference between the grand mean and the group mean specific to group k. In other words,

---

[1] In fact, there is a 2-sample t-test that is completely analogous to this ANOVA model. It is called the pooled t-test and is discussed in Chapter 7.

5.1. THE ONE-WAY MODEL: COMPARING GROUPS

$$\hat{\alpha}_k = \bar{y}_k - \hat{\mu} = \bar{y}_k - \bar{y}$$

where $\hat{\alpha}_k$ is the estimated value of the group effect for the kth group, $\bar{y}_k$ is the mean of the observations in group k, and $\hat{\mu}$ is the grand mean. If the alternative hypothesis is true, then the model is $Y = \mu + \alpha_k + \epsilon$ and the predicted value for any observation will depend on which group it is in and will be $\hat{y} = \bar{y} + (\bar{y}_k - \bar{y}) = \bar{y}_k$, the mean from the relevant group.

Finally, we need to estimate the values for the error terms in the model. Remember that the difference between the observed value and the predicted value is called the **residual**. We will compute the residuals in the same way as in the earlier chapters, using the ANOVA (or alternative) model.

$$\hat{\epsilon} = y - \hat{y} = y - \hat{\mu} - \hat{\alpha}_k = y - \bar{y}_k$$

We now put this all together.

Parameter estimates

The values used to estimate the ANOVA model are

$$\hat{\mu} = \bar{y}$$

$$\hat{\alpha}_k = \bar{y}_k - \bar{y}$$

where $\bar{y}$ is the mean of all observations and $\bar{y}_k$ is the mean of all observations in group k. Based on these estimates we compute the residuals as

$$\hat{\epsilon} = y - \bar{y}_k$$

In practice, we rely on software to compute these estimates. However, we do some of the necessary computations in Example 5.2 to illustrate the relationships between the various parameter estimates.

Example 5.2: *Fruit flies (continued)*

Recall that Table 5.1 listed the mean lifetimes for each of the five groups of fruit flies and the overall mean. From this table we see that the average lifetime of all of the male fruit flies was 57.44 days. Symbolically, this means $\hat{\mu} = 57.44$.

If we concentrate on the group of males who were housed with 1 virgin female (the group that we have denoted as group 4), the average lifetime was 56.76 days. The group effect for this group is

Treatment	Lifetime	$\hat{y}$	$y - \hat{y}$
None	61	63.56	-2.56
None	59	63.56	-4.56
1 Pregnant	70	64.80	5.20
1 Pregnant	56	64.80	-8.80
1 Virgin	60	56.76	3.24
8 Virgin	33	38.72	-5.72

Table 5.2: Computing residuals for a subset of 5 fruit flies

denoted by $\hat{\alpha}_4$ and is computed as the difference between the mean lifetime of those in group 4 and the overall mean lifetime:

$$\hat{\alpha}_4 = \bar{y}_4 - \bar{y} = 56.76 - 57.44 = -0.68$$

Finally we illustrate computing residuals for individual cases. We start with one particular fruit fly. It was assigned to live with 8 pregnant females and had a lifetime of 65 days. The mean lifetime for the group that our chosen fruit fly belonged to was 63.36 days so its residual was

$$residual = 65 - 63.36 = 1.64$$

Table 5.2 shows examples of calculated residuals for a few more of our fruit flies. The first column gives the treatment group that the observation belonged to, the second column gives the actual lifetime of that particular fruit fly, the third column gives the predicted lifetime for that particular fruit fly (the mean for its group), and the last column gives the residual. ◊

How Do We Use Residuals?

Once again, we start with the general idea, using the fruit flies as our guide, and then move on to the specifics behind the calculations.

In the null model, the response (ignoring random variation) is the same for all of the groups so, as we noted above, we just use the overall mean value of the responses as the predicted value for any given observation. In the alternative (or ANOVA) model, we think that the groups may have different response values so we use the group means as predicted values for observations in the individual groups.

To compare the two models, let's compute the residuals for each observation for each model. That is, for the null model we look at the difference between each observation and the mean for all observations. For the ANOVA model, we compute the difference between each observation and the mean for the observations in its group. If the ANOVA model is better, then we would expect the residuals to be smaller, in general, than those of the null model. If the ANOVA model is not better,

5.1. THE ONE-WAY MODEL: COMPARING GROUPS

then the residuals should be about the same as those from the null model.

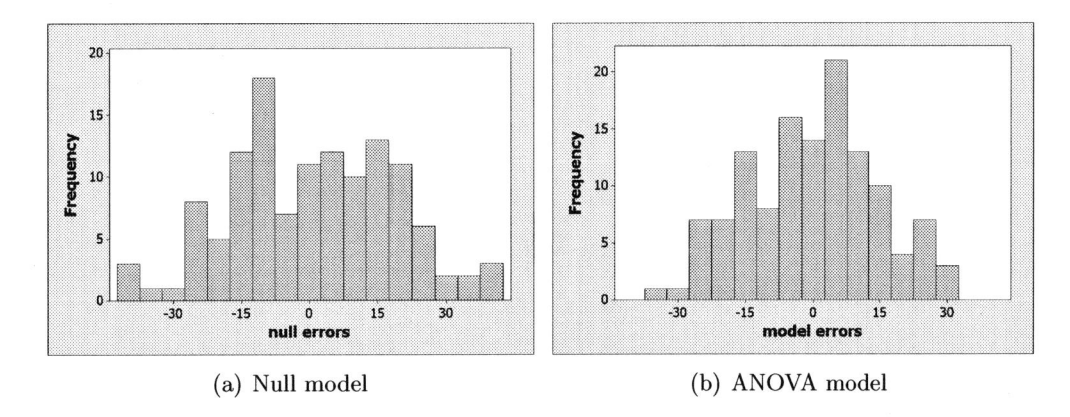

(a) Null model

(b) ANOVA model

Figure 5.2: Histograms of error terms for two models of lifespans for fruitflies

When we apply this idea to the fruit fly data, we get the graphs shown in Figure 5.2. Figure 5.2(a) gives a histogram of the residuals for the null model (all group means are the same) and Figure 5.2(b) gives a histogram of the residuals for the alternative model (at least one group mean is different from the rest). Notice that both histograms use the same scale on both axes. From these graphs, it appears that the ANOVA model has errors that are somewhat smaller. So we continue to suspect that there might be something to the idea that at least one group has a different mean. Now let's move on to the specifics with respect to the errors.

Variance of the Error Terms = Mean Square Error

As discussed above, we want to find a way to compare residuals from the null and alternative models in an appropriate way. What we will look for is less spread to the error terms of the ANOVA (alternative) model than those from the model just using the grand mean (null model). This will require that we compute the variance of the residuals to get a sense of their spread.

Computing the variance of the residuals in the ANOVA setting is similar to what we used in the regression chapters. Recall that $\Sigma (y - \hat{y})^2$ is what we called SSE and we divided that by the appropriate degrees of freedom to obtain the MSE. We will do the same thing here, but need to adjust the degrees of freedom to reflect the ANOVA model. To understand how this works requires us to examine the triple decomposition of the observations.

The triple decomposition

In Section 2.2, when we first discussed partitioning variance, we broke down the overall variability into two parts: the part explained by the model, and the part due to error. The basic idea is the same here. The difference is in the model itself and, therefore, how we calculate the SSModel term (called SSGroups in this setting).

Table 5.3 shows the same subset of the fruit fly data used in Table 5.2 with each response value split into the three components of the ANOVA model (the triple decomposition of observations). In this display, the last column (the residuals) shows the variability from one *individual* to the next (in essence, this is the variability within groups and is analogous to the residual variability in the regression setting). But we are also interested in the amount of variability from one *group* to the next (the variability between the groups, representing the model portion of the total variability). This is quantified in the next to last column (labeled group effect). As discussed earlier these values are the differences between the group means and the grand mean, and act like the residuals did for the individual observations, only now at the group level rather than the individual level. The following table gives an idea about the variability at the group level.

	Observed Value		Grand Mean		Group Effect		Residual
None	61.00	=	57.44	+	6.12	+	-2.56
None	59.00	=	57.44	+	6.12	+	-4.56
1 Pregnant	70.00	=	57.44	+	7.36	+	5.20
1 Pregnant	56.00	=	57.44	+	7.36	+	-8.80
1 Virgin	60.00	=	57.44	+	-0.68	+	3.24
8 Virgin	38.72	=	57.44	+	-18.72	+	-5.72
					↑ Variability from group to group		↑ Variability from one individual to the next

Table 5.3: ANOVA model applied to fruit flies data subset

	Group Mean	=	Grand Mean	+	Group Effect
None	63.56	=	57.44	+	6.12
1 Pregnant	64.80	=	57.44	+	7.36
8 Pregnant	63.36	=	57.44	+	5.92
1 Virgin	56.76	=	57.44	+	-0.68
8 Virgin	38.72	=	57.44	+	-18.72
Total	287.20	=	287.2	+	0.00

Table 5.4: Breakdown of Group Means in ANOVA model

Note that in this example, as seen in Table 5.4, the sum of the group effects is 0. This will be true in general if all groups have the same number of observations. Finally, the third part of the decomposition is the grand mean, that portion of the model that is common to all observations.

5.1. THE ONE-WAY MODEL: COMPARING GROUPS

To actually measure the two pieces of variability, the part explained by the model (groups) and the part explained by the error term (individuals), we compute the sum of squares (SS) for each of the pieces of the triple decomposition. (Remember that Table 5.3 only breaks down 5 of the 125 observations. To compute the sum of squares, you need to square the relevant numbers for *all* observations, not just the 5 in our table. The complete set of computations is just too large for us to show you.) The numbers given below are the sums of squares for each part of the equation for all 125 fruit flies and were found with software.

$$
\begin{array}{ccccccc}
\text{Observed} & = & \text{Grand Mean} & + & \text{Group Effect} & + & \text{Residual} \\
450672.00 & = & 412419.20 & + & \underbrace{11939.30}_{SSGroups} & + & \underbrace{26313.50}_{SSE}
\end{array}
$$

$$SSTotal$$

In practice, as with regression, we focus on the total variability ($SSTotal$) which is defined to be the sum of the variability "explained" by the model (in this case $SSGroups$) and the residual variability remaining within each group (SSE). The grand mean component does not contribute to the total variability, since adding a constant to each observation would not change the amount of variability in the data set. This leads us to the one-way ANOVA identity

$$SSTotal = SSGroups + SSE.$$

For the fruit flies data $SSTotal = 11939.30 + 26313.50 = 38252.80$.

It is worth noting that if we move the sum of squares of the grand mean from the right hand side of the equation to the left hand side, using some algebra the left hand side of the equation can be rewritten as $\sum (y - \bar{y})^2$. What is left on the other side of the equation is $SSTotal$. This means that $SSTotal = \sum (y - \bar{y})^2$. There are two things to note about this. First, this is the familiar form of the $SSTotal$ from Chapter 2. But also, it is a measurement of the amount of variability left after fitting the model with only the grand mean in it. Let's take a look at the decomposition of the observations again, this time focusing on the individual pieces.

$$
\begin{array}{ccccccc}
\text{Observed} & - & \text{Grand Mean} & = & \text{Group Effect} & + & \text{Residual} \\
& \underbrace{y - \bar{y}} & & = & \underbrace{\bar{y}_k - \bar{y}} & + & \underbrace{y - \bar{y}_k} \\
& & & & \hat{\alpha} & + & \hat{\epsilon}
\end{array}
$$

Grand Mean model variability   ANOVA model variability

Once again, taking the sum of squares of both sides we see that both sides of this equation now represent *SSTotal*, but they represent two different ways of thinking about the variability in the data. The left-hand side measures the variability assuming that the model with only the Grand Mean (the null model) is fit. The right-hand side is a finer model, allowing the data to come from groups with possibly different means. This represents a way of partitioning the same variability into pieces due to the (possibly) different group means and the leftover residual error. That is, the right-hand side represents the ANOVA model.

Now, regardless of how we think about the sums of squares components, we need to convert them into mean square components. To convert sums of squares to average variability for each part, we divide by the degrees of freedom as we did in Chapter 2 when we first saw an ANOVA table. The degrees of freedom for each portion of the model are given as follows:

$$\text{Observed} - \text{Grand Mean} = \text{Group Effect} + \text{Residual}$$

$$\# \text{ obs} - 1 = (\# \text{ groups -1}) + (\# \text{ obs - } \# \text{ groups})$$

$$\underbrace{125 - 1}_{df_{Total}} = \underbrace{\underbrace{4}_{df_{Groups}} + \underbrace{120}_{df_{Error}}}_{df_{Total}}$$

Note again that the *df* for total is the sum of the two components ($4 + 120 = 124$ in this case) and also the degrees of freedom for the observed values minus the one degree of freedom for the estimated grand mean ($125 - 1 = 124$).

5.2 Assessing and Using the Model

We now turn our attention to assessing the fit of the ANOVA model and making basic conclusions about the data, when we have a model that is appropriate. We start this section by a brief discussion of how we draw conclusions from this model. In Section 5.1 we set up the model as one that uses variability to determine if the means of several groups are the same, or if one or more are different. We continue with that theme here.

Using Variation to Determine if Means Are Different?

The issue that we want to address is how to actually determine if the model is significant. In other words, we want to ask the very basic inferential question: Does it appear that there is a difference between the means of the different groups? For this question, we need to return to the ANOVA table itself. As with ANOVA for regression, when we divide the sum of squares by the appropriate degrees of freedom, we call the result the Mean Square (MS):

5.2. ASSESSING AND USING THE MODEL

$$\text{Mean Square} = \text{Sum of Squares} \div \text{Degrees of Freedom}$$
$$\text{(MS)} \qquad\qquad \text{(SS)} \qquad\qquad \text{(df)}$$

We compute the Mean Square for both the Groups and the Error terms and put them in the ANOVA table as we did with regression. We then compare the values of the MSGroups and MSE by computing the ratio

$$F = \frac{\text{MSGroups}}{\text{MSE}}$$

This ratio, called the F-ratio, is then compared to the appropriate F-distribution (using degrees of freedom for the Groups and Error, respectively) to find the p-value. In practice, software does the calculations and provides a summary in the form of an ANOVA table.

ANOVA model

To test for differences between K group means, the hypotheses are

$H_0: \alpha_1 = \alpha_2 ... = \alpha_K = 0$
$H_a:$ At least one $\alpha_k \neq 0$

and the **ANOVA table** is

Source	Degrees of freedom	Sum of Squares	Mean Square	F statistic
Model	K-1	SSGroups	MSGroups	$F = \frac{MSGroups}{MSE}$
Error	$n - K$	SSE	MSE	
Total	$n - 1$	SSTotal		

If the proper conditions hold, the p-value is calculated using the upper tail of an F-distribution with $K - 1$ and $n - K$ degrees of freedom.

Example 5.3: *Fruit flies (continued)*

For the fruit flies data the computer gives us:

```
One-way ANOVA: Longevity versus Treatment

Source      DF    SS      MS     F      P
Treatment    4  11939   2985  13.61  0.000
Error      120  26314    219
Total      124  38253
```

The last column in this table gives the p-value for the **F-test** which in this case is approximately 0, using 4 numerator and 120 denominator degrees of freedom. This p-value is quite small, and so we conclude that the treatment effect is significantly different from zero. The length of time that male fruit flies live depends on the type and number of females that the male is living with. Of course we can't really trust the p-values yet. We need to assess the fit of the model first and we tackle that next. ◇

Now that we know how to compute the model and determine significance, we need to take a careful look at the four conditions required by the ANOVA model. We will first take each of these conditions, one-by-one, and discuss how to evaluate whether they are met or not. We then go on to give some suggestions about transforming the data in some cases in which the conditions are not met.

Remember that the four conditions on the ANOVA model are that the error terms have a Normal distribution, each group of error terms has the same variance, the error terms have mean zero, and the error terms are independent of each other. To check each of these conditions we compute the residuals as our estimates of the error terms and apply the condition checks to them.

Normal Error Distribution?

We begin by checking the Normality condition. In this case we suggest using a normal probability plot of the residuals (as discussed in the regression setting in Section 1.3). We illustrate below with a continuation of our fruit flies data set.

Example 5.4: *Fruit flies (continued)*

In Examples 5.1 and 5.2 we discussed fitting an ANOVA model to the fruit fly longevity data. But before we can proceed to inference with this model, we need to check the conditions required by ANOVA procedures. The first one that we tackle is the normality of the errors. The graph shown below is the Normal probability plot of the residuals from fitting the model to the fruit fly data. The points follow the line quite closely so we are comfortable with the idea that the error terms are, at least approximately, Normally distributed. ◇

5.2. ASSESSING AND USING THE MODEL

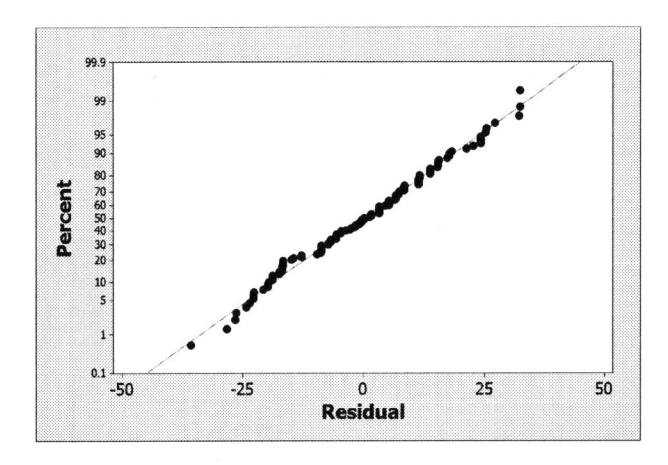

Figure 5.3: Normal Probability Plot of residuals for fruit flies

Equal Group Variances?

Next we consider the equal variance condition. To evaluate whether the standard deviations are the same across groups we have two choices: (a) Plot the residuals versus the fitted values and compare the distributions. Are the spreads roughly the same? (b) Compare the largest and smallest group standard deviations by computing the ratio S_{max}/S_{min}. Ideally this value would be 1 (indicating that all standard deviations are identical). Of course, it is unlikely that we will ever meet an ideal data set where samples in each group give exactly the same standard deviation. So we are prepared to accept values somewhat higher than 1 as being consistent with the idea that the *populations* have the same variance. While we do not like "rules of thumb" we suggest that a value of 2 or less is perfectly acceptable in this case. And, as the number of groups gets larger and the sample sizes in those groups gets smaller, we may even be willing to accept a ratio somewhat larger than 2. This is particularly true when the sample sizes across groups are equal.[2]

Equal variance condition

There are two ways to check the equal variance condition:

- Residuals versus fits plots: Check that the spreads are roughly the same

- Compute S_{max}/S_{min}: Assuming that comparative dotplots show no anomalies, is this value less than or equal to 2?

---

[2]A third option, Levene's test, is presented in Chapter 7.

Example 5.5: *Fruit flies (continued)*

We begin to evaluate the equal variances condition by plotting the residuals versus the fitted values. We see from the graph in Figure 5.4 that all 5 groups (note that two are almost overlapped on the right-hand side) have very similar spreads.

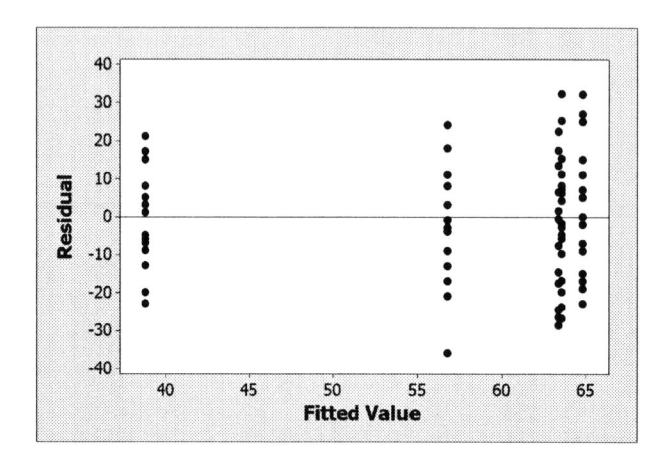

Figure 5.4: Residuals versus fits for fruit flies

The second way to evaluate this condition is to compute the standard deviations for each of the five groups and then compute the ratio of the largest to the smallest. In this case the standard deviations are

```
Variable    Treatment      StDev
Longevity   none           16.45
            pregnant - 1   15.65
            pregnant - 8   14.54
            virgin   - 1   14.93
            virgin   - 8   12.10
```

The largest of these is 16.45 and the smallest is 12.10. This means that the ratio is $16.45/12.10 = 1.36$, which is just a little bigger than 1 and certainly of a size that allows us to feel comfortable with the equal variance condition. ◇

Independent Errors with Mean Zero?

Finally, we deal with both the independence of the errors, and the fact that they should have mean zero, together. Both of these conditions will be assessed by taking a close look at how the data were collected.

5.2. ASSESSING AND USING THE MODEL

As with regression models, we will typically rely on the context of the data set to determine if the residuals are independent. This means that we need to think about how the data were collected to determine whether it is reasonable to believe that the error terms are independent or not. For example, if the data are a random sample, it is likely that the observations are independent of each other and therefore, the errors will also be independent of each other.

Unfortunately, the procedure we use for estimating the parameters based on the mean for each group will guarantee that the mean of all the residuals for each group (and overall) will be zero. This means that all data sets will suggest that the error terms have mean zero. In fact, we note that having a grand mean term in the model makes it relatively easy to insist that the average residual is zero. If the mean residual happened to be something different, say 0.6, we could always add that amount to the estimate of the grand mean and the new residuals would have mean zero. So once again, we rely on how the data were collected to assure ourselves that there was no bias present which would cause the observations to be representative of a population with errors which did not have mean zero.

Example 5.6: *Fruit flies (continued)*

The fruit flies were randomly allocated to one of 5 treatment groups. Because of the randomization, we feel comfortable assuming that the lifetimes of the fruit flies were independent of one another. Therefore, the condition of independent errors seems to be met. Also, because of this random allocation, we believe that there has been no bias introduced, so the mean of the errors should be zero. ⋄

Once we have determined that all of the conditions are met, then we can proceed to the inference stage of this model.

> **Checking conditions**
>
> The error term conditions should be checked as follows:
>
> - Zero mean: Always holds for residuals based on estimated group means; consider how the data were collected.
>
> - Same standard deviation condition:
>
>   - Plot the residuals versus the fitted values; is the spread similar across groups?
>   - Compute the ratio of S_{max}/S_{min}; is it 2 or less?
>
> - Normal distribution: Use a normal probability plot or normal quantile plot of the residuals; does it show a reasonably straight line?
>
> - Independence: Consider how the data were collected.

Transformations

Sometimes, one or more of the conditions for the ANOVA model are not met. In those cases, if the conditions not met involve either the normality of the errors or the equal standard deviation of the errors, we might consider a transformation as we have done in previous chapters. The next example illustrates the process.

Example 5.7: *Diamonds*

Diamonds have several different characteristics that people consider before buying them. Most think, first and foremost, about the number of carats in a particular diamond, and probably also the price. But there are other attributes that a more discerning buyer will look for, such as cut, color, and clarity. The file **Diamonds2** contains several variables measured on 307 randomly selected diamonds. (This is a subset of the 351 diamonds listed in the data set **Diamonds** used in Chapter 3. This set contains all of those diamonds with color D, E, F, and G — those colors for which there are many observations.) Among the measurements made on these diamonds are the color of the diamond and the number of carats in that particular diamond. A prospective buyer who is interested in diamonds with more carats might want to know if a particular color of diamond is associated with more or fewer carats.

CHOOSE

The model that we start with is the following:

$$\text{Number of carats} = \text{Grand Mean} + \text{Color Effect} + \text{Error Term}$$
$$Y = \mu + \alpha_k + \epsilon$$

5.2. ASSESSING AND USING THE MODEL

In this case there are four colors, so $K = 4$.

FIT

We use the following information to compute the grand mean and the treatment effects.

```
Variable  Color    N    Mean    StDev
Carat     D       52   0.8225  0.3916
          E       82   0.7748  0.2867
          F       87   1.0569  0.5945
          G       86   1.1685  0.5028

   Overall       307   0.9731  0.4939
```

First, we note that not all group sizes are equal in this data set, as they were in the fruitfly data set. But this method does not require equal sample sizes so we can proceed. The estimate for the grand mean will just be the mean of all observations. Here that is 0.9731. To compute the estimates for the treatment effects, we subtract the overall mean from each of the treatment means.

$\hat{\alpha}_1$ = -0.1506 (D)
$\hat{\alpha}_2$ = -0.1983 (E)
$\hat{\alpha}_3$ = 0.0838 (F)
$\hat{\alpha}_4$ = 0.1954 (G)

ASSESS

We begin by checking the four conditions for the errors of the ANOVA model: they add to zero, they are independent of each other, and they all come from a normal distribution with the same standard deviation.

We cannot directly check that the error terms add to zero since our method of fitting the model guarantees that the residuals will add to zero. And we cannot directly check that the error terms are independent of each other. What we can do is go back to how the data were collected and see if there is reason to doubt either of these conditions. Since these diamonds were randomly selected, we feel reasonably comfortable believing that these two conditions are met.

The next condition is that the amount of variability is constant from one group to the next. For this data set, we might accept this condition, though we also might not. The standard deviations for the four groups are listed in the output given above. From both Figure 5.5, which shows a graph of the residuals, and the fact that the ratio of the maximum standard deviation to the minimum standard deviation is greater than two, we could conclude that our data do not meet this condition. But since the ratio is quite close to two and the group with the smallest sample size did not have

216   CHAPTER 5. ANALYSIS OF VARIANCE

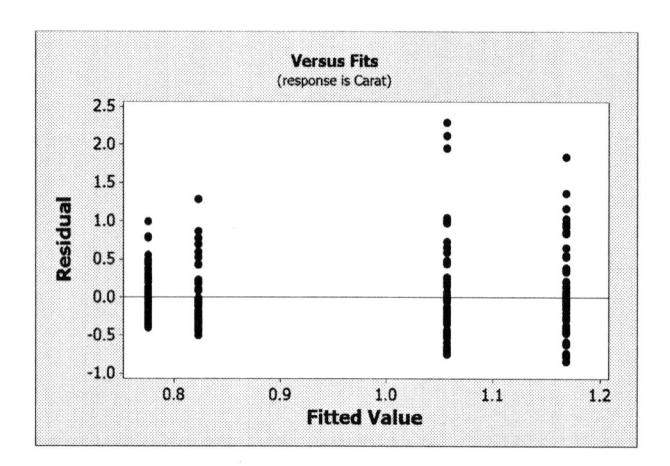

Figure 5.5: Plot of residuals versus fits

an extreme standard deviation, we might also be willing to proceed with caution.

$\frac{S_{max}}{S_{min}} = \frac{0.5945}{0.2867} = 2.07$

The last condition is that the error terms are normally distributed. As seen in Figure 5.6(a), this condition is clearly not met. So because of the fact that this condition is not met and we were on the fence for the last condition, our only conclusion can be that ANOVA is not an appropriate model for the number of carats in these diamonds based on color.

(a) Normal probability plot of residuals   (b) Dotplot of diamond residuals

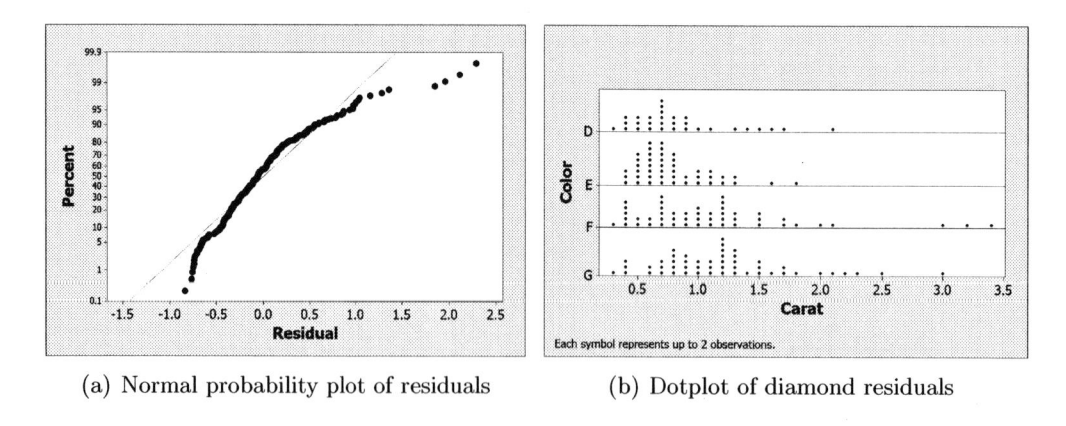

Figure 5.6: Residuals from the diamond data

CHOOSE (again)

Since the model above did not work, but we are still interested in answering the question of relationship between color and carats, we take a closer look at the original data. Figure 5.6(b) gives a dotplot of the numbers of carats for the four different color diamonds. Notice that for all four colors,

5.2. ASSESSING AND USING THE MODEL

the distribution of carats seems to be right skewed. This suggests that a natural log transformation might be appropriate here. The new model, then, is

$$\log(\text{Number of carats}) = \text{Grand Mean} + \text{Color Effect} + \text{Error Term}$$
$$\log(Y) = \mu + \alpha_k + e$$

ASSESS (again)

Taking the natural log of the number of carats does not affect the first two conditions, but may help us with the second two conditions. In fact, the ratio of the maximum standard deviation to the minimum standard deviation is now 1.56, well below our cutoff value of 2. And the normal probability plot of the residuals looks much better as well. Both the normal probability and residual versus fits plots are given in Figure 5.7. This model does now satisfy all of the conditions necessary for analysis.

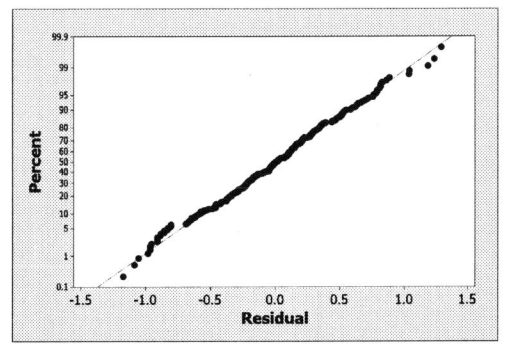

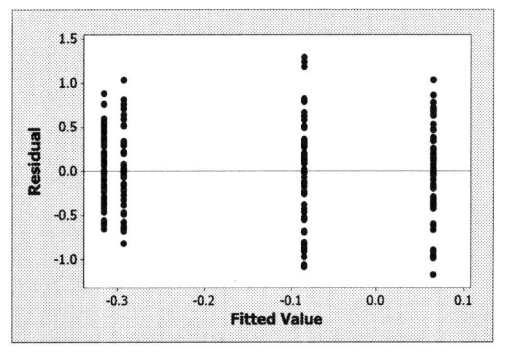

(a) Normal probability plot of residuals   (b) Scatterplot of residuals versus fits

Figure 5.7: Plots to assess the fit of the ANOVA model using log(Carats) as the response variable

Now that we have a model where the conditions are satisfied, we move on to inference to see if, in fact, there is a difference among the means of the natural log of the number of carats based on the color of the diamonds. So what is the conclusion? Do the different colors have a different number of log(carats)? Software gives us the following ANOVA table:

```
One-way ANOVA: log(carat) versus Color

Source   DF      SS      MS      F      P
Color     3   7.618   2.539  12.74  0.000
Error   303  60.382   0.199
Total   306  68.000
```

The p-value given in the last column is near zero, so we reject the null hypothesis here. There is a color effect for the number of log(carats).   ◇

5.3 Scope of Inference

The final step in the ANOVA procedure is to come to a scientific conclusion. The last column in the ANOVA table gives the p-value for the F-test, and we use this to decide whether we have enough evidence to conclude that at least one group mean is different from another group mean. But how far can we stretch our conclusion? Can we generalize to a group larger than just our sample? We usually hope so, but this depends on how the data were collected.

At this point we have seen two different examples: the fruit flies and the diamonds. Notice that the data for these two examples were collected in different ways. The fruit fly longevities were responses from an experiment. The researchers randomly allocated the fruit flies to different treatment groups and looked for the effect on their lifetimes. The diamond data, however, were the result of an observational study. Diamonds of different colors were randomly selected and the number of carats measured. If we find significant differences between groups in a data set that comes from a randomized experiment, because the treatments were randomly assigned to the experimental units, we can typically infer cause and effect. If, however, we have data from an observational study, what we can conclude depends on how the data were collected. If the observational units (in Example 5.7 these are the diamonds) are randomly chosen from the various populations, then significant differences found in the ANOVA F-test can be extended to the populations. The diamonds in our example were, in fact, randomly chosen so we feel comfortable concluding that, since we found significant differences among the natural log of the carats in our samples, there are differences between the carats among these four colors of diamonds in general.

Why Randomization Matters

All of our examples in this chapter have involved randomization to some extent. Why is that? In every data set that we analyze, we recognize that there will be variation present. Some of that variation will be between the groups and is, in fact, the signal that we are hoping to detect. However, there are other sources of variability that occur and can make our job of analyzing the data more difficult. By using randomization whenever possible, we hope to control those other sources of variability so that the resulting data are easier to analyze.

For example, think about the diamond data that we examined earlier. In this case the diamonds were selected at random from each of the colors of diamonds available. What if we had, instead, selected all of the diamonds available from one particular wholesaler? If we found that there was a difference in mean carats between the different colors, we might wonder whether that was true of all diamonds, or if that particular wholesaler had a fondness for a particular color diamond and therefore had larger diamonds of that color than the others. By randomizing the diamonds that we choose, we hope to eliminate selection bias (which in this case would be based on the wholesaler's

5.3. SCOPE OF INFERENCE

preference, but could just as easily be subconscious).

In experiments, randomly allocating the experimental units to treatments performs much the same function. It makes sure that there are no biases in the allocation that would result in a treatment seeming to be better (or worse) simply because of the way that the experimental units were divided up among the treatments. In the fruit fly experiment, if the researchers had mistakenly put all of the smallest fruit flies in one treatment group and then that group had a longer (or shorter) mean lifespan than the other groups, we wouldn't know if it was due to the treatment or just the fact that they were smaller than other fruit flies.

We also hope that by using randomization the variability between experimental units (or observational units) will behave in a chance-like way. This is what allows us to determine whether there is a *significant* difference between groups. We know that there will always be some difference. And if the variability acts in a chance-like way, we know how much variability to expect even when there are not real differences between the populations. If we see variability over and above the amount we would expect by chance, and we are fairly certain that the variability is driven by a chance-like mechanism, then we can say that there is a significant difference between populations.

In essence, by using randomization we are trying to accomplish two things: Reduce the risk of bias, and make the variability in experimental units behave in a chance-like way. We will discuss these ideas in much greater detail in Chapter 8.

Inference about Cause

The main example that we have been following throughout this chapter has been the experiment involving fruit flies. As we think about what kinds of generalizations we can make from these data, we emphasize that they come from an experiment. That is, the researchers randomly allocated the fruit flies to one of the 5 treatment groups, treated them (subjected them to one of 5 sets of living conditions), and then measured the response which in this case was lifetime. Because the fruit flies were randomly allocated to the groups, and we applied the treatments to the groups, then any differences we find that are large enough to be statistically significant, are likely due to the treatments themselves. In other words, since we actively applied the treatments, and we used randomization to equalize all lurking variables, we can feel confident making a conclusion of causality.

Inference about Populations

Now we know what kinds of conclusions we can make from data that come from a well-designed experiment. What can we do with data that come from an observational study when we have several populations that we wish to compare? Again, we have to pay attention to how the data were collected. Do the data constitute random samples from those populations? If so, and if the conditions for ANOVA are met, then we can generalize our results from the samples to the populations in question.

CHAPTER 5. ANALYSIS OF VARIANCE

Example 5.8: *Cancer survivability*

In the 1970's doctors wondered if giving terminal cancer patients a supplement of ascorbate would prolong their lives. They designed an experiment to compare cancer patients who received ascorbate to cancer patients who did not receive the supplement. The result of that experiment was that, in fact, ascorbate did seem to prolong the lives of these patients. But then a second question arose. Was the effect of the ascorbate different when different organs were affected by the cancer? The researchers took a second look at the data. This time they concentrated only on those patients who received the ascorbate and divided the data up by which organ was affected by the cancer. They had 5 different organs represented among the patients (all of whom only had one organ affected): Stomach, bronchus, colon, ovary, and breast. In this case, since the patients were not randomly assigned to which type of cancer they had, but were instead a random sample of those who had such cancers, we are dealing with an observational study. So let's analyze these data (found in the file **CancerSurvival**).

CHOOSE

The model that we start with is

$$\text{Survival time} = \text{Grand Mean} + \text{Organ Effect} + \text{Error Term}$$
$$Y = \mu + \alpha_k + e$$

In this example we have five groups (stomach, bronchus, colon, ovary, and breast) so $K = 5$.

FIT

A table of summary statistics for survival time broken down by each type of organ contains estimates for the mean in each group. The overall data give the estimate for the grand mean, in this case 558.6 days.

```
Variable  Organ      N    Mean    StDev
Survival  Breast     11   1396    1239
          Bronchus   17   211.6   209.9
          Colon      17   457     427
          Ovary      6    884     1099
          Stomach    13   286.0   346.3

          Overall    64   558.6   776.5
```

We can then compare each group mean to the grand mean to get the following estimates:

5.3. SCOPE OF INFERENCE

$$\hat{\alpha}_1 = 837.4 \quad \text{(Breast)}$$
$$\hat{\alpha}_2 = -347.0 \quad \text{(Bronchus)}$$
$$\hat{\alpha}_3 = -101.6 \quad \text{(Colon)}$$
$$\hat{\alpha}_4 = 325.4 \quad \text{(Ovary)}$$
$$\hat{\alpha}_5 = -272.6 \quad \text{(Stomach)}$$

ASSESS

Next we need to check the conditions necessary for the ANOVA model. In this case, we are confident that the data were collected in such a way that the error terms are independent and have mean zero. So we move on to checking the constant variance and normality of the error terms. The graphs in Figure 5.8 show both the residuals versus fits and the normal probability plot of the residuals. Both graphs have issues. In fact, we could have predicted that Figure 5.8(a) would have a problem since we have already seen that the maximum standard deviation is 1239 days and the minimum is 209.9. The ratio between these two is 5.9, which is far greater than 2.

(a) Residuals versus Fits   (b) Normal probability plot of Residuals

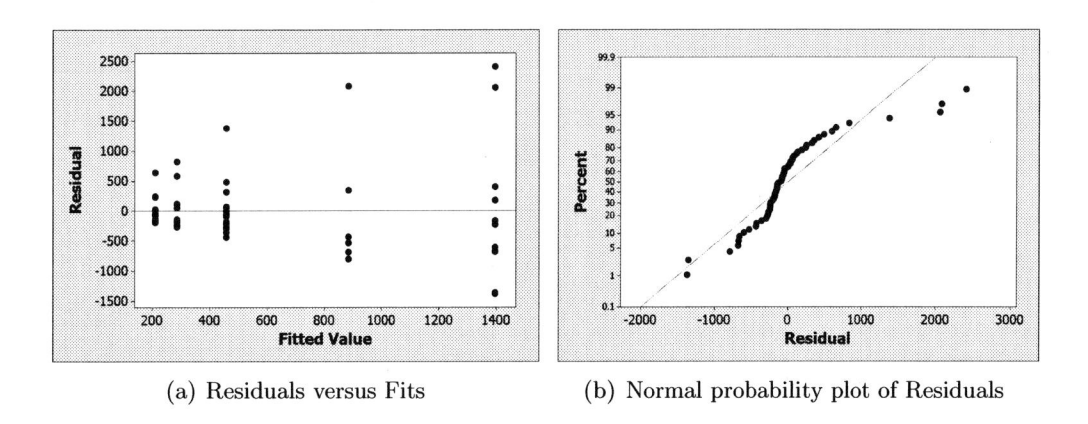

Figure 5.8: Residual plot and normal probability plot to assess conditions

But we have seen a case like this before where a transformation worked to solve these issues. We will do that again here by taking the natural log of the survival times.

CHOOSE (again)

Our new model is:

$$\log(\text{Survival time}) = \text{Grand Mean} + \text{Organ Effect} + \text{Error Term}$$
$$Y = \mu + \alpha_k + e$$

FIT (again)

The new group statistics are:

```
Variable       Organ      N   Mean   StDev
log(survival)  Breast     11  6.559  1.648
               Bronchus   17  4.953  0.953
               Colon      17  5.749  0.997
               Ovary       6  6.151  1.257
               Stomach    13  4.968  1.250

               Overall    64  5.556  1.314
```

Already this looks better. We see that the maximum standard deviation is 1.648 and the minimum is 0.953 which gives us a ratio of 1.73, which is acceptable.

ASSESS (again)

Since the issue before was with the constant variance and the normality of the errors, we go right to the graphs to assess these conditions. Figure 5.9 gives both plots and this time we do not have concerns either about the constant variance or the normality.

(a) Residuals versus Fits

(b) Normal probability plot of residuals

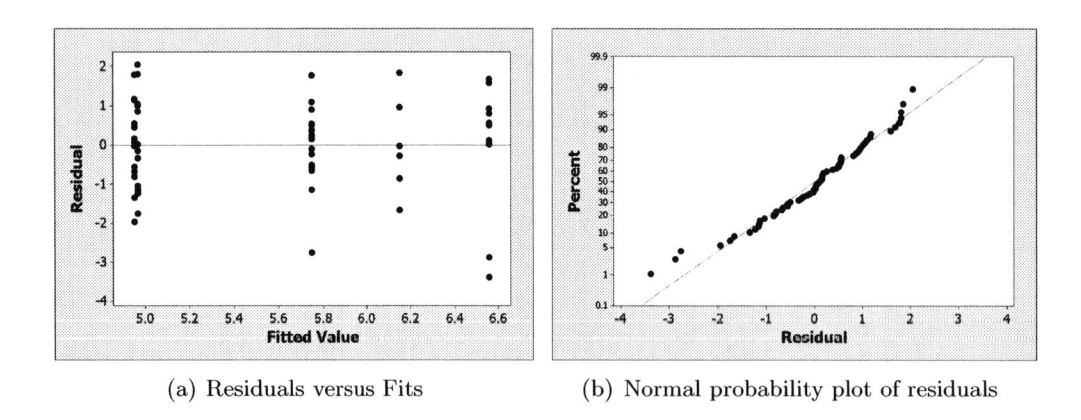

Figure 5.9: Residual plot and normal probability plot to assess conditions on transformed data

5.3. SCOPE OF INFERENCE

The next step is to have the computer fit the model and report the ANOVA table.

```
Source  DF      SS     MS     F      P
Organ    4   24.49   6.12  4.29  0.004
Error   59   84.27   1.43
Total   63  108.76

S = 1.195   R-Sq = 22.52%   R-Sq(adj) = 17.26%
```

With a p-value of 0.004, we find that we have enough evidence to conclude that the natural log of the survival time in days is different for people with cancer of one organ in comparison to people with cancer in another organ. At least one such difference exists.

USE

Now we come to the question of what generalization is appropriate. Since we can think of the people involved in this study as random samples of people who had a diagnosis of terminal cancer involving exactly one of these various organs, we can generalize to the populations of people with terminal cancer involving these organs. The original study found that ascorbate prolonged the lives of people with these types of terminal cancers. This analysis shows that the amount of increase in survival time depends on which organ is involved in the cancer. ◇

What You See Is All There Is

We do, of course, have to be careful how we extend our conclusions to a broader group of potential observations. The example that we have been working most carefully with throughout this chapter is an example of an experiment. In this case we feel comfortable suggesting that our conclusions are applicable to all fruit flies similar to those in the study. We also saw, in the cancer example, that we can often apply the methods of ANOVA to an observational study and make generalizations to the broader population, even though we cannot make causal inferences since the condition of organ is not randomly allocated. We caution you, however, that it is not always possible to make either causal inference or generalizations to a broader population. And we illustrate this with the following example.

Example 5.9: *Hawks*

Can you tell hawk species apart just from their tail lengths? Students and faculty at Cornell College in Mount Vernon, Iowa, collected the data over many years at the hawk blind at Lake MacBride near Iowa City, Iowa. The data set that we are analyzing here is a subset of the original data set, using only those species for which there were more than 10 observations. Data were collected on random samples of three different species of hawks: Red-tailed, Sharp-shinned, and Cooper's hawks. In this example we will concentrate on the tail lengths of these birds and how they might vary from one species to the next.

CHOOSE

The model that we start with is

$$\text{Tail Length} = \text{Grand Mean} + \text{Species Effect} + \text{Error Term}$$
$$Y = \mu + \alpha_k + e$$

In this example we have three groups (Cooper's Hawks, Red-tailed Hawks, and Sharp-shinned Hawks) so $K = 3$.

FIT

A table of summary statistics for tail length broken down by each type of hawk contains estimates for the mean in each group. The overall data give the estimate for the grand mean, in this case 198.83 centimeters.

```
Variable  Species           N    Mean   StDev
Tail      Cooper's         70  200.91   17.88
          Red-tailed      577  222.15   14.51
          Sharp-shinned   261  146.72   15.68

          Overall         908  198.83   36.82
```

Comparing each group mean to the grand mean we get the estimates of the treatment effects
$\hat{\alpha}_1 = 2.08$ (Cooper's)
$\hat{\alpha}_2 = 23.32$ (Red-Tailed)
$\hat{\alpha}_3 = -52.11$ (Sharp-shinned)

ASSESS

First we need to check our conditions.

The independence condition says, in effect, that the value of one error term is unrelated to the others. For the **Hawks** data this is almost surely the case, because the tail length of any one randomly selected bird probably does not depend on the tail lengths of any of the other birds. We also believe from the way the sample was collected that we have not introduced any bias into the data set, so the error terms should have mean 0.

5.3. SCOPE OF INFERENCE

Is it reasonable to assume that the variability of tail lengths is the same within each of the types of hawks? Figure 5.10 shows a plot of residuals versus the fitted values (in this case the three group means). The amount of variability is similar from one group to the next. Just for completeness, we find ratio of SDmax/SDmin.

$$\frac{S_{max}}{S_{min}} = \frac{17.90}{14.51} = 1.23$$

This value is less than 2 so an equal variance condition appears to be reasonable.

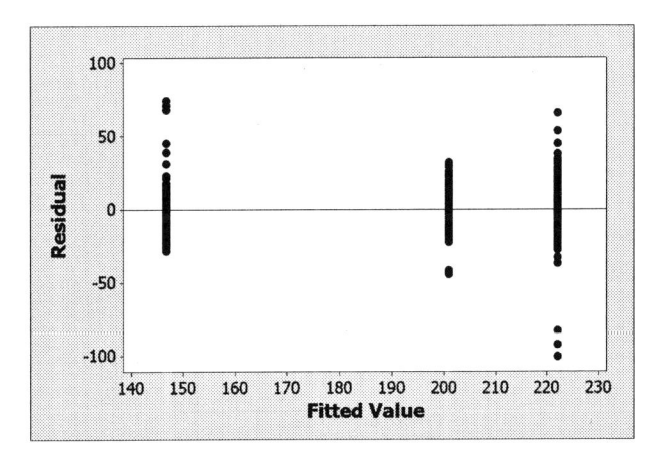

Figure 5.10: Residual plot for ANOVA

Finally, to check normality we use the normal probability plot in Figure 5.11. The error terms clearly are not normally distributed, so we have a cause for concern for the inference in this model. To investigate what is going on here we graph the tail lengths by species in the dotplot shown in Figure 5.12.

Immediately we see what the problem is. Since birds are randomly chosen from the three populations, we would like to assume that all other variables that might affect tail length have been randomized out. However, when we look at the dotplots in Figure 5.12 we discover that both the Sharp-shinned and Cooper's Hawks appear to have a bimodal distribution. This suggests that we have not randomized all other variables out. While we do not show the graphs here, it turns out that there is a sex effect as well, with females having a longer tail than the males.

So we have concluded that the conditions necessary for the ANOVA model have *not* been met. What can we do? In this case, all we can do is use descriptive statistics in our analysis. We saw in Figure 5.12 that there is a difference in the tail length between the birds in our sample of these different species. Specifically, the Sharp-shinned Hawks in our sample had shorter tail lengths. Unfortunately this is as far as we can take the analysis. We cannot conclude that these differences

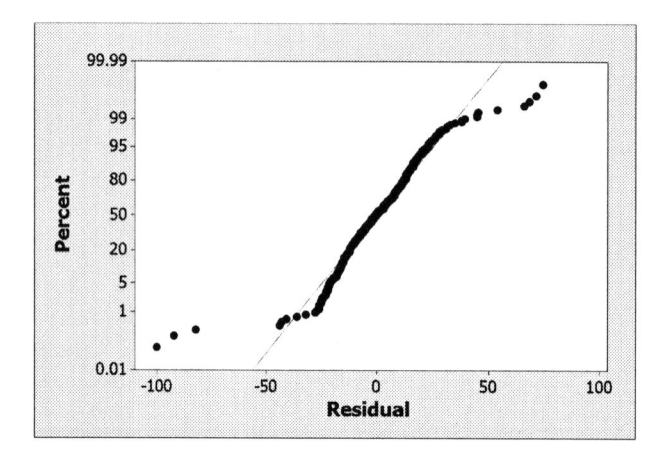

Figure 5.11: Normal probability plot of residuals

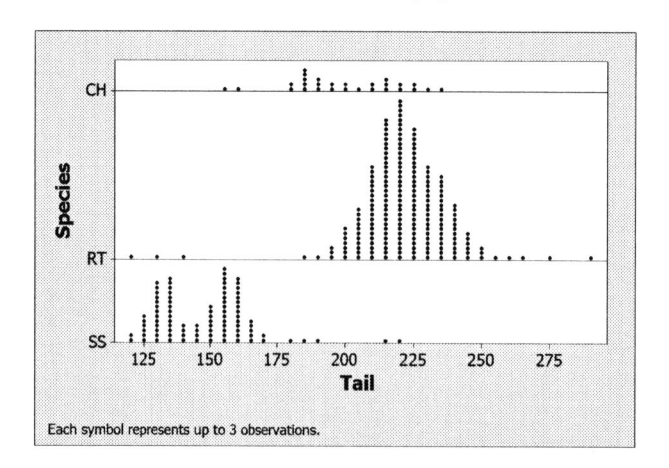

Figure 5.12: Dotplot of tail lengths by species

will also be found in the general population of these three species of hawks. ◇

5.4 Fisher's Least Significant Difference

Up to this point we have concerned ourselves with simply determining whether any difference between the means of several groups exists. While this is certainly an important point in our analysis, often it does not completely answer the research question at hand. Once we have determined that the population means are, in fact, different, it is natural to want to know something about those differences. There will even be some times when, despite the fact that the F-test is not significant, we will still want to test for some specific differences. How we test for those differences will depend on when we identified the differences of interest and whether they were specifically part of the original research question or not.

5.4. FISHER'S LEAST SIGNIFICANT DIFFERENCE

The simplest case to start with is when we want to compare all or some of the groups to each other, two at a time. A *comparison* is a difference of a pair of group means. In a given study, we might be interested in one comparison (we are only interested in two of the groups), several comparisons, or all possible comparisons. When we are interested in more than one comparison, we refer to the setting as *multiple comparisons*. Of course it may be true that we will want to compare more than two group means at one time. These situations are discussed fully in Chapter 7.

Comparisons

We start our discussion by returning to the fruit fly data. Let's consider just one specific comparison. One question that the researchers might have asked is whether the lifetimes of those living with 8 virgins is different from those living with no females. In this case the hypotheses are:

$H_0 : \mu_{8v} = \mu_{\text{none}}$
$H_a : \mu_{8v} \neq \mu_{\text{none}}$

Alternatively, the null hypothesis can be written as $\mu_{8v} - \mu_{\text{none}} = 0$. We estimate this difference in parameters with the difference in sample means: $\bar{y}_{8v} - \bar{y}_{\text{none}}$. Of course this looks like the beginning of a typical two-sample t-test and it is tempting to continue down that path. And, eventually, we will, though it will be modified somewhat. We first have to consider error rates.

Error Rates

Before simply continuing on with a t-test, we need to address why we used ANOVA in the first place and didn't just compute the usual confidence intervals or two-sample t-tests for each pair of differences from the very beginning. First, we note here that we can interchangeably discuss tests for a difference in means between two groups with confidence intervals for the difference. Even though the ANOVA table conducts a hypothesis test, in the setting of multiple comparisons after ANOVA the approach using intervals is more common, so that is how we will proceed.

> **Equivalence of Confidence Intervals and Tests for a Difference in Two Means**
>
> A pair of means can be considered significantly different at a 5% level $\iff$ a 95% confidence interval for a difference in the two means fails to include zero.

Recall that when we compute a 95% confidence interval for the difference between two means, what we are doing is using a method that in 95% of cases will capture the true difference. If we find that 0 is not in our interval, then, knowing that this method produces "good" results in 95% of samples, we feel justified in concluding that in this particular case the means are not the same. Obviously we can modify the relationship to account for different significance or confidence levels.

However, if we were to compute many different 95% intervals (for example, comparing all pairs of means for a factor with 7 levels requires 21 intervals) then we might naturally expect to see at least one interval showing a significant difference just by random chance, even if there were no real differences among any of the groups we were comparing. Remember that in the language of significance testing, concluding that a pair of groups have different means when actually there is no difference produces a Type I error. Setting a 5% significance level limits the chance of a Type I error to 5% for each individual test, but we are repeating that 5% chance over multiple tests. Thus the overall error rate, the chance of finding a significant difference for at least one pair when actually the are no differences among the groups, can be much larger than 5%. While each interval or test has its own individual error rate, the error rate of the intervals or tests taken together is called the *family-wise error rate*. It is this family-wise error rate that gives us pause when considering computing many confidence intervals to investigate differences among groups.

Error rates

There are two different types of error rates to consider when computing multiple comparisons: the **individual error rate** and the **family-wise error rate**.

- **Individual error rate**: The likelihood of rejecting a true null hypothesis when considering just one interval or test.

- **Family-wise error rate** The likelihood of rejecting at least one of the null hypotheses in a series of comparisons, when, in fact, all K means are equal.

As we have seen earlier in this chapter, ANOVA is a way of comparing several different means as a group to see if there are differences among them. It does so by looking at the amount of variability between means and comparing that variability to the amount of variability between observations. This procedure, by definition, does only one test, and therefore only has to deal with an individual error rate. The problem with *multiplicity* comes when we have decided that there is likely at least one difference and we want to compute several confidence intervals to find the difference(s). Now we have to be aware of the issue of the family-wise error rate.

There are several different ways that statisticians have developed to control the family-wise error rate, each with its own pros and cons. In this chapter we introduce you to one, Fisher's Least Significant Difference (LSD). In Chapter 7 we will take a closer look at this issue and introduce you to two more methods: Bonferroni method and Tukey's Honestly Significant Difference (HSD).

Fisher's LSD

A common approach to dealing with multiple comparisons is to produce confidence intervals for a difference in two means of the form

$$\bar{y}_i - \bar{y}_j \pm \text{ margin of error}$$

where $\bar{y}_i$ and $\bar{y}_j$ are the sample means from the two groups that we are comparing. The various methods differ in the computation of the margin of error. For example, we can generalize the usual two-sample t-interval for a difference in means to the ANOVA setting using

$$\bar{y}_i - \bar{y}_j \pm t^* \sqrt{MSE\left(\frac{1}{n_i} + \frac{1}{n_j}\right)}$$

where MSE is the mean square error from the ANOVA table. Under the conditions of the ANOVA model, the variance is the same for each group and the MSE is a better estimate of the common variance than the pooled variance from the two-sample, pooled t-test. Under this approach, we use the degrees of freedom for the error term $(n - K)$ when determining a value for t^*.

But how can we adjust for the problem of multiplicity? Different methods focus either on the individual error rate, or the family-wise error rate. When we have already found a significant F-ratio in the ANOVA table, we can apply **Fisher's Least Significant Difference (LSD)** which focuses only on the individual error rate. This method is one of the most liberal, producing intervals that are more likely to identify differences (either real or false) than other methods. Fisher's procedure uses the same form of the confidence interval as the regular two-sample t-interval except that it uses the MSE in the standard error for the difference of the means. It uses a t^* based on 95% confidence, regardless of the number of comparisons. This method has a larger family-wise error rate than other methods, but, in its favor, it has a smaller chance of missing actual differences that exist. Since we have already determined that there are differences (the ANOVA F-test was significant), we are more comfortable in stating that we have found a real difference.

> **Fisher's Least Significant Difference**
>
> To compare all possible group means using Fisher's LSD perform the following steps:
>
> a. Perform the ANOVA
>
> b. If the ANOVA F-test is not significant, stop.
>
> c. If the ANOVA F-test is significant, then compute the pairwise comparisons using the confidence interval formula:
>
> $$\bar{y}_i - \bar{y}_j \pm t^* \sqrt{MSE\left(\frac{1}{n_i} + \frac{1}{n_j}\right)}$$

Why do we call it LSD? The *least significant difference* aspect of Fisher's procedure arises from a convenient way to apply it. Recall that we conclude that two sample means are significantly different when the confidence interval for the difference does not include zero. This occurs precisely when the margin of error that we add and subtract is less than the magnitude (ignoring sign) of the difference in the groups means. Thus the margin of error represents the minimum point at which a difference can be viewed as significant. If we let

$$LSD = t^* \sqrt{MSE\left(\frac{1}{n_i} + \frac{1}{n_j}\right)}$$

we can conclude that a pair of means are significantly different if and only if $|\bar{y}_i - \bar{y}_j| > LSD$. This provides a convenient way to quickly identify which pairs of means show a significant difference, particularly when the sample sizes are the same for each group and the same LSD value applies to all comparisons.

Example 5.10: *Fruit flies (continued)*

There are actually several different ways that the researchers might have approached this data set. We started this section by asking if there is a difference between the mean lifetimes of those male fruit flies who had lived with 8 virgin females and those who had not lived with any females. We first need to compute the F-ratio in the ANOVA table to see if we reject the hypothesis that all group means are the same. Using computer software, we get the table listed below:

5.4. FISHER'S LEAST SIGNIFICANT DIFFERENCE

```
One-way ANOVA: Longevity versus Treatment

Source      DF      SS      MS      F       P
Treatment   4       11939   2985    13.61   0.000
Error       120     26314   219
Total       124     38253

S = 14.81   R-Sq = 31.21%   R-Sq(adj) = 28.92%
```

The F-ratio, with 4 and 120 degrees of freedom, is 13.61 and the p-value is approximately 0. We are, therefore, justified in deciding that at least one of the groups has a different mean lifetime than at least one other group. So, according to Fisher's LSD method, we now compute a confidence interval for the difference between the mean lifetime of the male fruit flies living with 8 virgins and the mean lifetime of the male fruit flies living alone.

As noted above, the confidence interval is:

$$\bar{y}_{8v} - \bar{y}_{none} \pm t^* \sqrt{MSE \left(\frac{1}{n_{8v}} + \frac{1}{n_{none}} \right)}$$

From table 5.1 we see that $\bar{y}_{8v} = 38.72$ and $\bar{y}_{none} = 63.56$. The MSE from the ANOVA table is 219 and each group has 25 observations. Since t^* has 120 degrees of freedom (which comes from the MSE), for a 95% confidence interval, $t^* = 1.9799$ so the interval is:

$$(38.72 - 63.56) \pm 1.9799 \sqrt{219 \left(\frac{1}{25} + \frac{1}{25} \right)} = -24.84 \pm 8.287 = (-33.127, -20.553)$$

Since that interval does not contain 0, we conclude that there is a difference between the lifetimes of male fruit flies who live with 8 virgins and male fruit flies who live alone. Since the interval is negative and we subtracted the mean of those who lived alone from the mean of those who lived with 8 virgins, we conclude that it is likely that the male fruit flies who live alone do live longer. ◇

A more typical situation occurs when we find that there is at least one difference among the groups, but we don't have any reason to suspect a particular difference. In that case we would like to look at all possible differences between pairs of groups. Again, if the F-ratio in the ANOVA table is significant, we can apply Fisher's LSD, this time to all pairs of differences.

Example 5.11: *Fruit flies (one last time)*

The researchers in this case may well have had more questions than just whether living with 8 virgins was different from living alone. They did, after all, test three other environments as well: 1

virgin, 8 pregnant, and 1 pregnant. Applying Fisher's LSD to all 10 pairs of means, Minitab gives us the following output:

```
Fisher 95% Individual Confidence Intervals
All Pairwise Comparisons among Levels of Treatment

Simultaneous confidence level = 71.79%


Treatment = none subtracted from:

Treatment       Lower    Center   Upper   -------+---------+---------+---------+--
pregnant - 1    -7.05     1.24     9.53                  (----*---)
pregnant - 8    -8.49    -0.20     8.09                  (---*---)
virgin - 1     -15.09    -6.80     1.49             (----*---)
virgin - 8     -33.13   -24.84   -16.55   (----*---)
                                          -------+---------+---------+---------+--
                                               -20         0        20        40


Treatment = pregnant - 1 subtracted from:

Treatment       Lower    Center   Upper   -------+---------+---------+---------+--
pregnant - 8    -9.73    -1.44     6.85                  (---*---)
virgin - 1     -16.33    -8.04     0.25              (---*---)
virgin - 8     -34.37   -26.08   -17.79   (---*---)
                                          -------+---------+---------+---------+--
                                               -20         0        20        40


Treatment = pregnant - 8 subtracted from:

Treatment       Lower    Center   Upper   -------+---------+---------+---------+--
virgin - 1     -14.89    -6.60     1.69              (---*---)
virgin - 8     -32.93   -24.64   -16.35   (---*---)
                                          -------+---------+---------+---------+--
                                               -20         0        20        40


Treatment = virgin - 1 subtracted from:

Treatment       Lower    Center   Upper   -------+---------+---------+---------+--
```

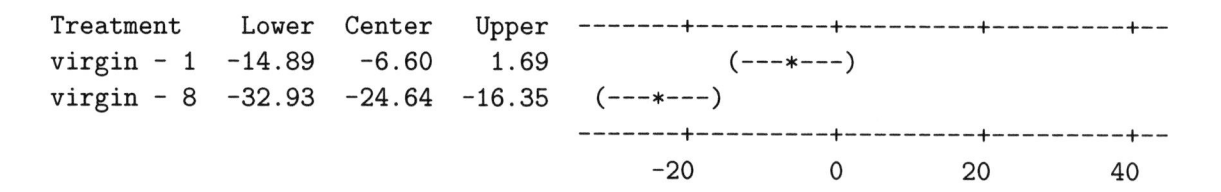

```
virgin - 8  -26.33  -18.04   -9.75         (---*---)
                                   -------+---------+---------+---------+--
                                        -20         0        20        40
```

So what can we conclude? First we look for all those intervals which do not contain 0. From the first grouping, we see that living with no females is only significantly different from living with 8 virgins. From the second grouping, we see that living with 1 pregnant female is significantly different from living with 8 virgins. From the third grouping, we see that living with 8 pregnant females is significantly different from living with 8 virgins. And finally, in the last interval, we see that living with 1 virgin female is significantly different from living with 8 virgins. In other words, we have discovered that living with 8 virgins significantly lowered the lifespan of the male fruit flies, but the life spans of the fruit flies in all of the other treatments cannot be declared to differ from on another. ◇

5.5 Chapter Summary

In this chapter we consider a **one-way analysis of variance (ANOVA)** model for comparing the means of a single quantitative response variable grouped according to a single categorical predictor. The categories of the predictor are indexed by $k = 1, 2, ..., K$ and the model is

$$Y = \mu_k + \epsilon$$

The estimates $\hat{\mu}_k$, obtained by the method of least squares, are the group means $\bar{y}_k$. For each response value there is a corresponding residual equal to the observed value (y) minus the fitted value $(\hat{\mu}_k)$.

The one-way model is often rewritten in terms of an overall mean μ and group effects $\alpha_1, \alpha_2, ..., \alpha_k$:

$$Y = \mu_k + \alpha_k + \epsilon$$

For this version of the model, least squares gives estimates

$$\hat{\mu} = \text{overall mean} = \bar{y}$$

$$\hat{\alpha}_k = \text{group mean} - \text{overall mean} = \bar{y}_k - \bar{y}$$

$$\hat{\epsilon} = \text{observed} - \text{fitted} = y - \bar{y}_k$$

Squaring individual terms and adding gives the corresponding **sums of squares**:

$$SSObs = \text{sum of all } \left(y^2\right) \text{ values}$$

$$SSGrand = \text{sum of all } \left(\hat{\mu}^2\right) \text{ values}$$

$$SSGroups = \text{sum of all } \left(\hat{\alpha}_k^2\right) \text{ values}$$

$$SSE = \text{sum of all } \left(\hat{\epsilon}^2\right) \text{ values}$$

Moreover, the SSs add in the same way as the estimates in the model:

$$Y = \hat{\mu} + \hat{\alpha}_k + \hat{\epsilon}$$

$$SSObs = SSGrand + SSGroups + SSE$$

To focus on variation, we also define

$$SSTotal = SSObs - SSGrand = SSGroups + SSE$$

Degrees of freedom (df) add according to the same pattern as for the SSs.

$$df_{Obs} = df_{Grand} + df_{Groups} + df_{Error}$$

where

$$df_{Obs} = \text{number of units}$$

$$df_{Grand} = 1$$

$$df_{Groups} = \text{number of groups} - 1$$

$$df_{Error} = \text{number of units} - \text{number of groups}$$

$$df_{Total} = df_{Obs} - df_{Grand} = df_{Groups} + df_{Error}$$

This completes the "triple decomposition" of ANOVA: observed values add, sum of squares add, and degrees of freedom add.

The one-way ANOVA model is often used to compare two competing hypotheses: a null hypothesis of no group differences ($H_0 : \alpha_1 = \alpha_2 = \alpha_3 = ... = \alpha_K = 0$) versus the alternative (H_a: At least one $\alpha_k \neq 0$). To test H_0, the error terms must satisfy the same four conditions as for regression models: zero means, equal standard deviations, independence, and normal shape.

If these required conditions are satisfied, then we can use the triple decomposition to test H_0 by comparing variation between groups and within groups. We measure variation using mean squares ($MS = SS/df$):

$$MSGroups = MSBetween = SSGroups/(\text{number of groups} - 1)$$

$$MSE = MSWithin = SSE/(\text{number of units} - \text{number of groups})$$

A large F-ratio, $F = MSGroups/MSE$, is evidence against H_0; the p-value tells the strength of that evidence: the lower the p-value, the stronger the evidence.

5.5. CHAPTER SUMMARY

As always, tests and intervals require that certain conditions be met. **Checking these conditions** is essentially the same for regression and ANOVA models. Two conditions, zero mean and independence, cannot ordinarily be checked numerically or graphically. Instead, you need to think carefully about how the data were collected. To check normality, use a normal plot of the residuals. To check for equal SDs, you have both a graphical method (does a plot of residuals versus fitted values show uniform vertical spread?) and a numerical method (compute group SDs; is S_{max}/S_{min} large, or near 1?).

If either of normality or equal SDs fails, a **transformation** to a new scale (for example, logs) may offer a remedy.

As always, the **scope of your conclusions** depends on how the data were produced. If treatments were randomly assigned to units, then inference about cause is justified. If units come from random sampling, inference about population is justified. If there was no randomization, "what you see is all there is."

Confidence intervals and multiple comparisons A confidence interval for the difference of two group means has the form

$$\bar{y}_i - \bar{y}_j \pm (\text{margin of error})$$

For the usual t-interval, the margin of error is

$$t^* \sqrt{MSE\left(\frac{1}{n_i} + \frac{1}{n_j}\right)}$$

where n_i and n_j are the group sizes and t^* is a percentile from the t-distribution with $df = df_{Error}$.

When you make several comparisons, it is useful to distinguish between the **individual error rate** for a single interval, and the **family-wise error rate**, equal to the probability that at least one interval fails to cover its true value.

Fisher's LSD is one strategy for controlling the family-wise error rate: First use the F-ratio to test for group differences; if not significant stop. If significant, form all pairwise comparisons:

$$\bar{y}_i - \bar{y}_j \pm t^* \sqrt{MSE\left(\frac{1}{n_i} + \frac{1}{n_j}\right)}$$

Declare two group means to be significantly different if and only if 0 is not in the corresponding interval.

5.6 Exercises

Conceptual Exercises

5.1 *Two-sample t-test?* True or False: When there are only two groups, an ANOVA model is equivalent to a 2-sample pooled, t-test.

5.2 *Random selection.* True or False: Randomly selecting units from populations and performing an ANOVA, allows you to generalize from the samples to the populations.

5.3 *No random selection.* True or False: In data sets in which there was no random selection, you can still use ANOVA results to generalize from samples to populations.

5.4 *Independence transformation?* True or False: If the data set does not meet the independence condition for the ANOVA model, a transformation might improve the situation.

Exercises 5.5 to 5.7 are multiple choice. Choose the answer that best fits.

5.5 *Why ANOVA?* The purpose of an ANOVA in the setting of this chapter is to learn about:

   a. The variances of several populations.

   b. The mean of one population.

   c. The means of several populations.

   d. The variance of one population.

5.6 *Not a condition?* Which of the following statements is not a condition of the ANOVA model?

   a. Error terms have the same standard deviation.

   b. Error terms are all positive.

   c. Error terms are independent.

   d. Error terms follow a normal distribution.

5.6. EXERCISES

5.7 *ANOVA plots.* The two best plots to assess the conditions of an ANOVA model are:

a. normal probability plot of residuals and dotplot of residuals.

b. scatterplot of response versus predictor and dotplot residuals.

c. bar chart of explanatory groups and normal probability plot of residuals.

d. normal probability plot of residuals and scatterplot of residual versus fits.

5.8 *Which sum of squares?* Match the measures of variability with the sum of square terms:

Groups Sum of Squares	Variability of observations about grand mean
Error Sum of Squares	Variability of group means about grand mean
Total Sum of Squares	Variability of observations about group means

5.9 *Student survey.* You gather data on the following variables from a sample of 75 undergraduate students on your campus:

- Major

- Sex

- Class year (first year, second year, third year, fourth year, other)

- Political inclination (liberal, moderate, conservative)

- Sleep time last night

- Study time last week

- Body mass index

- Total amount of money spent on textbooks this year

a. Assume that you have a quantitative response variable and that all of the above are possible explanatory variables. Classify each variable as quantitative or categorical. For each categorical variable, assume that it is the explanatory variable in the analysis and determine whether you could use a two-sample t-test or whether you would have to use an ANOVA.

b. State three research questions pertaining to these data for which you could use ANOVA. [Hint: For example, one such question would be to investigate whether men and women differ with regard to average sleep time.]

5.10 *Car ages.* Suppose that you want to compare ages of cars among faculty, students, administrators, and staff at your college or university. You take a random sample of 200 people who have a parking permit for your college or university, and then you ask them how old their primary car is. You ask several friends if it's appropriate to conduct ANOVA on the data, and you obtain the following responses. Indicate how you would respond to each.

  a. "You can't use ANOVA on the data because there are four groups to compare."

  b. "You can't use ANOVA on the data because the response variable is not quantitative."

  c. "You can't use ANOVA on the data because the sample sizes for the four groups will probably be different."

  d. "You can do the calculations for ANOVA on the data, even though the sample sizes for the four groups will probably be different, but you can't generalize the results to the populations of all people with a parking permit at your college/university."

5.11 *Comparing fonts.* Suppose that an instructor wants to investigate whether the font used on an exam affects student performance as measured by final exam score. She uses four different fonts (times, courier, helvetica, comic sans) and randomly assigns her 40 students to one of those four fonts.

  a. Identify the explanatory and response variables in this study.

  b. Is this an observational study or a randomized experiment? Explain how you know.

  c. Even though the subjects in this study were not randomly selected from a population, it's still appropriate to conduct an analysis of variance. Explain why.

5.12 *Comparing fonts (continued).* Refer to exercise 5.11. Determine each of the entries that would appear in the "degrees of freedom" column of the ANOVA table.

5.13 *All false.* Reconsider the previous exercise. Now suppose that the p-value from the ANOVA F-test turns out to be 0.003. All of the following statements are FALSE. Explain why each one is false.

  a. The probability is 0.003 that the four groups have the same mean score.

  b. The data provide very strong evidence that all four fonts produce different mean scores.

  c. The data provide very strong evidence that the comic sans font produces a different mean score than the other fonts.

  d. The data provide very little evidence that at least one of these fonts produces a different mean score than the others.

5.6. EXERCISES

e. The data do not allow for drawing a cause-and-effect conclusion between font and exam score.

f. Conclusions from this analysis can be generalized to the population of all students at the instructor's school.

5.14 *Can this happen?* You conduct an ANOVA to compare 3 groups.

a. Is it possible that all of the residuals in one group are positive? Explain how this could happen, or why it can not happen.

b. Is it possible that all but one of the residuals in one group are positive? Explain how this could happen, or why it can not happen.

c. If you and I are two subjects in the study, and if we are in the same group, and if your score is higher than mine, is it possible that your residual is smaller than mine? Explain how this could happen, or why it can not happen.

d. If you and I are two subjects in the study, and if we are in different groups, and if your score is higher than mine, is it possible that your residual is smaller than mine? Explain how this could happen, or why it can not happen.

5.15 *Schooling and political views.* You want to compare years of schooling among American adults who describe their political viewpoint as liberal, moderate, or conservative. You gather a random sample of American adults in each of these three categories of political viewpoint.

a. State the appropriate null hypothesis, in symbols and in words.

b. What additional information do you need about these three samples in order to conduct ANOVA to determine if there is a statistically significant difference among these three means?

c. What additional information do you need in order to assess whether the conditions for ANOVA are satisfied?

5.16 *Schooling and political views (continued).* Now suppose that the three sample sizes in the previous exercise are 25 for each of the three groups, and also suppose that the standard deviations of years of schooling are very similar in the three groups. Assume that all three populations do, in fact, have the same standard deviation. Suppose that the three sample means turn out to be 11.6, 12.3, and 13.0 years.

a. Without doing any ANOVA calculations, state a value for the standard deviation that would lead you to reject H_0. Explain your answer, as if to a peer who has not taken a statistics course, without resorting to formulas or calculations.

b. Repeat part (a), but state a value for the standard deviation that would lead you to fail to reject H_0.

Guided Exercises

5.17 *Life spans.* The *World Almanac* and *Book of Facts* lists notable people of the past in various occupational categories; it also reports how many years each person lived. Do these sample data provide evidence that notable people in different occupations have different average lifetimes? To investigate this question, we recorded the lifetimes for 973 people in various occupation categories. Consider the following ANOVA output:

```
Source       DF    SS      MS     F      P
Occupation                2749          0.000
Error        968   195149  202
Total        972   206147
```

a. Fill in the three missing values in this ANOVA table. Also show how you calculate them.

b. How many different occupations were considered in this analysis? Explain how you know.

c. Summarize the conclusion from this ANOVA (in context).

5.18 *Meth labs.* Nationally, the abuse of methamphetamine has become a concern, not only because of the effects of drug abuse, but also because of the dangers associated with the labs that produce them. A stratified random sample of a total of 12 counties in Iowa (stratified by size of county — small, medium or large) produced the following ANOVA table relating the number of methamphetamine labs to the size of the county. Use this table to answer the following questions.

One-way ANOVA: meth labs versus type

```
Source  DF   SS      MS    F
type         37.51
Error
Total        70.60
```

a. Fill in the values missing from the table.

b. What does the MS for county type tell you?

c. Find the p-value for the F-test in the table.

d. Describe the hypotheses tested by the F-test in the table, and using the p-value from part (b), give an appropriate conclusion.

5.6. EXERCISES

5.19 *Palatability.* A food company was interested in how texture might affect the palatability of a particular food. They set up an experiment in which they looked at two different aspects of the texture of the food: the concentration of a liquid component (low or high) and the coarseness of the final product (coarse or fine). The experimenters randomly assigned each of 16 groups of 50 people to one of the four treatment combinations. The response variable was a total palatability score for the group. For this analysis you will focus *only* on how the coarseness of the product affected the total palatability score. The data collected resulted in the following ANOVA table. Use this table to answer the following questions.

```
One-way ANOVA: Score versus Coarseness

Source      DF      SS      MS      F
Coarseness
Error               6113
Total               16722
```

a. Fill in the values missing from the table.

b. What does the MS for the coarseness level tell you?

c. Find the p-value for the F-test in the table.

d. Describe the hypotheses tested by the F-test in the table, and using the p-value from part (b), give an appropriate conclusion.

5.20 *Child poverty.* The same data set used in Exercise 5.18 also had information about the child poverty rate in those same Iowa counties. Below is the ANOVA table relating child poverty rate to type of county.

```
One-way ANOVA: child poverty versus type

Source  DF       SS        MS       F      P
type     2  0.000291  0.000145   0.22  0.810
Error    9  0.006065  0.000674
Total   11  0.006356
```

a. What hypotheses are being tested in this ANOVA table? Give the hypotheses both in words and in symbols.

b. Given in Figure 5.13 is the dotplot of the child poverty rates by type of county. Does this dotplot raise any concerns for you with respect to the ANOVA? Explain.

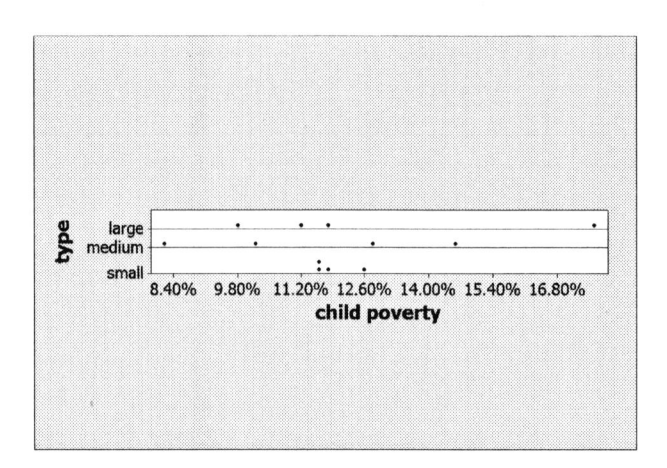

Figure 5.13: Child poverty rates in Iowa by size of county

5.21 *Palatability (continued).* In Exercise 5.19 you analyzed whether the coarseness of the product had an effect on the total palatability score of a food product. In the same experiment, the level of concentration of a liquid component was also varied between a low level and a high level. The following ANOVA table analyzes the effects of the concentration on the total palatability score.

```
One-way ANOVA: SCORE versus LIQUID
Source  DF    SS     MS    F      P
LIQUID   1   1024   1024  0.91  0.355
Error   14  15698   1121
Total   15  16722
```

a. What hypotheses are being tested in this ANOVA table? Give the hypotheses both in words and in symbols.

b. Given in Figure 5.14 is the dotplot of the palatability score for each of the two concentration levels. Does this dotplot raise any concerns for you with respect to the ANOVA? Explain.

5.22 *Fantasy Baseball.* A group of friends who participate in a "fantasy baseball" league became curious about whether some of them take significantly more or less time to make their selections in the "fantasy draft" through which they select players. The following table reports the times (in seconds) that each of the eight friends (identified by their initials) took to make their 24 selections in 2008 (the data are also available in the data file **FantasyBaseball**):

5.6. EXERCISES

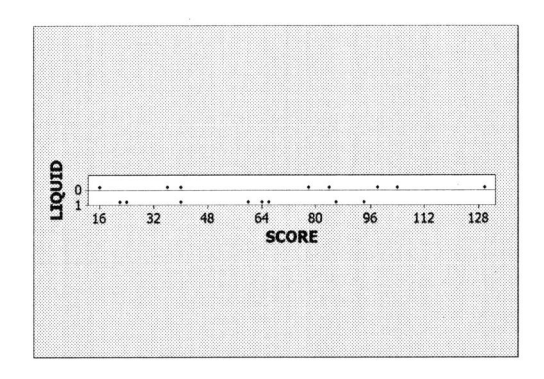

Figure 5.14: Palatability scores by level of concentration of liquid component

Round	DJ	AR	BK	JW	TS	RL	DR	MF
1	42	35	49	104	15	40	26	101
2	84	26	65	101	17	143	43	16
3	21	95	115	53	66	103	113	88
4	99	41	66	123	6	144	16	79
5	25	129	53	144	6	162	113	48
6	89	62	80	247	17	55	369	2
7	53	168	32	210	7	37	184	50
8	174	47	161	164	5	36	138	84
9	105	74	25	135	14	118	102	163
10	99	46	60	66	13	112	21	144
11	30	7	25	399	107	17	55	27
12	91	210	69	219	7	65	62	1
13	11	266	34	436	75	27	108	76
14	93	7	21	235	5	53	23	187
15	20	35	26	244	19	120	94	19
16	108	61	13	133	25	13	90	40
17	95	124	9	68	5	35	95	171
18	43	27	9	230	5	52	72	3
19	123	26	13	105	6	41	32	18
20	75	58	50	103	13	38	57	86
21	18	11	10	40	8	88	20	27
22	40	10	119	39	6	51	46	59
23	33	56	20	244	6	38	13	41
24	100	18	27	91	11	23	31	2

a. Produce boxplots and calculate descriptive statistics to compare the selection times for each participant. Comment on what they reveal. Also identify (by initials) which participant took the longest and which took the shortest time to make their selections.

b. Conduct a one-way ANOVA to assess whether the data provide evidence that averages as far apart as these would be unlikely to occur by chance alone if there really were no differences among the participants in terms of their selection times. For now, assume that all conditions are met. Report the ANOVA table, test statistic, and p-value. Also summarize your conclusion.

c. Use Fisher's LSD procedure to assess which participants' average selection times differ significantly from which others.

5.23 *Fantasy Baseball (continued).*

a. In Exercise 5.22 part (a), you produced boxplots and descriptive statistics to assess whether an ANOVA model was appropriate for the fantasy baseball selection times of the various members of the league. Now produce the normal probability plot of the residuals for the ANOVA model in Exercise 5.22 and comment on the appropriateness of the ANOVA model for these data.

b. Transform the selection times using the natural log. Repeat your analysis of the data and report your findings.

5.24 *Fantasy Baseball (continued).* Continuing with your analysis in Exercise 5.23, use Fisher's LSD to assess which participants' average selection times differ significantly from which others.

5.25 *Fantasy Baseball (continued).* Reconsider the data from Exercise 5.22. Now disregard the participant variable, and focus instead on the round variable. Perform an appropriate ANOVA analysis of whether the data suggest that some rounds of the draft tend to have significantly longer selection times than other rounds. Use a transformation if necessary. Write a paragraph or two describing your analysis and summarizing your conclusions.

5.26 *Fenthion.* Fenthion is a pesticide used against the olive fruit fly in olive groves. It is toxic to humans so it is important that there be no residue left on the fruit or in olive oil that will be consumed. One theory was that if there is residue of the pesticide left in the olive oil, it would dissipate over time. Chemists set out to test that theory by taking a random sample of small amounts of olive oil with fenthion residue and measuring the amount of fenthion in the oil at 3 different times over the year — day 0, day 281 and day 365.

a. Two variables given in the data set **Olives** are *fenthion* and *time*. Which variable is the response variable and which variable is the explanatory variable? Explain.

b. Check the conditions necessary for conducting an ANOVA to analyze the amount of fenthion present in the samples. If the conditions are met, report the results of the analysis.

c. Transform the amount of fenthion using the exponential. Check the conditions necessary for conducting an ANOVA to analyze the exponential of the amount of fenthion present in the samples. If the conditions are met, report the results of the analysis.

5.6. EXERCISES

5.27 *Hawks.* The data set on hawks was used in Example 5.9 to analyze the length of the tail based on species. Other response variables were also measured in **Hawks**. We now consider the the *weight* of the hawks as a response variable.

   a. Create dotplots to compare the values of *weight* between the three species. Do you have any concerns about the use of ANOVA based on the dotplots? Explain.

   b. Compute the standard deviation of the weights for each of the three species groups. Do you have any concerns about the use of ANOVA based on the standard deviations? Explain.

5.28 *Blood pressure.* A person's systolic blood pressure can be a signal of serious issues in their cardiovascular system. Are there differences between average systolic blood pressure based on smoking habits? The data set **Blood1** has the systolic blood pressure and the smoking status of 500 randomly chosen adults.

   a. Perform a 2-sample t-test, using the assumption of equal variances, to determine if there is a significant difference in systolic blood pressure between smokers and non-smokers.

   b. Compute an ANOVA table to test for differences in systolic blood pressure between smokers and non-smokers. What do you conclude? Explain.

   c. Compare your answers to parts (a) and (b). Discuss the similarities and differences between the two methods.

5.29 *North Carolina births.* The file **NCbirths** contains data on a random sample of 1450 birth records in the state of North Carolina in the year 2001. This sample was selected by John Holcomb, based on data from the North Carolina State Center for Health and Environmental Statistics. One question of interest is whether the distribution of birth weights differs among mothers' racial groups. For the purposes of this analysis we will consider four racial groups: white, black, Hispanic, and other (including Asian, Hawaiian, and Native American). Use the variable *MomRace* which gives the races with descriptive categories. (The variable *RaceMom* uses only numbers to describe the races.)

   a. Produce graphical displays of the birth weights (in ounces) separated by mothers' racial group. Comment on both similarities and differences that you observe in the distributions of birth weight among the races.

   b. Report the sample sizes, sample means, and sample standard deviations of birth weights for each racial group.

   c. Explain why it's not sufficient to examine the four sample means, note that they all differ, and conclude that all races do have birth weight distributions that differ from each other.

5.30 *North Carolina births (continued).* Return to the data discussed in the previous exercise.

a. Comment on whether the conditions of the ANOVA procedure and F-test are satisfied with these data.

b. Conduct an ANOVA. Report the ANOVA table, and interpret the results. Do the data provide strong evidence that mean birth weights differ based on the mothers' racial group? Explain.

5.31 *North Carolina births (continued).* We return to the birth weights of babies in North Carolina one more time.

a. Apply Fisher's LSD to investigate which racial groups differ significantly from which others. Summarize your conclusions, and explain how they follow from your analysis.

b. This is a fairly large sample so even relatively small differences in group means might yield significant results. Do you think that the differences in mean birth weight among these racial groups are important in a practical sense?

5.32 *Blood pressure (continued).* The data set used in Exercise 5.28 also measured the size of people using the variable *Overwt*. This is a categorical variable that takes on the values 0=Normal, 1=Overweight, and 2=Obese. Is the mean systolic blood pressure different for these three groups of people?

a. Why should we not use 2-sample t-tests to see what differences there are between the means of these three groups?

b. Compute an ANOVA table to test for differences in systolic blood pressure between normal, overweight, and obese people. What do you conclude? Explain.

c. Use Fisher's LSD to find any differences that exist between these three group mean systolic blood pressures. Comment on your findings.

5.33 *Salary.* A researcher wanted to know if the mean salaries of men and women are different. She chose a stratified random sample of 280 people from the 2000 U.S. Census consisting of men and women from New York State, Oregon, Arizona, and Iowa. The researcher, not understanding much about statistics, had Minitab compute an ANOVA table for her. It is shown below.

```
One-way ANOVA: salary versus sex

Source    DF          SS           MS       F      P
sex        1   8190848743   8190848743   12.45  0.000
Error    278   1.82913E+11    657958980
Total    279   1.91103E+11

S = 25651    R-Sq = 4.29%    R-Sq(adj) = 3.94%
```

5.6. EXERCISES

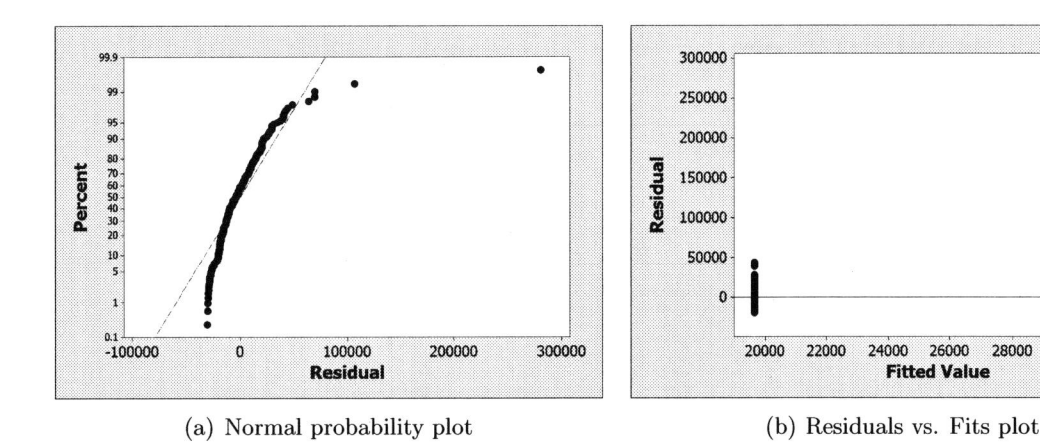

(a) Normal probability plot	(b) Residuals vs. Fits plot

Figure 5.15: Normal probability plot and Residual plot for Salary data

---

a. Is a person's sex significant in predicting their salary? Explain your conclusions.

b. What value of R^2 value does the ANOVA model have? Is this good? Explain.

c. The researcher did not look at residual plots. They have been produced for you in Figure 5.15. What conclusions do you reach about the ANOVA after examining these plots? Explain.

Open-ended Exercises

5.34 *Hawks (continued).* The data set on hawks was used in Example 5.9 to analyze the length of the tail based on species. Other response variables were also measured in the **Hawks** data. Analyze the length of the *culmen* (a measurement of beak length) for the three different species represented. Report your findings.

5.35 *Sea slugs.* Sea slugs, common on the coast of southern California, live on vaucherian seaweed. The larvae from these sea slugs need to locate this type of seaweed to survive. A study was done to try to determine whether chemicals that leach out of the seaweed attract the larvae. Seawater was collected over a patch of this kind of seaweed at 5-minute intervals as the tide was coming in and, presumably, mixing with the chemicals. The idea was that as more seawater came in, the concentration of the chemicals was reduced. Each sample of water was divided into 6 parts. Larvae were then introduced to this seawater to see what percentage metamorphosed. Is there a difference in this percentage over the 6 time periods? Open the data set **SeaSlugs**, analyze it, and report your findings.

5.36 *Auto pollution.* In a 1973 testimony before the Air and Water Pollution Subcommittee of the Senate Public Works Committee, John McKinley, President of Texaco discussed a new filter that had been developed to reduce pollution. Questions were raised about the effects of this filter on other measures of vehicle performance. The data set **AutoPollution** gives the results of an experiment on 36 different cars. The cars were randomly assigned to get either this new filter or a standard filter and the noise level for each car was measured. Is the new filter better or worse than the standard? The variable *Type* takes the value 1 for the standard filter and 2 for the new filter. Analyze the data and report your findings.

5.37 *Auto pollution (continued).* The experiment described in Exercise 5.36 actually used 12 cars each of three different sizes (small=1, medium=2 and large=3). These cars were, presumably, chosen at random from many cars of these sizes.

- a. Is there a difference in noise level among these three sizes of cars? Analyze the data and report your findings.

- b. Regardless of significance (or lack thereof) how must your conclusions for Exercise 5.36 and this exercise be different and why?

Chapter 6

Multifactor ANOVA

In Chapter 5 you saw how to explain the variability in a quantitative response variable using a single categorical explanatory factor (with K different levels). What if we wanted to use more than one categorical variable as explanatory factors? For example, are there differences in mean grade point average (GPA) based on type of major (Natural Science, Humanities, Social Science) or year in school (First Year, Sophomore, Junior, Senior)? How does the growth of a plant depend on the acidity of the water (None, Some, Lots) and the distance (Near, Moderate, Far) from a light source? While we could consider each factor separately, we might gain more insight to factors affecting GPA or plant growth by considering two factors together in the same model. In the same way that we moved from a simple linear model in Chapters 1 and 2 to multiple regression in Chapter 3, we can generalize the ANOVA model to handle differences in means among two (or more) categorical factors. In Section 6.1 we consider the most basic extension to include effects for a second categorical factor. In Section 6.2 we introduce the idea of interaction in the ANOVA setting. In Section 6.3 we define terms to account for possible interactions between the two factors. And in Section 6.4 we apply the new models to a case study.

6.1 The Two-way Additive Model (Main Effects Model)

Example 6.1: *Frantic fingers*

Are you affected by caffeine? What about chocolate? Long ago, scientists Scott and Chen published research that compared the effects of caffeine with those of theobromine (a similar chemical found in chocolate) and with those of a placebo. Their experiment used four human subjects, and took place over several days. Each day each subject swallowed a tablet containing one of caffeine, theobromine, or the placebo. Two hours later they were timed while tapping a finger in a specified manner (that they had practiced earlier, to control for learning effects). The response is the number of taps in a fixed time interval and the data are shown in Table 6.1.

A critical feature of this study was the way the pills were assigned to subjects. Each subject was given one pill of each kind, so that over the course of the experiment, each drug was given to all

250   CHAPTER 6. MULTIFACTOR ANOVA

Subject	Placebo	Caffeine	Theobromine	Mean
I	11	26	20	19
II	56	83	71	70
III	15	34	41	30
IV	6	13	32	17
Mean	22	39	41	34

Table 6.1: Rate of finger tapping by four trained subjects

four subjects. In order to protect against possible bias from carryover effects, subjects were given the drugs in an order determined by chance, with a separate randomized order for each subject. (This is an example of a randomized complete block design. Chapter 8 presents more detail.)

Another feature of this study is that there are two categorical variables, the drugs and the subjects. However, the two factors have different roles in the research. The goal of the study was to compare the effects of the drugs and drug is the primary factor of interest. The other factor, subjects, is not of particular interest. In fact, differences between subjects create a nuisance, in that they make it harder to detect the effects of the drugs. If you look at the subject averages in the right-most column of Table 6.1, you can see how much they differ. Subjects I and IV were very slow; Subject II was fast, and Subject III was in between.

A look at the drug averages (bottom row) shows that on average, the effects of caffeine and theobromine were quite similar, and that subjects tapped about twice as fast after one of the drugs as they did after the placebo. By now you know that we should ask, "Is the observed difference 'real'? That is, is it too big to be due just to coincidence?" For an answer, we need to make an inference from a model.

CHOOSE

We'll begin with a one-way ANOVA, using drugs (our factor of interest) as the lone categorical predictor even though we know that there are two categorical predictors. As you will see, using the wrong model here will lead to wrong conclusions.

One-way ANOVA: Drug versus Tapping

```
          Df Sum Sq Mean Sq F value Pr(>F)
Drug       2    872  436.00  0.6754  0.533
Residuals  9   5810  645.56
```

If we take the output ($P > 0.5$) at face value, the message is that there is no evidence of a drug effect. The variation due to drugs, as measured by the mean square of 436.0, is smaller than the

6.1. THE TWO-WAY ADDITIVE MODEL (MAIN EFFECTS MODEL)

variation in the residuals (MSE=645.56), and there is a greater than 50/50 chance of getting results like this purely by chance (P=0.533). This conclusion will turn out to be quite wrong, but it will take a two-factor model to reveal it.

Perhaps you've figured out why the one-way ANOVA fails to detect a drug effect. We know there are huge differences between subjects, and in our one-way analysis, those subject differences are part of the residual errors, making the residual sum of squares much too large. For this data set, testing for drug differences using a one-way analysis is like looking for microbes with a magnifying glass. To see what's there, we need a more powerful instrument, in this case, a two-way ANOVA, which accounts for both factors in the same model. ◊

In general, suppose that we have two categorical factors, call them Factor A and Factor B, that might be related to some quantitative response variable Y. Assume that there are K levels $(1, 2, \ldots, K)$ for Factor A (just as in one-way ANOVA of the previous chapter) and J levels $(1, 2, \ldots, J)$ for Factor B. The one-way ANOVA model for Y based on a single factor can be extended to account for effects due to the second factor.

The Two-way Additive Model for ANOVA

The **additive model** for a quantitative response Y based on main effects for two categorical factors A (row factor) and B (column factor) is

$$y = \mu + \alpha_k + \beta_j + \epsilon$$

where

$\mu =$ the grand mean for Y
$\alpha_k =$ the effect for the kth level of Factor A
$\beta_j =$ the effect for the jth level of Factor B
$\epsilon =$ the random error term

for $k = 1, 2, \ldots, K$ and $j = 1, 2, \ldots, J$.

As with inference for one-way ANOVA and regression we generally require that $\epsilon \sim N(0, \sigma_\epsilon)$ and are independent.

Now we have more parameters to estimate. The grand mean plays the same role as in the one-way ANOVA model and the effects for each factor have a similar interpretation, although now we allow for adjustments due to effects of levels of either factor in the same model. Before moving to the details for estimating these effects to fit the model we need to consider ways that the data might be organized.

The structure of two-way ANOVA data

Table 6.1 shows the two-way structure of the fingers data visually. Rows correspond to one factor, *Subject.* Columns correspond to a second factor, *Drug.* This format allows us to refer to "row factors" and "column factors," a terminology that evokes the two-way structure.

Subject	Treatment	Rate
I	Placebo	11
I	Caffeine	26
I	Theobromine	20
II	Placebo	56
II	Caffeine	83
II	Theobromine	71
III	Placebo	15
III	Caffeine	34
III	Theobromine	41
IV	Placebo	6
IV	Caffeine	13
IV	Theobromine	32

Table 6.2: Rate of finger tapping by four trained subjects displayed in "case by variables" format

However, this format differs from the generic "cases by variables" format used for many data sets. Table 6.2 shows the same data in this generic format. For data in this format, it is not clear which is the "row" factor, and which is the "column" factor. Instead, we refer to "Factor A" and "Factor B." Both formats for two-way data are useful, as are both ways of referring to factors. Statisticians use both, and so you should become familiar with both.

FIT

Once we have chosen a tentative model, we need to estimate its parameters. For the simplest two-way ANOVA model, we need to estimate the effects of both factors, use these to compute fitted values, and use the fitted values to compute residuals. For data sets with the structure of the fingers study, estimating the parameters in the two-way model is just a matter of computing averages and subtracting. The grand average of the observed values $\bar{y}$ estimates the grand mean μ. Each effect for the row factor (subjects) equals the row average minus the grand average, and each effect for the column factor equals the column average minus the grand average. Each fitted value equals the sum of the grand average, row effect, and column effect. Note that it doesn't matter which factor is represented by columns and which factor is represented by rows. As always, residual equals observed minus fitted.

Now that we have two factors, we have to think about row means (Factor A), column means (Factor B), cell means, and an overall mean. We need some new notation to indicate which mean we have

6.1. THE TWO-WAY ADDITIVE MODEL (MAIN EFFECTS MODEL)

in mind. We can't just use $\bar{y}$ for all of those different kinds of means. We will adopt what is commonly referred to as dot notation.

Dot Notation for Sample Means

For a response variable from a data set with two factors A and B, where A has K levels and B has J levels, we will use the following notation:

$\bar{y}_{kj} =$ the cell mean of the observations at the kth level of A and jth level of B
$\bar{y}_{k.} =$ the mean of all observations at the kth level of A regardless of the level of B
$\bar{y}_{.j} =$ the mean of all observations at the jth level of B regardless of the level of A
$\bar{y}_{..} =$ the mean of all of the observations (the grand mean)

In this notation, a "dot" means to take the mean over all levels of the factor for which there is a dot.

Using the dot notation we define the estimated effects as:

$$\text{Grand average} = \hat{\mu} = \bar{y}_{..}$$
$$\text{Row (Factor A) effects} = \hat{\alpha}_k = \bar{y}_{k.} - \bar{y}_{..}$$
$$\text{Column (Factor B) effects} = \hat{\beta}_j = \bar{y}_{.j} - \bar{y}_{..}$$

In the formulas above, $\bar{y}_{k.}$ represents the sample mean for the kth level of Factor A and $\bar{y}_{.j}$ is for the jth level of Factor B. We often refer to these as row means and columns means if the data are arranged in a two-way table as in Table 6.1. It is important to consider these in the context of the individual factors since a value like $\bar{y}_2$ is confusing on its own.

For the data on finger tapping, the grand average is 34; the drug averages are 22, 39, and 41; and the subject averages are 19, 70, 30, and 17. Thus the estimated effects for the levels of the each of the factors are as follows:

Drug (Column)	Estimated effect		
Placebo (1)	$\hat{\alpha}_1 = 22 - 34$	=	-12
Caffeine (2)	$\hat{\alpha}_2 = 39 - 34$	=	5
Theobromine (3)	$\hat{\alpha}_3 = 41 - 34$	=	7
Subject (Row)	Estimated effect		
Subject 1	$\hat{\beta}_1 = 19 - 34$	=	-15
Subject 2	$\hat{\beta}_2 = 70 - 34$	=	36
Subject 3	$\hat{\beta}_3 = 30 - 34$	=	-4
Subject 4	$\hat{\beta}_4 = 17 - 34$	=	-17

For two-way data, a useful way to show this is given in Table 6.3.

Subject	Placebo	Caffeine	Theobromine	Mean	Effect
I	11	26	20	19	-15
II	56	83	71	70	36
III	15	34	41	30	-4
IV	6	13	32	17	-17
Mean	22	39	41	34	
Eff	-12	5	7		

Table 6.3: Rate of finger tapping with estimated effects

These estimates allow us to rewrite each observed value as a grand average plus a sum of three components of variability, one each for the row effect (Factor A = *Subject*), the column effect (Factor B = *Drug*), and the residual. For example, the observed value of 11 for Subject I, Placebo is equal to

$$Obs = Grand + Subj + Drug + Res$$
$$11 = 35 + (-15) + (-12) + 4$$

As usual, the residual is computed as observed minus fitted: $Res = Obs - Fit$.

By squaring these components and adding, as shown in Table 6.4, we get the sums of squares we need for an ANOVA table.

Subj	Drug	Obs	=	Grand	+	Subj Effect	+	Drug Effect	+	Res
I	Pl	11	=	34	+	(-15)	+	(-12)	+	4
II	Pl	36	=	34	+	36	+	(-12)	+	(-2)
⋮	⋮	⋮		⋮		⋮		⋮		⋮
IV	Th	32	=	34	+	(-17)	+	7	+	8
Sum of Squares:		20554	=	13872	+	5478	+	872	+	332
		SS_{Obs}	=	SS_{Grand}	+	SS_{Rows}	+	SS_{Cols}	+	SS_{Res}

Table 6.4: Partitioning observed values

This format shows how each observed value y is partitioned. A useful alternative is to show how the deviations $(y - \bar{y})$ are partitioned. Table 6.5 shows this alternative. Here, also, by squaring and adding, we get the sums of squares we need.

Don't allow the details to distract you from the main point here: The two partitions are equivalent. Both are useful. Both are important to understand. These estimates allow us to rewrite each observed value as a grand average plus a sum of three components of variability, one each for

6.1. THE TWO-WAY ADDITIVE MODEL (MAIN EFFECTS MODEL)

Subj	Drug	Obs -Grand	=	Dev	=	Subj Effect	+	Drug Effect	+	Res
I	Pl	11 - 34	=	-23	=	(-15)	+	(-12)	+	4
II	Pl	36 - 34	=	22	=	36	+	(-12)	+	(-2)
⋮	⋮	⋮		⋮		⋮		⋮		⋮
IV	Th	32 - 34	=	-2	=	(-17)	+	7	+	8
Sum of Squares:		20554 - 13872	=	6682	=	5478	+	872	+	332
		$SS_{Obs} - SS_{Grand}$	=	SS_{Tot}	=	SS_{Rows}	+	SS_{Cols}	+	SS_{Res}

Table 6.5: Partitioning deviations and total variation

the row effect (Factor A = *Subject*), the column effect (Factor B = *Drug*), and the residual. By squaring these components and adding, we get the sum of squares we need for an ANOVA table. We start with total variability.

$$\text{Total Variability} = \text{Factor A} + \text{Factor B} + \text{Error}$$
$$y - \bar{y} = (\bar{y}_k - \bar{y}) + (\bar{y}_k - \bar{y}) + (y - \bar{y}_k - \bar{y}_j + \bar{y})$$

Squaring the components and adding leads us to the two-way ANOVA sums of squares identity

$$\sum (y - \bar{y})^2 = \sum (\bar{y}_k - \bar{y})^2 + \sum (\bar{y}_j - \bar{y})^2 + \sum (y - \bar{y}_k - \bar{y}_j + \bar{y})^2$$
$$SSTotal = SSA + SSB + SSE.$$

For the fingers data we get:

Obs	-	Grand Ave	=	Subj Effect	+	Drug Effect	+	Res
11	-	34	=	-15	+	-12	+	4
56	-	34	=	36	+	-12	+	-2
15	-	34	=	-4	+	-12	+	-3
6	-	34	=	-17	+	-12	+	1
26	-	34	=	-15	+	5	+	2
83	-	34	=	36	+	5	+	8
34	-	34	=	-4	+	5	+	-1
13	-	34	=	-17	+	5	+	-9
20	-	34	=	-15	+	7	+	-6
71	-	34	=	36	+	7	+	-6
41	-	34	=	-4	+	7	+	4
32	-	34	=	-17	+	7	+	8
SS: 20554	-	13872	=	5478	+	872	+	332

As you can see by comparing the estimated effects, the variation between subjects is much larger than the variation between drugs, and the variation from each of these sources is large compared to the residual variation. These comparisons are summarized by the mean squares in the ANOVA table.

Two-way ANOVA Table for Additive Model

For a two-way data set with K levels of Factor A, J levels of Factor B, and one observation for each combination of the two factors, the ANOVA table for the two-way main effects (additive) model of

$$y = \mu + \alpha_k + \beta_j + \epsilon$$

is shown below.

Source	df	Sum of Squares	Mean Square	F statistic	p-value
Factor A	$K-1$	SSA	MSA=$\frac{SSA}{K-1}$	$\frac{MSA}{MSE}$	$F_{K-1,(K-1)(J-1)}$
Factor B	$J-1$	SSB	MSB=$\frac{SSB}{J-1}$	$\frac{MSB}{MSE}$	$F_{J-1,(K-1)(J-1)}$
Error	$(K-1)(J-1)$	SSE	MSE=$\frac{SSE}{(K-1)(J-1)}$		
Total	$KJ-1$	SSTotal			

The two hypotheses being tested are
$H_0: \alpha_1 = \alpha_2 = \cdots \alpha_K = 0$ (Factor A) and $H_0: \beta_1 = \beta_2 = \cdots \beta_J = 0$ (Factor B)
$H_a:$ Some $\alpha_k \neq 0$ $\qquad\qquad\qquad\qquad\qquad\qquad$ $H_a:$ Some $\beta_j \neq 0$

In the ANOVA table below, the mean square for Drug is 436, the mean square for subjects is much larger, at 1826, and the mean square for residuals is comparatively tiny, at a mere 55.33.

```
          Df Sum Sq Mean Sq  F value Pr(>F)
Drug       2    872     436    7.880  .0210
Subjects   3   5478    1826   33.000  .0004
Residuals  6    332   55.33
```

The F-tests and p-values confirm that both the drug differences (F=7.880, P=0.210) and subject differences (F=33.00, P=0.004) are "real." Before we jump to conclusions based on the numbers alone, however, we should check the conditions.

6.1. THE TWO-WAY ADDITIVE MODEL (MAIN EFFECTS MODEL)

ASSESS

For now we'll look only at the standard plots you learned in the chapters on regression analysis and one-way ANOVA. (There are additional graphs designed especially for two-way ANOVA, but we'll save these for the supplementary exercises.) Figure 6.1(b) shows a plot of residuals versus fitted values, and Figure 6.1(a) shows a normal quantile plot. Both plots show the sorts of patterns we would expect to see when a model conditions are satisfied, points of the normal plot fall along a line and points of the residual plot form a directionless oval blob.

(a) Normal quantile plot of residuals   (b) Scatterplot of residuals versus fits

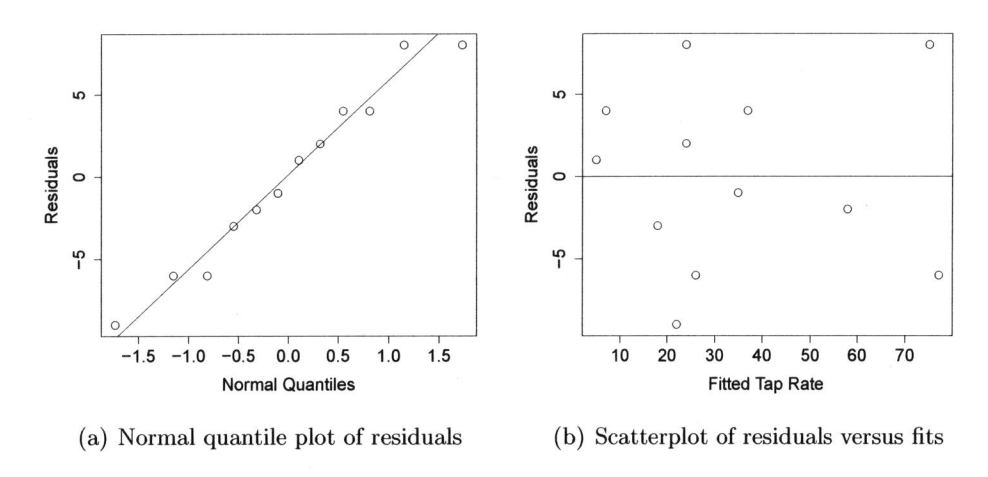

Figure 6.1: Plots to assess the fit of the ANOVA model

USE

A first use of the fitted model might be simply to test for differences between drugs. Here we find that the p-value of 0.02 is small, confirming that there are "real" differences. A look back at the original data (Table 6.1) reminds us that the averages for caffeine and theobromine are nearly equal, at 39 and 41, and both are roughly double the placebo average of 22. To provide more context for these comparisons, we will compute confidence intervals and effect sizes, but first, a bit more about the F-tests.

The two-way ANOVA provides tests for both factors, and even though the researchers cared much more about possible drug differences than about subject differences, the F-test and p-value for subjects confirms that there are indeed "real" differences between subjects.

Scope of inferences

Although the ANOVA model and ANOVA table treat the two factors equally, the meaning of the F-tests is quite different for the two factors, because of the differences in the ways the two factors were treated. One of the two factors, drugs, was experimental. It was controlled by the experimenters, and was assigned to subjects using chance. The other factor was observational, and in a strict sense, there was no randomization involved in choosing the subjects. (They were not chosen at random from some larger population.) For the factor of interest, drugs, the random assignment allows us to conclude that the significant difference is in fact caused by the drugs themselves, and not by some other, hidden cause.

For the observational factor, the scope of our inference is in some ways much more limited. Because the subjects were a convenience sample, we may be on shaky ground in trying to generalize to a larger population. We can be confident that the experiment demonstrates a causal relationship for the subjects in this particular study, and surely the results must apply to some population of similar subjects, but the study itself does not tell us who they are.

Example 6.1 is an example of a randomized complete block design. Chapter 8 discusses experimental design in detail, but here is a preview. The randomized complete block design has one factor of interest (the drugs), and one nuisance factor, the subjects, or blocks. Each possible combination of levels of the two factors occurs exactly once in the design. In Example 6.1, each subject gets all three drugs, and each drug is assigned to all four subjects. The assignment of drugs to subjects is randomized, that is, done using chance. Designs of this sort are used to reduce the residual variability, and make it easier to detect differences between drugs. To see how this works, consider a different plan, called a completely randomized design. In the complete block design, each subject is used three times, once with each drug. With a completely randomized design, each subject would get only one of the drugs, so to get four placebo observations, you'd need four subjects. To get four caffeine observations, you'd need another four subjects, and for theobromine, yet another four, so twelve in all. For data like this, you'd do a one-way ANOVA, and your residuals would include differences between subjects. Put differently, in order to compare caffeine and placebo using the one-way design, you have to compare different groups of subjects. With the block design, you can compare caffeine and placebo on the same subjects.

The same two-way additive ANOVA model used for Example 6.1 is appropriate for situations other than randomized block data with a single data value for each combination of the two factors. Suppose we have two factors A and B, one of which divides the data into K different groups while the other divides the same data into J different groups. We can still estimate a grand mean ($\hat{\mu} = \bar{y}_{..}$), effects for each level of Factor A ($\hat{\alpha}_k = \bar{y}_{k.} - \bar{y}_{..}$) and effects due to Factor B ($\hat{\beta}_j = \bar{y}_{.j} - \bar{y}_{..}$). The variabilities explained by the two factors (SSA and SSB) are computed as in a one-way ANOVA and the error variability is whatever is left to make the sum equal the total variability. Our two-way ANOVA identity can be written as

$$SSE = SSTotal - SSA - SSB.$$

The degrees of freedom for the ANOVA table remain $K-1$, $J-1$ and $KJ-1$ for Factor A, Factor B and the total variability, respectively. The degrees of freedom associated with the error term is then $KJ - K - J + 1$ (which simplifies to $(K-1)(J-1)$). This value also serves as the degrees

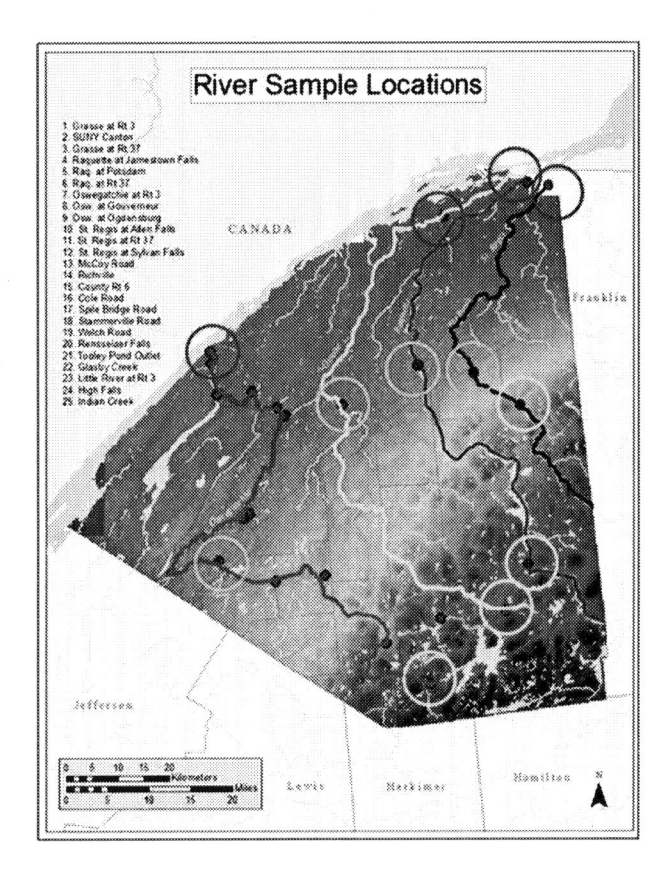

6.1. THE TWO-WAY ADDITIVE MODEL (MAIN EFFECTS MODEL)

Figure 6.2: Map of the rivers in upstate New York

of freedom for the t-distribution when doing inferences after the ANOVA based on $\sqrt{MSE}$ as an estimate of the common standard deviation which we discuss after the next example.

Example 6.2: *River iron*

Some geologists were interested in the water chemistry of rivers in upstate New York. They took water samples at three different locations in four rivers (Grasse, Oswegatchie, Raquette, and St. Regis). The sampling sites were chosen to investigate how the composition of the water changes as it flows from the source to the mouth of each river. The sampling sites were labeled as upstream, midstream, and downstream. Figure 6.2 shows the locations of the rivers and sites on a map.

Each water sample was analyzed to record the levels of many different elements in the water. Some of the data from this study are contained in the file **RiverElements**. Because chemists typically work with concentrations in the log scale (think pH) Table 6.6 shows the log (to base 10) of the iron concentration in parts per million (ppm).

	Observed					
	Grasse	Oswegatchie	Raquette	St. Regis		Average
Up	2.9750	2.9345	2.0334	2.8756		2.7046
Mid	2.7202	2.3598	1.5563	2.7543		2.3477
Down	2.5145	2.1139	1.4771	2.5441		2.1624
Ave	2.7366	2.4696	1.6889	2.7247		2.4049

Table 6.6: The observed river iron data transformed by the log base 10

As you can see from looking down the right-most column, the log concentrations decrease as you go downstream from source (upstream) to mouth (downstream), and the same pattern holds for each of the four individual rivers. We can show this pattern in a plot (Figure 6.3). Such plots, called "interaction plots," will be explained at great length in the next section.

Figure 6.3: The horizontal axis shows *Site*, from source (Upstream = left) to mouth (Downstream = right). The vertical axis shows iron concentration. There are four plotting symbols, one for each river. Each set of line segments shows a downward pattern from left to right: iron concentration decreases as you go from source to mouth.

For this data set, as for the previous one, we choose a two-way additive model[1] and fit it using the same method as in Example 6.1. Each *River* effect equals the average for that river minus the grand average, and each *Site* effect equals the average for that site minus the grand average. Each fitted value equals the grand average plus a *River* effect plus a *Site* effect. To assess the fit of the model, we look at a normal quantile plot of residuals and a residual versus fit plot in Figure 6.4.

---

[1] For the iron data in the original scale of ppm, without logs, the additive model is not appropriate. (See the supplementary exercises for more details.)

6.1. THE TWO-WAY ADDITIVE MODEL (MAIN EFFECTS MODEL)

(a) Normal quantile plot of residuals

(b) Scatterplot of residuals versus fits

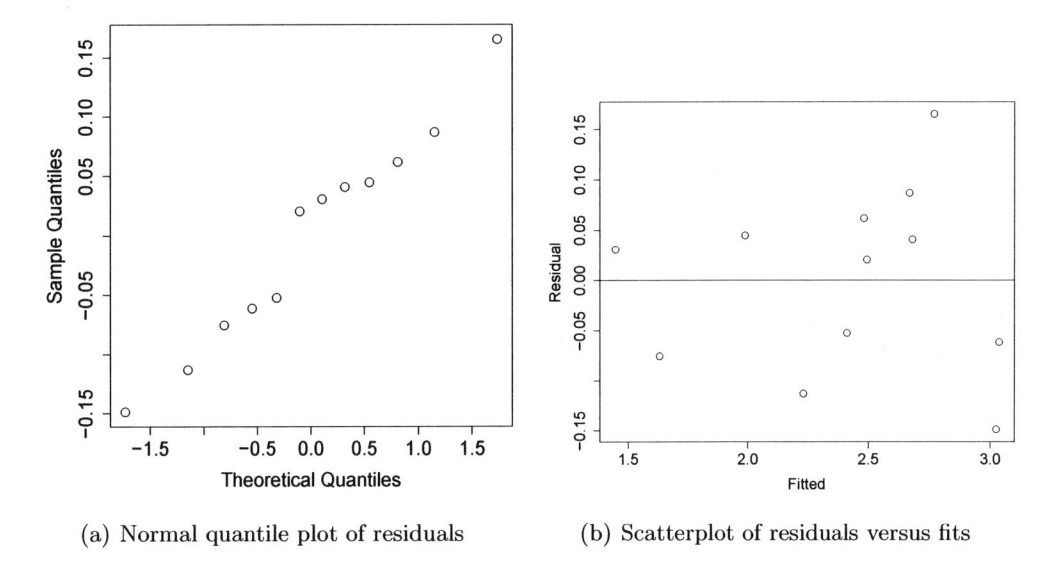

Figure 6.4: Plots to assess the fit of the ANOVA model

Both plots look reasonable, and so we move on, to use the model to test for significance. Here is the ANOVA table:

	Df	Sum Sq	Mean Sq	F value	Pr(>F)	
River	3	2.18703	0.72901	48.153	0.0001366	***
Site	2	0.60765	0.30383	20.069	0.0021994	**
Residuals	6	0.09084	0.01514			

Since we have such small p-values for both factors (P=0.000136 and P=0.0021994), we conclude that there are differences in mean log iron content between rivers and between sites that are unlikely to be due to chance alone. ◇

Inference after Two-way ANOVA

Example 6.3: *River iron (continued)*

Whenever the F-tests of an ANOVA indicate that differences between groups are significant, it makes sense to ask which groups are different, and by how much? Since the ANOVA was significant for both factors, we can apply the same reasoning as in the one-way case (see Section 5.4) to do multiple comparisons using confidence intervals of the form

$$\bar{y}_i - \bar{y}_j \pm t^* \sqrt{MSE \left(\frac{1}{n_i} + \frac{1}{n_j} \right)}$$

where i and j refer to different levels of the factor of interest. These apply equally well to compare means between rivers or between sites (but not to compare a *Site* mean to a *River* mean). Using **Fisher's LSD** we find t^* using the error degrees of freedom and an appropriate confidence level. For a randomized block design such as the river data the margin of error in the confidence interval is the same for comparing any two means for the same factor since $n_k = J$ for Factor A and $n_j = K$ for Factor B. Thus we can assess the differences efficiently by computing two LSD values

$$LSD_A = t^* \sqrt{\frac{2 \cdot MSE}{J}} \quad \text{and} \quad LSD_B = t^* \sqrt{\frac{2 \cdot MSE}{K}}$$

When comparing sites with Fisher's LSD at a 5% level we use $LSD_{Site} = 2.45\sqrt{2*0.0151/4} = 0.213$ and when we compare rivers we use $LSD_{River} = 2.45\sqrt{2*0.0151/3} = 0.246$. This indicates that upstream (mean = 2.7046) is different in average iron content from both midstream (mean = 2.3477) and downstream (mean = 2.1624), but the latter two sites aren't significantly different from each other. For the rivers, the Raquette (mean = 1.689) is significantly different from the other three, and the Oswegatchie (mean = 2.469) is significantly different from the Grasse (mean = 2.737) and St. Regis (mean = 2.725), but these last two are not significantly different from each other.

We can also compute effect sizes in the usual way, by dividing each difference in means by the estimated standard deviation for a single observation, equal to the square root of the MSE. For the river iron data, the estimated standard deviation is $s = \sqrt{.0151} = .123$, which gives the following:

Site effects sizes:

$$\text{Up - Mid} \quad \frac{2.7046 - 2.3477}{0.123} = 2.90$$

$$\text{Up - Down} \quad \frac{2.7046 - 2.1624}{0.123} = 4.41$$

$$\text{Mid - Down} \quad \frac{2.3477 - 2.1624}{0.123} = 1.51$$

Scope of inference. Both factors are observational: Neither the rivers nor the sites were chosen by random sampling to represent some larger population, and so we should be cautious about generalizing to other rivers. However, it is reasonable to think that the sites chosen were typical of sites that might have been chosen upstream, midstream, and downstream on those particular rivers, and it is reasonable to think that the water samples collected at each site are representative of other samples that might have been collected, and so there is good reason to regard the results observed as typical for those kinds of sites on the four rivers that were studied. ⋄

6.2 Interaction in the Two-way Model

In both examples so far the model assumed that effects are additive: the effect of a drug on tapping rate is the same for all four subjects; the effect of *Site* on the log of the iron concentration is the same for all four rivers. Although this assumption of additive effects may often be reasonable, it should not be taken for granted. Often the effect on one factor depends on the other factor: the two factors interact.

Consider first the everyday meaning of interaction. You are probably familiar with warnings about drug interactions: Don't mix sleeping pills and alcohol. Don't take certain antibiotics with food. (The interaction will keep the drug from working.) Avoid grapefruit if you're taking certain cough medicines. Bananas and aged cheese can be very dangerous if you are taking certain antidepressants. In ecology also, interaction can be a matter of life and death. For example, lichens consist of two interacting organisms, a fungus, and a partner organism that carries out photosynthesis. Neither organism can live without the other. Meteorologists identify an entire class called "optical phenomena" created by the interaction of light and matter. Rainbows are the most familiar of these, caused by the interaction of light and water droplets in the air. In cognitive psychology, perception often depends on the interaction of stimulus and context. Optical illusions are striking examples. You might even say that sociology is an entire academic field devoted to the study of interactions.

Now consider a more quantitative way to think about **interaction** in the context of two-way ANOVA. The key idea can be put in a nutshell: *an interaction is a difference of differences*. Note first that the effect of a factor is a difference, specifically, the difference between a group average and the overall grand average. For example, the estimated effect of caffeine is 5, the difference between the caffeine average of 39 and the grand average of 34. The estimated effect of the placebo is -12, the difference between the placebo average of 22 and the grand average of 34. The additive (no interaction) model requires that the effect of the drug must be the same for all four subjects. If in fact, different subjects react in different ways, then there is an interaction: a difference of differences.

The following example illustrates this idea at its simplest, with each factor at just two levels. We first show the difference of differences numerically, then visually using a plot designed to show interaction.

Example 6.4: *Feeding pigs*

Why is it that animals raised for meat are often given antibiotics in their food? The following experiment illustrates one reason, and at the same time illustrates a use of two-way ANOVA. A scientist in Iowa was interested in additives to standard pig chow that might increase the rate at which the pigs gained weight. For example, he could add antibiotics or he could add vitamin B12. He could also include both additives in a diet or leave both out (using just standard pig chow as a control). If we let Factor A be yes or no depending on whether the diet has *Antibiotics* and let

Factor B keep track of the presence or absence of $VitaminB12$, we can summarize the four possible diets in a two-way table. To perform the experiment, the scientist randomly assigned 12 pigs, three to each of the diet combinations. Their daily weight changes (in hundredths of a pound over 1.00) are summarized in the left panel of Table 6.7 and stored in the file called **PigFeed**.

We begin by computing the average weight gain for each of the four diets as in the right panel of Table 6.7.

Weight Gain for Individual Pigs				Average Weight Gain for Each Diet			
		Factor B	*Vitamin B12*			Factor B	*Vitamin B12*
		No	Yes			No	Yes
Factor A	No	30, 19, 8	26, 21, 19	Factor A	No	19	22
Antibiotics	Yes	5, 0, 4	52, 56, 54	*Antibiotics*	Yes	3	54

Table 6.7: Data for pig feed experiment. The left panel shows individual response values; the right panel shows diet averages.

Now consider the effect of *Vitamin B12*. For the pigs who did not get antibiotics (top row on the right), adding vitamin B12 to the diet increased the gain by 22 - 19 = 3 units (0.03 pounds per week) over those who did not get B12. But for the pigs that did get antibiotics, the difference due to adding vitamin B12 was 54 - 3 = 51 units (0.51 pounds per week). In short, the differences are different:

		Factor B	*Vitamin B12*	Difference
		No	Yes	due to vitamin B12
Factor A	No	19	22	3
Antibiotics	Yes	3	54	51

Notice that you can make a similar comparison within columns. For the pigs who got no vitamin B12, the difference due to adding antibiotics was negative: 3 - 19 = -16 units. But for the pigs who got vitamin B12, the difference was positive: 54 - 22 = 36 units[2]. Once again, the differences are different:

		Factor B	*Vitamin B12*
		No	Yes
Factor A	No	19	22
Antibiotics	Yes	3	54
Difference due to *Antibiotics*		-16	32

---

[2]Notice that if you compute the difference of differences both ways (by rows, and by columns) you get the same answer. For rows, the differences are 51 and 3, and their difference is 48. For columns, the differences are -16 and 32, and their difference is also 48. This number is a measure of the size of the interaction.

6.2. INTERACTION IN THE TWO-WAY MODEL

Interaction graphs

For the simplest two-way ANOVAs like the pig example, computing the difference of differences can be useful, but for more complicated data sets, the number of differences can be so large that patterns are hard to find, and a graphical approach allows the eye to focus quickly on what matters most. A two-way interaction graph plots the individual group means (diets, in the pig example) putting levels of one factor on the x-axis, the response values on the y-axis, and plotting a line segment for each level of the second factor.

In Figure 6.5, the two levels of *Vitamin B12* (no and yes) are shown on the x-axis. Each cell average is plotted as a point, with weight gain on the y-axis. The points are joined to form two line segments, one for each level of the *Antibiotics* factor. For each segment, the change in y-value equals the difference in weight gain due to vitamin B12 for that level of *Antibiotics*, absent or present.

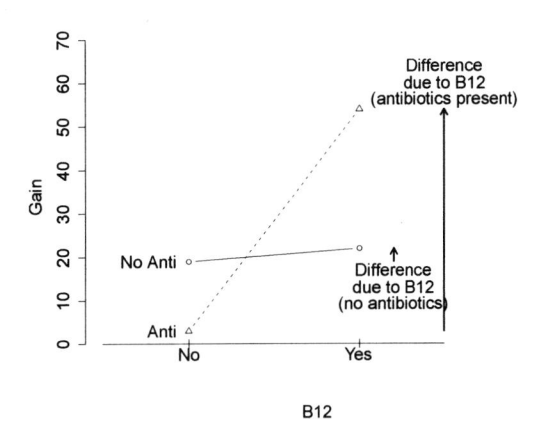

Figure 6.5: An interaction is a difference of differences. The vertical change of each segment equals the difference (Yes - No) due to *Vitamin B12*. Interaction is present: The two differences are different, and the segments are not parallel.

With an interaction plot you always have a choice of which factor to put on the x-axis, and so there are two possible plots. In theory, it doesn't matter which you choose, but in practice, for some data sets, one of the two may be easier to read. (The next example will illustrate this.) Figure 6.5 is the plot with *Vitamin B12* on the x-axis; Figure 6.6 shows the plot with *Antibiotics* on the x-axis.

To reinforce the idea that vertical change corresponds to the difference due either to *Vitamin B12* or to *Antibiotics*, compare the previous two graphs with the corresponding graphs for fitted values from the additive (no interaction) model (Figure 6.7). The table below shows those fitted values, computed using the method of Section 6.1.

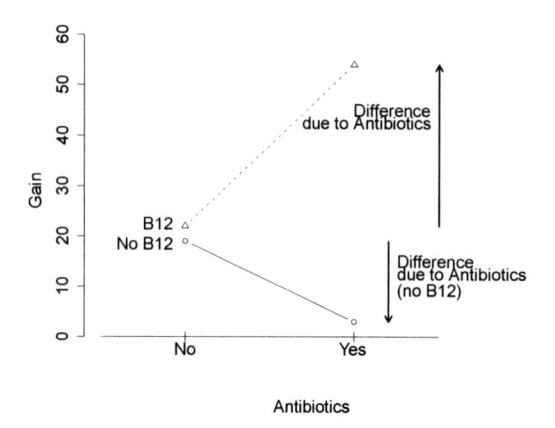

Figure 6.6: The other interaction plot for the PigFeed data. For this plot, the vertical change of each segment equals the difference (Yes - No) due to *Antibiotics*, with one segment for each level of the other factor, *Vitamin B12*.

$$\begin{array}{ccccccc}
& & & & \text{Additive Fit} & & \\
\text{Grand Ave} & + & \text{Row Effect} & + & \text{Col Effect} & = & \text{Additive Fit} \\
\begin{bmatrix} 24.5 & 24.5 \\ 24.5 & 24.5 \end{bmatrix} & + & \begin{bmatrix} -4 & -4 \\ 4 & 4 \end{bmatrix} & + & \begin{bmatrix} -13.5 & 13.5 \\ -13.5 & 13.5 \end{bmatrix} & = & \begin{bmatrix} 7 & 34 \\ 15 & 52 \end{bmatrix}
\end{array}$$

| **Fitted Values** | | Factor B | *Vitamin B12* | Difference |
Additive Model		No	Yes	due to *Vitamin B12*
Factor A	No	7	34	27
Antibiotics	Yes	15	42	27
Difference due to *Antibiotics*		8	8	0

As you can see, the difference due to *Vitamin B12* is the same for both levels of the *Antibiotics* factor; the difference due to *Antibiotics* is the same for both levels of *Vitamin B12*. To estimate interaction effects, we compute cell averages and subtract the additive fit.

$$\begin{array}{ccccc}
& & & \text{Interaction} & \\
\text{Cell Ave} & - & \text{Additive Fit} & = & \text{Interaction} \\
\begin{bmatrix} 19 & 22 \\ 3 & 54 \end{bmatrix} & - & \begin{bmatrix} 7 & 34 \\ 15 & 52 \end{bmatrix} & & \begin{bmatrix} 12 & -12 \\ -12 & 12 \end{bmatrix}
\end{array}$$

◇

When factors have more than just two levels, the plots are more complicated, but the key idea remains the same: if an additive model fits, segments will be parallel, apart from minor departures due to error.

6.2. INTERACTION IN THE TWO-WAY MODEL

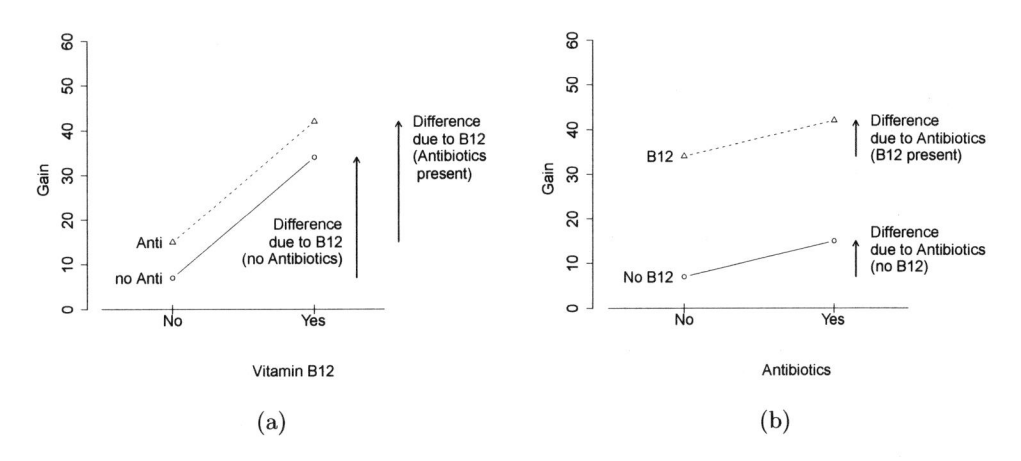

Figure 6.7: Interaction graphs for values from the additive (no interaction) model. When there is no interaction, the difference due to *Vitamin B12* (left panel) is the same for both levels of *Antibiotics*, and the change in y is the same for both line segments, and the segments are parallel. Similarly, the difference due to *Antibiotics* (right panel) is the same for both levels of *Vitamin B12*, and the segments are parallel.

Example 6.5: *River iron: Reading an interaction plot*

The data from New York rivers has two factors, *River* and *Site*, with four and three levels respectively. To help you practice reading these plots, we have created a set of several hypothetical versions of the river iron data, with and without *River* effects, with and without *Site* effects, with and without *River-by-Site* interaction. We start with a plot that gives an overview, then back up to consider features one at a time. The numbers in this example are hypothetical, chosen to emphasize the important features of a plot.

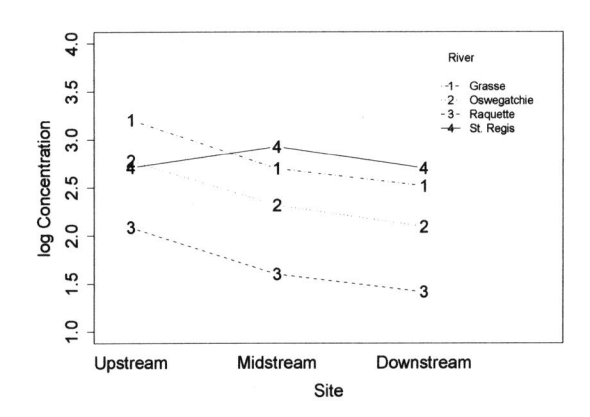

Figure 6.8: River iron interaction plot for hypothetical data

Figure 6.8 shows an interaction plot for hypothetical data. One factor, *Site*, is plotted along the x-axis, with levels in a natural order, from upstream on the left to downstream on the right. Iron concentrations are shown on the y-axis. There is one line for each of the four rivers.

Scanning an interaction plot from left to right, we get visual evidence of whether the first factor has an effect (Are the lines flat? Yes = no effect) and whether the second factor has an effect (Are the lines close together? Yes = no effect). But the main benefit from the plot is that we can check for interaction: Are the line segments roughly parallel? If not, we have evidence of an interaction.

We next consider a sequence of plots in Figure 6.9. Each plot uses a different hypothetical data set for the river iron data to illustrate a different type of situation.

Figure 6.9(a) shows no effect for either factor, so the simple model $Y = \mu + \epsilon$ would fit well.

Figure 6.9(b) shows an effect for *River*, but no *Site* effect, so the model would be $Y = \mu + \alpha_k + \epsilon$.

Figure 6.9(c) shows an effect from *Site*, but no *River* effect, so the model would be $Y = \mu + \beta_j + \epsilon$.

Figure 6.9(d) shows effects from *River* and *Site*, but the line segments are parallel so there is no evidence of interaction and the model would be $Y = \mu + \alpha_k + \beta_j + \epsilon$.

Figure 6.9(e) shows one set of segments which are not parallel with the others, which indicates that *River* and *Site* interact in their effects on Y, leading to the model $Y = \mu + \alpha_k + \beta_j + \gamma_{jk} + \epsilon$.

Finally, note that in theory, the choice to put *Site* on the horizontal axis and to use *River* as the "plotted" variable is arbitrary. If we switch the roles of *Site* and *River* then Figure 6.9(e) becomes Figure 6.9(f).

We again see non-parallel lines, indicating an interaction. Moreover, just as before, it is the combination of River=St. Regis and Site=Upstream for which the log iron concentration does not fit the pattern given in the rest of the data.

Although either choice for an interaction plot is correct, there are sometimes reasons to prefer one over the other. Such is the case here, because one factor, *Site*, has a natural ordering, from upstream to downstream, while the other factor, *River*, has no ordering. For a structure like this one, with one ordered and one unordered factor, it generally gives a more useful plot if you put the ordered factor levels on the x-axis. ◇

These graphs show what two-way data with interaction looks like in a graph. The next question is: How should you analyze such data? That is the topic of the next section.

6.2. INTERACTION IN THE TWO-WAY MODEL

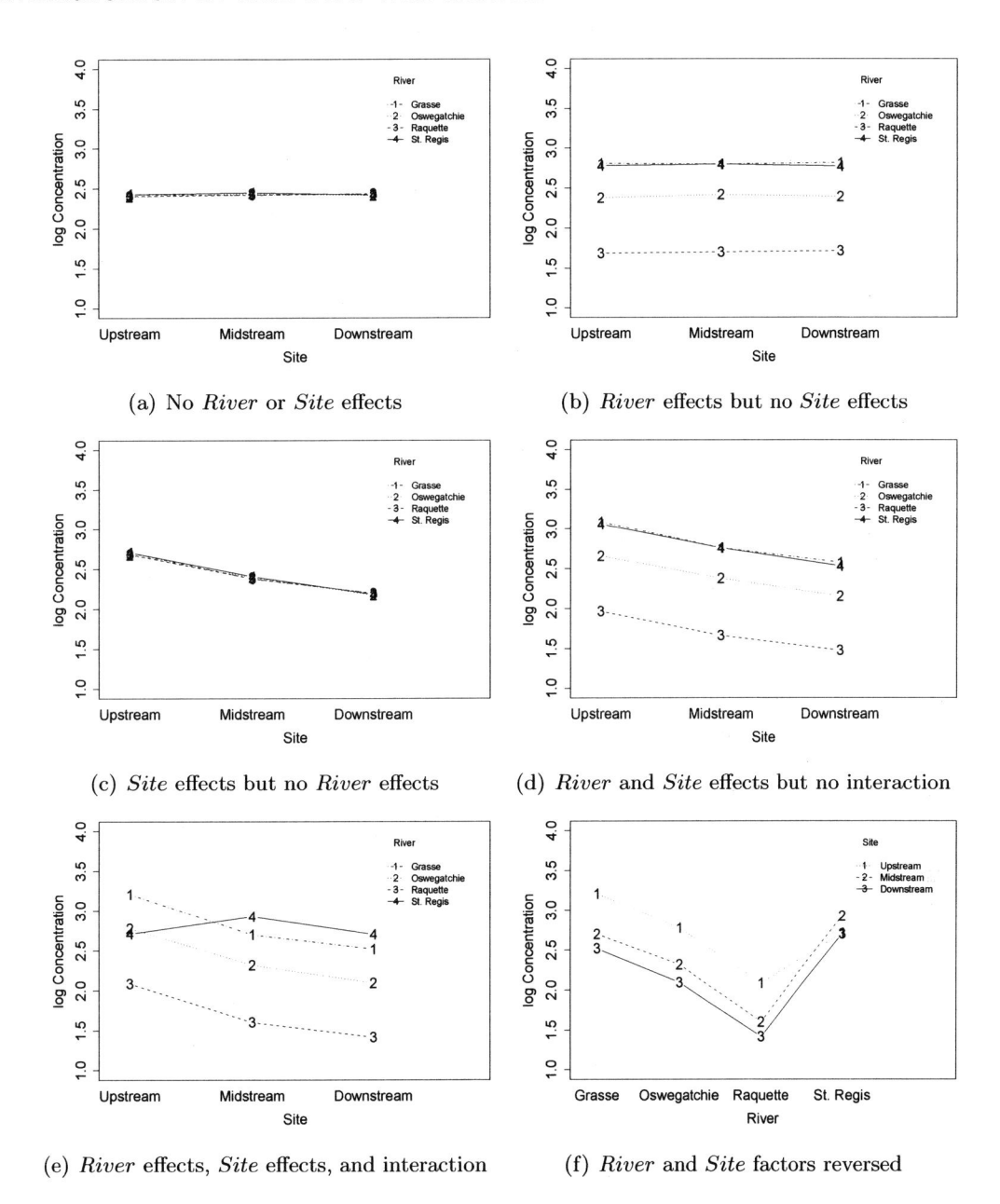

(a) No *River* or *Site* effects

(b) *River* effects but no *Site* effects

(c) *Site* effects but no *River* effects

(d) *River* and *Site* effects but no interaction

(e) *River* effects, *Site* effects, and interaction

(f) *River* and *Site* factors reversed

Figure 6.9: Hypothetical river iron interaction plots

6.3 The Two-way Non-additive Model (Two-Way ANOVA with Interaction)

In the last section you saw an example of a two-factor experiment with a pronounced interaction. In this section the main goal is to show you how to include interaction terms in the two-way model, how to fit the resulting model, how to assess the fit, and how to use the model for inference. First, however, by way of introduction, two preliminary issues: (1) Why interaction matters, and (2) How to tell if interaction is present.

a. Why interaction matters. There are myriad reasons, but two stand out.

i. Interaction changes the meaning of main effects. For the additive model, each main effect is the same for all levels of the other factor. Informally, we might say one size fits all. If interaction is present, each main effect depends on the level of the other factor. To be concrete, think about the pig data. According to the additive model (that is, without interaction), the effect of adding vitamin B12 to the diet appears to be to raise the average weekly weight gain by 17 units, either from 7 to 34 (no antibiotics), or from 15 to 42 (antibiotic present). One size (+17) apparently applies regardless of whether the diet contains antibiotics. In fact, and contrary to this additive model, the average effect of *Vitamin B12* is a paltry +3 (from 19 to 22) if the diet is antibiotic-free, but a substantial + 51 (from 3 to 54) if the diet contains antibiotics. The bottom line is: If interaction is present, it doesn't make much sense to talk about the effect of *Vitamin B12* alone, because there is more than one size. That is, its effect is different for different levels of the other factor.

ii. Interaction makes the F-tests of the additive model untrustworthy.

Here are the results for running the two-way main effects ANOVA model in R for these data.

```
            Df  Sum Sq  Mean Sq  F value  Pr(>F)
Antibiotic   1   192.00   192.00   0.8563  0.37892
Vitamin B12  1  2187.00  2187.00   9.7537  0.01226 *
Residuals    9  2018.00   224.22
```

The two tests would indicate that adding vitamin B12 has a significant effect (p-value =0.012), but there is not a significant difference in average weight change due to the antibiotic (p-value=0.379). In fact, with the correct, non-additive model, the effect of *Antibiotics* is borderline significant on its own (p-value=0.0504), but, more important,

6.3. THE TWO-WAY NON-ADDITIVE MODEL (TWO-WAY ANOVA WITH INTERACTION)

the interaction is highly significant (p-value=0.0001), so it is clear that *Antibiotics* have an effect. (We will return to this analysis later in the chapter.)

b. How can you tell whether interaction is present? Here are three things to consider.

i. *Designs with blocks.* Decades of experience with data suggests that often, though not always, when you have data from a complete block design, with one factor of interest and block as a nuisance factor, it will often be the case that the effects of the factor of interest are roughly the same in every block, and that the additive two-way model offers a good fit. The finger tapping example is typical: Each subject is a block, *Drug* is the factor of interest, and all four subjects respond in approximately the same way to the drugs. However, for the river iron data in the original scale (ppm), before transforming to logs, the additive model is not suitable. For more see (b) below, and the Supplementary Exercises.

ii. *Diagnostic plots.* For data sets like Examples 6.1 (fingers) and 6.2 (rivers) — two-way designs with only one observation per cell — it is not possible to get separate estimates for interaction effects and residuals: they are one and the same. (Another way to think about this is to think of the residual as "observed - cell mean" and recognized that when you have only one observation per cell, "observed" and "cell mean" are one and the same.)

Fortunately, for such data sets, there is a special kind of plot that looks for patterns in the residuals that might indicate the need to transform. The Supplementary Exercises show how to make such a plot. For the fingers data, the plot tells you that no transformation is needed, but for the rivers data, interaction is present for the original data, but not for a suitable transformation of the data, such as log concentrations.

iii. *Multiple observations per cell.* If you have just one observation per cell there is no way to get separate estimates of interaction and error, and so there can be no ANOVA F-test for the presence of interaction[3]. However, with more than one observation per cell, as with the pig data, it is possible to include interaction terms in the model, possible to get separate estimates for interaction terms and errors, and so possible to do an F-test for the presence of interaction.

CHOOSE

We now extend the two-way additive ANOVA model to create a two-way ANOVA model with interaction that allows us to estimate and assess interaction.

---

[3] There is a specialized method, Tukey's single degree of freedom for interaction, that will allow you to test for a specialized form of interaction, namely, the kind that can be removed by changing to a new scale.

> **Two-way ANOVA Model with Interaction**
>
> For two categorical factors A and B and a quantitative response Y, the ANOVA model with both main effects and the interaction between the factors is
>
> $$y = \mu + \alpha_k + \beta_j + \gamma_{kj} + \epsilon$$
>
> where
>
> $\mu = $ the grand mean for Y
> $\alpha_k = $ the effect for the kth level of Factor A
> $\beta_j = $ the effect for the jth level of Factor B
> $\gamma_{kj} = $ the interaction effect for the kth level of A with the jth level of B
> $\epsilon = $ the random error term
> $j = 1, 2, ..., J$ and $k = 1, 2, ..., K$.
>
> As with inference for the other ANOVA and regression models, we generally require that $\epsilon \sim N(0, \sigma_\epsilon)$ and are independent.

The new γ_{kj} terms in the model provide a mechanism to account for special effects that only occur at particular combinations of the two factors. For example, suppose that we have data on students at a particular school and are interested in the relationship between Factor A: *Year* in school (First year, Sophomore, Junior, Senior), Factor B: *Major* (Mathematics, Psychology, Economics, etc.) and grade point average (*GPA*). If students at this school tend to have a hard time adjusting to being away from home so that his or her first year grades suffer, we would have a main effect for *Year* ($\alpha_{FY} < 0$). If the Biology department has mostly faculty who are easy graders, we might find a main effect for *Major* ($\beta_{Bio} > 0$). Now suppose that Math majors at this school generally do pretty well, but the Junior year when they first encounter abstract proof courses is a real challenge for his or her grades. This would represent an interaction ($\gamma_{Math,Junior} < 0$); an effect that only occurs at a combination for the two factors and is not consistent across all the levels of one factor.

Now let's return to the example of pig diets to see how we can estimate the new interaction component of this model.

FIT

Fitting the non-additive model takes place in three main steps: (1) computing cell means and residuals; (2) fitting an additive model to the cell means; and (3) computing interaction terms. A final, fourth step puts the pieces together.

6.3. THE TWO-WAY NON-ADDITIVE MODEL (TWO-WAY ANOVA WITH INTERACTION)

Step 1: Cell means and residuals.

This is essentially the same as for the one-way ANOVA of Chapter 5. There, you wrote each observed value as a group mean plus a residual. Here, you regard each cell as a group and apply the same logic.

$$y = \bar{y}_{kj\cdot} + (y - \bar{y}_{kj\cdot})$$

Obs = Cell mean + Residual

For the upper left cell in Table 6.7 ($k = 1$, $j = 1$; No Antibiotics, No Vitamin B12) the cell mean is 19, and we get

Antibiotics	Vitamin B12	Observed Value		Cell Mean		Residual Error
No	No	30	=	19	+	11
No	No	19	=	19	+	0
No	No	8	=	19	+	-11

Step 2. Additive model for cell means.

To apply the two-way main effects ANOVA model we find the mean for diets with and without *Antibotics* ($No = 20.5, Yes = 28.5$), with and without $Vitamin B12$ ($No = 11, Yes = 38$) and the grand mean weight gain for all 12 pigs ($\bar{y} = 24.5$). From these values we can estimate the effects for both factors.

Factor A:
    Antibiotics (No) $\hat{\alpha}_1 = -4.0$
    Antibiotics (Yes) $\hat{\alpha}_2 = +4.0$
Factor B:
    Vitamin B12 (No) $\hat{\beta}_1 = -8.5$
    Vitamin B12 (Yes) $\hat{\beta}_2 = +8.5$

Step 3. Interaction terms.

While estimating the main effects for each factor is familiar from the additive model, this step introduces new ideas to estimate the interaction. Since the residual error term, ϵ, has mean zero, the mean produced by the two-way (with interaction) ANOVA model for values in the $(k, j)^{th}$ cell is

$$\mu_{kj} = \mu + \alpha_k + \beta_j + \gamma_{kj}$$

To avoid the difficulties we saw in estimating the cell means using only the main effects, we can use the cell mean from the sample, call it $\bar{y}_{kj\cdot}$, as an estimate for μ_{kj} in the population. We already know how to estimate the other components of the equation above so, with a little algebra substituting each of the estimates into the model, we can solve for an estimate of the interaction effect γ_{kj} for the $(k,j)^{th}$ cell.

$$\hat{\gamma}_{kj} = \bar{y}_{kj.} - \bar{y}_{k..} - \bar{y}_{.j.} + \bar{y}_{...}$$

Thus each interaction term is the just difference between the observed cell mean and the fitted mean from the main effects model in Step 2.

For example, consider again the table of cell, row, column, and grand means for the **PigFeed** data.

Cell means			Factor B	Vitamin B12	Row
			No	Yes	Mean
Factor A	No		19.0	22.0	20.5
Antibiotics	Yes		3.0	54.0	28.5
Column	Mean		11.0	38.0	24.5

The estimated interaction effect for the first cell is

$$\gamma_{11} = 19 - 20.5 - 11.0 + 24.5 = 12.0$$

You can check that the other interaction effects in this 2×2 situation are also ± 12 just as the interaction effect that we calculated was; this is a special feature when each factor has just two levels and we have an equal sample size in each cell. Note that when comparing the tables earlier showing the cell means and estimates based on the main effects model for this example, there was a difference of ± 12 in every cell. That is precisely the discrepancy that the interaction term in the model is designed to account for.

The interaction terms allow us to decompose each cell mean into a sum of four pieces:

Antibiotics	Vitamin B12	Cell mean		Grand Average		Antibiotics Effect		Vitamin B12 Effect		Interaction
No	No	19	=	24.5	+	-4	+	-13.5	+	12
No	Yes	22	=	24.5	+	-4	+	13.5	+	-12
Yes	No	3	=	24.5	+	4	+	-13.5	+	-12
Yes	Yes	54	=	24.5	+	4	+	13.5	+	12

Step 4. Decomposition.

Writing each observed value as a cell mean $\bar{y}_{jk}$ plus residual $y - \bar{y}_{jk}$ and then decomposing the cell means allows us to decompose each observed value into a sum of five pieces, or equivalently, to decompose each deviation $y - \bar{y}$ as a sum of four pieces. Squaring these pieces and adding gives the sums of squares needed for the ANOVA table.

6.3. THE TWO-WAY NON-ADDITIVE MODEL (TWO-WAY ANOVA WITH INTERACTION)

Anti-biotics	Vitamin B12	Observed Value		Grand Average		Antibiotics Effect		Vitamin B12 Effect		Inter-action		Resid-ual
No	No	30	-	24.5	=	-4	+	-13.5	+	12	+	11
No	No	19	-	24.5	=	-4	+	-13.5	+	12	+	0
No	No	8	-	24.5	=	-4	+	-13.5	+	12	+	-11
No	Yes	26	-	24.5	=	-4	+	13.5	+	-12	+	4
No	Yes	21	-	24.5	=	-4	+	13.5	+	-12	+	-1
No	Yes	19	-	24.5	=	-4	+	13.5	+	-12	+	-3
Yes	No	5	-	24.5	=	4	+	-13.5	+	-12	+	2
Yes	No	0	-	24.5	=	4	+	-13.5	+	-12	+	-3
Yes	No	3	-	24.5	=	4	+	-13.5	+	-12	+	1
Yes	Yes	52	-	24.5	=	4	+	13.5	+	12	+	-2
Yes	Yes	56	-	24.5	=	4	+	13.5	+	12	+	2
Yes	Yes	54	-	24.5	=	4	+	13.5	+	12	+	0
	SS	3077			=	192	+	2187	+	1728	+	290
	df	11			=	1	+	1	+	1	+	8

If we've added another term to the ANOVA model, we should also expand our ANOVA table to allow us to assess whether the interaction term is actually useful for explaining variability in the response. While a table such as the one above is a useful teaching tool for understanding how the model works, in practice we generally rely on technology to automate these calculations. Fortunately this is generally easy to accomplish with statistical software. For example, here are the results using R. Notice that R does not use the typical "multiplication" notation of the interaction, but instead uses a colon.

```
               Df  Sum Sq  Mean Sq  F value   Pr(>F)
Antibiotic      1  192.00   192.00   5.2966  0.050359 .
B12             1 2187.00  2187.00  60.3310  5.397e-05 ***
Antibiotic:B12  1 1728.00  1728.00  47.6690  0.000124 ***
Residuals       8  290.00    36.25
```

Now we have three different tests:

$H_0 : \alpha_1 = \alpha_2 = 0$ (Main effect for Factor A)
$H_a : \alpha_1 \neq 0$ or $\alpha_2 \neq 0$

$H_0 : \beta_1 = \beta_2 = 0$ (Main effect for Factor B)
$H_a : \beta_1 \neq 0$ or $\beta_2 \neq 0$

$H_0: \gamma_{11} = \gamma_{12} = \gamma_{21} = \gamma_{22} = 0$ (Interaction effect for A·B)
$H_a:$ Some $\gamma kj \neq 0$

We see here in the ANOVA table that the interaction term is quite significant, along with the main effect for *Vitamin B12*, while the main effect for *Antibiotic* appears more significant than in the main effects only model. Once again, we see that by including a new component that successfully accounts for variability in the response, we can make the the other tests more sensitive to differences.

Two-way ANOVA Calculations for Balanced Complete Factorial Data

Balanced Complete Factorial Design - Two Factors

Suppose that A and B are two categorical factors with K levels for Factor A and J levels for Factor B. In a **complete factorial design** we have sample values for each of the KJ possible combinations of levels for the two factors. We say the data is **balanced** if the sample sizes are the same for each combination of levels.

If we let c denote the sample size in each cell, the **PigFeed** data is a 2×2 balanced complete factorial design with $c = 3$ entries per cell. A randomized block design, such as the river data in Example 6.2, is a special case of the balanced complete factorial design where $c = 1$. As we show shortly we can't include an interaction term for randomized block data since there is no variability in each cell to use as an estimate of the error term. With just one entry per cell, if the predicted value is the cell mean in every case, a two-way model with interaction would predict each data case perfectly. It would also exhaust all of the degrees of freedom of the model, essentially adding just enough parameters to the model to match the number of data values in the sample.

So let's assume we have balanced complete factorial data with at least $c > 1$ entries per cell. Start by writing down a partition based on the deviations for each component of the two-way ANOVA model with interaction.

$$y - \bar{y}_{...} = (\bar{y}_{k..} - \bar{y}_{...}) + (\bar{y}_{.j.} - \bar{y}_{...}) + (\bar{y}_{kj.} - \bar{y}_{k..} - \bar{y}_{.j.} + \bar{y}_{...}) + (y - \bar{y}_{kj})$$

Total = Factor A + Factor B + Interaction + Error

There's a lot in this equation but if we break it down in pieces we see the familiar estimates of the main effects and the new estimate of the interaction term. To make the equation balance, the residual error term is $y - \bar{y}_{kj}$. Note that this is just the usual *observed − predicted* value since the "predicted" value in this model is just the cell mean, $\bar{y}_{kj}$. Once we have the partition, we compute sums of squares of all the terms (or most commonly we leave it up to the computer to compile all the sums). Just as for the additive model, if we have the same number of observed values per cell,

6.3. THE TWO-WAY NON-ADDITIVE MODEL (TWO-WAY ANOVA WITH INTERACTION)

we can get the ANOVA sums of squares from the partition by squaring and adding. This leads us to the general two-way ANOVA sum of squares identity for partitioning the variability with a new term, $SSAB$, representing the variability explained by the interaction.

$$SSTotal = SSA + SSB + SSAB + SSE.$$

This all goes into the ANOVA table to develop tests for the interaction and main effects.

Two-way ANOVA Table with Interaction - Balanced Complete Factorial Design

For a balanced complete factorial design with K levels for Factor A, J levels for Factor B and $c > 1$ samples values per cell, the two-way ANOVA model with interaction is

$$y = \mu + \alpha_k + \beta_j + \gamma_{kj} + \epsilon$$

where $\epsilon \approx N(0, \sigma_\epsilon)$ and independent.

The ANOVA table is

Source	D.F.	S.S.	M.S.	F statistic	p-value
Factor A	$K-1$	SSA	MSA=$\frac{SSA}{K-1}$	$\frac{MSA}{MSE}$	$F_{K-1, KJ(c-1)}$
Factor B	$J-1$	SSB	MSB=$\frac{SSB}{J-1}$	$\frac{MSB}{MSE}$	$F_{J-1, KJ(c-1)}$
A × B	(K-1)(J-1)	SSAB	MSB=$\frac{SSAB}{(K-1)(J-1)}$	$\frac{MSAB}{MSE}$	$F_{(K-1)(J-1), KJ(c-1))}$
Error	$KJ(c-1)$	SSE	MSE=$\frac{SSE}{KJ(c-1)}$		
Total	$n-1$	SSTotal			

The usual three hypotheses to test are:
$H_0 : \alpha_1 = \alpha_2 = \cdots \alpha_K = 0$ (Factor A) $H_0 : \beta_1 = \beta_2 = \cdots \beta_J = 0$ (Factor B)
$H_a :$ Some $\alpha_k \neq 0$ $H_a :$ Some $\beta_j \neq 0$

$H_0 :$ All $\gamma_{kj} = 0$ (Interaction)
$H_a :$ Some $\gamma_{kj} \neq 0$

Note that the degrees of freedom for the interaction component is the same as the degrees of freedom for the error term when we have a main effects model for a randomized block design ($c = 1$). In that case the interaction component is precisely the unexplained variability in the sample.

For the sake of completeness, we include here the old-fashioned short-cut formulas, left over from times B.C. (Before Computers), but these formulas are definitely "moldy oldies."

$$\begin{aligned} SSA &= \sum_k Jc(\bar{y}_{k.} - \bar{y}_{..})^2 & \text{Factor A} \\ SSB &= \sum_j Kc(\bar{y}_{.j} - \bar{y}_{..})^2 & \text{Factor B} \\ SSAB &= \sum_{k,j} c(\bar{y}_{kj} - \bar{y}_{k.} - \bar{y}_{.j} + \bar{y}_{..})^2 & \text{Interaction} \\ SSE &= \sum (y - \bar{y}_{kj})^2 & \text{Error} \\ SSTotal &= \sum (y - \bar{y}_{..})^2 & \text{Total} \end{aligned}$$

ASSESS

Checking conditions for the residual errors is the same as before: Look at a residual plot and a normal probability plot. For the pig data, neither plot shows anything unusual (see Figure 6.10), although for the case study of the next section, that will not be so.

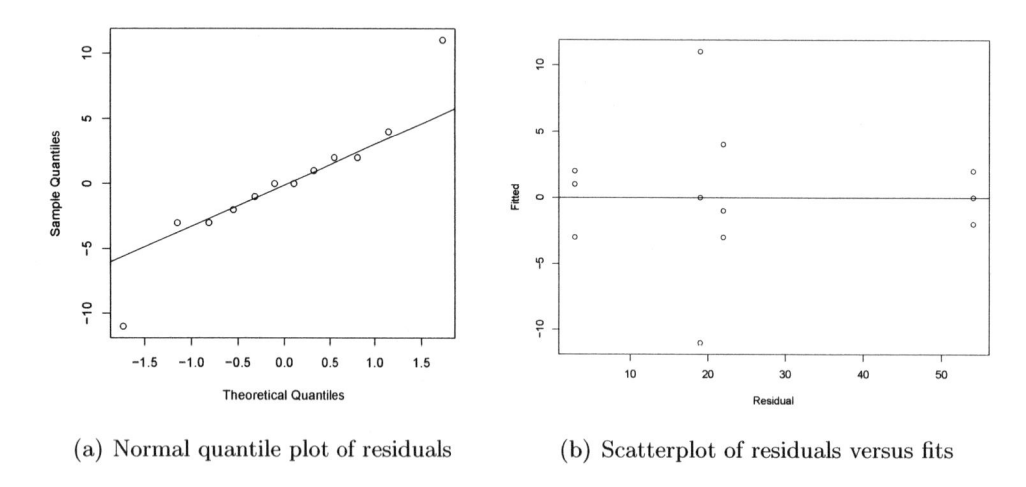

(a) Normal quantile plot of residuals   (b) Scatterplot of residuals versus fits

Figure 6.10: Plots to assess the fit of the ANOVA model

USE

Because there is no evidence that the pig data values violate the requirements of the model, we can base conclusions on the F-tests, and construct confidence intervals. Thus, for example, we can compare the mean weight gain for the control diet (no antibiotics, no vitamin B12) with the mean

6.4. CASE STUDY

gain for the diet with both using the same method as used for the river iron data of Example 6.3. The interval for $\hat{\mu}_{22} - \hat{\mu}_{11}$ is

$$\bar{y}_{22} - \bar{y}_{11} \pm t * \sqrt{\frac{2 * MSE}{c}}$$

where c is the number of observations per cell.

For the pigs, this works out to

$$54 - 19 \pm (2.3060)\sqrt{\frac{2 * 36.25}{3}}$$

or 35 ± 11.3.

Scope of inference. Because the diets were assigned to pigs at random, we are safe in concluding that it was the diets that caused the differences in weight gains. Although the pigs were not chosen through random sampling from a larger population, it is reasonable to think that other pigs would respond in a similar way to the different diets.

6.4 Case Study

We conclude this chapter with a case study illustrating the use of the ANOVA model from start to finish.

Overview: This particular analysis is based on a two-way data set that calls for transforming to a new scale. The original two-way ANOVA turns out to be misleading, in part because the data in the original scale fail to satisfy the conditions required for inference. Once a suitable scale has been chosen, the two-way model with interaction (7 df) can be replaced by a much simpler model (2 df).

Background: Many biochemical reactions are slowed or prevented by the presence of oxygen. For example, there are two simple forms of fermentation, one which converts each molecule of sugar to two molecules of lactic acid, and a second which converts each molecule of sugar to one each of lactic acid, ethanol, and carbon dioxide. The second form is inhibited by oxygen. The particular experiment that we consider here was designed to compare the inhibiting effect of oxygen on the metabolism of two different sugars, glucose and galactose, by Streptococcus bacteria. In this case there were four levels of oxygen that were applied to the two kinds of sugar. The data from the experiment appear in the following table.

	\multicolumn{2}{c}{Oxygen concentration}						
	\multicolumn{2}{c	}{0}	\multicolumn{2}{c	}{46}	\multicolumn{2}{c	}{92}	\multicolumn{2}{c}{138}
Glucose	59	30	44	18	22	23	12	13
Galactose	25	3	13	2	7	0	0	1

Choose, fit, and assess

The two-way structure, with more than one observation per cell, suggests a two-way ANOVA model with interaction as a tentative first model, although it will turn out that choosing a model is not entirely straightforward.

As usual, it is good practice to begin an analysis with a set of plots. We deliberately present these plots with minimal comments, in order to offer our readers the chance to practice. Try to anticipate the analysis to come!

For this data set, because one of the two factors (oxygen concentration) is quantitative, a scatterplot of response (ethanol concentration) versus the quantitative factor offers a good first plot (Figure 6.11).

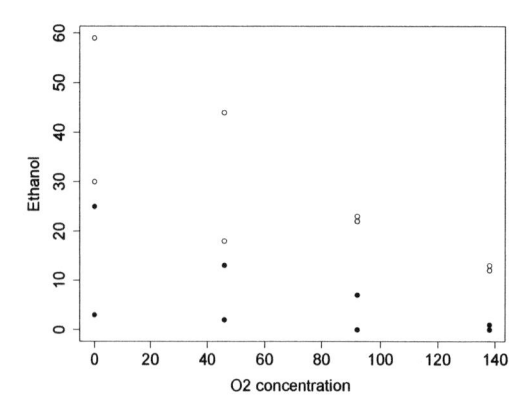

Figure 6.11: Plot of Ethanol concentration versus oxygen concentration. Solid circles: Galactose. Open circles: Glucose.

Next, an interaction plot (Figure 6.12) can show the pattern of the cell means without the distraction from the individual measurements.

As a tentative first analysis, we fit a two-way ANOVA with interaction. Here, just as with the graphs, we wait to comment so that you can think about what the numbers suggest.

```
             Df  Sum Sq  Mean Sq  F value  Pr(>F)
Sugar         1  1806.25  1806.25  13.2935  0.006532 **
Oxygen        3  1125.50   375.17   2.7611  0.111472
Sugar:Oxygen  3   181.25    60.42   0.4446  0.727690
Residuals     8  1087.00   135.87
```

6.4. CASE STUDY

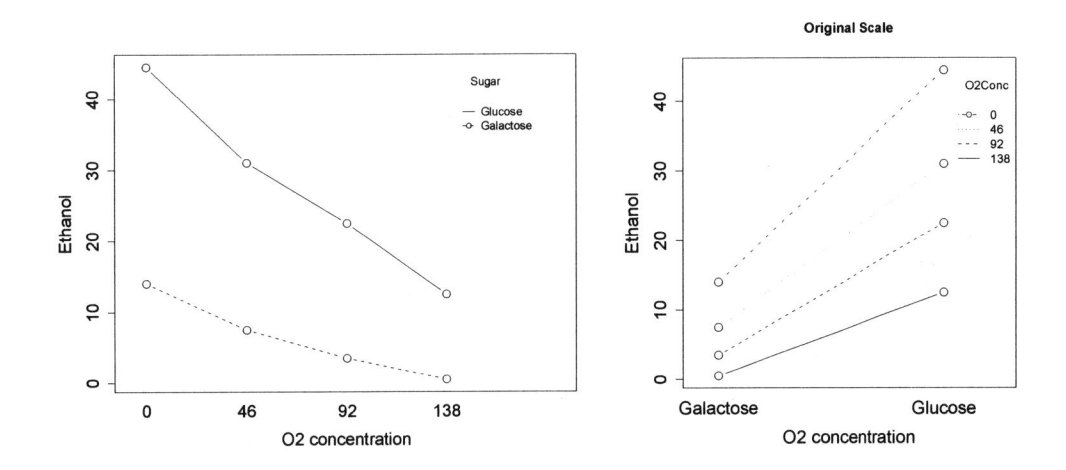

Figure 6.12: Interaction plot for the Sugar Metabolism data. The interaction plot in the left panel is better than the one to it's right, because that first plot makes the patterns easier to see.

After fitting the two-way model we get the residual versus fit plot in Figure 6.13(a) and normal quantile plot in Figure 6.13.

(a) Residuals versus fits plot

(b) Normal quantile plot

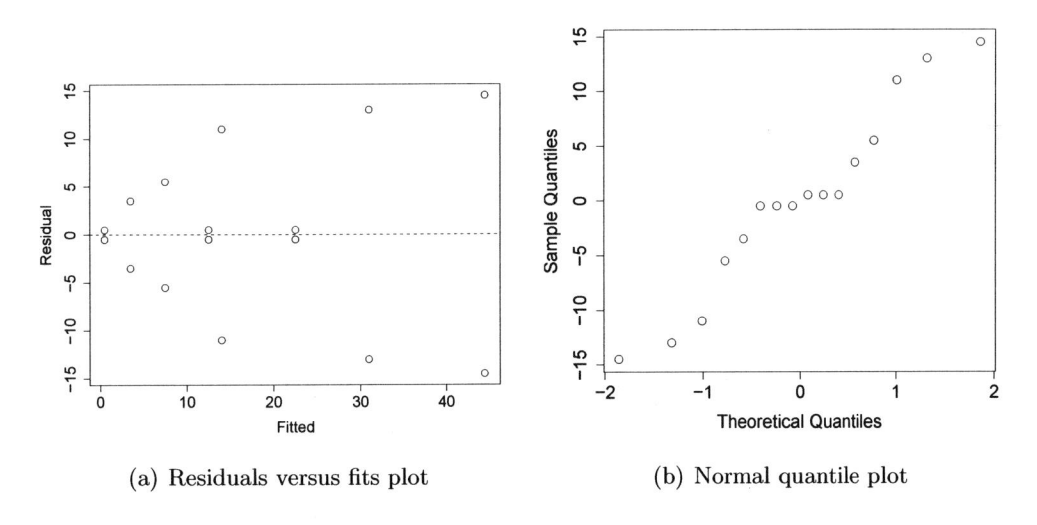

Figure 6.13: Residual plots for the Sugar Metabolism data

Taken together, the dot plot, interaction plot, residual plot, and normal quantile plot show several important patterns:

a. Ethanol production goes down as the oxygen concentration goes up, and the rate of decrease is greater (curves are steeper) at the lower O2 concentrations. Perhaps a transformation to a new scale can straighten the curves.

b. Interaction is present. Ethanol production is much greater from glucose than galactose, and the difference for the two sugars is larger at lower oxygen concentrations. In other words, the inhibiting effect of oxygen is greater for glucose than for galactose. (The curve for glucose is steeper.) Perhaps transforming to a new scale can make a simpler, additive model fit the data.

c. Variation is not constant: Spread is positively related to the size of the observed and fitted values, with larger spreads being many times bigger than the smaller spreads. (See Figure 6.13(a).) Perhaps a transformation to a new scale can make the spreads more nearly equal.

d. The normal quantile plot looks reasonably straight. (The flat stretch of six points in the middle of the plot is due to the three pairs of observed values that differ by only 1. These pairs create residuals of $\pm\frac{1}{2}$.)

e. According to the ANOVA, the observed difference between sugars is too big to be due to chance-like variation, but the effect of increasing oxygen concentration is not pronounced enough to qualify as statistically significant, and the same is true of the interaction. (Preview: A re-analysis of the transformed data will show that the effect of oxygen does register as significant after all.)

The first three patterns suggest transforming to a new scale.

Choose, fit, and assess (continued): Looking for a transformation

The supplementary exercises extend the analysis of the sugar metabolism data by describing new kinds of plots that you can use to find a suitable transformation for a data set like this one. These plots suggest transformations like square roots, cube roots, or logarithms. In what follows, we show you all three.

Figure 6.14 shows interaction plots for the original data along with three transformed versions. We are hoping to find a transformation that does two things at once: (1) It straightens the curved relationships between ethanol concentration and oxygen concentration, and (2) it makes the resulting lines close to parallel, so that we can use an additive model, with no need for interaction terms.

Which plots suggest fitted lines that are closest to parallel? (Any difference in slopes suggests interaction.) Notice that in the original scale (power = 1), the fitted lines get closer together as you go from left to right. For square roots (power = 1/2) and cube roots (power = 1/3), the lines are roughly parallel, and for logarithms (which corresponds to power = 0 – see supplementary exercises) the lines get farther apart as you go from left to right. According to these plots, the log transformation (shown on the bottom right) "goes too far": in the log scale, the line for galactose is steeper than the line for glucose. The cube roots and square roots do the best job of eliminating the interaction, and making the lines nearly parallel. These same two transformations also result in the most nearly linear effect of oxygen concentration.

6.4. CASE STUDY

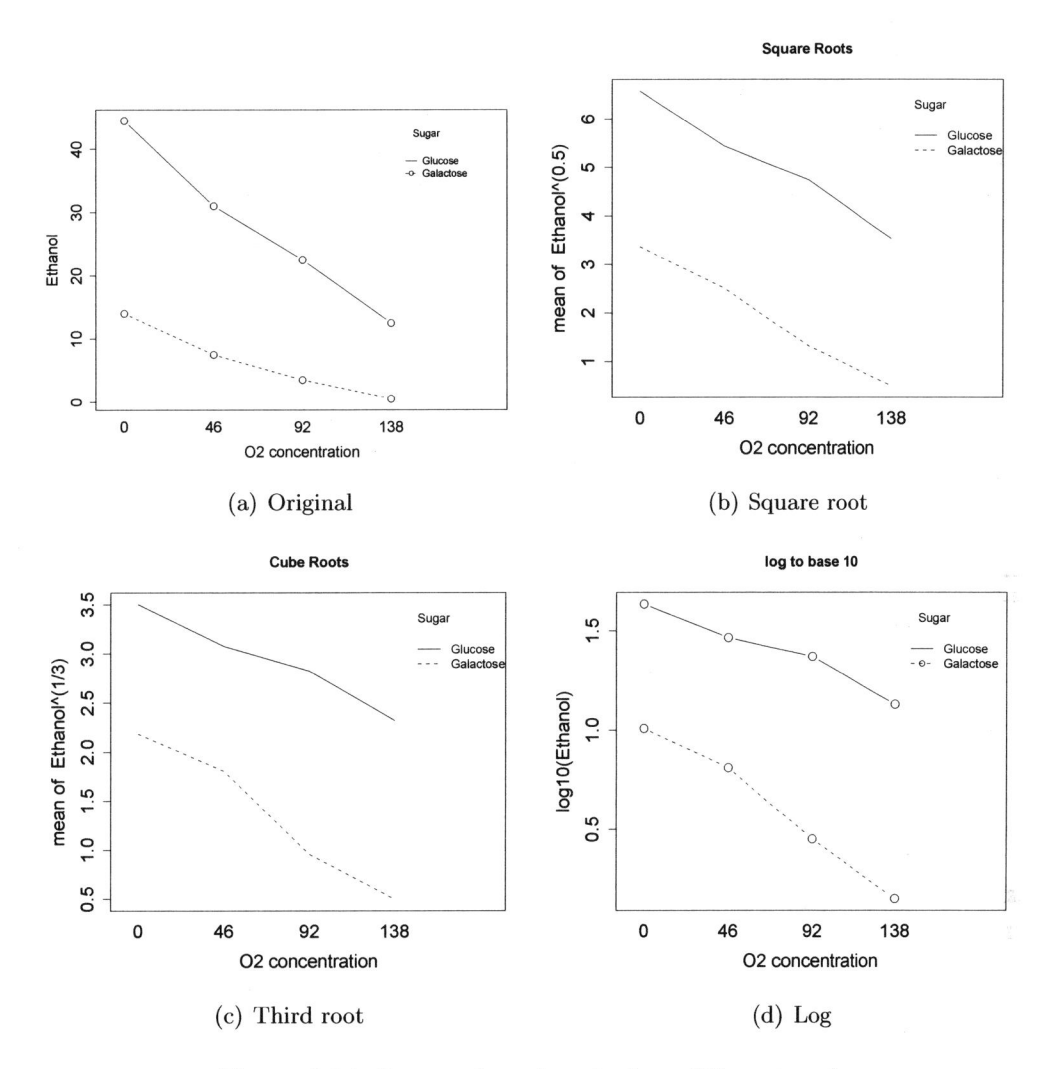

(a) Original	(b) Square root
(c) Third root	(d) Log

Figure 6.14: Interaction plots in four different scales

The interaction plots, especially for the cube roots and square roots, suggest two major simplifications of our original two-way model: (1) We can leave out the interaction terms, and (2) we can treat oxygen concentration as a quantitative predictor instead of a categorical factor. Put differently, we can choose, as our revised model, two parallel lines, one for glucose, and one for galactose.

Here is computer output from fitting this model to data in four scales: original scale (ethanol concentration), square roots, cube roots, and logs.

Original:

```
Residuals:
    Min     1Q  Median     3Q     Max
-14.550  -5.363  -1.400   5.050  20.200

Coefficients:
             Estimate Std. Error t value Pr(>|t|)
(Intercept)    6.3750     3.5133   1.815 0.092730 .
SugarGlucose  21.2500     4.9685   4.277 0.000901 ***
LinearO2      -0.1620     0.0483  -3.353 0.005192 **
---
Signif. codes:  0 '***' 0.001 '**' 0.01 '*' 0.05 '.' 0.1 ' ' 1

Residual standard error: 9.937 on 13 degrees of freedom
Multiple R-squared: 0.6944,    Adjusted R-squared: 0.6473
F-statistic: 14.77 on 2 and 13 DF,  p-value: 0.0004507
```

Square roots:

```
Residuals:
     Min      1Q   Median      3Q     Max
-1.66362 -1.01748  0.05486  1.08597  1.60433

Coefficients:
              Estimate Std. Error t value Pr(>|t|)
(Intercept)   1.924696   0.406507   4.735 0.000390 ***
SugarGlucose  3.149074   0.574887   5.478 0.000106 ***
LinearO2     -0.021318   0.005589  -3.814 0.002148 **
---
Signif. codes:  0 '***' 0.001 '**' 0.01 '*' 0.05 '.' 0.1 ' ' 1

Residual standard error: 1.15 on 13 degrees of freedom
Multiple R-squared: 0.7741,    Adjusted R-squared: 0.7394
F-statistic: 22.28 on 2 and 13 DF,  p-value: 6.311e-05
```

6.4. CASE STUDY

Cube roots:

Residuals:
```
    Min      1Q  Median      3Q     Max
-1.1190 -0.5498  0.1315  0.3601  0.8359
```

Coefficients:
```
             Estimate Std. Error t value Pr(>|t|)
(Intercept)  1.361307   0.220769   6.166 3.4e-05 ***
SugarGlucose 1.568442   0.312215   5.024 0.000233 ***
LinearO2    -0.010533   0.003035  -3.470 0.004145 **
---
Signif. codes:  0 '***' 0.001 '**' 0.01 '*' 0.05 '.' 0.1 ' ' 1
```

Residual standard error: 0.6244 on 13 degrees of freedom
Multiple R-squared: 0.7414,   Adjusted R-squared: 0.7017
F-statistic: 18.64 on 2 and 13 DF, p-value: 0.0001519

Logs:

Residuals:
```
    Min      1Q  Median      3Q     Max
-1.1328 -0.5617  0.1050  0.2413  1.0792
```

Coefficients:
```
             Estimate Std. Error t value Pr(>|t|)
(Intercept)  1.394331   0.249878   5.580 8.92e-05 ***
SugarGlucose 1.830292   0.353380   5.179 0.000177 ***
LinearO2    -0.011371   0.003436  -3.310 0.005641 **
---
Signif. codes:  0 '***' 0.001 '**' 0.01 '*' 0.05 '.' 0.1 ' ' 1
```

Residual standard error: 0.7068 on 13 degrees of freedom
Multiple R-squared: 0.744,   Adjusted R-squared: 0.7046
F-statistic: 18.89 on 2 and 13 DF, p-value: 0.0001424

Several features of this output are worth noting.

- First, the results are strikingly similar for all four ANOVA tables. In particular, the effects of both sugar and oxygen are highly significant.

- Second, the simplification in the model has reduced the model degrees of freedom from 7 in the original model to just 2 in the current model. The extra 5 degrees of freedom are now part of the residual sum of squares. Informally, our analysis suggests that the extra 5 degrees of freedom were not linked to differences, but just to additional error. In different words, the remaining 2 degrees of freedom in our model are linked to "signal," and, in a suitable scale, all the other degrees of freedom are associated with "noise." As a rule, other things being equal, the larger the residual degrees of freedom, the better, provided these degrees of freedom are associated with noise and not with signal.

- Third, the values of R-squared (adjusted) are nearly the same for all four versions of the model: 64.7% (original), 73.9% (square roots), 70.2% (cube roots), and 70.5% (logs). A slight edge goes to square roots, but any of the four models seems acceptable.

The normal probability plot and residual-versus-fit plots look pretty much the same in all four scales, so we show only one of each, for the cube roots:

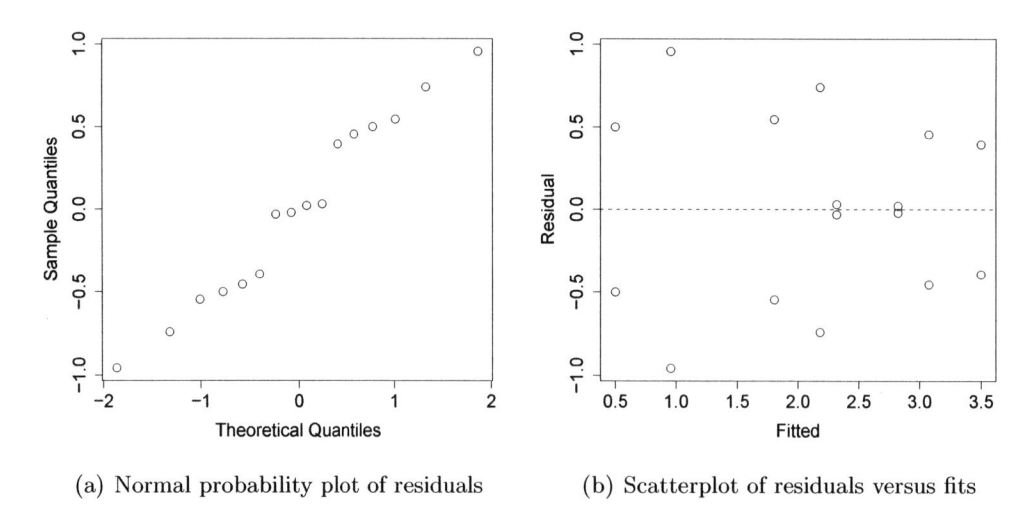

(a) Normal probability plot of residuals (b) Scatterplot of residuals versus fits

Figure 6.15: Plots to assess the fit of the ANOVA model using the third root of the ethanol measurement as the response variable

No single one of the three transformations (square root, cube root and log) stands out as clearly superior to the other two, although the interaction plots look slightly better for cube roots and square roots than for logs. Any of the three transformations does better than the original scale: the

6.5. CHAPTER SUMMARY

variation is more nearly constant, the effect of oxygen is more nearly the same for the two sugars, and (unlike the case with the original scale) the effect of oxygen registers as statistically significant in a two-way ANOVA.

The choice among the three scales is somewhat arbitrary. In the cube roots scale, the fitted model tell us that the cube root of the ethanol concentration is about 1.568 higher for glucose than for galactose, at all oxygen concentrations, and decreases by about 0.01 for every increase of one in oxygen concentration.[4]

6.5 Chapter Summary

In this chapter we introduced two different **multi-factor ANOVA** models. We concentrated specifically on **two-way models** (models with two factors) but the ideas from this chapter can be extended to three or more factors.

The first model considered was the **two-way additive model** (also known as the **main effects model**)

$$y = \mu + \alpha_k + \beta_j + \epsilon$$

where $\epsilon \sim N(0, \sigma_\epsilon)$ and the errors are independent of one another. The variability is now partitioned into three parts representing Factor A, Factor B, and error. The resulting **ANOVA table** has three lines, one for each factor and one for the error term. The table allows us to test the effects of each of the two factors (known as the **main effects**) using F-tests.

As in the situation of one-way ANOVA, we can use **Fisher's LSD** for any factors found to be significant. These intervals, based on the common value of the MSE, help us decide which groups are different and by how much.

If there is only one observation per **cell** (combination of levels from both factors) then the additive model is the only model we can fit. If, however, there is more than one observation per cell, we often start by considering the second model we introduced, the **two-way non-additive model** (also known as the **two-way ANOVA with interaction**). For ease of calculation, we only consider models that are **balanced**, that is, models that have the same number of observations per cell.

We say an **interaction** between two factors exists if the relationship between one factor and the response variable changes for the different levels of the second factor. **Interaction plots** can give a good visual sense of whether an interaction may be present or not.

If an interaction appears to be present, this should be accounted for in the model so we should use

---
[4]The log transformation, although it doesn't produce quite so nice an interaction plot, offers ease of interpretation, and the familiarity that comes from its frequent use to express concentrations.

the **two-way non-additive model**

$$y = \mu + \alpha_k + \beta_j + \gamma_{kj} + \epsilon,$$

where $\epsilon \sim N(0, \sigma_\epsilon)$ and the errors are independent of one another. The γ_{kj} term represents the interaction in this model.

The **ANOVA table** now partitions the variability into four sources: Factor A, Factor B, interaction, and error. This table leads to three F-tests, one for each factor, and one for the interaction. If there is a significant interaction effect, interpreting significant main effects can be difficult. Again, **interaction plots** can be useful.

Finally, a case study leads us through the process from start to finish and points out some ways to incorporate both analysis of variance and regression analysis.

6.6 Exercises

Conceptual Exercises

Exercises 6.1 to 6.6. **Factors and levels.** For each of the following studies: (a) give the response, (b) the name of the two factors, (c) for each factor, tell whether it is observational or experimental and the number of levels it has, and (d) tell whether the study is a compete block design.

6.1 *Bird calcium.* Ten male and ten female robins were randomly divided into groups of five. Five birds of each sex were given a hormone in their diet; the other five of each sex were given a control diet. At the end of the study, the researchers measured the calcium concentration in the plasma of each bird.

6.2 *Crabgrass competition.* In a study of plant competition, two species of crabgrass (*Digitaria sanguinalis (D.s.)* and *Digitatria ischaemum (D.i.)*) were planted together in a cup. In all, there were twenty cups, each with 20 plants. Four cups held 20 D.s. each; another four held 15 D.s. and 5 D.i., another four held 10 of each species, still another four held 5 D.s. and 15 D.i., and the last four held 20 D.i. each. Within each set of four cups, two were chosen at random to receive nutrients at normal levels; the other two cups in the set received nutrients at low levels. At the end of the study the plants in each cup were dried and the total weight recorded.

6.3 *Behavior therapy for stuttering.* Thirty-five years ago the journal *Behavior Research and Therapy*[5] reported a study that compared two mild shock therapies for stuttering. There were 18 subjects, all of them stutterers. Each subject was given a total of three treatment sessions, with the order randomized separately for each subject. One treatment administered a mild shock *during* each moment of stuttering, another gave the shock *after* each stuttered word, and the third treatment was a control, with no shock. The response was a score that measured a subject's adaptation.

6.4 *Noise and ADHD.* It is now generally accepted that children with attention deficit and hyperactivity disorder tend to be particularly distracted by background noise. About 20 years ago a study was done to test this hypothesis.[6] The subjects were all second graders. Some had been diagnosed as hyperactive; the other subjects served as a control group. All the children were given sets of math problems to solve, and the response was their score on the set of problems. All the children solved problems under two sets of conditions, high noise and low noise. (Results showed that the controls did better with the higher noise level, whereas the opposite was true for the hyperactive children.)

6.5 *Running dogs.* In a study conducted at the University of Florida (*Science News*, July 20, 2002) investigators compared the effects of three different diets on the speed of racing dogs. (The investigators wanted to test the common belief among owners of racing dogs that giving their dogs

---

[5] "Rate of stuttering adaptation under two electro-shock conditions," 1967, pp. 49-54.
[6] S.S. Zentall and J.H. Shaw (1980), "Effects of classroom noise on performance and activity of second-grade hyperactive and control children," J. Ed. Psych., 72, pp. 830-840.

large doses of vitamin C will cause them to run faster.) In the University of Florida study, each of five dogs got all three diets, one at a time, in an order that was determined using a separate randomization for each dog. (To the surprise of the scientists, the results showed that when the dogs ate the diet high in vitamin C, they ran slower, not faster.)

6.6 *Fat rats.* Is there a magic shot that makes dieting easy? Researchers investigating appetite control measured the effect of two hormone injections, leptin and insulin, on the amount eaten by rats (*Science News*, July 20, 2002). Male rats and female rats were randomly assigned to get one hormone shot or the other. (The results showed that for female rats, leptin lowered the amount eaten, compared to insulin; for male rats, insulin lowered the amount eaten, compared to leptin.)

6.7 *F-ratios.* Suppose you fit the two-way main effects ANOVA model for the river data of Example 6.2, this time using the log concentration of copper (instead of iron) as your response, and then do an F-test for differences between rivers.

a. If the F-ratio is near 1, what does that tell you about the differences between rivers?

b. If the F-ratio is near 10, what does that tell you?

6.8 *Degrees of freedom.* If you carry out a two-factor ANOVA (main effects model) on a data set with Factor A at four levels and Factor B at five levels, with one observation per cell, how many degrees of freedom will there be for

a. Factor A?

b. Factor B?

c. Interaction?

d. Error?

6.9 *More degrees of freedom.* If you carry out a two-factor ANOVA (with interaction) on a data set with Factor A at four levels and Factor B at five levels, with three observations per cell, how many degrees of freedom will there be for

a. Factor A?

b. Factor B?

c. Interaction?

d. Error?

6.10 *Fill in the blank.* If your data set has two factors and you carry out a one-way ANOVA, ignoring the second factor, your SSE will be too _____ (small, large) and you will be _____ (more, less) likely to detect real differences than would a two-way ANOVA.

6.6. EXERCISES

6.11 *Fill in the blank, again.* If you have two-way data with one observation per cell the only model you can fit is a main effects model and there is no way to tell whether interaction is present. If in fact there is interaction present, your SSE will be too _____ (small, large) and you will be _____ (more, less) likely to detect real differences due to each of the factors.

6.12 *Interaction.* Is interaction present in the following data? How can you tell?

	Heart	Soul
Democrats	2, 3	10, 12
Republicans	8, 4	11, 8

6.13 *Interaction, again.* Is interaction present in the following data? How can you tell?

	Blood	Sweat	Tears
Males	5, 10	10, 20	15, 15
Females	15, 20	20, 30	25, 25

Exercises 6.14 to 6.17 are True or False exercises. If the statement is false, explain why it is false.

6.14 If interaction is present, it is not possible to describe the effects of a factor using just one set of estimated main effects.

6.15 In a randomized complete block study, at least one of the factors must be experimental.

6.16 The conditions for the errors for the two-way additive ANOVA model are the same as the conditions for the errors of the one-way ANOVA model.

6.17 The conditions for the errors of the two-way ANOVA model with interaction are the same as the conditions for the errors of the one-way ANOVA model.

Guided Exercises

6.18 *Burning calories.* If you really work at it, how long does it take to burn 200 calories on an exercise machine? Does it matter whether you use a treadmill or a rowing machine? An article in *Medicine and Science in Sports and Exercise* (August, 2001) reported average times to burn 200 calories for men and for women using a treadmill and a rowing machine for heavy exercise. The results:

Average minutes to burn 200 calories

	Men	Women
Treadmill	12	17
Rowing machine	14	16

a. Draw both interaction graphs.

b. If you assume that each person in the study used both machines, in a random order, is the two-way ANOVA model appropriate? Explain.

c. If you assume that each subject used only one machine, either the treadmill or the rower, is the two-way ANOVA model appropriate? Explain.

6.19 *Drunken teens, part 1.* A survey was done to find the percentage of 15-year-olds, in each of 18 European countries, who reported having been drunk at least twice in their lives. Here are the results, for boys and girls, by region. (Each number is an average for 6 countries.)

	Male	Female
Eastern	24.17	42.33
Northern	51.00	51.00
Continental	24.33	33.17

Draw an interaction plot, and discuss the pattern. Relate the pattern to the context. (Don't just say "The lines are parallel, so there is no interaction," or "The lines are not parallel, so interaction is present.")

6.20 *Happy face: interaction.* Researchers at Temple University[7] wanted to know: if you work waiting tables and you draw a happy face on the back of your customers' checks, will you get better tips? To study this burning question at the frontier of science, they enlisted the cooperation of two servers at a Philadelphia restaurant. One was male, the other female. Each server recorded his or her tips for their next 50 tables. For 25 of the 50, following a predetermined randomization, they drew a happy face on the back of the check. The other 25 randomly chosen checks got no happy face. The response was the tip, expressed as a percentage of the total bill. The averages for the male server were 18% with a happy face, 21% with none. For the female server, the averages were 33% with a happy face, 28% with none.

a. Regard the data set as a two-way ANOVA, which is the way it was analyzed in the article. Name the two factors of interest, tell whether each is observational or experimental, and tell the number of levels.

b. Draw an interaction graph. Is there evidence of interaction? Describe the pattern in words, using the fact that an interaction, if present, is a difference of differences.

---

[7]B. Rind and P. Bordia (1996). "Effect on restaurant tipping of male and female servers drawing a happy face on the backs of customers' checks," *J. Soc. Psych*, 26, pp. 215-225.

6.6. EXERCISES

6.21 *Happy face: ANOVA (continued).* A partial ANOVA table is given below. Fill in the missing numbers.

Source	df	SS	MS	F
Face (Yes/No)				
Gender (M/F)			2,500	
Interaction		400		
Residuals			100	
Total		25,415		

6.22 *River iron.* This is an exercise with a moral: Sometimes, the way to tell that your model is wrong requires you to ask "Do the numbers make sense in the context of the problem?" Consider the New York river data of Example 6.2, with iron concentrations in the original scale of parts per million:

	Grasse	Oswegatchie	Raquette	St. Regis	Mean
upstream	944	860	108	751	665.75
midstream	525	229	36	568	339.50
downstream	327	130	30	350	209.25
Mean	598.7	406.3	58.0	556.3	404.83

a. Fit the two-way additive model $FE = River + Site + Error$.

b. Obtain a normal probability plot of residuals. Is there any indication from this plot that the normality condition is violated?

c. Obtain a plot of residuals versus fitted values. Is there any indication from the shape of this plot that the variation is not constant? Are there pronounced clusters? Is there an unmistakable curvature to the plot?

d. Finally, look at the leftmost point, and estimate the fitted value from the graph. Explain why this one fitted value strongly suggests that the model is not appropriate.

6.23 *Iron deficiency.* In developing countries, roughly one fourth of all men and half of all women and children suffer from anemia due to iron deficiency. Researchers[8] wanted to know whether the trend away from traditional iron pots in favor of lighter, cheaper aluminum could be involved in this most common form of malnutrition. They compared the iron content of twelve samples of three Ethiopian dishes: one beef, one chickpea, and one vegetable casserole. Four samples of each dish were cooked in aluminum pots, four in clay pots, and four in iron pots. Given below is a parallel dot plot of the data.

---

[8] A.A. Adish, et al. (1999). "Effect of food cooked in iron pots on iron status and growth of young children: a randomized trial," *The Lancet*, 353, pp. 712-716.

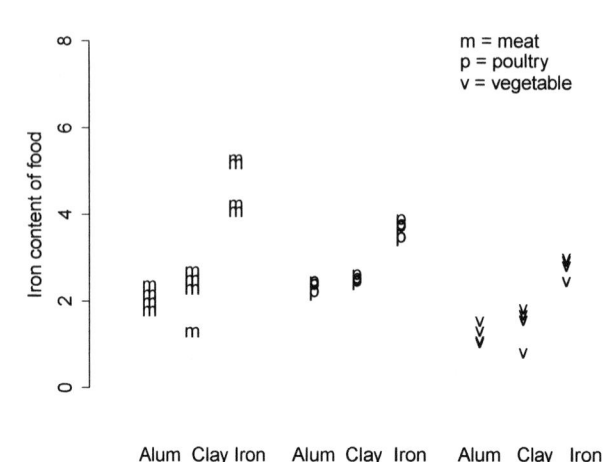

Describe what you consider to be the main patterns in the plot. Cover the usual features keeping in mind that in any given plot, some features deserve more attention than others: how are the group averages related (to each other and to the researchers' question)? Are there gross outliers? Are the spreads roughly equal? If not, is there evidence that a change of scale would tend to equalize spreads?

6.24 *Alfalfa sprouts.* Some students were interested in how an acidic environment might affect the growth of plants. They planted alfalfa seeds in 15 cups and randomly chose five to get plain water, five to get a moderate amount of acid (1.5M HCl), and five to get a stronger acid solution (3.0M HCl). The plants were grown in an indoor room so the students assumed that the distance from the main source of daylight (a window) might have an affect on growth rates. For this reason, they arranged the cups in five rows of three with one cup from each *Acid* level in each row. These are labeled in the data set as *Row*: a= farthest from the window through e=nearest to the window. Each cup was an experimental unit and the response variable was the average height of the alfalfa sprouts in each cup after four days (*Ht*4). The data are shown in Table 6.8 and stored in the **Alfalfa** file.

Treatment\Cup	a	b	c	d	e
water	1.45	2.79	1.93	2.33	4.85
1.5 HCL	1.00	0.70	1.37	2.80	1.46
3.0 HCL	1.03	1.22	0.45	1.65	1.07

Table 6.8: Four-day alfalfa growth

a. Find the means for each row of cups (a, b, ... e) and each treatment (water, 1.5HCL, 3.0HCl). Also find the average and standard deviation for the growth in all 15 cups.

6.6. EXERCISES

b. Construct a two-way main effects ANOVA table for testing for differences in average growth due to the acid treatments using the rows as a blocking variable.

c. Check the conditions required for the ANOVA model.

d. Based on the ANOVA would you conclude that there is a significant difference in average growth due to the treatments? Explain why or why not.

e. Based on the ANOVA would you conclude that there is a significant difference in average growth due to the distance from the window? Explain why or why not.

6.25 *Alfalfa sprouts (continued).* Refer to the data and two-way ANOVA on alfalfa growth in Exercise 6.24. If either factor is significant, use Fisher's LSD (at a 5% level) to investigate which levels are different.

6.26 *Unpopped popcorn.* Lara and Lisa don't like to find unpopped kernels when they make microwave popcorn. Does the brand make a difference? They conducted an experiment to compare Orville Redenbacher's Light Butter Flavor vs. Seaway microwave popcorn. They made 12 batches of popcorn, 6 of each type, cooking each batch for four minutes. They noted that the microwave oven seemed to get warmer as they went along so they kept track of six trials and randomly chose which brand would go first for each trial. For a response variable they counted the number of unpopped kernels and then adjusted the count for Seaway for having more ounces per bag of popcorn (3.5 vs 3.0). The data are shown below and stored in **Popcorn**.

Brand \ Trial	1	2	3	4	5	6
Orville	26	35	18	14	8	6
Seaway	47	47	14	34	21	37

Table 6.9: Unpopped Popcorn by *Brand* and *Trial*

a. Find the mean number of unpopped kernels for the entire sample and estimate the effects (α_1 and α_2) for each brand of popcorn.

b. Run a two-way ANOVA model for this randomized block design. (Remember to check the required conditions.)

c. Does the brand of popcorn appear to make a difference in the mean number of unpopped kernels? what about the trial?

6.27 *Swahili attitudes.* Hamisi Babusa, a Kenyan scholar, administered a survey to 480 students from Pwani and Nairobi provinces about their attitudes towards the Swahili language. In addition, the students took an exam on Swahili. From each province, the students were from 6 schools (3 girls schools and 3 boys schools) with 40 students sampled at each school, so half of the students from each province were males and the other half females. The survey instrument contained 40 statements about attitudes towards Swahili and students rated their level of agreement to each. Of these questions, 30 were positive questions and the remaining 10 were negative questions. On an individual question the most positive response would be assigned a value of 5 while the most negative response would be assigned a value of 1. By summing (adding) the responses to each question, we can find an overall *Attitude Score* for each student. The highest possible score would be 200 (an individual who gave the most positive possible response to every question). The lowest possible score would be 40 (an individual who gave the most negative response to every question). The data are stored in **Swahili**.

a. Investigate these data using *Province* (Nairobi or Pwani) and *Sex* to see if attitudes towards Swahili are related to either factor or an interaction between them. For any effects that are significant, give an interpretation that explains the direction of the effect(s) in the context of this data situation.

b. Do the normality and equal variance assumptions look reasonable for the model you chose in (a)? Produce a graph (or graphs) and summary statistics to justify your answers.

c. The **Swahili** data also contains a variable coding the school for each student. There are 12 schools in all (labeled A, B, ..., L). Despite the fact that we have an equal sample size from each school, explain why an analysis using *School* and *Province* as factors in a two-way ANOVA would NOT be a balanced complete factorial design.

Supplementary Exercises

A: Transforming to equalize standard deviations.

When group standard deviations are very unequal (S_{max}/S_{min} is large), you can sometimes find a suitable transformation as follows:

Step 1: Compute the average and standard deviation for each group.

Step 2: Plot $\log(s)$ versus $\log(\text{ave})$.

Step 3: Fit a line by eye and estimate the slope.

Step 4: Compute $p = 1 - \text{slope}$. p tells us the transformation. For example, $p = 0.5$ means the square root, $p = 0.25$ means the fourth root, and $p = -1$ means the reciprocal. (For technical

6.6. EXERCISES

reasons, $p = 0$ means the logarithm.)

6.28 *Simple illustration.* This exercise was invented to show the method described above at work using simple numbers. Consider a data set with four groups and three observations per group:

Group	Observed values
A	0.9, 1.0, 1.1
B	9, 10, 11
C	90, 100, 110
D	900, 1000, 1100

Notice that you can think of each set of observed values as $m - s$, m, and $m + s$. Find the ration s/m for each group and notice that the "errors" are constant in *percentage* terms. For such data, a transformation to logarithms will equalize the standard deviations, so applying steps 1-4 should show that transforming is needed, and that the right transformation is $p = 0$.

a. Compute the means and standard deviations for the four groups. (Don't use a calculator. Use a short cut instead: check that if the response values are $m - s$, m, and $m + s$, then the mean is m and the standard deviation is s.)

b. Compute S_{max}/S_{min}. Is a transformation called for? itemPlot $\log_{10}(s)$ versus $\log_{10}(m)$, and fit a line by eye. (Note that the fit is perfect: the right transformation will make $S_{max}/S_{min} = 1$ in the new scale.)

c. What is $p = 1 - $ slope? What transformation is called for?

d. Use a calculator to transform the data and compute new group means and standard deviations.

e. Check S_{max}/S_{min}. Has changing scales made the standard deviations more nearly equal?

6.29 *Sugar metabolism.* Figure 6.16 shows the plot of the log(s) versus log(ave) for the data from this chapter's case study.

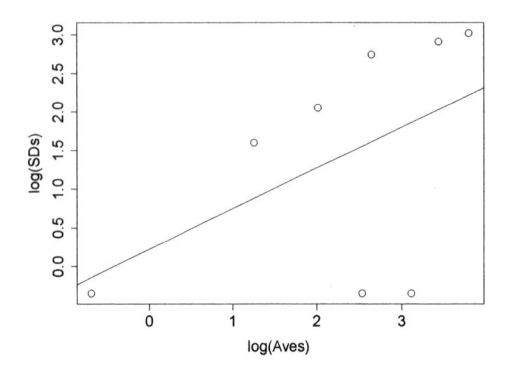

Figure 6.16: Graph of log(standard deviations) versus log(averages) for the eight groups of the sugar metabolism study.

a. Fit a line by eye to all eight points, estimate the slope, and compute $p = 1 - \text{slope}$. What transformation is suggested?

b. If you ignore the two outliers, the remaining six points lie very close to a line (see Figure 6.17). Estimate its slope and compute $p = 1 - \text{slope}$. What transformation is suggested?

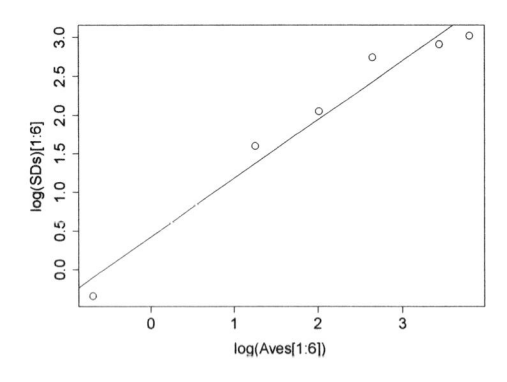

Figure 6.17: Graph of log(standard deviations) versus log(averages) for six groups of the sugar metabolism study.

6.30 *Diamonds.* Here are the means, standard deviations, and their logs for the diamond data of Example 5.7.

6.6. EXERCISES

Color	Mean	st. dev.	log(Mean)	log(st. dev.)
D	0.8225	0.3916	-0.0849	-0.4072
E	0.7748	0.2867	-0.1108	-0.5426
F	1.0569	0.5945	0.0240	-0.2258
G	1.1685	0.5028	0.0676	-0.2986

a. Plot $\log(s)$ versus $\log(\text{ave})$ for the four groups. Do the points suggest a line?

b. Fit a line by eye and estimate its slope.

c. Compute $p = 1 - \text{slope}$. What transformation, if any, is suggested?

B: Transforming for additivity.

For many two-way data sets, the additive (no interaction) model does not fit when the response is in the original scale. For some of these data sets, however, there is a transformed scale for which the additive model does fit well. When such a transformation exists, you can find it as follows:

Step 1: Fit the additive model to the cell means, and write the observed values as a sum:

$$\text{obs} = \text{Grand Ave} + \text{Row Effect} + \text{Col Effect} + \text{Res}$$

Step 2: Compute a "comparison value" for each cell:

$$\text{Comp} = (\text{Row eff})(\text{Col effect})/\text{Grand}$$

Step 3: Plot residual versus comparison values.

Step 4: Fit a line by eye and estimate its slope.

Step 5: Compute $p = 1 - \text{slope}$; p tells the transformation as in A above.

6.31 *Sugar metabolism.* Figure 6.18 shows the plot of residuals versus comparison values for the data from this chapter's case study. Fit a line by eye and estimate its slope. What transformation is suggested?

6.32 *River iron.* Here is the river iron data in the original scale

Site	Grasse	Oswegatchie	Raquette	St. Regis
Up	944	860	108	751
Mid	525	229	36	568
Down	327	130	30	350

A decomposition of the observed values gives a grand average of 404.83, site effects of 260.92, -65.33 and -195.58, and river effects of 193.83, 1.50, and -346.83. The residuals are:

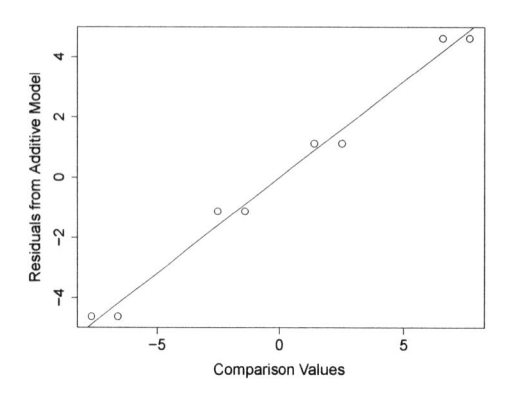

Figure 6.18: Graph of the residuals versus the comparison values for the sugar metabolism study.

Site	Grasse	Oswegatchie	Raquette	St. Regis
Up	84.42	192.75	-210.92	-66.25
Mid	-8.33	-112.00	43.33	77.00
Down	-76.08	-80.75	167.58	-10.75

a. Use a calculator or spreadsheet to compute the comparison values.

b. Plot the residuals versus the comparison values.

c. Fit a line by eye and estimate its slope. What transformation is suggested?

Chapter 7

Additional ANOVA Topics

In Chapters 5 and 6 we introduced you to what we consider to be the basics of analysis of variance. Those chapters contain the topics we think you must understand to be equipped to follow someone else's analysis and to be able to adequately perform an ANOVA on your own data set. This chapter takes a more in-depth look at the subject of ANOVA and introduces you to ideas that, while not strictly necessary for a beginning analysis, will substantially strengthen an analysis of data.

We prefer to think of the sections of this chapter as topics rather than subdivisions of a single theme. Each topic stands alone and they can be read in any order. Because of this, the exercises at the end of the chapter have been organized by topic.

7.1 Topic: Levene's Test for Homogeneity of Variances

When we introduced the ANOVA procedure in Chapter 5, we discussed the conditions that are required in order for the model to be appropriate, and how to check them. In particular, we discussed the fact that all error terms need to come from a distribution with the same variance. That is, if we group the error terms by the levels of the factor, all groups of errors need to be from distributions that have a common variance. In Chapter 5 we gave you two ways to check this: making a residual plot and comparing the ratio S_{max}/S_{min} to 2. Neither of these guidelines is very satisfying. The value 2 is just a "rule of thumb" and the usefulness of this rule depends on the sample sizes (when the sample sizes are equal, the value 2 can be replaced by 3 or more). Moreover the residual plot also carries a lot of subjectivity with it. No two groups will have exactly the same spread. So how much difference in spread is tolerable and how much is too much?

In this section we introduce you to another way to check this condition: **Levene's test for homogeneity of variances**. You may be wondering why we didn't introduce you to this test in Chapter 5 when we first discussed checking conditions. The answer is simple: Levene's test is a form of an ANOVA itself, which means that we couldn't introduce it until we had finished introducing the ANOVA model.

Levene's test is designed to test the hypotheses

$H_0: \sigma_1^2 = \sigma_2^2 = \sigma_3^2 = ... = \sigma_K^2$ (no differences in variances)
H_a : Not all variances are equal

Notice that for our purposes, what we are hoping to find is that the data are *consistent* with H_0, rather than the usual hope of rejecting H_0. This means we hope to find a fairly large p-value rather than a small one.

Levene's test first divides the data points into groups based on the level of the factor they are associated with. Then the median value of each group is computed. Finally the absolute deviation between each point and the median of its group is calculated. In other words, we calculate |observed − Median$_k$| for each point using the median of the group that the observation belongs to. This give us a set of estimated error measurements based on the grouping variable. These absolute deviations are not what we typically call residuals (observed - predicted) but they are similar in that they give an idea of the amount of variability within each group. Now we want to see if the average absolute deviation is the same for all groups or if at least one group differs from the others. This leads us to employ an ANOVA model.

> **Levene's Test for Homogeneity of Variances**
>
> Levene's Test for Homogeneity of Variances tests the null hypothesis of equality of variances by computing the absolute deviations between the observations and their group medians, and then applying an ANOVA model to those absolute deviations. If the null hypothesis is *not* rejected, then the condition of equality of variances required for the original ANOVA analysis can be considered to be met.

The procedure, as described above, sounds somewhat complicated if we need to compute every step, but most software will perform this test for you from the original data, without you having to make the intermediate calculations. We now illustrate this test with several data sets that you have already seen.

Example 7.1: *Checking equality of variances in the fruit fly data*

The data for this example are in the file **FruitFlies**. Recall from Chapter 5 that we applied both of our previous methods for checking the equality of the variances in the various groups of fruit flies. Figure 7.1 is the dotplot of the residuals for this model. We observed in Chapter 5 that the spreads seem to be rather similar in this graph.

We also computed the ratio of the extreme standard deviations to be $16.45/12.10 = 1.36$ which we observed was closer to 1 than to 2 and so seemed to be perfectly acceptable.

7.1. TOPIC: LEVENE'S TEST FOR HOMOGENEITY OF VARIANCES

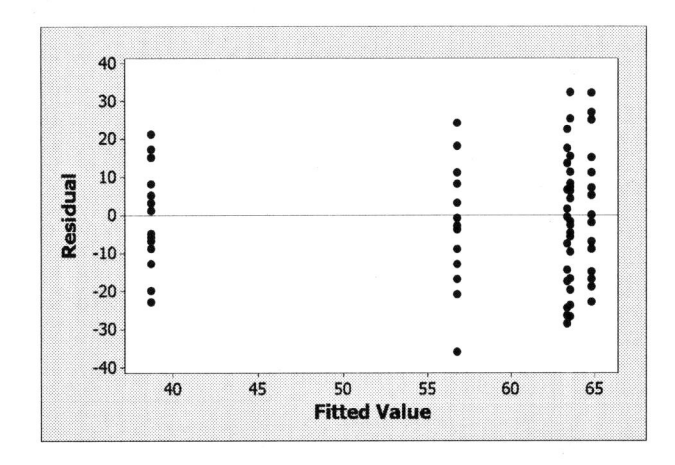

Figure 7.1: Residuals versus fitted values for fruit flies

Now we apply Levene's test to these data. Computer output is given below.

```
Levene's Test (Any Continuous Distribution)
Test statistic = 0.49, p-value = 0.742
```

Notice that the p-value is 0.742 which is a rather large p-value. We do not have enough evidence to reject the null hypothesis. Or, put another way, our data are consistent with the null hypothesis. That is, our data are consistent with the idea that the (population) variances are the same among all 5 groups. So our decision to proceed with the ANOVA model in Chapter 5 for these data is justified. ◊

Now recall the cancer survivability data set from Chapter 5 (Example 5.9). In that case, when we looked at the original data we had some concerns about equality of the variances.

Example 7.2: *Checking equality of variances in the cancer data set*

For this data set, found in the file **CancerSurvival**, first presented in Section 5.3, we saw immediately that the condition of equality of variances was violated. Figure 7.2 is the residual plot for the cancer data. It is quite clear from this plot that there are differences in the variances among the various forms of cancer. Also, when we calculated the ratio of the maximum standard deviation to the minimum standard deviation we obtained 5.9, which is far greater than 2.

What does Levene's test tell us? We would hope that it would reject the null hypothesis in this case. And, in fact, as the output below shows, it does.

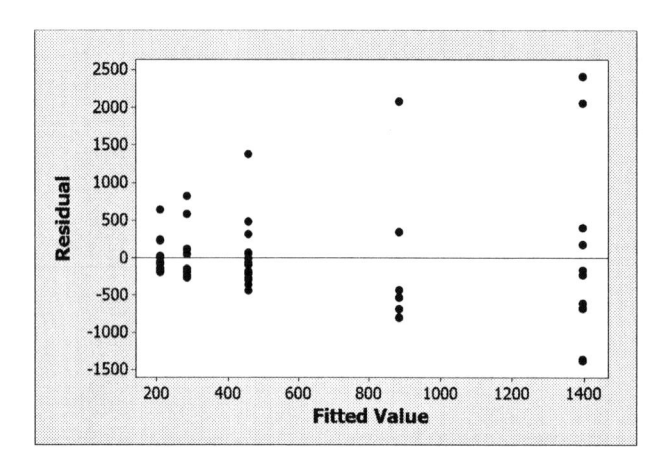

Figure 7.2: Residuals versus fitted values for cancer survival times

```
Levene's Test (Any Continuous Distribution)
Test statistic = 4.45, p-value = 0.003
```

With a p-value of 0.003, there is strong evidence that the condition of equal variances is violated. Our suggestion in Chapter 5 was to transform the survival times using the natural log. Here is Levene's test when using the natural log of the survival times.

```
Levene's Test (Any Continuous Distribution)
Test statistic = 0.67, p-value = 0.616
```

This time the p-value is computed to be 0.616 which is large enough that we can feel comfortable using a model of equal variances. If the rest of the conditions are met (and we decided in Chapter 5 that they were) we can safely draw conclusions from an ANOVA model fit to the natural log of the survival times. ◊

Up to this point all of our examples have been for one-way ANOVA models. The obvious question is: Can we use Levene's test for two-way or higher models as well? The answer is: Yes. Note, however, that we apply Levene's test to groups of data formed by the cells (that is, the combinations of levels of factors), rather than to groups formed by levels of a single factor. In both Minitab and R this is accomplished by designating the combination of factors that are in the model. Our final example illustrates this using the PigFeed data set first encountered in Chapter 6.

Example 7.3: *Checking equality of variances in the pig data set*

This data set, found in the file **PigFeed**, from Section 6.2, presents the weight gains of pigs based on the levels of two different factors: receiving antibiotics (yes, no) and receiving vitamin B12 (yes, no). Thus, Levene's test operates on 4 groups (cells) of data, corresponding to the 4 combinations of factor levels. The output from the test is shown below. With a p-value of 0.223 we fail to reject the null hypothesis. We find our data to be consistent with the condition that the population variances are equal.

```
Levene's Test (Any Continuous Distribution)
Test statistic = 1.81, p-value = 0.223
```

◇

7.2 Topic: Multiple Tests

What happens to our analysis when we have concluded that there is at least one difference among the groups in our data? One question to ask would be: Which groups are different from which other groups? But this may involve doing many comparisons. We begin by summarizing what we presented in Section 5.4, and then we expand on the options for dealing with this situation.

Why Worry About Multiple Tests?

Remember that when we compute a 95% confidence interval we use a method that, if used many times, will result in a confidence interval that contains the appropriate parameter in 95% of cases. If we only compute one confidence interval, we feel pretty good about our interval as an estimate for the parameter because we know that most of the time such intervals contain the parameter.

But what happens if we compute lots of such intervals? Although we may feel good about any one interval individually, we have to recognize that among those many intervals there is likely to be at least one interval that doesn't contain its parameter. And the more intervals we compute, the more likely it is that we will have at least one interval that doesn't capture its intended parameter.

Family-wise Error Rate

What we alluded to in the above discussion is the difference between individual error rates and family-wise error rates. If we compute 95% confidence intervals, then the individual error rate is 5% for each of them, but the family-wise error rate (the likelihood that at least one interval among the group does not contain its parameter) increases as the number of intervals increases.

There are quite a few approaches to dealing with multiple comparisons. Each one deals with the two different kinds of error rates in different ways. In Chapter 5 we introduced you to one such method: Fisher's LSD. Here we review Fisher's LSD and introduce you to two more methods: the Bonferroni adjustment and Tukey's HSD. We also discuss the relative merits of all three methods.

We start by noting that all three methods produce confidence intervals of the form

$$\bar{y}_i - \bar{y}_j \pm \text{ margin of error}$$

What differs from method to method is the definition of the margin of error. In fact, for all three methods, the margin of error will be

$$cv\sqrt{MSE\left(\frac{1}{n_i} + \frac{1}{n_j}\right)}$$

where cv stands for the critical value. What differs between the methods is the critical value.

Fisher's LSD: A Liberal Approach

Recall that in Section 5.4 we introduced you to **Fisher's Least Significant Difference (LSD)**. This method is the most liberal of the three we will discuss, producing intervals that are often much narrower than those produced by the other two methods and thus is more likely to identify differences (either real or false). The reason that its intervals are narrower is that it focuses only on the individual error rate. We employ this method only when the F-test from the ANOVA is significant. Since we have already determined that there are differences (the ANOVA F-test was significant), we feel comfortable identifying differences with the liberal LSD method. Because LSD only controls the individual error rate, this method has a larger family-wise error rate. But, in its favor, it has a small chance of missing actual differences that exist.

The method starts with verifying that the F-test is significant. If it is, then we compute the usual 2-sample confidence intervals using the MSE as our estimate of the sample variance and using as the critical value a t^* with the MSE degrees of freedom, $n - K$, and the individual α-level of choice.

Fisher's LSD

To compute multiple confidence intervals comparing pairs of means using Fisher's LSD:

a. Verify that the F-test is significant.

b. Compute the intervals

$$\bar{y}_i - \bar{y}_j \pm t^* \sqrt{MSE\left(\frac{1}{n_i} + \frac{1}{n_j}\right)}.$$

Use an individual α-level (e.g. 0.05) and the MSE degrees of freedom, $n - K$, for finding t^*.

Bonferroni Adjustment: A Conservative Approach

On the other end of the spectrum is the **Bonferroni method**. This is one of the most simple methods to understand. Whereas Fisher's LSD places its emphasis on the individual error rate (and only controls the family-wise error rate through the ANOVA F-test), the Bonferroni method places its emphasis solely on the family-wise error rate. It does so by using a smaller individual error rate for each of the intervals.

In general, if we want to make m comparisons we replace the usual α with α/m and make the corresponding adjustment in the confidence levels of the intervals. For example, suppose that we are comparing means for $K = 5$ groups, which require $m = 10$ pairwise comparisons. To assure a family-wise error rate of at most 5% we use $0.05/10 = 0.005$ as the significance level for each individual test or, equivalently, we construct the confidence intervals using the formula given above with a t^* value to give 99.5% confidence. Using this procedure, in 95% of data sets drawn from populations for which the means are all equal, our *entire set* of conclusions will be correct (no differences will be found significant). In fact, the family-wise error rate is often smaller than this, but we cannot easily compute exactly what it is. That is why we say the Bonferroni method is *conservative*. The actual family-wise confidence level is at least 95%. And, since we have accounted for the family-wise error rate, it is less likely to incorrectly signal a difference between two groups that actually have the same mean.

The bottom line is that the Bonferroni method is easy to put in place and provides an upper bound on the family-wise error rate. These two facts make it an attractive option.

Bonferroni Method

To compute multiple confidence intervals comparing two means using the Bonferroni method:

a. Choose an α-level for the family-wise error rate.

b. Decide how many intervals you will be computing. Call this number m.

c. Find t^* for $100 - \frac{\alpha}{2m}\%$ confidence intervals, using the MSE degrees of freedom, $n - K$.

d. Compute the intervals

$$\bar{y}_i - \bar{y}_j \pm t^* \sqrt{MSE\left(\frac{1}{n_i} + \frac{1}{n_j}\right)}.$$

Tukey's "Honest Significant Difference": A Moderate Approach

Fisher's LSD and Bonferroni lie towards the extremes of the conservative to liberal spectrum among multiple comparison procedures. There are quite a few more moderate options that statisticians often recommend. Here we discuss one of those: **Tukey's honest significant difference**.

Once again, the confidence intervals are of the form

$$\bar{y}_i - \bar{y}_j \pm cv \sqrt{MSE\left(\frac{1}{n_i} + \frac{1}{n_j}\right)}$$

7.2. TOPIC: MULTIPLE TESTS

in which *cv* stands for the appropriate critical value. The question is what to use for *cv*. In both Fisher's LSD and Bonferroni, the *cv* was a value from the t-distribution. For Tukey's HSD the critical value depends on a different distribution called the *studentized range distribution*.

Like Bonferroni, Tukey's HSD method concerns itself with the family-wise error rate but it is designed to create intervals that are somewhat narrower than Bonferroni intervals. The idea is to find an approach that controls the family-wise error rate while retaining the usefulness of the individual intervals.

The idea behind Tukey's HSD method is to think about the case where all group means are, in fact, identical, and all sample sizes are the same. Under these conditions we would like the confidence intervals to include 0 most of the time (so that the intervals indicate no significant difference betweeen means). In other words, we would like $|\bar{y}_i - \bar{y}_j| \leq$ margin of error for most pairs of sample means.

To develop a method that does this we start by thinking about $(\bar{y}_{\max} - \bar{y}_{\min})$, the difference between the largest and smallest group means. If $(\bar{y}_{\max} - \bar{y}_{\min}) \leq$ margin of error then all differences $|\bar{y}_i - \bar{y}_j|$ will be less than or equal to the margin of error and all intervals will contain 0. Tukey's HSD chooses a critical value so that $(\bar{y}_{\max} - \bar{y}_{\min})$ will be less than the margin of error in 95% of data sets drawn from populations with a common mean. So, in 95% of data sets in which all population means are the same and all sample sizes are the same, all confidence intervals for pairs of differences in means will contain 0. This is how Tukey's HSD controls for the family-wise error rate. Although the idea behind Tukey's HSD depends on having equal sample sizes, the method can be used when the sample sizes are unequal.

Tukey's HSD

To compute multiple confidence intervals comparing 2 means using Tukey's HSD method:

a. Choose an α-level for the family-wise error rate.

b. Find the value of q from the studentized range distribution based on the number of intervals, α, and the MSE degrees of freedom.

c. Compute the intervals

$$\bar{y}_i - \bar{y}_j \pm \frac{q}{\sqrt{2}} \sqrt{MSE \left(\frac{1}{n_i} + \frac{1}{n_j} \right)},$$

Although tables of the studentized range distribution are available, we will rely on software to compute these intervals. We also note that while the family-wise confidence level is exact for cases in which the sample sizes are the same, Tukey's HSD is conservative for cases in which the sample sizes are different. That is, if sample sizes are different and a 95% level is used, the actual family-wise confidence level is somewhat higher than 95%.

A comparison of all three methods

The Bonferroni method, Tukey's HSD, and Fisher's LSD are different not only in their respective family-wise error rates, but also in the settings in which we might choose to use them. When we use Bonferroni or Tukey's HSD, we have tacitly decided that we want to make sure that the overall Type I error rate is low. In other words, we have decided that if we falsely conclude that two groups are different, this is worse than failing to find actual differences. In practice, we are likely to use Bonferroni or Tukey when we have specific differences in mind ahead of time.

When we use Fisher's LSD, we are not quite so worried about the family-wise error rate and therefore are willing to take a slightly higher risk of making a Type I error. We think that missing actual differences is a bigger problem than finding the occasional difference that doesn't actually exist. Perhaps we are doing exploratory data analysis and would like to see what differences might exist. Using Fisher's LSD with $\alpha = 0.05$ means that we only find a "false" difference about once in every 20 intervals. So most differences we find will, in fact, be true differences, since the F-test has already suggested that differences exist.

We have called Fisher's LSD a more liberal approach, Tukey's HSD a moderate approach, and Bonferroni a more conservative approach. This is, in part, because of the lengths of the intervals they produce. If we compare the critical values from all three, we find that Fisher's cv $\leq$ Tukey's cv $\leq$ Bonferroni's cv and equality only holds when the number of groups is 2. So the Fisher's LSD intervals will be the narrowest and most likely to find differences, followed by Tukey's HSD, and finally Bonferroni, which will have the widest intervals and the least likelihood of finding differences.

Example 7.4: *Fruit flies (multiple comparisons)*

The data for this example are found in the file **FruitFlies**. Recall from Chapter 5 that we are wondering if the lifetime of male fruit flies differs depending on what type and how many females they are living with. Here we present all three ways of computing multiple intervals, though in practice one would use only one of the three methods.

Fisher's LSD

In Chapter 5 we found that the F-test was significant, allowing us to compute intervals using Fisher's LSD. Table 7.1 gives a summary of the confidence intervals that we found using this method earlier.

Note that the conclusion here is that living with 8 virgins does significantly reduce the life span of male fruit flies in comparison to all other living conditions tested, but none of the other conditions are significantly different from each other.

7.2. TOPIC: MULTIPLE TESTS

Group Difference	Confidence Interval	Contains 0?
1 preg - none	(-7.05, 9.53)	yes
8 preg - none	(-8.49, 8.09)	yes
1 virgin - none	(-15.09, 1.49)	yes
8 virgin - none	(-33.13, -16.55)	no
8 preg - 1 preg	(-9.73, 6.85)	yes
1 virgin - 1 preg	(-16.33, 0.25)	yes
8 virgin - 1 preg	(-34.37, -17.79)	no
1 virgin - 8 preg	(-14.89, 1.69)	yes
8 virgin - 8 preg	(-32.93, -16.35)	no
8 virgin - 1 virgin	(-26.33, -9.75)	no

Table 7.1: Fisher's LSD confidence intervals

Bonferroni

Computer output is given below for the 95% Bonferroni intervals. For this example we note that the conclusions are the same as for those found using the Fisher's LSD intervals. That is, the life span of male fruit flies living with 8 virgins is significantly shorter than that of all other groups, but none of the other groups are significantly different from each other.

```
Bonferroni 95.0% Simultaneous Confidence Intervals
Response Variable Longevity
All Pairwise Comparisons among Levels of Treatment
Treatment = 1 pregnant   subtracted from:

Treatment    Lower   Center   Upper  ----+---------+---------+---------+-
1 virgin    -20.02    -8.04    3.94          (----*----)
8 pregnant  -13.42    -1.44   10.54             (---*----)
8 virgin    -38.06   -26.08  -14.10  (----*---)
none        -13.22    -1.24   10.74             (----*---)
                                     ----+---------+---------+---------+-
                                       -25         0        25        50

Treatment = 1 virgin   subtracted from:

Treatment    Lower   Center    Upper  ----+---------+---------+---------+-
8 pregnant   -5.38    6.60   18.578                (----*---)
8 virgin    -30.02  -18.04   -6.062   (----*----)
none         -5.18    6.80   18.778                (----*----)
                                      ----+---------+---------+---------+-
                                        -25         0        25        50
```

312   CHAPTER 7. ADDITIONAL ANOVA TOPICS

```
Treatment = 8 pregnant   subtracted from:

Treatment     Lower   Center   Upper  -----+---------+---------+---------+-
8 virgin     -36.62   -24.64  -12.66  (----*----)
none         -11.78     0.20   12.18              (----*----)
                                      -----+---------+---------+---------+-
                                         -25        0        25        50

Treatment = 8 virgin   subtracted from:

Treatment     Lower   Center   Upper  -----+---------+---------+---------+-
none          12.86    24.84   36.82                          (----*----)
                                      -----+---------+---------+---------+-
                                         -25        0        25        50
```

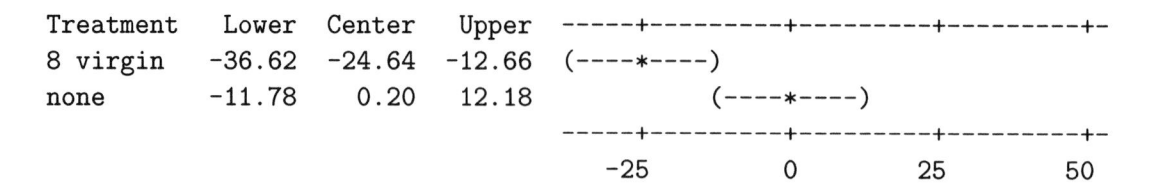

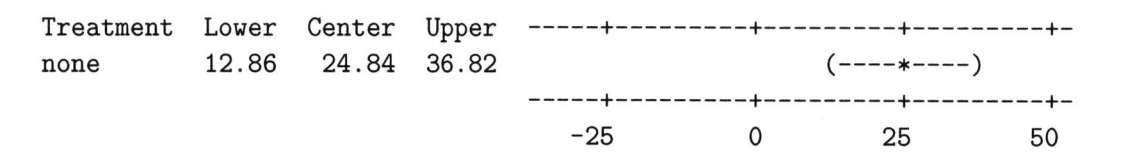

Tukey's HSD

Finally, we present the 95% intervals computed using Tukey's HSD method. Again we note that the conclusions are the same for this method as they were for the other two.

```
Tukey 95.0% Simultaneous Confidence Intervals
Response Variable Longevity
All Pairwise Comparisons among Levels of Treatment
Treatment = 1 pregnant   subtracted from:

Treatment     Lower   Center   Upper  -----+---------+---------+---------+-
1 virgin     -19.65    -8.04    3.57          (----*---)
8 pregnant   -13.05    -1.44   10.17             (---*----)
8 virgin     -37.69   -26.08  -14.47  (----*---)
none         -12.85    -1.24   10.37             (----*---)
                                      -----+---------+---------+---------+-
                                         -25        0        25        50
```

7.2. TOPIC: MULTIPLE TESTS

```
Treatment = 1 virgin   subtracted from:

Treatment    Lower   Center    Upper   -----+---------+---------+---------+-
8 pregnant   -5.01    6.60    18.210                  (----*---)
8 virgin    -29.65  -18.04    -6.430   (----*---)
none         -4.81    6.80    18.410                  (----*---)
                                       -----+---------+---------+---------+-
                                          -25         0        25        50


Treatment = 8 pregnant   subtracted from:

Treatment    Lower   Center    Upper   -----+---------+---------+---------+-
8 virgin   -36.25  -24.64   -13.03    (---*----)
none       -11.41    0.20    11.81                (----*----)
                                       -----+---------+---------+---------+-
                                          -25         0        25        50


Treatment = 8 virgin   subtracted from:

Treatment    Lower   Center    Upper   -----+---------+---------+---------+-
none         13.23   24.84    36.45                         (----*----)
                                       -----+---------+---------+---------+-
                                          -25         0        25        50
```

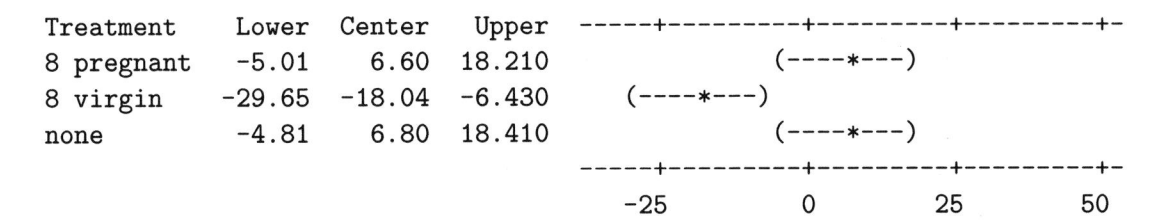

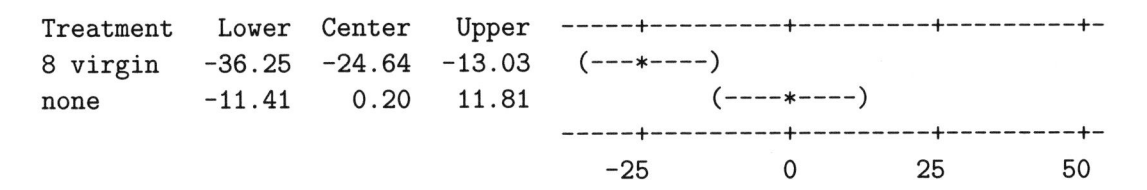

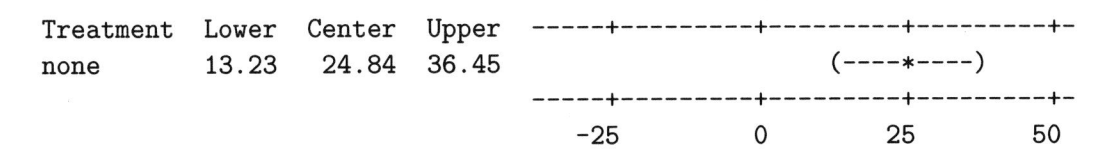

We also draw your attention to the relative lengths of the intervals. As an example, consider the lengths of the 8 virgins - none intervals. For Fisher's LSD, the interval is (-33.13, -16.55), which has a length of $|-33.13 + 16.55| = 16.58$ days. The Bonferroni interval is (-36.82, -12.86), which has a length of $|-36.82 + 12.86| = 23.96$. Finally, the Tukey's HSD interval is (-36.45, -13.23) which has a length of $|-36.45 + 13.23| = 23.22$. As predicted, Fisher's LSD results in the shortest intervals. This allows it to find more differences to be significant. Bonferroni's intervals are the longest, reflecting the fact that it is the most conservative of the three methods. Tukey's HSD fits in between, though for this example, since we have a relatively small number of intervals and small sample sizes, it is similar to Bonferroni. ◇

7.3 Topic: Comparisons and Contrasts

We start this topic by returning to the main example from Chapter 5: the fruit flies. Remember that the researchers were interested in the life spans of male fruit flies and how they were affected by the number and type of females that were living with each male. In Chapter 5 we performed a basic ANOVA and learned that there are significant differences in mean lifetime between at least two of the treatment groups. We also introduced the idea of comparisons where we compared the treatments to each other in pairs.

But there are other types of analyses that the researchers might be interested in. For instance, they might want to compare the mean lifetimes of those fruit flies who lived with virgins (either one or eight) to those who lived with pregnant females (again, either one or eight). Notice that this is still a comparison of two ideas (pregnant versus virgin), but the analysis would involve more than two of the treatment groups. In this case it would involve four out of the five treatment groups.

For this situation, we introduce the idea of a *contrast*. In fact, strictly speaking, a contrast is used any time we want to compare two or more groups. But because we are so often interested specifically in comparing two groups, we give these contrasts a special name: *comparisons*.

Contrasts and Comparisons

When we have planned comparisons for our data analysis we will use the concept of **contrasts** which in special cases are also called **comparisons**.

- **Contrast**: A comparison of two ideas that uses 2 or more of the K possible groups.

- **Comparison**: The special case of a contrast when we compare only two of the K possible groups.

Comparing Two Means

If we have two specific groups that we would like to compare to each other, as discussed in Sections 5.4 and 7.2, we employ a typical two-sample t-test or confidence interval. The only modification that we make in the ANOVA setting is that we use the MSE from the ANOVA model as our estimate of the variance. This makes sense because we assume, in the ANOVA model, that all groups have the same variance and the MSE is an estimate of that variance using information from all groups.

In Sections 5.4 and 7.2 we relied on confidence intervals for these comparisons, as is the usual course of action for comparisons. Here, however, we present this same analysis as a hypothesis test. We

7.3. TOPIC: COMPARISONS AND CONTRASTS

will build on this idea when we move to the more general idea of contrasts, where hypothesis tests are more common.

Example 7.5: *Comparison using fruit flies*

The data for this example are found in the file **FruitFlies**. In Chapter 5 we argued that researchers would be interested in testing

$H_0 : \mu_{8v} - \mu_{none} = 0$
$H_a : \mu_{8v} - \mu_{none} \neq 0$

The appropriate test statistic is

$$t = \frac{\bar{y}_{8v} - \bar{y}_{none} - 0}{\sqrt{MSE\left(\frac{1}{n_i} + \frac{1}{n_j}\right)}} = \frac{38.72 - 63.56 - 0}{\sqrt{219\left(\frac{1}{25} + \frac{1}{25}\right)}} = -5.93$$

There are 120 degrees of freedom for this statistic and the p-value is approximately 0. Because the p-value is so small we conclude that there is a significant difference in the mean lifetime of fruit flies who live with eight virgins and fruit flies who live alone. ◊

Note that here we are doing only one test. Since this is the one test of interest we do not need to worry about the family-wise error rate; nor do we necessarily need the original ANOVA to be significant. If, however, we plan to compare every pair of treatments we do need to be concerned with the family-wise error rate. We discussed Fisher's LSD in Chapter 5 as one way to deal with this problem. In Topic 7.2 we introduced you to two other methods commonly used in this situation.

The Idea of a Contrast

But how do we proceed if our question of interest involves more than 2 groups? For example, in the fruit flies example there are three questions we might like to answer that fit this situation. We have already identified one of them: Is the lifespan for males living with pregnant females different from that of males living with virgins? The researchers thought so. Their assumption was that the pregnant females would not welcome advances from the males and so living with pregnant females would result in a different mean lifespan than living with virgins.

In fact, the researchers thought that living with pregnant females would be like living alone. This would lead us to test the null hypothesis that the fruit flies living with pregnant females would have the same lifetime as those living alone. In this case we would be working with 3 groups: those living alone, those with one pregnant female, and those with 8 pregnant females.

Combining the previous two ideas could lead us to ask a final question in which we ask if living with virgins (either 1 or 8, it doesn't matter) is different from living alone. Again we would focus on three groups, using the two groups living with virgins together to compare to the group living alone.

Linear Combinations of Means

When we analyze contrasts we need a way of considering the means of more than two groups. The simplest situation involves the case where some (or all) of the groups can be divided into two classifications and we want to compare these two classifications. A perfect example of this is comparing male fruit flies who lived with virgins (no matter how many) to those who lived with pregnant females (again, no matter how many). In this case, each classification has two groups in it. Now we need a way to combine the information from all groups within a classification. It makes sense to take the mean of the means. Since we want to compare those living with virgins to those living with pregnant females, this leads us to the following hypothesis for the fruit flies:

$$H_0 : \frac{1}{2}(\mu_{1v} + \mu_{8v}) = \frac{1}{2}(\mu_{1p} + \mu_{8p})$$

That is, the average life span for males who lived with virgin females is equal to the average life span for males who lived with pregnant females, no matter how many of the respective type of female the male lived with. Notice that the hypothesis can be rewritten as

$$H_0 : \frac{1}{2}(\mu_{1v} + \mu_{8v}) - \frac{1}{2}(\mu_{1p} + \mu_{8p}) = 0$$

to make it clear that we want to evaluate the difference between these two kinds of treatments. Of course, to actually evaluate this difference we need to estimate it from our data. For this example we would calculate

$$\frac{1}{2}(\bar{y}_{1v} + \bar{y}_{8v}) - \frac{1}{2}(\bar{y}_{1p} + \bar{y}_{8p}) = \frac{1}{2}\bar{y}_{1v} + \frac{1}{2}\bar{y}_{8v} - \frac{1}{2}\bar{y}_{1p} - \frac{1}{2}\bar{y}_{8p}$$

Contrasts

In general we will write contrasts in the form

$$c_1\mu_1 + c_2\mu_2 + \ldots + c_k\mu_k$$

where

$$c_1 + c_2 + \ldots + c_k = 0$$

and some c_i might be 0. The contrast is estimated by substituting sample means for the population means:

$$c_1\bar{y}_1 + c_2\bar{y}_2 + \ldots + c_k\bar{y}_k$$

In the example above c_{1v} and c_{8v} are both $\frac{1}{2}$, c_{1p} and c_{8p} are both $-\frac{1}{2}$, and c_{none} is 0.

7.3. TOPIC: COMPARISONS AND CONTRASTS

Notice that what we call a comparison is just a simple case of a contrast. That is, the statistic of interest for a comparison is $\bar{y}_i - \bar{y}_j$. In this case one group has a coefficient of 1, a second group has a coefficient of -1, and all other groups have coefficients of 0.

What about the case where there are differing numbers of groups within the two classifications? We will use the same ideas here. We start by comparing the mean of the relevant group means and proceed as above. The following example illustrates the idea.

Example 7.6: *Fruit flies (virgins versus alone)*

In this case we want to compare the life spans of those living with either one or eight virgin females to those living alone. This leads us to the null hypothesis

$$H_0 : \frac{1}{2}(\mu_{1v} + \mu_{8v}) = \mu_{none}$$

or

$$H_0 : \frac{1}{2}(\mu_{1v} + \mu_{8v}) - \mu_{none} = 0$$

The estimate of the difference is

$$\frac{1}{2}\bar{y}_{1v} + \frac{1}{2}\bar{y}_{8v} - \bar{y}_{none}$$

Notice that the sum of the coefficients is 0. ◇

The Standard Error for a Contrast

Now that we have calculated an estimate for the difference that the contrast measures, we need to decide how to determine if the difference we found is significantly different from 0. This means we have to figure out how much variability there would be from one sample to the next in the value of the estimate.

Let's start with a simple case that we already know about. If we compare just two groups to each other, as we did in Chapter 5, we estimate the variability of our statistic, $\bar{y}_1 - \bar{y}_2$, with $\sqrt{MSE\left(\frac{1}{n_1} + \frac{1}{n_2}\right)}$. Recall that the actual standard deviation of $\bar{y}_1 - \bar{y}_2$ is $\sqrt{\frac{\sigma_1^2}{n_1} + \frac{\sigma_2^2}{n_2}}$ but we only use the ANOVA model when all groups have the same variance. So we call that common variance σ^2 and factor it out, getting $\sqrt{\sigma^2\left(\frac{1}{n_1} + \frac{1}{n_2}\right)}$. Of course, we don't know σ^2 so we estimate it using the MSE.

Now we need to consider the more general case where our contrast is $c_1\bar{y}_1 + c_2\bar{y}_2 + \ldots + c_k\bar{y}_k$. Note that in the above discussion, when we were comparing just two groups, $c_1 = 1$ and $c_2 = -1$. As we consider the more general case we will continue to use the MSE as our estimate for the common variance but now we need to take into consideration the sample sizes of all groups involved and the coefficients used in the contrast.

> **Standard error of a contrast**
>
> The general formula of the standard error for contrasts is
> $$\sqrt{MSE \sum_{i=1}^{k} \frac{c_i^2}{n_i}}$$

Example 7.7: *Fruit flies (virgins versus pregnant)*

The main question we have been considering is whether the mean life span of the male fruit flies living with virgins is different from that of those living with pregnant females. The contrast that we are using is $\frac{1}{2}\bar{y}_{1v} + \frac{1}{2}\bar{y}_{8v} - \frac{1}{2}\bar{y}_{1p} - \frac{1}{2}\bar{y}_{8p}$. All groups consist of 25 fruit flies and all coefficients are $\pm\frac{1}{2}$ so the standard error is

$$\sqrt{219\left(\frac{\left(\frac{1}{2}\right)^2}{25} + \frac{\left(\frac{1}{2}\right)^2}{25} + \frac{\left(-\frac{1}{2}\right)^2}{25} + \frac{\left(-\frac{1}{2}\right)^2}{25}\right)} = \sqrt{219\left(\frac{\frac{1}{4}}{25} + \frac{\frac{1}{4}}{25} + \frac{\frac{1}{4}}{25} + \frac{\frac{1}{4}}{25}\right)} = \sqrt{219\left(\frac{1}{25}\right)} = 2.9597$$

◇

Example 7.8: *Fruit flies (virgins versus alone)*

We can apply the same idea to the hypothesis presented in Example 7.6. Recall that the difference here is that not all of the coefficients are the same in absolute value. Recall that our estimate was $\frac{1}{2}\bar{y}_{1v} + \frac{1}{2}\bar{y}_{8v} - \bar{y}_{\text{none}}$. So the standard error is

$$\sqrt{219\left(\frac{\left(\frac{1}{2}\right)^2}{25} + \frac{\left(\frac{1}{2}\right)^2}{25} + \frac{(-1)^2}{25}\right)} = \sqrt{219\left(\frac{1.5}{25}\right)} = 3.6249$$

◇

The t-test for a Single Contrast

At this point we have a statistic (the estimated contrast), its hypothesized value (typically 0), and its estimated standard error. All that is left to do is to put these three pieces together to create a test statistic and compare it to the appropriate distribution.

7.3. TOPIC: COMPARISONS AND CONTRASTS

It should come as no surprise that the statistic is of the typical form that you no doubt saw in your statistics first course. That is,

$$\text{test statistic} = \frac{\text{estimate} - \text{hypothesized value}}{\text{standard error of the estimate}}$$

where the test statistic is the appropriate linear combination of sample means, the hypothesized value comes from the null hypothesis of interest, and the standard error is as defined above.

We have already seen in the case of a comparison (where we are just comparing two groups to each other) that since we use the MSE as our estimate of the common group variance, the test statistic has a t-distribution with the same degrees of freedom as the MSE, $n - K$. Thankfully this result generalizes to the case of a contrast in which we have more than two groups involved. Once again the test statistic has a t-distribution, and, since we continue to use the same estimate (the MSE) for the common group variance, the degrees of freedom remains that of the MSE.

Let's now put all of the pieces together and decide whether the researchers were right in their hypothesis that there is a (statistically) significant difference between the life spans of fruit flies who lived with virgins and the life spans of fruit flies who lived with pregnant females.

Example 7.9: *Fruit flies (virgins versus pregnant)*

We determined earlier that the relevant contrast is $\frac{1}{2}\bar{y}_{1v} + \frac{1}{2}\bar{y}_{8v} - \frac{1}{2}\bar{y}_{1p} - \frac{1}{2}\bar{y}_{8p}$. From Chapter 5 we know that $\bar{y}_{1v} = 56.76$, $\bar{y}_{8v} = 38.72$, $\bar{y}_{1p} = 64.80$, and $\bar{y}_{8p} = 63.36$. From Example 7.7 we know that the standard error of this contrast is 2.9597 so the test statistic is

$$t = \frac{\frac{1}{2}(56.76) + \frac{1}{2}(38.72) - \frac{1}{2}(64.80) - \frac{1}{2}(63.36) - 0}{2.9597} = -5.52$$

Comparing this to a t-distribution with 120 degrees of freedom, we find that the p-value is approximately 0 and we conclude that there is a significant difference between the life spans of those fruit flies that lived with virgins compared to those fruit flies that lived with pregnant females. ⋄

We leave the remaining tests for the fruit flies to the exercises and we conclude this topic with one final example that puts all of the pieces together.

Example 7.10: *Walking babies*

As a rule, it takes about a year before a baby takes his or her first steps alone. Scientists wondered if they could get babies to walk sooner by prescribing a set of special exercises. They decided to compare babies given the special exercises to a control group of babies. But the scientists recognized that just showing an interest in the babies and their parents could cause a placebo effect, and it could be that any exercise would affect walking age. The final experimental design that they settled on included four groups of babies and used treatments listed below. The researchers had a total of 24 babies to use in this experiment so they randomly assigned them, six to a group. The data can be found in the file **WalkingBabies**.

- *Special exercises*: Parents were shown the special exercises and encouraged to use them with their children. They were phoned weekly to check on their child's progress.

- *Exercise control:* These parents were not shown the special exercises, but they were told to make sure their babies spent at least 15 minutes a day exercising. This control group was added to see if any type of exercise would help or if the special exercises were, indeed, special.

- *Weekly report:* Parents in this group were not given instructions about exercise. Like the parents in the treatment group, however, they received a phone call each week to check on progress: "Did your baby walk yet?" This control group would help the scientists discover if just showing an interest in the babies and their parents affected age at first walking.

- *Final report:* These parents were not given instructions about exercise or weekly phone calls. They reported at the end of the study. This final control group was meant to measure the age at first walking of babies with no intervention at all.

To start the analysis we ask the simple question: Is there a difference in the mean time to walking for the four groups of babies? So we start by evaluating whether ANOVA is an appropriate analysis tool for this data set by checking the necessary conditions.

The method of fitting the model guarantees that the residuals always add to zero, so there's no way to use residuals to check the condition that the mean error is zero. Essentially, the condition says that the equation for the structure of the response hasn't left out any terms. Since babies are randomly assigned to the four treatment groups, we hope that all other variables that might affect walking age have been randomized out.

The independence of the errors condition says, in effect, that the value of one error is unrelated to the others. For the walking babies this is almost surely the case, because the time it takes one baby to learn to walk doesn't depend on the times it takes other babies to learn to walk.

The next condition says that the amount of variability is the same from one group to the next. One way to check this is to plot residuals versus fitted values and compare columns of points. Figure 7.3 shows that the amount of variability is similar from one group to the next. We can also compare the largest and smallest group standard deviations by computing the ratio SDmax/SDmin.

```
group      StDev
Control_E  1.898
Weekly     1.558
Final      0.871
Special    1.447
```

$$\frac{\text{SD}_{max}}{\text{SD}_{min}} = \frac{1.898}{0.871} = 2.18$$

7.3. TOPIC: COMPARISONS AND CONTRASTS

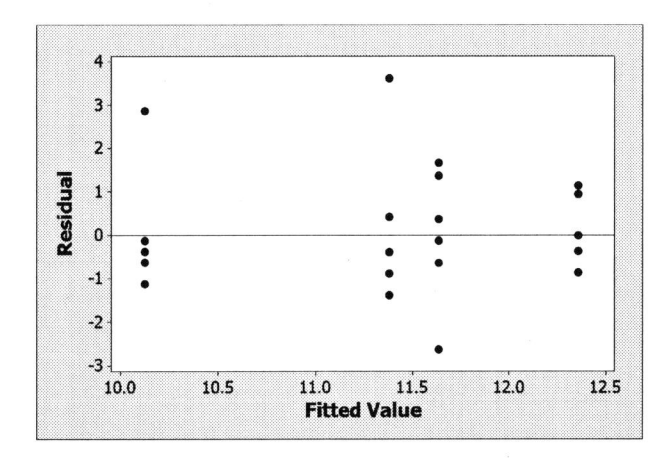

Figure 7.3: Residual plot for walking babies ANOVA model

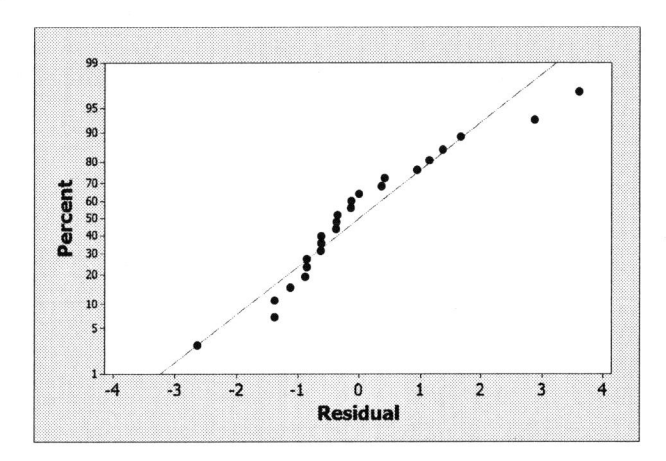

Figure 7.4: Normal probability plot of residuals

Although this ratio is slightly larger than 2, we are talking about 4 groups with only 6 observations in each group so we are willing to accept this ratio as being small enough to be consistent with the condition that the population variances are all the same.

Finally, the last condition says that the error terms should be normally distributed. To check this condition we look at a normal probability plot, which should suggest a line. Figure 7.4 looks reasonably straight so we are willing to accept the normality of the errors.

The end result for this example is that the conditions seem to be met and we feel comfortable proceeding with our analysis using ANOVA.

The ANOVA table is given below.

```
One-way ANOVA: age versus group

Source  DF   SS     MS    F     P
group    3  15.60  5.20  2.34  0.104
Error   20  44.40  2.22
Total   23  60.00
```

From the table we see that the F-statistic is 2.34 with 3 and 20 degrees of freedom and the p-value is 0.104. We also compute $R^2 = 1 - \frac{SSE}{SSTotal} = 1 - \frac{44.40}{60.00} = 1 - 0.74 = 0.26$. This tells us that the group differences account for 26% of the total variability in the data. And the test of the null hypothesis that the mean time to walking is the same for all groups of children is not significant. In other words, we do not have significant evidence that any group is different from the others.

However, notice in this case that we really have one treatment group and three kinds of control groups. Each control group is meant to ferret out whether some particular aspect of the treatment is in fact important. Is it the special exercise itself, or is it any exercise at all that will help? The first control group could help us decide this. Is it extra attention, not the exercise itself? This is where the second control group comes in. Finally is it all of the above? This is what the third control group adds to the mix.

While the F-test for this data set was not significant, the researchers had two specific questions from the outset that they were interested in. First was the question of whether there was something special to the exercises labeled "special." That is, when compared to the control group with exercises (treated in every way the same as the treatment group except for the type of exercise), is there a difference? As a secondary question, they also wondered whether there was a difference between children who used some type of exercise (any type) versus those who did not use exercise. The first question calls for a comparison, the second requires a more complicated contrast.

We start with the first question: Is there a difference between the group that received the special exercises and the group that were just told to exercise? The hypotheses here are

$$H_0: \mu_{\text{se}} - \mu_{\text{ce}} = 0$$
$$H_a: \mu_{\text{se}} - \mu_{\text{ce}} \neq 0$$

The group means are given in the table below

	Group Mean
Final	12.360
Weekly	11.633
Control_E	11.383
Exercise	10.125
Total	45.501

7.3. TOPIC: COMPARISONS AND CONTRASTS

Our estimate of the comparison is $\bar{y}_{se} - \bar{y}_{ce} = 10.125 - 11.383 = -1.258$. Since there are 6 babies in each group, the standard error of the comparison is

$$\sqrt{MSE\left(\frac{1^2}{6} + \frac{(-1)^2}{6}\right)} = \sqrt{2.22\left(\frac{1}{6} + \frac{1}{6}\right)} = 0.8602$$

This leads to a test statistic of

$$t = \frac{-1.258 - 0}{0.8602} = -1.46$$

Notice that this has 20 degrees of freedom, from the MSE. The p-value associated with this test statistic is 0.1598, so we do not have significant evidence of a difference between the special exercises and any old exercises.

Finally we test to see if using exercises (special or otherwise) gives a different mean time to walking for babies in comparison to no exercises. For this case, the hypotheses are

$$H_0: \frac{1}{2}(\mu_{se} + \mu_{ce}) - \frac{1}{2}(\mu_{rw} + \mu_{re}) = 0$$

$$H_0: \frac{1}{2}(\mu_{se} + \mu_{ce}) - \frac{1}{2}(\mu_{rw} + \mu_{re}) \neq 0$$

Our estimate of the contrast is $\frac{1}{2}(10.125 + 11.383) - \frac{1}{2}(11.633 + 12.360) = -1.2425$. For this case the standard error is

$$\sqrt{2.22\left(\frac{\left(\frac{1}{2}\right)^2}{6} + \frac{\left(\frac{1}{2}\right)^2}{6} + \frac{\left(\frac{-1}{2}\right)^2}{6} + \frac{\left(\frac{-1}{2}\right)^2}{6}\right)} = \sqrt{\frac{2.22}{6}} = 0.6083$$

This leads us to the test statistic

$$t = \frac{-1.2425 - 0}{0.6083} = -2.043$$

with 20 degrees of freedom and a p-value of 0.0544. Here we conclude that we have moderate evidence against the null hypothesis. That is, it appears that having babies take part in exercises may lead to earlier walking.

We end this example with the following note. You will have noticed that we did the analysis for the comparison and contrast of interest even though the F-test in the ANOVA table was not significant. This is because these two comparisons were ones that were planned at the outset of the experiment. These were the questions that the researchers designed the study to ask. Planned comparisons can be undertaken even if the overall F-test is not significant. ⋄

7.4 Topic: ANOVA and Regression with Indicators

In Chapters 5 and 6 we have considered several models under the general heading of ANOVA for Means. These include:

One-way ANOVA (single categorical factor):
$$Y = \mu + \alpha_k + \epsilon$$

Two-way ANOVA with main effects only:
$$Y = \mu + \alpha_k + \beta_j + \epsilon$$

Two-way ANOVA with interaction:
$$Y = \mu + \alpha_k + \beta_j + \gamma_{kj} + \epsilon$$

While we can estimate the effects in each of these models using sample means for various levels of the factors, it turns out that we can also fit them using ordinary multiple regression techniques that we used in Chapter 3 if we use indicator variables to identify the categories for each factor. In this section we examine these connections, first for a simple two sample situation, then for each of the ANOVA models listed above.

Two-sample Comparison of Means as Regression

We start with the simplest case for comparing means. This is the case in which we have just two groups. Here we start with the pooled 2-sample t-test. Then we continue by comparing that to the relevant ANOVA. Finally we illustrate the use of regression in this situation.

Example 7.11: *Fruit flies (continued for two categories)*

Consider the **FruitFlies** data from Chapter 5 where we examined the life span of male fruit flies. Two of the groups in that study were *8 virgins* and *none*, which we compare again here (ignoring, for now, the other three groups in the study). The 25 fruit flies in the *8 virgins* group had an average life span of 38.72 days, whereas the average for the 25 fruit flies living alone (in the *none* group) was 63.56 days.

Is the difference, 38.72 versus 63.56 statistically significant? Or could the two sample means differ by $38.72 - 63.56 = -24.84$ just by chance? Let's approach this question from three distinct directions: the pooled two-sample t-test of Chapter 0, one-way ANOVA for a difference in means as covered in Chapter 5, and regression with an indicator predictor as described in Chapter 3. ◇

Pooled two-sample t-test

Parallel dotplots, shown in Figure 7.5, give a visual summary of the data. The life spans for most of the fruit flies living alone are greater than those of most of the fruit flies living with 8 virgins.

7.4. TOPIC: ANOVA AND REGRESSION WITH INDICATORS

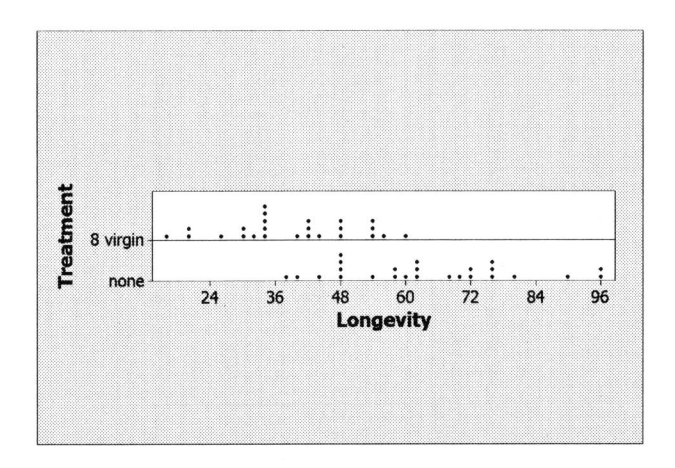

Figure 7.5: Life spans for 8 virgins and living alone groups

In the Minitab output for a pooled two-sample t-test the value of the test statistic is -6.08, with a p-value of approximately 0 based on a t-distribution with 48 degrees of freedom. This small p-value gives strong evidence that the average life span for fruit flies living with 8 virgins is smaller than the average life span for those living alone.

```
Two-sample T for Longevity

Treatment    N   Mean   StDev   SE Mean
8 virgin    25   38.7   12.1      2.4
none        25   63.6   16.5      3.3


Difference = mu (8 virgin) - mu (none)
Estimate for difference:  -24.84
95% CI for difference:  (-33.05, -16.63)
T-Test of difference = 0 (vs not =): T-Value = -6.08  P-Value = 0.000  DF = 48
Both use Pooled StDev = 14.4418
```

ANOVA with Two Groups

As we showed in Chapter 5, these data can also be analyzed using an ANOVA model. The output for the ANOVA table is given below.

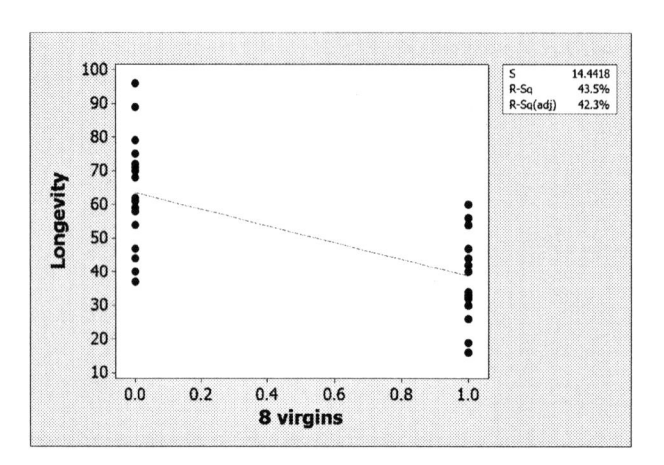

Figure 7.6: Life spans for fruit flies living alone and with 8 virgins

```
Source      DF     SS     MS      F       P
Treatment    1   7713   7713   36.98   0.000
Error       48  10011    209
Total       49  17724

S = 14.44   R-Sq = 43.52%   R-Sq(adj) = 42.34%


                              Individual 95% CIs For Mean Based on
                              Pooled StDev
Level      N    Mean  StDev   ------+---------+---------+---------+--
8 virgin  25   38.72  12.10   (-----*-----)
none      25   63.56  16.45                            (-----*----)
                              ------+---------+---------+---------+--
                                   40        50        60        70

Pooled StDev = 14.44
```

Notice that the p-value is approximately 0, the same that we found when doing the two-sample test. Notice also that the F-value is 36.98 which is -6.08^2 approximately. The only difference between the -6.08^2 and 36.98 is due to rounding error. In fact, the F-statistic will always be the square of the t-statistic and the p-value will be the same no matter which test is run.

7.4. TOPIC: ANOVA AND REGRESSION WITH INDICATORS

Regression with an Indicator

In Chapter 3 we introduced the idea of using an indicator variable to code a binary categorical variable as 0 or 1 in order to use it as a predictor in a regression model. What if that is the *only* predictor in the model? For the fruit flies example we can create an indicator variable, *V8*, to be 1 for the fruit flies living with 8 virgins and 0 for the fruit flies living alone. The results for fitting a regression model to predict life span using *V8* are shown below. Figure 7.6 shows a scatterplot of *Lifespan* versus *V8* with the least squares line. Note that the intercept for the regression line (63.56) is the mean life span for the sample of 25 *none* fruit flies ($V8 = 0$). The slope $\hat{\beta}_1 = -24.84$ shows how much the mean decreases when we move to the *8 virgin* fruit flies (mean=38.72). We also see that the t-test statistic, degrees of freedom, and p-value for the slope in the regression output are identical to the corresponding values in the pooled two-sample t-test.

```
The regression equation is
Longevity = 63.6 - 24.8 v8


Predictor     Coef    SE Coef      T       P
Constant    63.560     2.888    22.01   0.000
v8         -24.840     4.085    -6.08   0.000


S = 14.4418    R-Sq = 43.5%    R-Sq(adj) = 42.3%


Analysis of Variance

Source           DF       SS       MS       F       P
Regression        1   7712.8   7712.8   36.98   0.000
Residual Error   48  10011.2    208.6
Total            49  17724.0
```

One-way ANOVA for Means as Regression

What happens if we try regression using dummy indicator predictors for a categorical factor that has more than two levels? Suppose that Factor A has K different groups. We can construct K different indicator variables, one for each of the groups of Factor A.

$$A_1 = \begin{cases} 1 & \text{if Group} = 1 \\ 0 & \text{otherwise} \end{cases} \qquad A_2 = \begin{cases} 1 & \text{if Group} = 2 \\ 0 & \text{otherwise} \end{cases} \quad \cdots \quad A_K = \begin{cases} 1 & \text{if Group} = K \\ 0 & \text{otherwise} \end{cases}$$

However, if we try to include all of these indicator variables in the same multiple regression model we'll have a problem since any one of them is an exact linear function of the other $K-1$ variables. For example, $A_1 = 1 - A_2 - A_3 - \cdots - A_K$ since any data case in Group 1 is coded as zero for each of the other indicators and any case outside of Group 1 is coded as one for exactly one other indicator. When one predictor is exactly a linear function of other predictors in the model the problem of minimizing the sum of squared errors has no unique solutions. Most software packages will either produce an error message or automatically drop one of the predictors if we try to include them all in the same model.

Thus, to include a categorical factor with K groups in a regression model we use any $K-1$ of the indicator variables. The level that is omitted is known as the *reference* group. The reason for this term should become apparent in the next example.

Example 7.12: *Fruit flies (continued - five categories)*

We continue with the **FruitFlies** data from Chapter 5 where we have five categories: 8 virgins, 1 virgin, 8 pregnant, 1 pregnant, and none. The Minitab output for analyzing possible differences in mean life span for these five groups is reproduced here.

```
One-way ANOVA: Longevity versus Treatment

Source       DF      SS     MS      F      P
Treatment     4   11939   2985  13.61  0.000
Error       120   26314    219
Total       124   38253

S = 14.81   R-Sq = 31.21%   R-Sq(adj) = 28.92%


                             Individual 95% CIs For Mean Based on
                             Pooled StDev
Level        N    Mean  StDev  -------+---------+---------+---------+--
1 pregnant  25   64.80  15.65                              (-----*-----)
1 virgin    25   56.76  14.93                    (-----*-----)
8 pregnant  25   63.36  14.54                           (-----*-----)
8 virgin    25   38.72  12.10  (-----*-----)
none        25   63.56  16.45                             (-----*----)
                               -------+---------+---------+---------+--
                                     40        50        60        70

Pooled StDev = 14.81
```

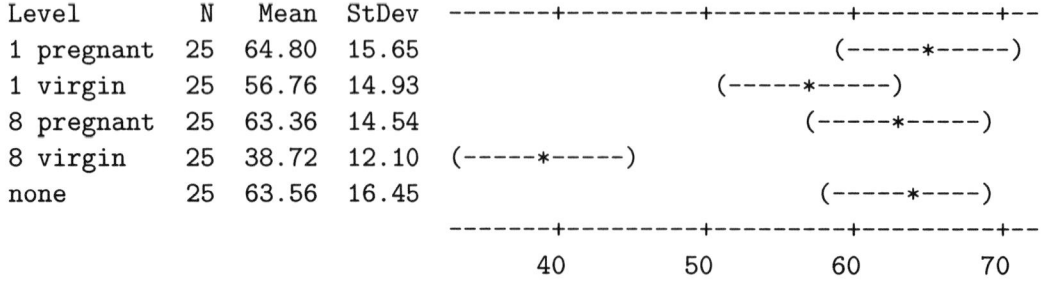

7.4. TOPIC: ANOVA AND REGRESSION WITH INDICATORS

The one-way ANOVA shows evidence for a significant difference in mean life span among these five groups.

To assess this situation with a regression model we create indicator variables for each of the five categories and then choose any four of them to include in the model. For brevity, we label the indicators as $v8$, $v1$, $p8$, $p1$, and *None*. Although the overall significance of the model does not depend on which indicator is left out, the interpretations of individual coefficients may be more meaningful in this situation if we omit the indicator for the *None* group. This was a control group for this experiment since those fruit flies lived alone. Thus our multiple regression model is

$$Longevity = \beta_0 + \beta_1 v8 + \beta_2 v1 + \beta_3 p8 + \beta_4 p1 + \epsilon$$

Here is some of the output for the model with four indicator variables:

```
The regression equation is
Longevity = 63.6 + 1.24 p1 - 6.80 v1 - 0.20 p8 - 24.8 v8


Predictor      Coef    SE Coef      T       P
Constant     63.560      2.962   21.46   0.000
p1            1.240      4.188    0.30   0.768
v1           -6.800      4.188   -1.62   0.107
p8           -0.200      4.188   -0.05   0.962
v8          -24.840      4.188   -5.93   0.000


S = 14.8081    R-Sq = 31.2%    R-Sq(adj) = 28.9%


Analysis of Variance

Source           DF       SS       MS      F       P
Regression        4  11939.3   2984.8  13.61   0.000
Residual Error  120  26313.5    219.3
Total           124  38252.8
```

Notice that the ANOVA table given for the multiple regression matches (up to roundoff) the ANOVA table in the one-way output. The indicators for the categories of the factor provide an overall "significant" regression model exactly when the one-way ANOVA indicates that there are significant differences in the mean responses among the groups. This helps us understand why we have $K-1$ degrees of freedom for the groups when the categorical factor has K groups (one for each of the $K-1$ indicator predictors in the regression model).

When interpreting the estimated coefficients in the model, note that the intercept, $\hat{\beta}_0 = 63.560$, equals the mean for the *none* group of fruit flies — the level that we didn't include as an indicator in the model. This makes sense since the values of all of the indicators in the model are zero for the "left out" group. When computing a predicted mean for one of the indicators in the model we simply add its coefficient to the constant term. For example, a life span for a fruit fly in the *8 virgins* group would be predicted to be $\hat{Longevity} = 63.56 - 24.84 = 38.72$, precisely the sample mean for that group. Thus, we can recover each of the group means from the fitted regression equation and each of the indicator coefficients shows how the sample mean for that group compares to the sample mean for the reference group (*None*). ◇

Two-way ANOVA for Means as Regression

It is easy to extend the idea of using dummy indicators to code a categorical factor in a multiple regression setting to handling a model for main effects with more than one factor. Just include indicators (for all but one level) of each of the factors. To illustrate this idea we return to our earlier example of feeding pigs antibiotics and vitamins; then explore how to also account for interactions with a regression model.

Example 7.13: *Feeding pigs (continued with dummy regression)*

In Example 6.4 we looked at a two-factor model to see how the presence or absence of *Antibiotics* and vitamin $B12$ might affect the weight gain of pigs. Recall that the data in **PigFeed** had two levels (yes or no) for each the two factors and three replications in each cell of the 2×2 factorial design, for an overall sample size of $n = 12$. Since $K = 2$ and $J = 2$ we need just one indicator variable per factor to create a regression model that is equivalent to the two-way ANOVA with main effects.

$$WgtGain = \beta_0 + \beta_1 A + \beta_2 B + \epsilon$$

where A is one for the pigs who received antibiotics in their feed (zero if not) and B is one for the pigs who received vitamin B12 (zero otherwise).

The two-way ANOVA (main effects only) and multiple regression output are shown below:

7.4. TOPIC: ANOVA AND REGRESSION WITH INDICATORS

```
Two-way ANOVA: WgtGain versus Antibiotic, B12

Source      DF    SS      MS       F      P
Antibiotic   1  2187  2187.00   9.75  0.012
B12          1   192   192.00   0.86  0.379
Error        9  2018   224.22
Total       11  4397

S = 14.97   R-Sq = 54.11%   R-Sq(adj) = 43.91%

                    Individual 95% CIs For Mean Based on
                    Pooled StDev
Antibiotic  Mean  --+---------+---------+---------+-------
No           11   (---------*---------)
Yes          38                         (---------*---------)
                  --+---------+---------+---------+-------
                    0        15        30        45

                    Individual 95% CIs For Mean Based on
                    Pooled StDev
B12   Mean        ---+---------+---------+---------+------
No    20.5        (-------------*------------)
Yes   28.5                   (-------------*------------)
                  ---+---------+---------+---------+------
                     10        20        30        40
```

```
Regression Analysis: WgtGain versus A, B

The regression equation is
WgtGain = 7.00 + 27.0 A + 8.00 B

Predictor    Coef   SE Coef      T      P
Constant    7.000     7.487   0.93  0.374
A          27.000     8.645   3.12  0.012
B           8.000     8.645   0.93  0.379

S = 14.9741   R-Sq = 54.1%   R-Sq(adj) = 43.9%
```

```
Analysis of Variance
Source          DF      SS      MS      F       P
Regression       2   2379.0  1189.5   5.31   0.030
Residual Error   9   2018.0   224.2
Total           11   4397.0

Source  DF  Seq SS
A        1  2187.0
B        1   192.0
```

Now we have to dig a little deeper to make the connections between the two-way ANOVA (main effects) output and the multiple regression using the two indicators. Obviously, we see a difference in the ANOVA tables themselves since the multiple regression combines the effects due to both A and B into a single component while they are treated separately in the two-way ANOVA. However, the regression degrees of freedom and SSModel terms are just the sums of the individual factors shown in the the two-way ANOVA model; for example, $2 = 1 + 2$ for the degrees of freedom and $2379 = 2187 + 192$ for the sum of squares. Furthermore, the "Seq SS" numbers given by software regression output show the contribution to the variability explained by the model for each factor. These sums of squares match those in the two-way ANOVA output. Furthermore, note that the p-values for the individual t-tests of the coefficients of the two indicators in the multiple regression match the p-values for each factor as main effects in the two-way model; this is a bonus that comes when there are only two levels in a factor.

Comparing group means in the two-way ANOVA output shows that the difference for *Antibiotic* is $38 - 11 = 27$ and for $B12$ is $28.5 - 20.5 = 8$, exactly matching the coefficients of the respective indicators in fitted regression. What about the estimated constant term of $\hat{\beta}_0 = 7.0$? Our experience tells us that this should have something to do with the no antibiotic, no B12 case ($A = B = 0$), but the mean of the data in that cell is $\bar{y}_{11} = 19$. Remember that the main effects model also had some difficulty predicting the individual cells accurately. In fact, you can check that the predicted means for each cell using the two-indicator regression match the values generated from the estimated effects in the main effects only ANOVA of Example 6.4. That is what led us to consider adding an interaction term to the model, so let's see how to translate the interaction model into a multiple regression setting.

Recall that in earlier regression examples (such as comparing two regression lines in Section 3.3 or the interaction model for perch weights in Example 3.10) we handled interaction by including a term that was a product of the two interacting variables. The same reasoning works for indicator variables. For the **PigFeed** data the appropriate model is

$$WgtGain = \beta_0 + \beta_1 A + \beta_2 B + \beta_3 A \cdot B + \epsilon$$

Output for the two-way ANOVA with interaction is shown below.

7.4. TOPIC: ANOVA AND REGRESSION WITH INDICATORS

```
Source         DF      SS      MS      F      P
Antibiotic      1   2187.0  2187.0  60.33  0.000
B12             1    192.0   192.0   5.30  0.050
Antibiotic*B12  1   1728.0  1728.0  47.67  0.000
Error           8    290.0    36.3
Total          11   4397.0

S = 6.02080   R-Sq = 93.40%   R-Sq(adj) = 90.93%

Means
Antibiotic  B12   N   WgtGain
No          No    3   19.000
No          Yes   3    3.000
Yes         No    3   22.000
Yes         Yes   3   54.000
```

Output for the multiple regression model with A, B and AB is shown below.

```
The regression equation is
WgtGain = 19.0 + 3.00 A - 16.0 B + 48.0 AB

Predictor    Coef   SE Coef     T      P
Constant   19.000     3.476  5.47  0.001
A           3.000     4.916  0.61  0.559
B         -16.000     4.916 -3.25  0.012
AB         48.000     6.952  6.90  0.000

S = 6.02080   R-Sq = 93.4%   R-Sq(adj) = 90.9%

Analysis of Variance
Source          DF      SS      MS      F      P
Regression       3   4107.0  1369.0  37.77  0.000
Residual Error   8    290.0    36.2
Total           11   4397.0

Source  DF   Seq SS
A        1   2187.0
B        1    192.0
AB       1   1728.0
```

Again, we see that the three components of the ANOVA model are combined in the multiple regression ANOVA output, but the separate contributions and degrees of freedom are shown in the sequential sum of squares section. Note, however, that in this case the p-values for the individual terms in the regression are not the same as the results of the F-tests for each component in the regression model. This should not be too surprising since the addition of the interaction product term introduces a new predictor that is obviously correlated with both A and B.

What about the coefficients in the fitted regression equation? With the interaction term present we see that the constant term, $\hat{\beta}_0 = 19.0$, matches the cell mean for the no antibiotic, no vitamin B12 condition ($A = B = 0$). The coefficient of A (3.0) says that the mean should go up by +3 (to 22.0) when we move to the antibiotic group but keep the B12 value at "no." This gives the sample mean for that cell. Similarly, the coefficient of B indicates that the mean should decrease by 16 when we move from the (no, no) cell to the (no, yes) cell, giving a mean mean of just 3.0. Finally, we interpret the interaction by starting at 19.0 in the reference cell, going up by +3 when adding the antibiotic, going down by 16 for including B12, but then going up by another +48 for the interaction effect when the antibiotic and B12 are used together. So we have $19 + 3 - 16 + 48 = 54$, the sample mean of the (yes, yes) cell. This form of the model makes it relatively easy to see that this interaction is important when deciding how to feed the piggies. ⋄

Example 7.14: *Ants on a sandwich*

As part of an article in the *Journal of Statistics Education*, Margaret Mackisack of Queensland University of Technology described an experiment conducted by one of her students. The student, Dominic, noticed that ants often congregated on bits of sandwich that were dropped on the ground. He wondered what kind of sandwiches ants preferred to eat, so he set up an experiment. Among the factors he considered were the *Filling* of the sandwich (vegemite, peanut butter, ham and pickle) and the type of *Bread* (rye, whole wheat, multi-grain, white) used. He also used butter on some sandwiches and not others, but we will ignore that factor for the moment. Dominic prepared 4 sandwich pieces for each combination of *Bread* and *Filling*, so 48 observations in total. Randomizing the order, he left a piece of sandwich near an anthill for five minutes, then trapped the ants with an inverted jar and counted how many were on the sandwich. After waiting for the ants to settle down or switching to a similar size anthill, he repeated the process on the next sandwich type. The data in **SandwichAnts** are based on the counts he collected.

Let Factor A represent the *Filling* with $K = 3$ levels and Factor B be the type of *Bread* with $J = 4$. We have a 3×4 factorial design with $c = 4$ values in each cell. The response variable is the number of *Ants* on each sandwich piece. Table 7.2 shows the means for each cell as well as the row means (*Filling*) and columns means (*Bread*).

The two-way ANOVA table with interaction for these data follows.

	Rye	WholeWheat	MultiGrain	White	Row mean
Vegemite	29.00	37.25	33.50	38.75	34.625
PeanutButter	40.25	49.50	37.00	38.25	40.375
HamPickles	57.50	49.50	58.50	56.50	55.50
Column mean	42.25	44.25	43.00	44.50	43.50

Table 7.2: Mean numbers of ants on sandwiches

```
Analysis of Variance for Ants

Source          DF        SS        MS       F       P
Filling          2    3720.5    1860.3   10.39   0.000
Bread            3      40.5      13.5    0.08   0.973
Filling*Bread    6     577.0      96.2    0.54   0.777
Error           36    6448.0     179.1
Total           47   10786.0
```

The two-way ANOVA indicates that the type of filling appears to make a difference, with the row means indicating the ants might prefer ham and pickle sandwiches. The type of bread does not appear to be a significant factor in determining sandwich preferences for ants and there also doesn't appear to be ant interaction between the filling and the type of bread it is in.

What do these results look like if we use multiple regression with indicators to identify the types of breads and fillings? Define indicator variables for each level of each factor and choose one to leave out for each factor. The choice of which to omit is somewhat arbitrary in this case, but statistical software often selects either the first or last level from an alphabetical or numerical list. For this example we include the indicator variables listed below in the multiple regression model.

Main effect for *Filling*: $A_1 = PeanutButter$, $A_2 = Vegemite$ (leave out *HamPickles*)
Main effect for *Bread*: $B_1 = Rye$, $B2 = White$, $B_3 = WholeWheat$ (leave out *MultiGrain*)
Interaction for *Filling · Bread*: $A_1B_1, A_1B_2, A_1B_3, A_2B_1, A_2B_2, A_2B_3$

Note that there are 2 degrees of freedom for *Filling*, 3 degrees of freedom for *Bread*, and 6 degrees of freedom for the interaction. The output from running this regression follows.

The regression equation is
Ants = 58.5 - 21.5 A1 - 25.0 A2 - 1.00 B1 - 2.00 B2 - 9.00 B3 + 4.3 A1B1
       + 3.3 A1B2 + 18.0 A1B3 - 3.5 A2B1 + 7.2 A2B2 + 12.8 A2B3

Predictor	Coef	SE Coef	T	P
Constant	58.500	6.692	8.74	0.000
A1	-21.500	9.463	-2.27	0.029
A2	-25.000	9.463	-2.64	0.012
B1	-1.000	9.463	-0.11	0.916
B2	-2.000	9.463	-0.21	0.834
B3	-9.000	9.463	-0.95	0.348
A1B1	4.25	13.38	0.32	0.753
A1B2	3.25	13.38	0.24	0.810
A1B3	18.00	13.38	1.34	0.187
A2B1	-3.50	13.38	-0.26	0.795
A2B2	7.25	13.38	0.54	0.591
A2B3	12.75	13.38	0.95	0.347

S = 13.3832   R-Sq = 40.2%   R-Sq(adj) = 22.0%

Analysis of Variance

Source	DF	SS	MS	F	P
Regression	11	4338.0	394.4	2.20	0.037
Residual Error	36	6448.0	179.1		
Total	47	10786.0			

Source	DF	Seq SS
A1	1	234.4
A2	1	3486.1
B1	1	25.0
B2	1	6.1
B3	1	9.4
A1B1	1	10.1
A1B2	1	68.1
A1B3	1	180.2
A2B1	1	155.0
A2B2	1	1.0
A2B3	1	162.6

As we saw in the previous example, the ANOVA F-test for the multiple regression model combines both factors and the interaction into a single test for the overall model. Check that the degrees of freedom (11=2+3+6) and sum of squares (4338=3720.5+40.5+577.0) are sums of the three model components in the two-way ANOVA table. In the multiple regression setting, the tests for the main effects for Factor A (coefficients of A_1 and A_2), Factor B (coefficients of B_1, B_2 and B_3), and the interaction effect (coefficients of the six indicator products) can be viewed as nested F-tests. For example, to test the interaction we sum the sequential SS values from the Minitab output, $10.1 + 68.1 + 180.2 + 155.0 + 1.0 + 162.6 = 576.9$, matching (up to roundoff) the interaction sum of squares in the two-way ANOVA table and having six degrees of freedom. In this way we can also find the sum of squares for Factor A (234.4+3486.1=3720.5) with two degrees of freedom and for Factor B (25.0+6.1+6.4=4.5) with three degrees of freedom.

Let's see if we can recover the cell means in Table 7.2 from the information in the multiple regression output. The easy case is the constant term, which represents the cell mean for the indicators that weren't included in the model (*HamPickles* on *MultiGrain*). If we keep the bread fixed at *MultiGrain* we see that the cell means drop by -21.5 ants and -25.0 ants as we move to *PeanutButter*, and *Vegemite*, respectively. On the other hand, if we keep *HamPickles* as the filling, the coefficients of coefficients of B_1, B_2, and B_3 indicate what happens to the cell means as we change the type of bread to *Rye*, *White*, and *WholeWheat*, respectively. For any of the other cell means we first need to adjust for the filling, then the bread, and finally the interaction. For example, to recover the cell mean for *Vegemite* (A_2) and *Rye* (B_1) we have $58.5 - 25.0 - 1.0 - 3.5 = 29.0$, the sample mean number of ants for a vegemite on rye sandwich. ⋄

With most statistical software, we can run the multiple regression version of the ANOVA model without explicitly creating the individual indicator variables, as the software will do this for us. Once we recognize that we can include any categorical factor in a regression model by using indicators, the next natural extension is to allow for a mixture of both categorical and quantitative terms in the same model. Although we have already done this on a limited scale with single binary categorical variables, the next topic explores these sorts of models more fully.

7.5 Topic: Analysis of Covariance

We now turn our attention to the setting in which we would like to model the relationship between a continuous response variable and a categorical explanatory variable, but we suspect that there may be another quantitative variable affecting the outcome of the analysis. Were it not for the additional quantitative variable we would use an ANOVA model as discussed in chapter 5 (assuming that conditions are met). But we may discover, often after the fact, that the experimental or observational units were different at the onset of the study, with the differences being measured by the additional continuous variable. For example, it may be that the treatment groups had differences between them even before the experimental treatments were applied to them. In this case any results from an ANOVA analysis become suspect. If we find a significant group effect, is it due to the treatment, the additional variable, or both? If we do not find a significant group effect, is there really one there that has been masked by the extra variable?

Setting the Stage

One method for dealing with both categorical and quantitative explanatory variables is to use a multiple regression model as discussed in Section 3.3. But in that model, both types of explanatory variables have equal importance with respect to the response variable. Here we discuss what to do when the quantitative variable is more-or-less a nuisance variable: We know it is there, we have measured it on the observations, but we really don't care about its relationship with Y other than how it interferes with the relationship between Y and the factor of interest. This type of variable is called a *covariate*.

> **Covariate**
>
> A continuous variable X_c not of direct interest, but that may have an effect on the relationship between the response variable Y and the factor of interest, is called a **covariate**.

The type of model that takes covariates into consideration is called an *analysis of covariance model*.

Example 7.15: *Grocery stores*

Grocery stores and product manufacturers are always interested in how well the products on the store shelves sell. An experiment was designed to test whether the amount of discount given on products affected the amount of sales of that product. There were three levels of discount, 5%, 10%, and 15%, and sales were held for a week. The total number of products sold during the week of the sale was recorded. The researchers also recorded the wholesale price of the item put on sale. The data are found in the file **Grocery**.

7.5. TOPIC: ANALYSIS OF COVARIANCE

CHOOSE

We start by analyzing whether the discount had any effect on sales using an ANOVA model.

FIT

The ANOVA table is given below:

```
One-way ANOVA: Sales versus Discount

Source     DF     SS      MS     F      P
Discount    2   1288     644   0.57  0.569
Error      33  37074    1123
Total      35  38363

S = 33.52    R-Sq = 3.36%    R-Sq(adj) = 0.00%
```

ASSESS

Figure 7.7 shows both the residual versus fits plot and the normal probability plot of the residuals for this model. Both indicate that the conditions of normality and equal variances are met. Levene's test (see Section 7.1) also confirms the consistency of the data with an equal variances model.

```
Levene's Test (Any Continuous Distribution)
Test statistic = 0.93, p-value = 0.406
```

Since the products were randomly assigned to the treatment groups, we are comfortable assuming that the observations are independent of one another and that there was no bias introduced.

USE

The p-value for the ANOVA model is quite large (0.569) so we would fail to reject the null hypothesis. In other words, the data do not suggest that average sales over the one-week period are different for products offered at different discount amounts. In fact, the R^2 value is very small at 3.36%. So the model just does not seem to be very helpful in predicting product sales based on the amount of discount. ◇

The conclusion of the last example seems straightforward and it comes from a typical ANOVA model. But we also need to be aware of the fact that we have information that may be relevant

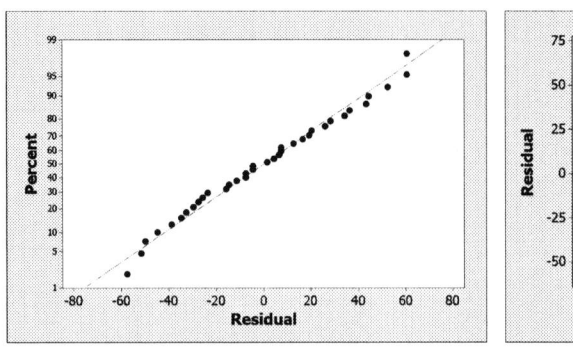

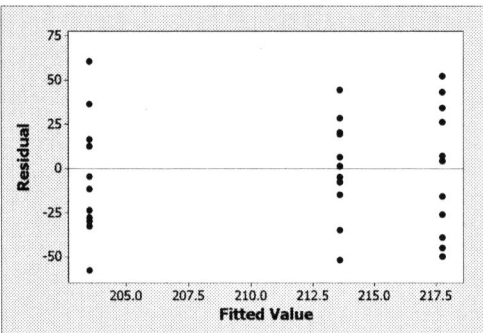

(a) Normal probability plot of residuals

(b) Scatterplot of residuals versus fits

Figure 7.7: Plots to assess the fit of the ANOVA model

to the question but which we have not used in our analysis. We know the wholesale price of each product as well. It could be that the prices of the products are masking a relationship between the amount of sales and the discount, so we will treat the wholesale price as a covariate and use an *ANCOVA model*.

What Is the ANCOVA Model?

The ANCOVA model is similar to the ANOVA model. Recall that the ANOVA model is

$$Y = \mu + \alpha_k + \epsilon$$

The ANCOVA model essentially takes the error term from the ANOVA model and divides it into two pieces: one piece takes into account the relationship between the covariate and the response while the other piece is the residual that remains.

ANCOVA model

The ANCOVA model is written as

$$Y = \mu + \alpha_k + \beta X_c + \epsilon_{adj}$$

where X_c is the covariate and ϵ_{adj} is the error term adjusted for the covariate.

Of course the ANCOVA model is only appropriate under certain conditions.

7.5. TOPIC: ANALYSIS OF COVARIANCE

Conditions

The conditions necessary for the ANCOVA model are:

- All of the ANOVA conditions are met for Y with the factor of interest.
- All of the linear regression conditions are met for Y with X_c.
- There is no interaction between the factor and X_c.

Note that the last condition is equivalent to requiring that the linear relationship between Y and X_c have the same slope for each of the levels of the factor.

We saw in Example 7.15 that the discount did not seem to have an effect on the sales of the products. Figure 7.8(a) gives a dotplot of the product sales broken down by amount of discount. As expected, the plot shows quite considerable overlap between the three groups. This plot is a good way to visualize the conclusion from the ANOVA.

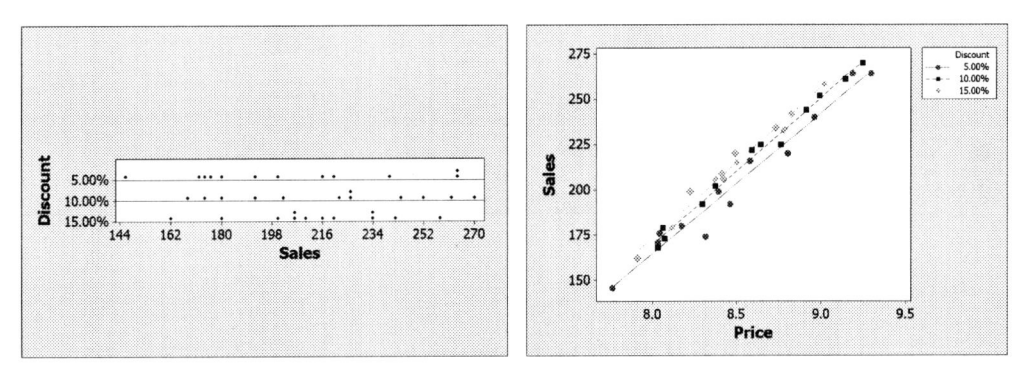

(a) Dotplot of sales loss by amount of discount (b) Scatterplot of sales against price by amount of discount

Figure 7.8: Plots to assess the fit of the ANCOVA model

Now consider Figure 7.8(b). This is a scatterplot of the product sales versus the product price with different symbols for the three different discount amounts. We have also included the regression lines for each of the three groups. This graph shows a more obvious difference between the three groups. If we collapse the y values across the x-axis (as the dotplot does) there is not much difference between the sales at the different discount levels. But, if we concentrate on specific values for the product price, there is a difference. The scatterplot suggests that the higher the discount, the more the product will sell. This suggests that an ANCOVA model may be appropriate.

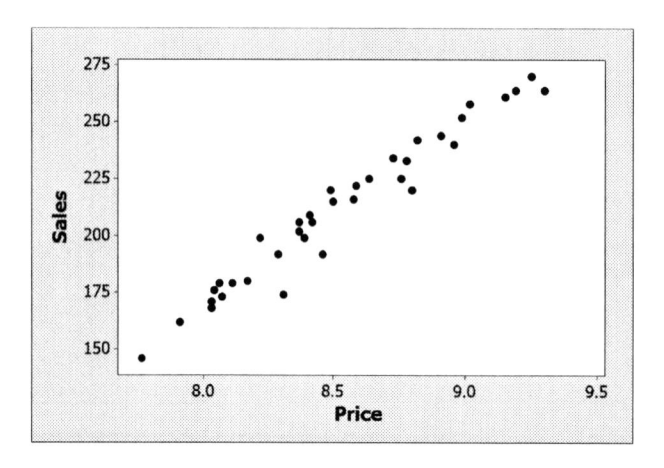

Figure 7.9: Scatterplot of sales versus price

Example 7.16: *Grocery — checking conditions*

In Example 7.15 we already discovered that the conditions for an ANOVA model with discount predicting the product sales were met, so we move on to checking the conditions for the linear regression of the sales on the price of the product. Figure 7.9 is a scatterplot of the response versus the covariate. There is clearly a strong linear relationship between these two variables. Next, we consider Figure 7.10 which gives both the normal probability plot of the residuals and the residuals versus fits plot. Both of these graphs are consistent with the conditions for linear regression (although the normal probability plot is not as linear as we would like). We also have no reason to believe that one product's sales will have affected any other product's sales so we can consider the error terms to be independent.

(a) Normal probability plot of residuals   (b) Scatterplot of residuals versus fits

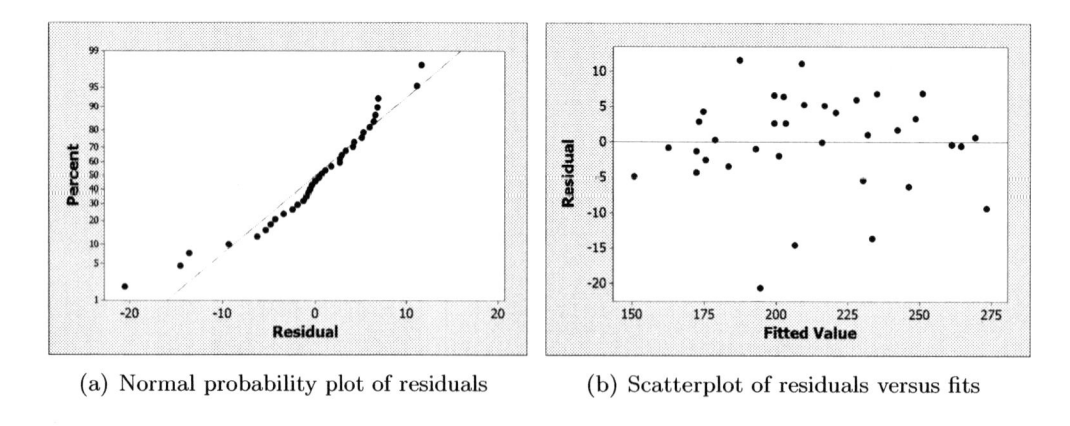

Figure 7.10: Plots to assess the fit of the linear regression model

Finally we need to consider whether the slopes for the regression of sales on price are the same for all three groups. Figure 7.8(b) shows that, while they are not identical, they are quite similar. In

7.5. TOPIC: ANALYSIS OF COVARIANCE

fact, because of variability in data, it would be very surprising if a data set resulted in slopes that were exactly the same.

It appears that all of the conditions for the ANCOVA model have been met and we can proceed with the analysis. ◇

At this point we have discussed the general reason for using an ANCOVA model, what the equation for the model looks like, and what the conditions are for the model to be appropriate. What's left to discuss is how to actually fit the model and how to interpret the results. We are not going to go into the details of how the parameters are estimated, but rather rely on computer software to do the computations for us and concentrate on how to interpret the results. However, we note that the basic idea behind the sums of squares is the same as the idea used in Chapters 2 and 5 to derive sums of squares from the definitions. In fact, the output looks very similar to that of the ANOVA model but with one more line in the ANOVA table corresponding to the covariate. When we interpret this output we typically continue to concentrate on the p-value associated with the factor, as we did in the ANOVA model. Specifically, we are interested in whether the effect of the factor is significant now that we have controlled for the covariate, and whether the significance has changed in moving from the ANOVA model to the ANCOVA model. If the answer to the latter question is yes, then the covariate was indeed important to the analysis.

Example 7.17: *Grocery using ANCOVA model*

In Example 7.16 we saw that the conditions for the ANCOVA model were satisfied. The computer output for the model is shown below.

```
Analysis of Variance for Sales, using Adjusted SS for Tests

Source    DF  Seq SS  Adj SS  Adj MS        F      P
Price      1   36718   36230   36230  1372.84  0.000
Discount   2     800     800     400    15.15  0.000
Error     32     844     844      26
Total     35   38363


S = 5.13714   R-Sq = 97.80%   R-Sq(adj) = 97.59%
```

Once again we check to see whether the F-test for the factor (discount rate) is significant. In this analysis, with the covariate (*Price*) taken into account, we see that the p-value for the discount rate is indeed significant. In fact, the software reports the p-value to be approximately 0. It seems quite clear that the sales amounts are different depending on the amount of discount on the products.

Even more telling is the fact that the R^2 value has risen quite dramatically from 3.36% in the ANOVA model to 97.8% in the ANCOVA model. We now conclude that at any given price level, discount rate is important, and looking back at the scatterplot in Figure 7.8(b) we see that the higher the discount rate, the more of the product that is sold.   ⋄

The grocery store data that we have just been considering shows one way that a covariate can change the relationship between a factor and a response variable. In that case, there did not seem to be a relationship between the factor and the response variable until the covariate was taken into consideration. The next example illustrates another way that a covariate can change the conclusions of an analysis.

Example 7.18: *Exercise and heart rate*

Does how much exercise you get on a regular basis affect how high your heart rate goes when you are active? This was the question that a Stat 2 instructor set out to examine. He had his students rate themselves on how active they were generally (1 = not active, 2 = moderately active, 3 = very active). This variable was recorded with the name "Exercise." He also measured several other variables that might be related to active pulse rate, including sex and whether they smoked or not. The last explanatory variable he had them measure was their resting pulse rate. Finally, he assigned them to one of two treatments (walk or run up and down a flight of stairs 3 times) and measure their pulse when they were done. This last pulse rate was their "active pulse rate" and was the response variable for the study. The data are found in the file **Pulse**.

CHOOSE

We start with the simplest model. That is, we start by looking to see what the ANOVA model tells us since we have a factor with three levels and a quantitative response.

FIT

The ANOVA table is given below:

```
One-way ANOVA: Active versus Exercise

Source     DF     SS     MS      F      P
Exercise    2  10523   5261  16.90  0.000
Error     229  71298    311
Total     231  81820

S = 17.64   R-Sq = 12.86%   R-Sq(adj) = 12.10%
```

7.5. TOPIC: ANALYSIS OF COVARIANCE

ASSESS

Figure 7.11 shows both the residual versus fits plot and the normal probability plot of the residuals for this model. Both indicate that the conditions of normality and equal variances are met. Levene's test (see Section 7.1) also confirms the consistency of the data with an equal variances model.

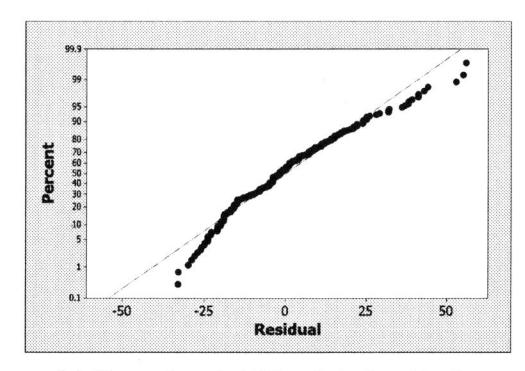

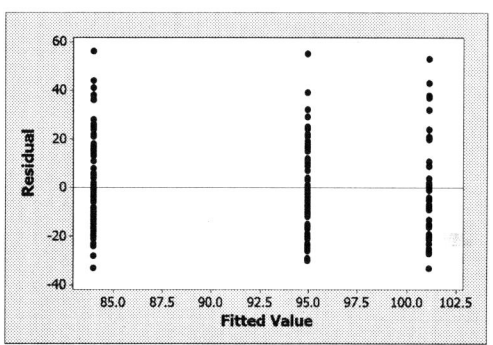

(a) Normal probability plot of residuals   (b) Scatterplot of residuals versus fits

Figure 7.11: Plots to assess the fit of the ANOVA model

```
Levene's Test (Any Continuous Distribution)
Test statistic = 2.11, p-value = 0.124
```

USE

In this case the p-value is quite small in the ANOVA table and therefore it seems that the amount of exercise a student gets in general does affect how high their pulse goes after exercise. But we also notice that the R^2 value is still pretty low at 12.86%, confirming that this model does not explain everything about the after-exercise pulse rate.

Now we use the fact that we do have a covariate that might affect the analysis: resting pulse rate.

CHOOSE

So we redo the analysis this time using analysis of covariance.

346   CHAPTER 7. ADDITIONAL ANOVA TOPICS

FIT

The ANCOVA table is given below:

Analysis of Variance for Active, using Adjusted SS for Tests

```
Source      DF   Seq SS   Adj SS   Adj MS       F      P
Rest         1    29868    19622    19622   86.57  0.000
Exercise     2      276      276      138    0.61  0.544
Error      228    51676    51676      227
Total      231    81820
```

S = 15.0549   R-Sq = 36.84%   R-Sq(adj) = 36.01%

ASSESS

We already assessed the conditions of the ANOVA model. Now we move on to assessing the conditions of the linear regression model for predicting active pulse rate from resting pulse rate. Figure 7.12 is a scatterplot of the response versus the covariate. There is clearly a strong linear relationship between these two variables. Next, we consider Figure 7.13, which gives both the normal probability plot of the residuals and the residuals versus fits plot. Although the residuals versus fits plot is consistent with the conditions for linear regression of equal variance and we have no reason to believe that one person's active pulse rate will affect any other person's active pulse rate, the normal probability plot of the residuals suggests that the residuals have a distribution that is right-skewed. So we have concerns about the ANCOVA model at this point.

For completeness we check whether the slopes for the regression of active pulse rate on resting pulse rate are the same for all three groups. Figure 7.14 shows that, while they are not identical, they are reasonably similar.

CHOOSE (again)

At this point we cannot proceed with the ANCOVA model because the normality condition of the residuals for the linear model has not been met. This leads us to try using a log transformation on both the active pulse rate and the resting pulse rate.

FIT (again)

Since we are transforming both our response variable and our covariate, we re-run both the ANOVA table and the ANCOVA table.

7.5. TOPIC: ANALYSIS OF COVARIANCE

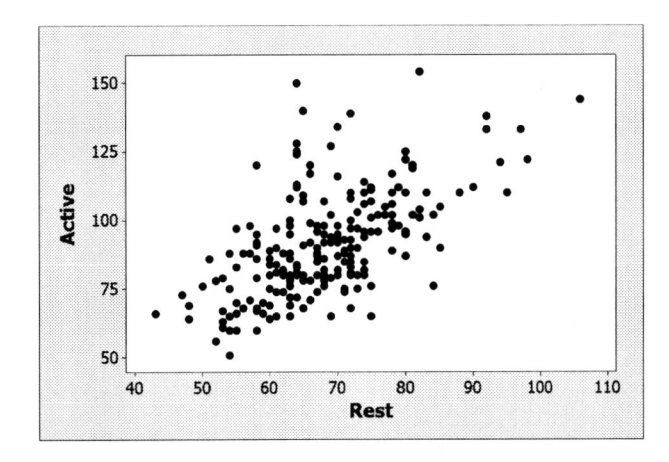

Figure 7.12: Scatterplot of active pulse rate versus resting pulse rate

(a) Normal probability plot of residuals (b) Scatterplot of residuals versus fits

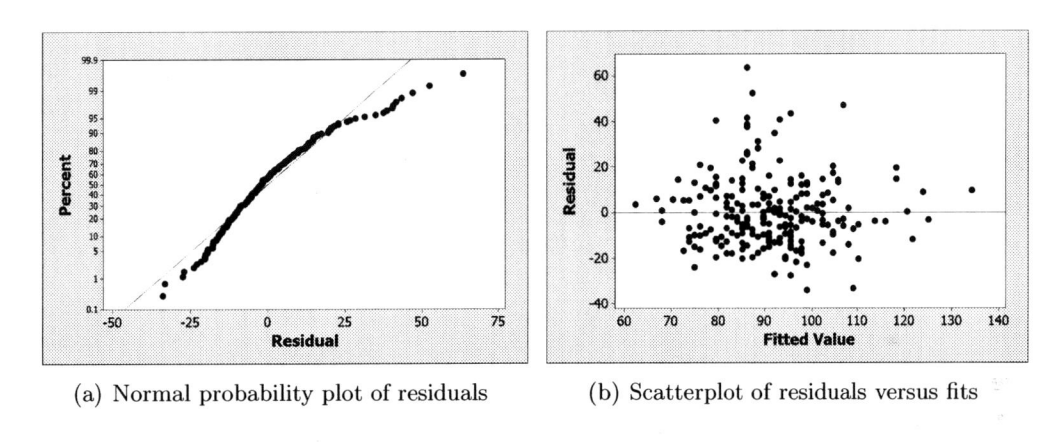

Figure 7.13: Plots to assess the fit of the linear regression model

First we display the ANOVA table:

```
One-way ANOVA: log(active) versus Exercise

Source     DF      SS      MS      F      P
Exercise    2  1.3070  0.6535  18.40  0.000
Error     229  8.1316  0.0355
Total     231  9.4386

S = 0.1884   R-Sq = 13.85%   R-Sq(adj) = 13.10%
```

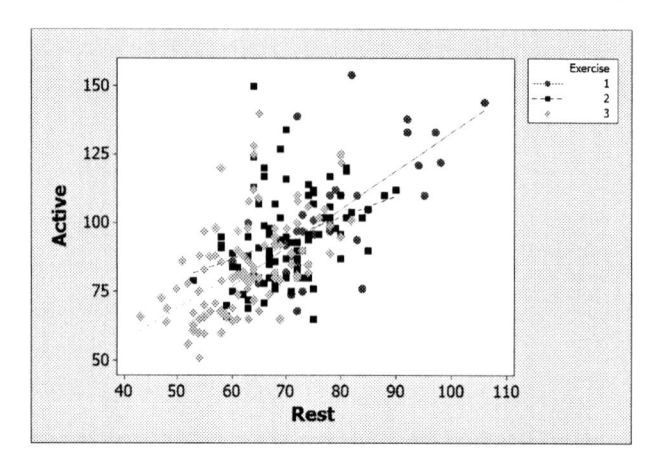

Figure 7.14: Scatterplot of active pulse rate versus resting pulse rate for each exercise level

---

Here we note that the factor is still significant (though we still need to re-check the conditions for this ANOVA) and the R^2 is a similar value at 13.85%.

Next we display the ANCOVA table:

```
Analysis of Variance for log(active), using Adjusted SS for Tests

Source       DF   Seq SS   Adj SS   Adj MS      F      P
log(rest)     1   3.5951   2.3305   2.3305   91.59  0.000
Exercise      2   0.0424   0.0424   0.0212    0.83  0.436
Error       228   5.8011   5.8011   0.0254
Total       231   9.4386


S = 0.159510   R-Sq = 38.54%   R-Sq(adj) = 37.73%
```

And now we note that the exercise factor is not significant, though, again, we still have to re-check the conditions.

ASSESS (again)

First we assess the conditions for the ANOVA model. Figure 7.15 shows both the residual versus fits plot and the normal probability plot of the residuals for this model. Both indicate that the

7.5. TOPIC: ANALYSIS OF COVARIANCE

conditions of normality and equal variances are met. Levene's test (see Section 7.1) also confirms the consistency of the data with an equal variances model.

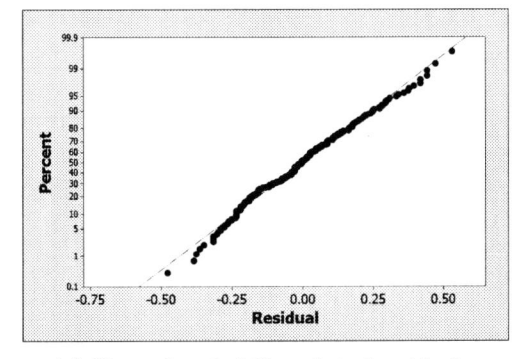

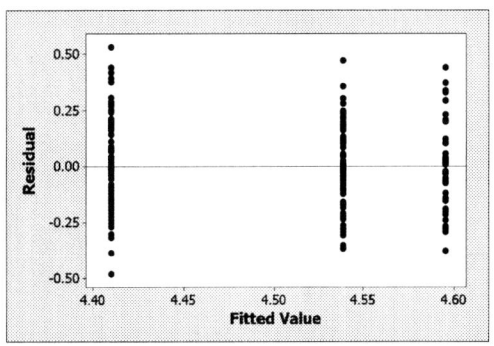

(a) Normal probability plot of residuals   (b) Scatterplot of residuals versus fits

Figure 7.15: Plots to assess the fit of the ANOVA model

```
Levene's Test (Any Continuous Distribution)
Test statistic = 1.47, p-value = 0.231
```

We have already decided that the independence condition holds for that data set so we do not need to check for that again.

Now we check the conditions for the linear regression model. Figure 7.16 is a scatterplot of the response versus the covariate. There is clearly a strong linear relationship between these two variables. Next, we consider Figure 7.17 which gives both the normal probability plot of the residuals and the residuals versus fits plot. This time we do not see anything that concerns us about the conditions.

Finally we again check to make sure the slopes are approximately the same, and Figure 7.18 shows us that they are.

USE:
It finally appears that all of the conditions for the ANCOVA model have been met and we can proceed with the analysis. Looking back at the ANCOVA table with the log of the active pulse rate for the response variable, we see that now the factor measuring average level of exercise is no longer significant, even though it was in the ANOVA table. This means that the typical level of exercise was really just another way to measure resting pulse rate and that we would probably be better off just running a simple linear regression model using resting pulse rate to predict active pulse rate.   ⋄

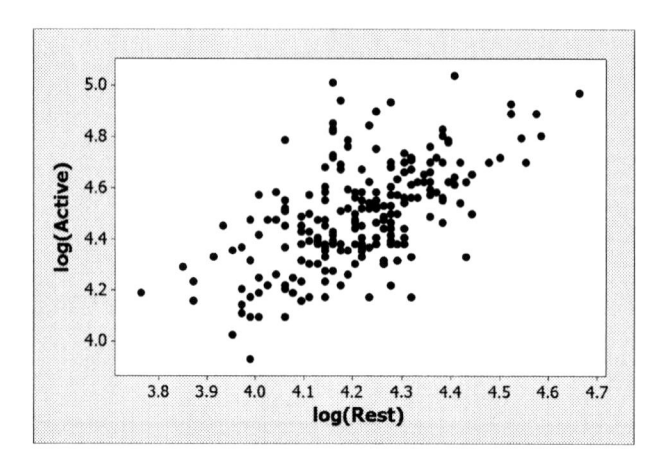

Figure 7.16: Scatterplot of active pulse rate versus resting pulse rate

(a) Normal probability plot of residuals

(b) Scatterplot of residuals versus fits

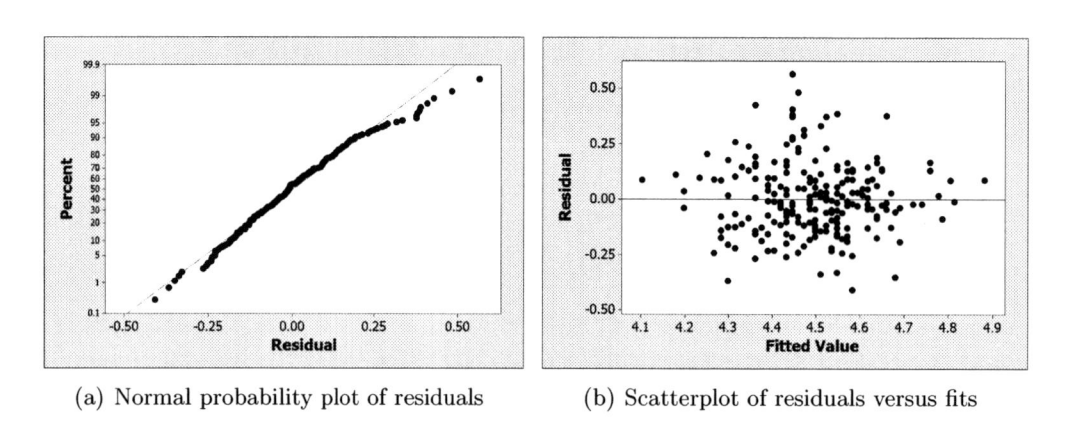

Figure 7.17: Plots to assess the fit of the linear regression model

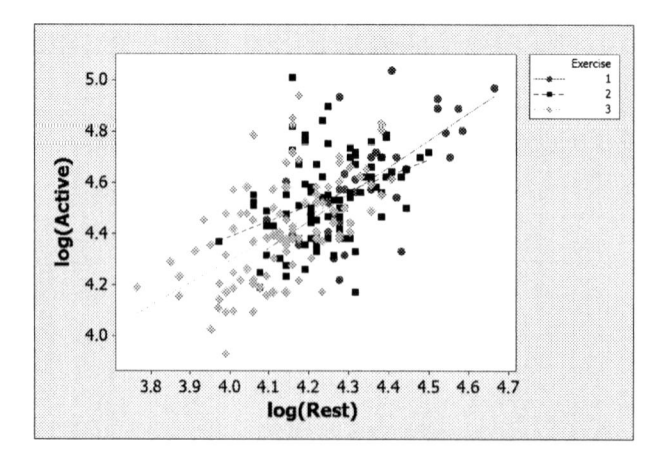

Figure 7.18: Scatterplot of active pulse rate versus resting pulse rate for each exercise level

7.6 Exercises

Topic 7.1 Exercises: Levene's Test for Homogeneity of Variances

7.1 *North Carolina births.* The file **NCbirths** contains data on a random sample of 1450 birth records in the state of North Carolina in the year 2001. This sample was selected by John Holcomb, based on data from the North Carolina State Center for Health and Environmental Statistics. One question of interest is whether the distribution of birth weights differs among mothers' racial groups. For the purposes of this analysis we will consider four racial groups as reported in the variable *MomRace*: white, black, Hispanic, and other (including Asian, Hawaiian, and Native American). Use Levene's test to determine if the condition of equality of variances is satisfied. Report your results.

7.2 *Blood pressure.* A person's systolic blood pressure can be a signal of serious issues in their cardiovascular system. Are there differences between average systolic blood pressure based on smoking habits? The data set **Blood1** has the systolic blood pressure and the smoking status of 500 randomly chosen adults. We would like to know if the mean systolic blood pressure is different for smokers and non-smokers. Use Levene's test to determine if the condition of equality of variances is satisfied. Report your results.

7.3 *Blood pressure (continued).* The data set used in Exercise 7.2 also measured the sizes of people using the variable *Overwt*. This is a categorical variable that takes on the values 0=Normal, 1=Overweight, and 2=Obese. We would like to know if the mean systolic blood pressure differs for these three groups of people. Use Levene's test to determine if the condition of equality of variances is satisfied. Report your results.

7.4 *Swahili attitudes.* Hamisi Babusa, a Kenyan scholar, administered a survey to 480 students from Pwani and Nairobi provinces about their attitudes towards the Swahili language. In addition, the students took an exam on Swahili. From each province, the students were from 6 schools (3 girls schools and 3 boys schools) with 40 students sampled at each school, so half of the students from each province were males and the other half females. The survey instrument contained 40 statements about attitudes towards Swahili and students rated their level of agreement to each. Of these questions, 30 were positive questions and the remaining 10 were negative questions. On an individual question the most positive response would be assigned a value of 5 while the most negative response would be assigned a value of 1. By adding the responses to each question, we can find an overall *Attitude Score* for each student. The highest possible score would be 200 (an individual who gave the most positive possible response to every question). The lowest possible score would be 40 (an individual who gave the most negative response to every question). The data are stored in **Swahili**.

  a. There are two explanatory variables of interest, *Province* and *Sex*. Use Levene's test for each factor by itself. Would either one-way ANOVA model satisfy the equal variances condition? Explain.

b. Use Levene's test with both factors together. Would a two-way ANOVA model be appropriate? Explain.

Supplemental Exercise

7.5 *Sea slugs.* Sea slug larvae need vaucherian seaweed, but the larvae from these sea slugs need to locate this type of seaweed to survive. A study was done to try to determine whether chemicals that leach out of the seaweed attract the larvae. Seawater was collected over a patch of this kind of seaweed at 5-minute intervals as the tide was coming in and, presumably, mixing with the chemicals. The idea was that as more seawater came in, the concentration of the chemicals was reduced. Each sample of water was divided into 6 parts. Larvae were then introduced to this seawater to see what percentage metamorphosed. The question of interest is whether or not there is a difference in this percentage over the 5 time periods. Open the data set **SeaSlugs**. We will use this data set to illustrate the way that Levene's test is calculated.

   a. Find the median percent metamorphosed larvae for each of the 6 time periods. Find the absolute deviation between those medians and the actual observations. Plot those deviations on one dotplot, grouped by time period. Does it look as if there is a difference in the average absolute deviation for the 6 different time periods? Explain.

   b. Compute an ANOVA table for the absolute deviations. What is the test statistic? What is the p-value?

   c. Run Levene's test on the original data. What is the test statistic? What is the p-value? How do these values compare to the values that you computed in part (b)?

Topic 7.2 Exercises: Multiple Tests

7.6 *Fantasy baseball.* Recall the data from Exercise 5.22. The data recorded the amount of time each of 8 "fantasy baseball" participants took in each of 24 rounds to make their selection. The data are listed in Exercise 5.22 and are available in the data file **FantasyBaseball**. In Chapter 5, Exercise 5.23 we asked you to transform the selection times using the natural log before continuing with your analysis because the residuals were not normally distributed. We ask you to do the same again here.

   a. Use Tukey's HSD to compute confidence intervals to identify the differences in average selection times for the different participants. Report your results.

   b. Which multiple comparisons method would you use to compute confidence intervals to assess which rounds' average selection times differ significantly from which other? Explain.

7.7 *North Carolina births.* The file **NCbirths** contains data on a random sample of 1450 birth records in the state of North Carolina in the year 2001. In Exercises 5.29- 5.31 we conducted an analysis to determine whether there was a race-based difference in birth weights. In Exercise 5.30

7.6. EXERCISES

we found that the test in the ANOVA table was significant. Use the Bonferroni method to compute confidence intervals to identify differences in birth weight for babies of moms of different races. Report your results.

7.8 *Blood pressure (continued).* The data set used in Exercise 7.2 also measured the sizes of people using the variable *Overwt*. This is a categorical variable that takes on the values 0=Normal, 1=Overweight, and 2=Obese. Are the mean systolic blood pressures different for these three groups of people?

a. Use Bonferroni intervals to find any differences that exist between these three group mean systolic blood pressures. Report your results

b. Use Tukey's HSD intervals to find any differences that exist between these three group mean systolic blood pressures. Report your results.

c. Were your conclusions in (a) and (b) different? Explain. If so, which would you prefer to use in this case and why?

7.9 *Sea slugs.* Sea slugs, common on the coast of southern California, live on vaucherian seaweed. But the larvae from these sea slugs need to locate this type of seaweed to survive. A study was done to try to determine whether chemicals that leach out of the seaweed attract the larvae. Seawater was collected over a patch of this kind of seaweed at 5-minute intervals as the tide was coming in and, presumably, mixing with the chemicals. The idea was that as more seawater came in, the concentration of the chemicals was reduced. Each sample of water was divided into 6 parts. Larvae were then introduced to this seawater to see what percentage metamorphosed. Is there a difference in this percentage over the 5 time periods? Open the data set **SeaSlugs**.

a. Use Fisher's LSD intervals to find any differences that exist between the percent of larvae that metamorphosed in the different water conditions.

b. Use Tukey's HSD intervals to find any differences that exist between the percent of larvae that metamorphosed in the different water conditions.

c. Were your conclusions to (a) and (b) different? Explain. If so, which would you prefer to use in this case and why?

Topic 7.3 Exercises: Comparisons and Contrasts

7.10 *Fruit flies.* Use the fruit fly data in **FruitFlies** to test the second question the researchers had. That is, is there a difference between the life spans of males living with pregnant females and males living alone?

a. What are the hypotheses that describe the test we would like to perform?

b. Write the contrast of interest in symbols and compute its estimated value.

c. What is the standard error of the contrast?

d. Perform the hypothesis test to test the alternative hypothesis that the mean life span of fruit flies living with pregnant females is different from that of fruit flies living alone. Be sure to give your conclusions.

7.11 *Blood pressure.* A person's systolic blood pressure can be a signal of serious issues in their cardiovascular system. Are there differences between average systolic blood pressure based on weight? The data set **Blood1** has the systolic blood pressure and the weight status of 300 randomly chosen adults. The categorical variable *Overwt* records the values 0=Normal, 1=Overweight, and 2=Obese for each individual. We would like to compare those who are of normal weight to those who are either overweight or obese.

a. What are the hypotheses that describe the test we would like to perform?

b. Write the contrast of interest in symbols and compute its estimated value.

c. What is the standard error of the contrast?

d. Perform the hypothesis test to test the alternative hypothesis that the mean systolic blood pressure is different for normal weight people as compared to overweight or obese people. Be sure to give your conclusions.

7.12 *Auto pollution.* The data set **AutoPollution** gives the results of an experiment on 36 different cars. The cars were randomly assigned to get either a new filter or a standard filter and the noise level for each car was measured. For this problem we are going to ignore the treatment itself and just look at the sizes of the cars (also given in this data set). The 36 cars in this data set consisted of 12 randomly selected cars in each of three sizes (1=small, 2=medium and 3=large). The researchers wondered if the large cars just generally produced a different amount of noise than the other two categories combined.

a. What are the hypotheses that describe the test we would like to perform?

b. Write the contrast of interest in symbols and compute its estimated value.

c. What is the standard error of the contrast?

d. Perform the hypothesis test to see if mean noise level is different for large cars as compared to small or medium-sized cars. Be sure to give your conclusions.

7.13 *Cancer survivability.* Example 5.8 discusses the data set from the following example. In the 1970's doctors wondered if giving terminal cancer patients a supplement of ascorbate would prolong their lives. They designed an experiment to compare cancer patients who received ascorbate to cancer patients who did not receive the supplement. The result of that experiment was that, in

7.6. EXERCISES

fact, ascorbate did seem to prolong the lives of these patients. But then a second question arose. Was the effect of the ascorbate different when different organs were affected by the cancer? The researchers took a second look at the data. This time they concentrated only on those patients who received the ascorbate and divided the data up by which organ was affected by the cancer. They had 5 different organs represented among the patients (all of whom only had one organ affected): stomach, bronchus, colon, ovary, and breast. In this case, since the patients were not randomly assigned to which type of cancer they had, but were, instead a random sample of those who had such cancers, we are dealing with an observational study. The data are available in the file **CancerSurvival**. In Example 5.8 we discovered that we needed to take the natural log of the survival times for the ANOVA model to be appropriate.

- a. We would like to see if the survival times for breast and ovary cancer are different from the survival times of the other three types. What are the hypotheses that describe the test we would like to perform?

- b. Write the contrast of interest in symbols and compute its estimated value.

- c. What is the standard error of the contrast?

- d. Perform the hypothesis test to test the alternative hypothesis that the mean survival time is different for breast and ovary cancers as compared to bronchus, colon, or stomach cancers. Be sure to give your conclusions.

Topic 7.4 Exercises: ANOVA and Regression with Indicators

7.14 *North Carolina births.* Exercises 5.29 and 5.30 asked for an ANOVA to determine whether there were differences between the birth weights of babies born to mothers of different races. We will repeat that analysis here using indicator variables and regression.

- a. Create indicator variables for the race categories as given in the *Momrace* variable. How many indicator variables will you use in the regression model?

- b. Run the regression model using the indicator variables and interpret the coefficients in the model. How do they relate to the ANOVA model? Explain.

- c. Compare the ANOVA table created by the regression model to the ANOVA table created by the ANOVA model? What conclusions do you reach with the regression analysis? Explain.

7.15 *Fenthion.* In Exercise 5.26 we introduced you to the problem of fruit flies in olive groves. Specifically, fenthion is a pesticide used against the olive fruit fly and it is toxic to humans so it is important that there be no residue left on the fruit or in olive oil that will be consumed. At one point in time, there was a theory that if there was residue of the pesticide left in the olive oil, it would dissipate over time. Chemists set out to test that theory by taking olive oil with fenthion residue and measuring the amount of fenthion in the oil at 3 different times over the year — day 0, day 281 and day 365. The data set for this problem is **Olives**.

a. Exercise 5.26 asked for an ANOVA to determine if there was a difference in fenthion residue over these three testing periods. For comparison sake, we ask you to run that ANOVA again here (using an exponential transformation of *fenthion* so that conditions are met).

b. Now analyze these data using regression, but treat the time period as a categorical variable. That is, create indicator variables for the time periods and use those in your regression analysis. Interpret the coefficients in the regression model.

c. Discuss how the results are the same between the regression analysis and the analysis in part (a).

d. In this case, the variable that we used as categorical in the ANOVA analysis really constitutes measurements at three different values of a continuous variable (*Time*). Treat *Time* as a continuous variable and use it in your regression analysis. Are your conclusions any different? Explain.

e. Which form of the analysis do you think is better in this case? Explain.

7.16 *Sea slugs.* Sea slugs, common on the coast of southern California, live on vaucherian seaweed, but the larvae from these sea slugs need to locate this type of seaweed to survive. A study was done to try to determine whether chemicals that leach out of the seaweed attract the larvae. Seawater was collected over a patch of this kind of seaweed at 5-minute intervals as the tide was coming in and, presumably, mixing with the chemicals. The idea was that as more seawater came in, the concentration of the chemicals was reduced. Each sample of water was divided into 6 parts. Larvae were then introduced to this seawater to see what percentage metamorphosed. Is there a difference in this percentage over the 5 time periods? Exercise 5.35 asked that the data set **SeaSlugs** be analyszed using ANOVA.

a. Repeat the analysis you did in Exercise 5.35 using regression with indicator variables. Interpret the coefficients in the regression model.

b. What are your conclusions from the regression analysis? Explain.

c. Now notice that the grouping variable is actually a time variable that we could consider to be continuous. Use regression again, but this time use *Time* as a continuous explanatory variable instead of using indicator variables. What are your conclusions? Explain.

d. Which form of the analysis do you think is better in this case? Explain.

7.17 *Auto pollution.* The data set **AutoPollution** gives the results of an experiment on 36 different cars. This experiment used 12 randomly selected cars each of three different sizes (small, medium, and large) and assigned half in each group randomly to receive a new air filter or a standard air filter. The response of interest was the noise level for each of the three cars.

7.6. EXERCISES

a. Run an appropriate two-way ANOVA on this data set. What are your conclusions? Explain.

b. Now run the same analysis except using indicator variables and regression. Interpret the coefficients in the regression model.

c. What are your conclusions from the regression analysis? Explain.

d. In what ways are your analyses (ANOVA and regression) the same and how are they different? Explain.

Topic 7.5 Exercises: Analysis of Covariance

7.18 *Weight loss.* Losing weight is an important goal for many individuals. An article in the *Journal of the American Medical Association* describes a study (Volpp, John, Troxel, et. al., 2008) in which researchers investigated whether financial incentives would help people lose weight more successfully. Some participants in the study were randomly assigned to a treatment group that was offered financial incentives for achieving weight loss goals, while others were assigned to a control group that did not use financial incentives. All participants were monitored over a four month period and the net weight change (*Before − After* in pounds) at the end of this period was recorded for each individual. Then the individuals were left alone for three months with a followup weight check at the seven-month mark to see whether weight losses persisted after the original four months of treatment. For both weight loss variables the data are the change in weight (in pounds) from the beginning of the study to the measurement time and positive values correspond to weight losses. The data are stored in the file **WeightLossIncentive**. Ultimately we don't care as much about whether incentives work during the original weight loss period, but rather, whether people who were given the incentive to begin with continue to maintain (or even increase) weight losses after the incentive has stopped.

a. Use an ANOVA model to first see if the weight loss during the initial study period was different for those with the incentive than for those without. Be sure to check conditions and transform if necessary.

b. Use an ANOVA model to see if the final weight loss was different for those with the incentive than for those without. Be sure to check conditions and transform if necessary.

c. It is possible that how much people were able to continue to lose (or how much weight loss they could maintain) in the second, unsupervised, time period might be related to how much they lost in the first place. Use the initial weight loss as the covariate and perform an ANCOVA. Be sure to check conditions and transform if necessary.

7.19 *Fruit flies.* Return to the fruit fly data used as one of the main examples in Chapter 5. Recall that the analysis that was performed in that chapter was to see if the mean life spans of male fruit flies were different depending on how many and what kind of females they were housed with. The result of our analysis was that there was, indeed, a difference in the mean life spans of the different groups of male fruit flies.

There was another variable measured, however, on these male fruit flies — the length of the thorax (variable name *Thorax*). This is a measure of the size of the fruit fly and may have an effect on its longevity.

   a. Use an ANCOVA model on this data, taking the covariate length of thorax into account. Be sure to check conditions and transform if necessary.

   b. Compare your ANCOVA model to the ANOVA model fit in Chapter 5. Which model would you use for the final analysis? Explain.

7.20 *Horse prices.* Undergraduate students at Cal Poly collected data on prices of 50 horses advertised for sale on the internet. The response variable of interest is the price of the horse and the explanatory variable is the sex of the horse and the data are in **HorsePrices**.

   a. Perform an ANOVA to answer the question of interest. Be sure to check the conditions and transform if necessary.

   b. Perform an ANCOVA to answer the question of interest using the height of the horse as the covariate.

   c. Which analysis do you prefer for this data set? Explain.

7.21 *Three car manufacturers.* Do prices of three different brands of cars differ? We have a data set of cars offered for sale at an internet site. All of the cars in this data set are either Porsches, Jaguars, or BMWs. We have data on their price, their age and their mileage as well as what brand car they are in **ThreeCars**.

   a. Start with an ANOVA model to see if there are different mean prices for the different brands of car. Be sure to check conditions. Report your findings.

   b. Now move to an ANCOVA model with age as the covariate. Be sure to check conditions and report your results.

   c. Finally use an ANCOVA model with mileage as the covariate. Be sure to check your conditions and report your results.

   d. Which model would you use in a final report as your analysis and why?

Supplemental Exercises

7.22 *Grocery store.* The data set used in Examples 7.15 through 7.17 is available in the data set **Grocery**. In those examples we were interested in knowing whether the amount that products are discounted influences how much of that product sells. In fact, the study looked at another factor

7.6. EXERCISES

as well: Where was the product displayed? There were three levels of this factor. Some products were displayed at the end of the aisle in a special area, some products had a special display but it was where the product was typically located in the middle of the aisle, and the rest were just located in the aisle as usual, with no special display at all.

  a. Use analysis of covariance (again using *Price* as the covariate) to see if the type of display affects the amount of the product that is sold. Be sure to check the necessary conditions.

  b. If you take both factors (*Display* and *Discount*) into consideration and ignore the covariate, this becomes a two-way ANOVA model. Run the two-way ANOVA model and report your results. Be sure to check the necessary conditions.

  c. Analysis of covariance can also be used with a two-way ANOVA model. The only difference in checking conditions is that we must be sure that the conditions for the two-way ANOVA are appropriate and that the slopes between the two continuous variables for observations in each combination of factor levels are the same. Check these conditions for this data set and report whether they are met or not.

  d. Run the analysis of covariance using *Price* as the covariate with the two-way model. Compare your results with the results of part (b) and report them.

Chapter 8

Overview of Experimental Design

Introduction

One of the most important but less obvious differences between regression and ANOVA is the role of statistical design. In statistics "design" refers to the methods and strategies used to produce data, together with the ways your choice of methods for data production will shape the conclusions you can and cannot make based on your data. As you'll come to see from the examples of this chapter, issues of design often make the difference between a sound conclusion and a blunder, between a successful study and a failure, and, sad but true, between life and death for people who suffer the consequences of conclusions based on badly designed studies.

The single most important strategy for design is randomization: that is, using chance to choose subjects from some larger group you want to know about, or using chance to decide which subjects get a new treatment and which ones get the current standard. Indeed, David Moore, one of our leading statisticians, has said that "The randomized controlled experiment is perhaps the single most important contribution from statistics during the 20th century."[1]

Randomization is one of the ways that regression and ANOVA data sets often differ. For regression data, the relationship between response and predictor is typically observed, but not assigned: You can observe the mileage and price of a used Porsche, but you can't pick a Porsche, assign it a particular mileage, and see what price results. For ANOVA data, you can often choose and assign the values of the predictor variable, and when you can, your best strategy is usually to use *chance* to decide, for example, which subjects get the blood pressure medicine and which ones get the inert pill. Using chance in the right way will make all possible choices equally likely, and — presto! — you have an out-of-the-box probability model for your data, with no error conditions to worry about. This simple principle of randomizing your choices turns out to have surprisingly deep consequences.

---

[1] David Moore, personal communication. Professor Moore has served as President of both the American Statistical Association and the International Association for Statistics Education.

This chapter presents an overview of the basic ideas of experimental design. Section 8.1 describes the two essential requirements of a randomized, controlled experiment: the need for comparison and the need to randomize. Section 8.2, on the randomization F-test, shows how you can analyze data from a randomized comparative experiment without having to rely on any of the error conditions of the ANOVA model. The next two sections describe two important strategies for making ANOVA experiments more efficient, called blocking and factorial crossing. You have already seen an example of blocking in the finger tapping experiment at the beginning of Chapter 6. For many situations, blocking can save you time and money, reduce residual variation, and increase the chances that your experiment will detect real differences. You have also seen factorial crossing in Chapter 6, in the context of the two-way ANOVA. The one-way and two-way designs of Chapters 5 and 6 are just a beginning: ANOVA designs can be three-way, four-way, and higher orders. In principle, there's no limit.

8.1 Comparisons and Randomization

The need for comparisons

To qualify as a comparative experiment, a study must compare two or more groups. The need for a comparison was a lesson learned slowly and painfully, especially in the area of surgery, where bad study designs proved to be particularly costly. It is only natural that the medical profession would resist the need for comparative studies. After all, if you are convinced that you have found a better treatment, aren't you morally obligated to offer it to anyone who wants it? Wouldn't it be unethical to assign some patients to a control treatment that you are sure is inferior? Sadly, the history of medicine is littered with promising-seeming innovations that turned out to be useless, or even harmful. Fortunately, however, it has become standard practice to compare the treatment of interest with some other group, often called a "control." Sometimes the control group gets a "placebo," a non-treatment created to appear as much as possible like the actual treatment.

Example 8.1: *Discarded surgeries*

(a) Internal mammary ligation. Internal mammary ligation is the name of an operation that used to be given to patients with severe chest pain related to heart disease. In this operation, a patient is given a local anesthetic and a shallow chest incision is made to expose the mammary artery, which is then tied off. Patients who were given this operation reported they could walk more often and longer without chest pain, and for a while the operation was in common use. That began to change after researchers began comparing the operation with a placebo operation. For the placebo, the patient was given the local anesthetic and the incision was made to expose the artery, but instead of tying it off, the surgeon just sewed the patient back up. It was sobering to discover that results from the fake operation were about as good as from the real one. Once the operation began to be compared with a suitable control group, it became clear that the operation was of no value.

8.1. COMPARISONS AND RANDOMIZATION

(b) The portacaval shunt. The livers of patients with advanced cirrhosis have trouble filtering the blood as it flows through the liver. In the hope of lightening the load on the liver, surgeons invented a partial bypass that shunted some of the blood past the liver. In 1966 the journal *Gastroenterology* published a summary of other articles about the operation. For 32 of these articles, there was no control group, and 75% of these articles were "markedly enthusiastic" about the operation, and the remaining 25% were "moderately enthusiastic." For the 19 articles that did include a control group, only about half (10 of 19) were markedly enthusiastic, and about one fourth (5 of 19) were unenthusiastic.

Control	*Enthusiasm*			Number
Group	Marked	Moderate	None	of studies
No	75%	25%	0%	32
Yes	53%	21%	26%	19

From these results it seems clear that not including a comparison group tended to make the operation look better than was justified.

◇

These two examples illustrate why it is so important to include a control group, and within the area of statistical design, one extreme position is that a study with no comparison is not even considered to be an experiment. *If there's no comparison group, conclusions are very limited.*

The need for randomization

Comparison is necessary, but it is not enough to guarantee a sound study. There are too many ways to botch the way you choose your comparison group. Experience offers a sobering lesson, one that goes against the grain of intuition: Rather than trust your own judgment to choose the comparison group, you should rely instead on blind *chance*. Sometimes a dumb coin toss is smarter than even the best of us. Imagine, for a moment you're a pioneering surgeon, you've invented a new surgical procedure, and you want to persuade other surgeons to start using it. You know that you should compare your new operation with the standard, and you have a steady stream of patients coming to you for treatment. How do you decide which ones get your new operation and which ones get the old standard? The answer: don't trust your own judgment. Put your trust in chance. For example, for each pair of incoming patients, use a coin flip to decide which one gets your wonderful new treatment, with the control going to the "loser" of the coin flip. The next example tells the stories of studies where judgment failed but chance would have succeeded. For these examples, the "winners" were the ones in the control groups.

Example 8.2: *The need to randomize*

(a) Cochran's rats. William G. Cochran (1909 - 1980) was one of the pioneers of experimental design, a subject that he taught at Harvard for many years. In his course on design, he told students about one of his own early experiences in which he decided, wrongly as it turned out, that it wasn't necessary to randomize. He and a coworker were doing a study of rat nutrition, comparing a new diet with a control. To decide which rats got the special diet and which got the control, the two of them simply grabbed rats one at a time from the cages and put them into one group or the other. They thought it was largely a matter of chance which rats went where. It later turned out, according to Professor Cochran himself, that he had been acting as though he felt sorry for the smaller rats. At any rate, without knowing it at the time, he had tended to put the smaller rats in the treatment group, and the larger rats in the control group. As a result, the special diet didn't look as special as theory had predicted. As Cochran told his classes, if only he had used chance to decide which diet each rat got, he would not have gotten such biased results.

(b) The portacaval shunt, continued. In Example 8.1b you saw data about 51 articles discussing the portacaval shunt operation. There the lesson was that not using a control group made the operation look better than could be justified based on the evidence from comparative studies. In fact, the results were even more sobering. Of the 19 studies with a control group, only four had used randomization to decide which patients got the operation and which went into the control group. None of those four reported marked enthusiasm for the operation, only one reported moderate enthusiasm, and three of those four reported no enthusiasm.

Method	Marked	*Enthusiasm* Moderate	None	Number of studies
Judgement	67%	20%	13%	15
Randomized	0%	25%	75%	4

The bottom line is that all the studies with no controls were positive about the operation, and 75% were markedly enthusiastic. Among the very few studies with randomized controls, none was markedly enthusiastic, and 75% were negative.

◇

The lesson from these examples is clear: randomization is an essential part of sound experimentation. *If you don't randomize, conclusions are limited.*

Reasons to randomize

There are three main reasons to randomize, which we first list and then discuss in turn:

a. Protect against bias.

b. Permit conclusions about cause.

c. Justify using a probability model.

8.1. COMPARISONS AND RANDOMIZATION

Randomize to protect against bias

Bias is due to systematic variation that makes the groups you compare different even before you apply the treatments. If you can't randomize the assignment of subjects or experimental units to the treatment groups, you have no guarantee that the groups are comparable. A particularly striking example comes from the history of medicine.

Example 8.3: *Thymus surgery: bias with fatal consequences*

The world-famous Mayo Clinic was founded by two brothers who were outstanding medical pioneers. Even the best of scientists can be misled by hidden bias, however, as this example illustrates. In 1914 Charles Mayo published an article in the *Annals of Surgery*, recommending surgical removal of the thymus gland as a treatment for respiratory obstruction in children. The basis for his recommendation was a comparison of two groups of autopsies, children who had died from respiratory obstruction and adults who had died from other causes. Mayo's comparison led him to discover that the children who had died of obstruction had much larger thymus glands than those in his control group, and this difference led him to conclude that the respiratory problems must be due to an enlarged thymus. In his article, he recommended removal of the thymus, even though the mortality rate for this operation was very high.

Sadly, it turned out that Mayo had been misled by hidden bias. Unlike most body parts, that get larger as we grow from childhood to adulthood, the thymus gland actually gets smaller as we age. Although Mayo didn't know it, the children's glands that he considered too large were in fact normal sized; they only seemed large in comparison to the smaller glands of the adults in the control group.

◇

Mayo was misled, despite his use of a control group, because a *hidden* source of systematic variation made his two groups different. Without randomized assignment, inference about cause and effect is risky if not impossible. *The best way to protect against bias from hidden sources is to use chance to create your groups.* To see how this works, carry out the following thought experiment.

Example 8.4: *A thought experiment with Cochran's rats*

Example 8.2a described a study in which rats were assigned to get either a special diet or a control diet. The two groups of rats were chosen in a biased way, with smaller rats being more likely to end up in the treatment group. The bias in this study made the special diet look less effective than in fact it was.

Imagine that you are going to repeat the study, and you have a group of rats, numbered from 1 to 20. To see why randomization is valuable, consider two different scenarios, one where you know the rat weights, and a second where you don't.

Scenario 1: You know the rat weights in advance. For the sake of the thought experiment, pretend that the weights (in grams) are as shown below.

Rat ID	Weight	Rat ID	Weight	Rat ID	Weight	Rat ID	Weight
1	18	6	19	11	20	16	21
2	18	7	19	12	20	17	22
3	18	8	19	13	21	18	22
4	18	9	20	14	21	19	22
5	19	10	20	15	21	20	22

Since you know the weights, you can see that some rats are large, some are smaller, and you can use this information in planning your experiment. One reasonable strategy would be to put rats into groups according to weight, with all four 18 gram rats in a group, all four 19 gram rats in a group, etc. Within each group, two rats will get the special diet, and the other two will get the control. This strategy ensures that the initial weights are balanced, and protects against bias due to initial weight, but it requires that you know the source of bias in advance. The main threat from bias comes from the sources you *don't* know about. For the sake of this thought experiment, imagine now that you don't know the rat weights.

Scenario 2: You don't know the rat weights. This scenario corresponds to situations in which there are hidden sources of systematic variation. Since you don't know what the systematic differences are, or how to create groups by matching as in Scenario 1, your best strategy is to form the groups using chance, and rely on the averaging process to even things out. One way to randomize would be to put the numbers 1 to 20 on cards, shuffle, and deal out 10. The numbers on those ten cards tell the ID numbers of the rats in the control group. Even though you don't know the rat weights, by randomizing you make it likely that your treatment and control groups will be similar. For this particular example, the histogram in Figure 8.1 shows the results for 10,000 randomly chosen control groups. As you can see, the mean weight of the rats in a group is almost never more than about 1 gram from the average of 20.

◇

Example 8.5: *Randomization, bias, and the portacaval shunt*

Here, once again, is the summary of controlled studies of the portacaval shunt operation:

Random-ized	Enthusiasm Marked	Moderate	None	Number of studies
No	67%	20%	13%	15
Yes	0%	25%	75%	4

It is impossible to know for sure why the operation looked so much better when the controls were not chosen by randomization. Nevertheless, it is not hard to imagine a plausible explanation. Advanced liver disease is a serious illness, and abdominal surgery is a serious operation. Almost surely

8.1. COMPARISONS AND RANDOMIZATION

Figure 8.1: Mean weight of rats in a randomly chosen control group, based on 10,000 randomizations

some patients would have been judged to be not healthy enough to make the operation a good risk. If only the healthier patients got the operation, and the sicker ones were used as the control group, it would not be surprising if the patients in the treatment group appeared to be doing better than the controls, even if the operation was worthless.

The proper way to evaluate the operation would have been to decide first which patients were healthy enough to be good candidates for the operation, and then use chance to assign these healthy candidates either to the treatment or the control group. This method ensures that the treatment and control groups differ only because of the random assignment process, and have no systematic differences.

◇

Randomize to permit conclusions about cause and effect

The logic of statistical testing relies on a process of elimination. If you randomize and group means are different, then there are only two possible reasons: either the observed differences are due to the effect of the treatments, or else they are due to the randomization. If your p-value is tiny, you know it is almost impossible to get the observed differences just from the randomization. If you can rule out the randomization as a believable cause, there is only one possibility left: the treatments must be responsible for at least part of the observed differences, for example, the treatments are effective.

Using chance to create your groups gives you a simple probability model that you can use to calibrate the likely size of group differences due to the assignment process. The process of random assignment is transparent and repeatable, and repetition lets you see what size differences are typical, and which differences are extreme.

Compare a randomized experiment with what happens if you don't randomize. If your group means are different, and you haven't randomized, you are stuck. As above, there are two possible reasons for the differences: either the differences are due to the effect of the treatments, or else they are due to the way the groups were formed. But unlike the mechanism of random assignment which is explicit and repeatable, the mechanism for non-randomized methods is typically hidden and not repeatable. There's no way to use repetition to see what's typical, and no way to know whether the differences were caused by the way the groups were formed. There's no way to know whether the observed differences are due to the features you know about, like the presence of respiratory obstruction, or the features you don't know about, like the shrinking of the human thymus over the course of the life cycle. Are the results of the rat feeding experiment due to the diet, or to Cochran's unconscious sympathy for the scrawnier rats?

Randomize to justify using a probability model

Randomized experiments offer a third important advantage, namely, the fact that you use a probability-based method to create your data means that you have a built-in probability model to use for your statistical analysis. The analyses you have seen so far, in both the chapters on regression and the chapters on ANOVA, rely on a probability model that requires independent normal errors with constant variance. Until recently, these requirements or conditions were often called "assumptions" because there is nothing you can do to guarantee that the conditions are in fact met and that model of normal errors is appropriate. The term "assumption" acknowledged that you had to cross your fingers and assume that the model was reasonable. For randomized experiments, you are on more solid ground. If you randomize the assignment of treatments to units, you automatically get a suitable probability model. This model, and how to use it for ANOVA, is the subject of the next section.

Randomized comparative experiments; treatments and units

The example below illustrates the structure of a typical randomized comparative experiment.

Example 8.6: *Calcium and blood pressure*

The purpose of this study was to see whether daily calcium supplements can lower blood pressure. The subjects were 21 men; each was randomly assigned either to a treatment group or to a control group. Those in the treatment group took a daily pill containing calcium. Those in the control group took a daily pill with no active ingredients. Each subject's blood pressure was measured at the beginning of the study, and again at the end. The response values below show the decrease in systolic blood pressure. Thus a negative value means that the blood pressure went up over the course of the study.

8.1. COMPARISONS AND RANDOMIZATION

												Mean
Calcium	7	-4	18	17	-3	-5	1	10	11	-2		5.000
Placebo	-1	12	-1	-3	3	-5	5	2	-11	-1	-3	-0.273

◇

Treatments and units. In the calcium experiment there are two conditions of interest to be compared, namely the calcium supplement and the placebo. Because these are under the experimenter's control, they are called **treatments**. The treatments in this study were assigned to subjects; each subject got either the calcium or the placebo. The subjects or objects that get the treatments are called **experimental units**.

Randomized Comparative Experiment

In a **randomized comparative experiment**, the experimenter must be able to assign the conditions to be compared. These assignable conditions are called **treatments**. The subjects, animals, or objects that receive the treatments are called **experimental units**. In a randomized experiment the treatments are assigned to units using chance.

In the calcium example, the treatments had the ordinary, everyday meaning as medical treatments, but that won't always be the case. As you read the following example, see if you can recognize the treatments and experimental units.

Example 8.7: *Rating Milgram*

One of the most famous and most disturbing psychological studies of the twentieth century took place in the laboratory of Stanley Milgram at Yale University. The unsuspecting subjects were ordinary citizens recruited by Milgram's advertisement in the local newspaper. The ad simply asked for volunteers for a psychology study, and offered a modest fee in return. Milgram's study of "obedience to authority" was motivated by the atrocities in Nazi Germany and was designed to see whether people off the street would obey when "ordered" to do bad things by a person in a white lab coat. If you had been one of Milgram's subjects, you would have been told that he was studying ways to help people learn faster, and that your job in the experiment was to monitor the answers of a "learner" and to push a button to deliver shocks whenever the learner gave a wrong answer. The more wrong answers, the more powerful the shock.

Even Milgram himself was surprised by the results: Every one of his subjects ended up delivering what they *thought* was a dangerous 300-volt shock to a slow "learner" as punishment for repeated wrong answers. Even though the "shocks" were not real and the "learner" was in on the secret, the results triggered a hot debate about ethics and experiments with human subjects. Some argued that the experiment itself was unethical, because it invited subjects to do something they would later regret and feel guilty about. Others argued that the experiment was not only ethically acceptable but important as well, and that the uproar was due not to the experiment itself, but to the results, namely that every one of the subjects was persuaded to push the button labeled "300 volts: XXX," suggesting that any and all of us can be influenced by authority to abandon our moral principles.

To study this aspect of the debate, Harvard graduate student Maryann de Mateo conducted a randomized comparative experiment. Her subjects were 37 high school teachers who did not know about the Milgram study. Using chance, Maryann assigned each teacher to one of three treatment groups:

- Group 1: Actual results. Each subject in this group read a description of Milgram's study, including the actual results that every subject delivered the highest possible "shock."

- Group 2: Many complied. Each subject read the same description given to the subjects in Group 1, except that the actual results were replaced by fake results, that many but not all subjects complied.

- Group 3. Most refused. For subjects in this group, the fake results said that most subjects refused to comply.

After reading the description, each subject was asked to rate the study according to how ethical they thought it was, from 1 (not at all ethical) to 9 (completely ethical.) Here are the results:

													Mean
Actual Results	6	1	7	2	1	7	3	4	1	1	1	6	3	3.308
Many Complied	1	3	7	6	7	4	3	1	1	2	5	5	5	3.846
Many Refused	5	7	7	6	6	6	7	2	6	3	6			5.545

◇

Here each high school teacher is an experimental unit, and the three treatments are the three descriptions of Milgram's study. (Check that the definition applies: the treatments are the conditions that get assigned; the units are the people, animals, or objects that receive the treatments.) In both examples so far, the experimental unit was an individual, which makes the units easy to recognize. In the example below, there are individual insects, but the experimental unit is not the individual. As you read the description of the study, keep in mind the definitions, and ask yourself: What conditions are being randomly assigned? What are they being assigned to?

8.1. COMPARISONS AND RANDOMIZATION

Example 8.8: *Leafhopper survival*

The goal of this study was to compare the effects of four diets on the lifespan of small insects called potato leafhoppers. One of the four was a control diet: just distilled water with no nutritive value. Each of the other three diets had a particular sugar added to the distilled water, one of glucose, sucrose, or fructose. Leafhoppers were sorted into groups of eight and each group was put into one of eight lab dishes.[2] Each of the four diets was added to two dishes, chosen using chance. The response for each dish was the time (in days) until half the leafhoppers in the dish had died.

Control	Glucose	Fructose	Sucrose
2.3	2.9	2.1	4.0
2.7	2.7	2.3	3.6

For the leafhopper experiment, there are four treatments (the diets) and 64 leafhoppers, but the leafhoppers are not the experimental units, because the diets were not assigned to individual insects. Each diet was assigned to an entire dish, and so the experimental unit is the dish.

◇

In each of the last three examples, there was one response value for each experimental unit, and you might be tempted to think that you can identify the units by making that pattern into a rule, but, unfortunately, you'd be wrong. Moreover, failure to identify the units is one of bigger and the more common blunders in planning and analyzing experimental data. As you read the next (hypothetical) example, try to identify the experimental units.

Example 8.9: *Comparing textbooks*

Two statisticians who teach at the same college want to compare two textbooks A and B for the introductory course. They arrange to offer sections of the course at the same time, and arrange also for the registrar to randomly assign each student who signs up for the course to one section or the other. In the end 60 students are randomly assigned, 30 to each section. The two instructors flip a coin to decide who teaches with book A, who with B. At the end of the semester all sixty students take a common final exam.

There are two treatments, the textbooks A and B. Although there are 60 exam scores, and although the 60 students were randomly assigned to sections, the experimental unit is not a student. The treatments were assigned not to individual students, but to entire sections. The unit is a section, and there are only two units. (So there are zero degrees of freedom for error. F-tests are not possible.)

◇

---

[2] Assume, for the sake of the example, that placement of the dishes was randomized.

This section has described the simplest kind of randomized controlled experiment, called a one-way completely randomized design. All the examples of randomized experiments in this section are instances of this one design. The next section presents a probability model for such designs, and shows how to use that model for testing hypotheses. Section 8.3 and 8.4 show two optional but important design strategies that can often lead to more efficient experiments.

8.2 Randomization F Test

As you saw in the last section, there are many reasons to "randomize the assignment of treatments to units," that is, to use chance to decide, for example, which men get calcium and which ones get the placebo. The three most important reasons are:

a. To protect against bias.

b. To permit inference about cause-and-effect.

c. To justify using a probability model.

This section is about the third reason and its consequences. If you use probability methods to assign treatments, you automatically get a built-in probability model to use for inference. Errors don't have to be normal, independence is automatic, and the size of the variation doesn't matter. The model depends *only* on the random assignment. If you randomize, there are no other conditions to check![3] This is a very different approach from the inference methods of Chapter 5 - 7; those methods do require that your data satisfy the usual four conditions.

The rest of this section has four main parts: the mechanics of a randomization test, the logic of the randomization test, and the randomization test and the usual F-test compared.

Mechanics of the Randomization F-Test

A concrete way to think about the mechanics of the randomization F-test is to imagine putting each response value on a card, shuffling the cards, dealing them into treatment groups, and computing the F-statistic for this random set of groups. The p-value is the probability that this randomly created F-statistic will be at least as large as the one from the actual groups. To estimate the p-value, shuffle, deal, and compute F, over and over, until you have accumulated a large collection of F-statistics, say 10,000. The proportion of times you get an F at least as large as the one from the actual data is an estimate of the p-value. (The more shuffles you do, the better your estimate. Usually, 10,000 is good enough.)

---

[3]In what follows, the randomization test uses the same F-statistic as before. Depending on conditions, a different statistic might work somewhat better for some data sets.

8.2. RANDOMIZATION F TEST

Randomization F-test for One-way ANOVA

Step 1: Observed F: Compute the value F_{obs} of the F-statistic in the usual way.

Step 2: Randomization distribution.

- Step 2a: Re-randomize: Use a computer to create a random re-ordering of the data.

- Step 2b: Compute F_{Rand}: Compute the F-statistic for the re-randomized data.

- Step 2c: Repeat and record: Repeat Steps 2a and 2b a large number of times (e.g., 10,000) and record the set of values of F_{Rand}.

Step 3: P-value. Find the proportion of F_{Rand} values from Step 2c that are greater than or equal to the value of F_{obs} from Step 1. This proportion is an estimate of the p-value.

Example 8.10: *Calcium*

For the data on calcium and blood pressure, we have two groups:

												Mean
Calcium	7	-4	18	17	-3	-5	1	10	11	-2		5.000
Placebo	-1	12	-1	-3	3	-5	5	2	-11	-1	-3	-0.273

Step 1: Value from the data F_{obs}.

A one-way ANOVA gives an observed F of 2.6703. This value of F is the usual one, computed using the methods of Chapter 5. The p-value in the computer output is also based on the same methods. For the alternative method based on randomization, use a different approach to compute the p-value. Ordinarily, the two methods give very similar p-values, but the logic is different, and the required conditions for the data are different.

```
          Df  Sum Sq Mean Sq F value Pr(>F)
Groups     1  145.63 145.628  2.6703 0.1187
Residuals 19 1036.18  54.536
```

Step 2: Create the randomization distribution.

- *Step 2a: Re-randomize.* The computer randomly reassigns the 21 response values to groups, with the result shown below.

374   CHAPTER 8. OVERVIEW OF EXPERIMENTAL DESIGN

												Mean
Calcium	18	-5	-1	17	-3	7	-2	3	-1	10		4.300
Placebo	11	-3	-1	-3	1	5	-5	12	-4	2	-11	-0.364

- *Step 2b: Compute F_{Rand}.* The F-statistic for the re-randomized data turns out to be 1.4011.

```
          Df  Sum Sq Mean Sq F value Pr(>F)
Groups     1   81.16  81.164  1.4011 0.2511
Residuals 19 1100.65  57.929
```

- *Step 2c: Repeat and record.* Here are the values of F for the first ten re-randomized data sets:

$$1.4011 \quad 0.8609 \quad 6.4542 \quad 0.0657 \quad 0.8872$$
$$0.5017 \quad 0.0892 \quad 0.04028 \quad 0.0974 \quad 0.2307$$

As you can see, only one of the ten randomly generated F values is larger than the observed value $F_{obs} = 2.6703$. So based on these ten repetitions, we would estimate the p-value to be one out of ten, or 0.10. For a better estimate, we repeat Steps 2a and 2b a total of 10,000 times, and summarize the **randomization distribution** with a histogram (Figure 8.2):

Step 3: P-value.

The estimated p-value is the proportion of F-values greater than or equal to Fobs = 2.6703 (Figure 8.3). For the calcium data, this proportion is 0.227. The p-value of 0.227 is not small enough to rule out chance variation as an explanation for the observed difference.

◇

Example 8.11: *Rating Milgram (continued)*

For the Milgram data, the high school teachers who served as subjects were randomly assigned to one of three groups. On average those who read the actual results gave the study the lowest ethical rating (3.3 on a scale from 1 to 7); those who read that most subjects refused to give the highest shocks gave the study a substantially higher rating of 5.5.

													Mean
Actual Results	6	1	7	2	1	7	3	4	1	1	1	6	3	3.308
Many Complied	1	3	7	6	7	4	3	1	1	2	5	5	5	3.846
Many Refused	5	7	7	6	6	6	7	2	6	3	6			5.545

8.2. RANDOMIZATION F TEST

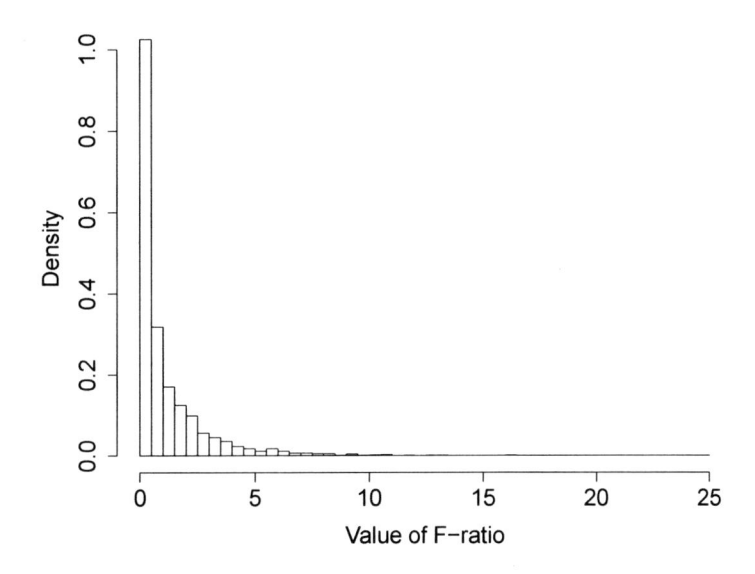

Figure 8.2: Distribution of F-statistics for 10,000 re-randomizations of the Calcium data.

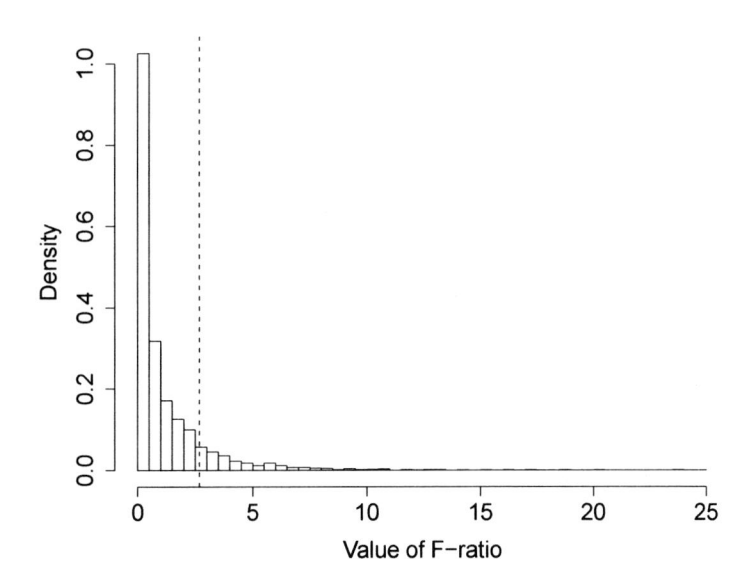

Figure 8.3: The p-value equals the area of the histogram to the right of the dashed line.

376   CHAPTER 8. OVERVIEW OF EXPERIMENTAL DESIGN

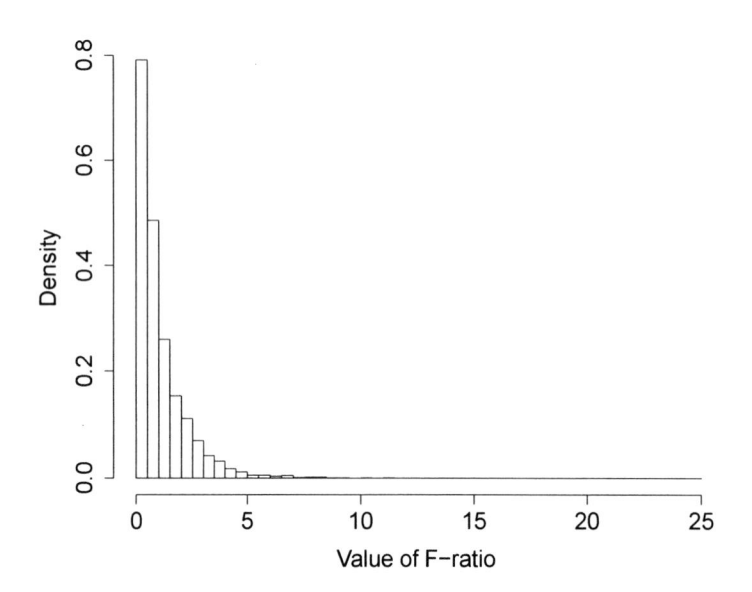

Figure 8.4: Distribution of F-statistics for 10,000 re-randomizations of the Milgram data.

To decide whether results like these could be due just to chance alone, we carry out a randomization F-test.

Step 1: F_{obs} value from the data.

For this data set, software tells us that the value of the F-statistic is $F_{obs} = 3.488$.

Step 2: Create the randomization distribution.

- *Steps 2a and 2b: Re-randomize and compute F_{Rand}.* We used the computer to randomly reassign the 37 response values to groups, with the result shown below. Notice that for this re-randomization, the groups means are very close together. The value of F_{Rand} is 0.001.

													Mean
Actual Results	6	3	1	6	2	1	3	5	1	7	7	7 4	4.077
Many Complied	3	2	7	1	7	5	6	7	6	5	1	1 6	4.385
Many Refused	6	3	1	4	7	6	1	6	1	7	3		4.091

- *Step 2c: Repeat and record.* Figure 8.4 shows the randomization distribution of F based on 10,000 repetitions.

8.2. RANDOMIZATION F TEST

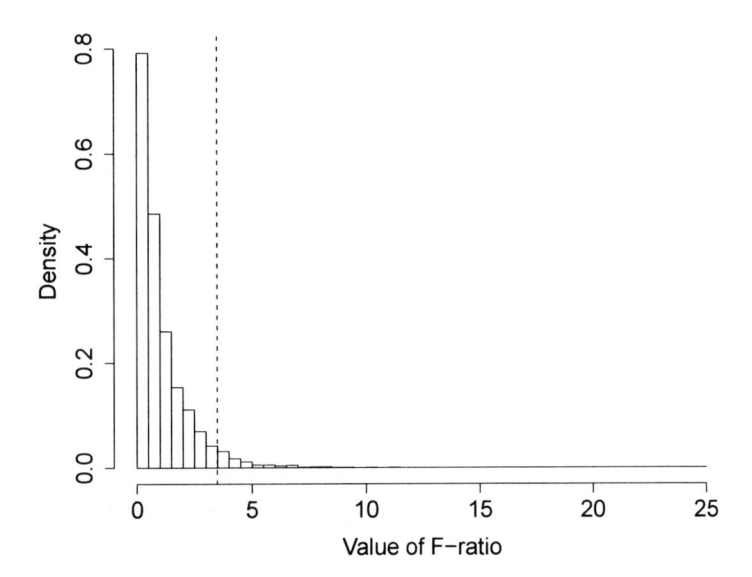

Figure 8.5: The p-value equals the area of the histogram to the right of the dashed line.

Step 3: P-value.

The estimated p-value is the proportion of F-values greater than or equal to F_{obs} (see Figure 8.5). For the Milgram data, this proportion is 0.044.

The p-value of 0.044 is below 0.05, and so we reject the null hypothesis and conclude that the ethical rating does depend on which version of the Milgram study the rater was given to read.

◇

We now turn from the mechanics of the randomization test to the logic of the test: What is the relationship between the randomization F-test and the F-test that comes from an ANOVA table? What is the justification for the randomization test, and how does it compare with the justification for the ordinary F-test?

The logic of the randomization F-test

The key to the logic of the randomization F-test is the question, *"Suppose there are no group differences. What kinds of results can I expect to get?"* The answer: If there are no group differences (no treatment effects) then the *only* thing that can make one group different from another is the random assignment. If that's true, then the data I got should look pretty much like other data sets I

378   CHAPTER 8. OVERVIEW OF EXPERIMENTAL DESIGN

might have gotten by random assignment, that is, data sets I get by re-randomizing the assignment of response values to groups. On the other hand, if my actual data set looks quite different from the typical data sets that I get by re-randomizing, then that difference is evidence that it is the treatments, and not just the random assignment, that is causing the differences to be big. To make this reasoning concrete, consider a simple artificial example based on the calcium study.

Example 8.12: *Calcium (continued)*

Consider the eleven subjects in the calcium study who got the placebo. Here are the decreases in blood pressure for these eleven men:

$$-1 \quad 12 \quad -1 \quad -3 \quad 3 \quad -5 \quad 5 \quad 2 \quad -11 \quad 1 \quad -3$$

The eleven men were treated exactly the same, so these values tell how these men would respond when there is no treatment effect. Suppose you use these eleven subjects for a pseudo-experiment, by randomly assigning each man to one of two groups, which you call "A" and "B", with six men assigned to Group A and five men assigned to B.

							Mean
Group A	-1	3	2	-11	1	-3	-1.5
Group B	12	-1	-3	-5	5		1.6

Although you have created two groups, there is no treatment, and no control. In fact, the men aren't even told which group they belong to, so there is absolutely no difference between the two groups in terms of how they are treated. True, the group means are different, but *the difference is due entirely to the random assignment.* This fact is a major reason for randomizing: If there are no treatment differences, then any observed differences can only be caused by the randomization. So, if the observed differences are roughly the size you would expect from the random assignment, there is no evidence of a treatment effect. On the other hand, if the observed differences are too big to be explained by the randomization alone, then we have evidence that some other cause is at work. Moreover, because of the randomization, the only other possible cause is the treatment.

We now carry out the randomization F-test to compare Groups A and B:

Step 1: Value from the data F_{obs}.

For this data set, the value of the F-statistic is $F_{obs} = 0.731$.

Step 2: Create the randomization distribution.

Figure 8.6 shows the distribution based on 10,000 repetitions.

8.2. RANDOMIZATION F TEST

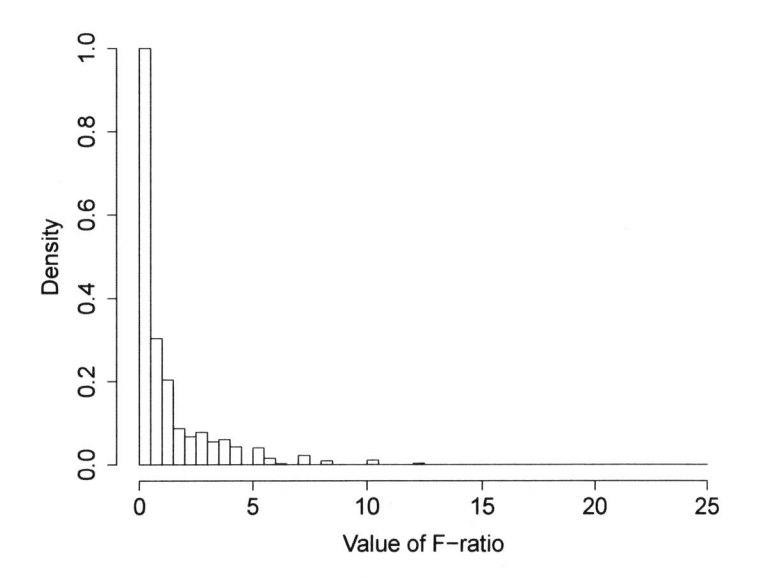

Figure 8.6: Distribution of F_{Rand} based on 10,000 re-randomizations

Step 3: P-value.

Because 4499 of 10,000 re-randomizations gave a value of F_{Rand} greater than or equal to $F_{obs} = 0.731$ (Figure 8.7), the estimated p-value is 0.45.

The p-value tells us that almost half (45%) of random assignments will produce an F-value of 0.731 or greater. In other words, the observed F-value looks pretty much like what we can expect to get from random assignment alone. There is no evidence of a treatment effect, which is reassuring, given that we know there was none.

◇

Isn't it possible to get a large value of F purely by chance, due just to the randomization? Yes, that outcome (a very large value of F) is always possible, but unlikely. The p-value tells us how unlikely. Whenever we reject a null hypothesis based on a small p-value, the force of our conclusion is based on "either/or": Either the null hypothesis is false, or an unlikely event has occurred. Either there are real treatment differences, or the differences are due to a very unlikely and atypical random assignment.

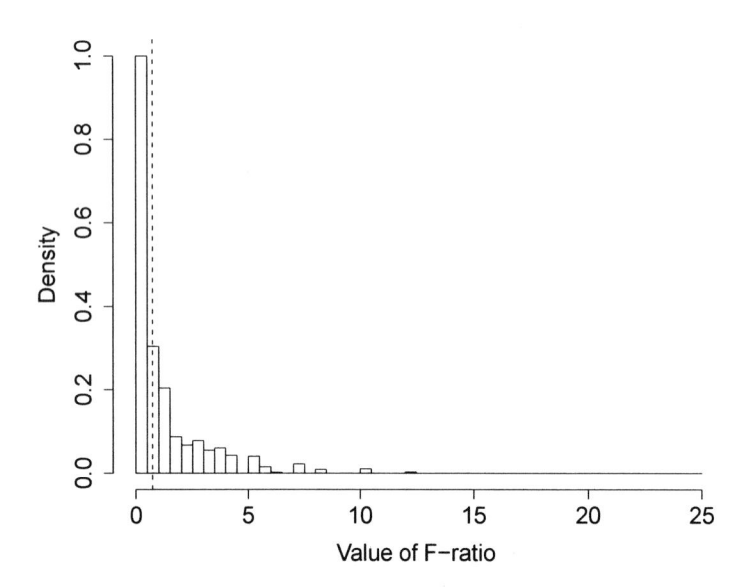

Figure 8.7: 4499 of 10,000 values of F_{Rand} were at least as large as $F_{obs} = 0.731$.

The randomization F-test and the ordinary F-test compared

Many similarities. In this sub-section we compare the randomization F-test with the "ordinary" F-test for one-way ANOVA. The two varieties of F-tests are quite similar in most ways.

- Both F-tests are intended for data sets with a quantitative response and one categorical predictor that sorts response values into groups.

- Both F-tests are designed to test whether observed differences in group means are too big to be due just to chance.

- Both F-tests use the same data summary, the F-statistic, as the test statistic.

- Both F-tests use the same definition of a p-value as a probability, namely, the probability that a data set chosen at random will give an F-statistic at least as large as the one from the actual data: p-value = $Pr(F_{Rand} \geq F_{obs})$.

- Both F-tests use a probability model to compute the p-value[4].

Different probability models. The last of these points of similarity is also the key to the main difference between the two tests: They use very different probability models. The randomization test uses a model of equally likely assignments; the ordinary F-test uses a model of independent normal errors.

---

[4]In practice, the ordinary F-test uses mathematical theory as a shortcut for finding the p-value.

8.2. RANDOMIZATION F TEST

- Equally likely assignment model for the randomization F-test:
  - All possible assignments of treatments to units are equally likely[5]
- Independent normal error model for the ordinary ANOVA F-test:
  - Each observed value equals mean plus error: $y = \mu + \epsilon$
    * Errors have mean zero: $E(\epsilon) = 0$
    * Errors are independent
    * Errors have constant standard deviation: $SD(\epsilon) = \sigma$
    * Errors follow a normal distribution: $\epsilon \sim N$

Assessing the models. These differences in the models have consequences for assessing their suitability. For the randomization test, assessing conditions is simple and straightforward: the probability model you use to compute a p-value comes directly from the method you use to produce the data.

- Checking conditions for the randomization model.
  - Were treatments assigned to units using chance? If yes, you can rely on the randomization F-test.

For the normal errors model, the justification for the model is not nearly so direct or clear. There is no way to know with certainty whether errors are normal, or whether standard deviations are equal. Checks are empirical, and subject to error.

- Checking conditions for the normal errors model.
  - Ideally, you should check all four conditions for the errors. In practice, this means at the very least looking at a residual plot and a normal probability plot or normal quantile plot.

With the normal errors model, if you are wrong about whether the necessary conditions are satisfied, you risk being wrong about your interpretation of small p-values.

Scope of inference. The randomization F-test is designed for data sets for which the randomization is clear from the way the data were produced. As you know, there are two main ways to randomize, and two corresponding inferences. If you randomize the assignment of treatments to

---

[5]If there are N response values, with group sizes $n_1, n_2, ..., n_J$ then the number of equally likely assignments is $N!/(n_1!n_2!...n_J!)$. In practice this number is so large that we randomly generate 10,000 assignments in order to estimate the p-value, but in theory we could calculate the p-value exactly by considering every one of these possible assignments.

units, your experimental design eliminates all but two possible causes for group differences, the treatments and the randomization. A tiny p-value lets you rule out the randomization, leaving the treatments as the only plausible cause. Thus when you have randomized the assignment of treatments to units, you are justified in regarding significant differences as evidence of cause and effect. The treatments caused the differences. If you could not randomize, or did not, then to justify an inference about cause you must be able to eliminate other possible causes that might have been responsible for observed differences. Eliminating these alternative causes is often tricky, because they remain hidden, and because the data alone can't tell you about them.

The second of the two main ways to randomize is by random sampling from a population; like random assignment, random sampling eliminates all but two possible causes of group differences: either the observed differences reflect differences in the populations sampled, or else they are caused by the randomization. Here, too, a tiny p-value lets you rule out the randomization, leaving the population differences as the only plausible cause of observed differences in group means.

Although the exposition in this section has been based on random assignment, you can use the same randomization F-test for samples chosen using chance[6]. As with random assignment, when you have used a probability-based method to choose your observational units, you have an automatic probability model that comes directly from your sampling method. Your use of random sampling guarantees that the model is suitable, and there is no need to check the conditions of the normal errors model. On the other hand, if you have not used random sampling, then you cannot automatically rule out selection bias as a possible cause of observed differences, and you should be more cautious about your conclusions from the data.

A surprising fact. As you have seen, the probability models for the two kinds of F-tests are quite different. Each of the models tells how to create random pseudo-data sets, and so each model determines a distribution of F-statistics computed from those random data sets. It would be reasonable to expect that if you use both models for the same data set, you would get quite different distributions for the F-statistic, and different p-values. Surprisingly, that doesn't often happen. Most of the time, the two methods give very similar distributions and p-values. Figure 8.8 compares the two F-distributions for each of the three examples of this section. For each panel, the histogram shows the randomization distribution, and the solid curve shows the distribution from the normal errors model. For each of the three examples, the two distributions are essentially the same.

The three panels of Figure 8.8 are typical in that the two F-distributions, one based on randomization, the other based on a model of normal errors, are often essentially the same. In practice this means that for the most part, it doesn't matter much which model you use to compute your p-value. What *does* matter, and often matters a great deal, is the relationship between how you produce your data and the scope of your inferences.

---

[6]The justification for the randomization test is different. For an explanation, see the article by Michael Ernst in *Statistical Science*.

8.3. DESIGN STRATEGY: BLOCKING

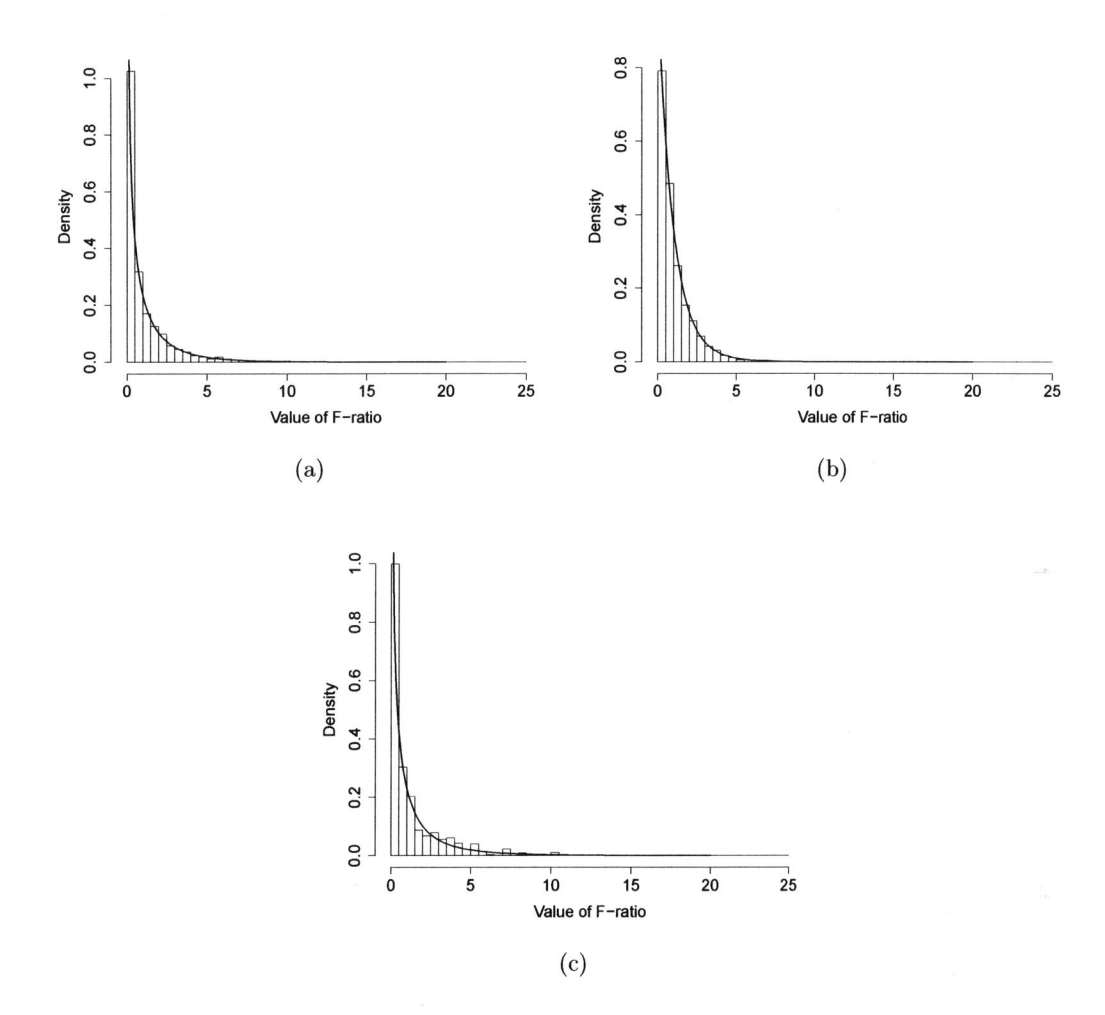

Figure 8.8: F-distributions from randomization model (histogram) and normal errors model (solid curve) for (a) Calcium study (b) Milgram study (c) Hypothetical Calcium study

8.3 Design Strategy: Blocking

Abstractly, a **block** is just a group of similar units, but until you have seen a variety of concrete examples, the abstract definition by itself may not be very meaningful. To introduce the strategy of blocking we rely on the finger tapping example from Chapter 6. The goal of the study was to compare the effects of two alkaloids, caffeine and theobromine, with each other and with the effect of a placebo, using the rate of finger tapping as the response. There were four subjects, and each subject got all three drugs, one at a time on separate days, in a random order that was decided separately for each subject. This design is an example of one type of block design called a repeated measures block design, what psychologists often call a within-subjects design.

> **Repeated Measures Block Design (Within-subjects Design)**
>
> - Each subject gets all the treatments, a different treatment in each of several time slots.
>
> - Each treatment is given once to each subject.
>
> - The order of the treatments is decided using chance, with a separate randomization for each subject.
>
> - For the repeated measures block design, the experimental unit is a time slot; each subject is a block of time slots.

The repeated measures block design relies on the strategy of **reusing** subjects in order to get several units from each person and the repeated measures part of the title refers to how blocks are created.

Not all situations lend themselves to reusing in this way. In agricultural studies, for example, although it might be possible to compare three varieties of wheat by reusing plots of farmland three times in order to grow each variety on each plot, one variety per year, there are practical reasons to prefer a different strategy. In particular, it would take three years to complete the experiment based on reusing plots of land. In this case we create blocks in a different way, by **subdividing** each large block of land into smaller plots to use as units.[7]

> **Blocks by Subdividing**
>
> - Divide each large block of land into equal-sized smaller plots, with the same number of plots per block, equal to the number of treatments.
>
> - Randomly assign a treatment to each plot in such a way that each block gets all the treatments, one per plot, and each treatment goes to some plot in every block.

The idea of the block design is due to Ronald Fisher, who published an example in his 1935 book *The Design of Experiments*. In Fisher's example there were five varieties of wheat to be compared, and 8 blocks of farmland. Each block was sub-divided into 5 equal-sized plots, which served

---

[7] You may have noticed that it is a bit odd to refer to human subjects as "blocks." The term block comes from the use of block designs in agriculture, which provided the first uses of the design.

8.3. DESIGN STRATEGY: BLOCKING

as the units. All five varieties were planted in each block, one variety per plot, using chance to decide which variety went to which plot, with a separate randomization for each of the eight blocks.

The strategy of creating blocks by subdividing is not limited to farmland and agricultural experiments. For example, a study of two treatments of glaucoma "subdivided" human subjects by using individual eyes as experimental units, so that each subject was a block of two units. A comparison of different patterns of tire tread might take advantage of the fact that a car has four wheels, using each car as a block of four units.

Example 8.13: *Bee stings: blocks by subdividing*

If you are stung by a bee, does that make you more likely to get stung again? Might bees leave behind a chemical message that tells other bees to attack you? To test this hypothesis, scientists used a **complete block design**. On nine separate occasions they dangled a 4x4 array of 16 muslin-wrapped cotton balls over a beehive. Eight of 16 had been previously stung; the other eight were fresh. (See Figure 8.9.)

F	S	F	S
S	F	S	F
F	S	F	S
S	F	S	F

Figure 8.9: *The four-by-four array used to tempt the bees.* Each square represents a cotton ball. Those labeled S (stung) had been previously stung; those labeled F (fresh) were pristine.

The response was the total number of new stingers left behind by the bees. The results:

Occasion	1	2	3	4	5	6	7	8	9	Mean
Stung	27	9	33	33	4	22	21	33	70	28
Fresh	33	9	21	15	6	16	19	15	10	16
Mean	30	9	27	24	5	19	20	24	40	22

As you can see, the average number of new stingers was much higher for the cotton balls with the old stingers in them.

◇

This is an example in which blocks, units, and random assignment are hard to recognize. You may be inclined to think of each individual cotton ball as a unit, but remind yourself that a unit is what the treatment gets assigned to. Here the treatments, Stung and Fresh, get assigned to whole sets of eight, so a unit is a set of eight.

Next, randomization: What is it that gets randomized? Each 4x4 square offers two units, two sets of eight balls. One unit gets "Stung" and the other gets "Fresh."[8]

Finally, what about blocks? If each 4x4 array consists of two units, made similar because they belong to the same occasion, then a block must be a 4x4 array, or equivalently, an occasion.

You've now seen blocks created by reusing and blocks created by subdividing. A third common strategy for creating blocks is by sorting units into groups, then using each group as a block.

> **Blocks by Grouping (Matched Subjects Design)**
>
> - Each individual (each subject) is an experimental unit.
> - Sort individuals into equal sized groups (blocks) of similar individuals, with group size equal to the number of treatments.
> - Randomly assign a treatment to each subject in such a way that each group gets all of the treatments, one per individual, and each treatment goes to some individual in every group.

In studies with lab rats, it is quite common to form blocks of littermates. Each rat is an experimental unit, and rats from the same litter are chosen to serve as a block. In learning experiments, subjects might be grouped according to a pre-test. In such studies, blocks come from reusing subjects, with two measurements, pre-test and post-test. Trying to separate genetic and environmental effects is often a challenge in science, and a main reason why twin studies are used. As the next example illustrates, each twin pair serves as a block, with individuals as units.

Example 8.14: *Radioactive twins: Blocks by grouping similar units*

In a study to compare the effect of living environment (rural or urban) on human lung function, researchers were of course unable to run a true experiment, because they were unable to control where people chose to live. Nevertheless, they were able to use the strategy of blocking in an observational study. They were able to locate seven pairs of twins with one twin in each pair living in the country, the other in a city.

To measure lung function, twins inhaled an aerosol of radioactive Teflon particles. By measuring the level of radioactivity immediately and then again after an hour, the scientists could measure

---
[8]There are other possible ways to decide which of the 16 cotton balls get "Stung" and which get "Fresh." Such designs are more complicated than the one described here.

8.3. DESIGN STRATEGY: BLOCKING

the rate of "tracheobronchial clearance." The percentage of radioactivity remaining in the lungs after an hour told how quickly subjects' lungs cleared the inhaled particles. Here are the results; lower percentages indicate healthier clearance rates:

Twin Pair	1	2	3	4	5	6	7	Mean
Rural	10.1	51.8	33.5	32.8	69.0	38.8	54.6	41.5
Urban	28.1	36.2	40.7	38.8	71.0	47.0	57.0	45.5
Mean	19.1	44.0	37.1	35.8	70.0	42.9	55.8	43.5

Notice that this study is observational, not experimental, because there is no way to randomly decide which twin in each pair lives in the country. Because there is no randomization, inference about cause and effect is uncertain. The data tell us that those with less healthy lungs tend to live in the city, but this study by itself cannot tell us whether that association means that city air is bad for lungs, or, on the other hand, for example, that people with less healthy lungs tend to choose to live in the city.

◇

Abstract definition of a block. You have now seen three different ways to get blocks: by reusing (repeated measures design), by subdividing, and by grouping (matched subjects design). Although the three designs differ in what you do to get the blocks, they are all block designs, they all lead to data sets that would be analyzed using the same ANOVA model as in Example 6.1, and abstractly all three designs are the same. To emphasize the underlying similarity, the definition of a block applies to all three designs.

> **Block**
>
> A **block** is a group of similar units.

The definition of a block as a group of similar units omits one essential piece of information, namely, what is it that makes units similar? The answer: Units are similar if they tend to give similar response values under control conditions. Whenever you have a plan for blocks that you think will make the units in a block similar in this way, using a block design is likely to be a good decision. In the example below, we compare the actual complete block design from the finger tapping study with a hypothetical alternative, using a completely randomized design to compare the same three drugs.

Example 8.15: *Finger tapping: the effectiveness of blocking*

Look again at the results of the finger tapping study.

Subject	Placebo	Caffeine	Theobromine	Mean
I	11	26	20	19
II	56	83	71	70
III	15	34	41	30
IV	6	13	32	17
Mean	22	39	41	34

It is clear that the differences between subjects are large, and that these large differences are not surprising: Some people are just naturally faster than others, but whether you are fast or slow, your tap rate today is likely to be similar to your tap rate tomorrow, and your tap rate the day after that. If you serve as a block of units, the three time slots you provide will tend to be similar, much more nearly the same than the rates for two different people. For a study like this one, blocking is a good strategy.

If we compare the analysis of the actual data with the analysis of a hypothetical data set, we can quantify both the effectiveness of the blocking and what it is that you give up as the cost of using a block design instead of a completely randomized design with no blocks. For the hypothetical data set, we use the same response values as above, but pretend they come from a design with no blocks. If we don't use subjects as blocks, we instead regard each subject as an experimental unit. For a completely randomized design, we assign treatments to units, one drug to each subject, using chance. Since there are three drugs, if we want four observations for each drug, we'll need a total of 12 subjects. Here are hypothetical results, using the response values from the actual data:

Subject	2	7	6	8	11	1	5	12	10	3	4	9
Drug	Placebo				Caffeine				Theobromine			
Tap rate	11	56	15	6	26	83	34	13	20	71	41	32
Mean	22				39				41			

Figure 8.10 shows side-by-side ANOVA tables for the two versions of the data set.

Taken together, the two analyses in Figure 8.10 give a good overview of the advantages and disadvantages of blocking.

a. Blocking can dramatically reduce residual variation. For the block design, differences between subjects go into the block sum of squares. Because these differences are not part of the residuals, the residual sum of squares is correspondingly lower. For the completely randomized design, differences between subjects are part of the residuals, because subjects are units, not blocks. If there are large differences between blocks, as in this example, the two analyses give very different sums of squares and mean squares for the residuals.

b. Blocking comes at a cost, however: For the block analysis on the left, the residual degrees of freedom are reduced from 9 to 6, because 3 degrees of freedom are shifted from error to

8.4. DESIGN STRATEGY: FACTORIAL CROSSING

Block design							No Blocks				
SOURCE	df	SS	MS	F	p	SOURCE	df	SS	MS	F	p
Drugs	2	872	436.0	7.88	0.0210	Drugs	2	872	436.0	0.68	0.5330
Subjects	3	5478	1826.0	33.00	0.0004	Error	9	5810	645.6		
Error	6	332	55.3								
Total	11	6682				Total	11	6682			

Figure 8.10: Two analyses for the Finger Tapping data. The analysis on the left, for the block design, has three sources of variation. The analysis on the right, for the completely randomized design, has only two sources. Notice that for both analyses, the bottom rows are the same: There are 11 df total, and the total SS is 6682. Moreover, for both analyses, there are 2 df for Drugs, and the SS for Drugs is the same, at 872. Finally, although the analysis on the right has no row for Subjects, the df and SS on the right for Error come from adding the df and SS for Subjects and Error on the left.

blocks. To see why this loss of degrees of freedom is a cost, imagine that there had been no real subject differences, but you had used a block design anyway. Then you would end up with a large residual sum of squares, but only 6 degrees of freedom for error. Your resulting F-test for *Drugs* would be substantially less powerful than the F-test from the completely randomized design.

c. Blocking by reusing often allows you to get better information with fewer subjects. On the other hand, such block designs take longer to run. In this case it takes three days with four subjects, compared with one day using twelve subjects. For some situations, reusing may not be practical. If your response is the time it takes an infant to learn to walk, or the time it takes an electronic component to fail, for example, you can't reuse the same infants or the same semiconductors. Blocking by subdividing takes less time than blocking by reusing, but not all situations lend themselves to subdividing. (Subdividing infants is not generally recommended.) Finally, blocking by matching or grouping is often possible when reusing and subdividing aren't practical, but you need a good basis for creating your groups.

◇

The strategy of blocking is a strategy for dealing with units and residual variation, and does not have much to do with choosing the treatments. In the next section we discuss another design strategy, factorial crossing, for choosing treatments.

8.4 Design Strategy: Factorial Crossing

This section, on the design strategy called factorial crossing, has three parts. The first explains what it is, the second explains why it is useful, and the third describes a few experiments that illustrate the strategy at work.

390   CHAPTER 8. OVERVIEW OF EXPERIMENTAL DESIGN

What is factorial crossing? You actually know about factorial crossing already. Although the concept is one you have seen, some of the vocabulary may be new.

In ANOVA, the categorical predictor variables are called **factors**, and their categories are called **levels of the factors**, or just **levels**. One-way ANOVA is the method of analysis for one-factor designs; two-way ANOVA is the method of analysis for two-factor designs. In ANOVA, two factors are **crossed** if all possible combinations of levels appear in the design. If we have the same number of observations for each combination of levels we say the design is **balanced**.[9]

The Basics of Factorial Crossing

- Each categorical variable is a **factor**. Each category is a **level** of the factor.

- Two factors are **crossed** if all combinations of levels of the two factors appear in the design.

- Each combination of factor levels is called a **cell**.

- A design is **balanced** if there are equal numbers of observations per cell.

- In order to estimate or test for the presence of interaction effects, you must have more than one observation per cell.

Example 8.16: *Fruit flies, river iron, and pigs*

(a) Fruit flies. The fruit fly data of Example 5.1 on page 198 is an example of a one-way design. Each male was assigned to one of five treatment groups (one of five levels of the treatment factor):

Group	Females	Status
0	0	—
1	1	Virgin
2	1	Pregnant
3	8	Virgin
4	8	Pregnant

---
[9]Balanced designs offer many advantages, but designs need not be balanced. For unbalanced designs, a multiple regression approach with indicator variables, of the sort described in Chapter 7, is often used to analyze the results.

8.4. DESIGN STRATEGY: FACTORIAL CROSSING

Although the design is a one-factor design, notice that it contains a two-way design within it. If you omit the control group (Group 0), the four groups that are left come from crossing two factors, *Females*, with two levels (0 or 8), and *Status*, also with two levels (Virgin or Pregnant):

		Factor 2: *Status*	
		Virgin	Pregnant
Factor 1:	1		
Females	8		

The 2x2 layout shows the reason for the term "crossed." There is one row for each level of the row factor, and one column for each level of the column factor. Each possible combination of levels corresponds to a place (called a **cell**) where a row and column cross.[10]

(b) River iron. For the design of Example 6.2 on page 259, there are two factors, *river*, with four levels, and *site* with three levels.

	River				
	Grasse	Oswegatchie	Raquette	St. Regis	Mean
Up	2.97	2.93	2.03	2.88	2.70
Mid	2.72	2.36	1.56	2.75	2.35
Down	2.51	2.11	1.48	2.54	2.16
Ave	2.74	2.47	1.69	2.72	2.40

The two factors are crossed: every combination of factor levels appears in the design. For this design, there is only one observation per cell.

(c) Pig feed. (Refer to Example 6.4 on page!263.) For this study, there are again two factors, *antibiotics* (0 mg or 40 mg), and *B12* (0 mg or 5 mg). The two factors are crossed, as in (b) above, but for this design there are three observations for each combination of factor levels:

		Factor B: *Vitamin B12*	
		No	Yes
Factor A:	No	30, 19, 8	26, 21, 19
Antibiotics	Yes	5, 0, 4	52, 56, 54

◇

For the river iron data, with only one observation per cell, it is not possible to separate interaction effects from residual error. For the pigs, because there is more than one observation per cell, it

---
[10] The two versions of the fruit fly experiment illustrate how the strategy of crossing can be useful even if you don't end up with a two-way design in the strict sense.

is possible to get separate estimates of the interaction effects and the residual errors. The moral: *If you think there may be interaction, be sure to assign more than one experimental unit to each combination of factor levels.*

When and why is factorial crossing a good strategy? Crossing offers two main advantages, which we can label *efficiency* and *the capacity to estimate interaction effects*.

Efficiency: "Buy one factor, get one almost free." Consider two versions of the Pig Feed study, planned by two ambitious but fictitious doctors, Dr. Hans Factor-Solo and Dr. Janus Two-Way, who are in hot competition for a Golden Trough Award, the Pig Science equivalent of a Nobel Prize. Dr. Factor-Solo has reason to think that adding vitamin B12 to the diets of pigs will make them gain weight faster. He consults a statistician who tells him he needs six pigs on each diet, so together they design a one-way completely randomized design that requires the use of twelve pigs:

One-way design	
0 mg B12	6 pigs
5 mg B12	6 pigs

Just down the hall, Dr. Janus Two-Way is ignorant of the subtleties of pig physiology, but statistically half-savvy, unscrupulous, and very lucky. He has secretly discovered Dr. Factor-Solo's vitamin B12 hypothesis, and also knows that certain intestinal bacteria can actually aid digestion, but he has no sense of interaction, and so he hypothesizes – wrongly – that adding antibiotics to pigs' diets, which will kill the beneficial bacteria, will in turn cause the pigs to gain weight more slowly. To beat Dr. Factor-Solo to the Trough, he wants to be able to report on both *Antibiotics* and *Vitamin B12*. He and his consulting statistician recognize that for the cost of a dozen pigs, they can study both factors in the same experiment, so they decide on a two-way completely randomized design that uses only twelve pigs:

Two-way design

		Factor B: *Vitamin B12*	
		No	Yes
Factor A	No	3 pigs	3 pigs
Antibiotics	Yes	3 pigs	3 pigs

This design allows the unscrupulous Dr. Two-Way to measure both factors for the price of one. For just twelve pigs, he can estimate both the main effect of *Vitamin B12*, and the main effect of *Antibiotics*.[11]

---

[11]There is a slight cost. Dr. Factor-Solo's one-way design has 10 degrees of freedom for error; Dr. Two-Way's design has only 8 degrees of freedom for error. This makes Dr. Factor-Solo's design slightly more powerful at detecting an effect due to *Vitamin B12*. However, when it comes to detecting the effect of antibiotics or interaction, Dr. Factor-Solo is completely powerless.

8.4. DESIGN STRATEGY: FACTORIAL CROSSING

When the results come in, however, Dr. Two-Way is in trouble. He doesn't know enough to know that when interaction is present, main effects are hard to interpret. The results of his pig experiment confuse him, and he is unsure what to do next. Fortunately for Dr. Two-Way, his postdoctoral researcher, Dr. Interaction, is able to recognize that, just by luck, her boss has used a design that allows him to get separate estimates for interaction and residual error. She tactfully points out that the capacity to estimate interaction effects is another advantage of his two-way design.

The first lesson from this porcine parable: *Often factorial crossing lets you estimate the effects of **two** factors for the price of one.* Dr. Interaction's second lesson follows.

The capacity to estimate interaction effects. *The main new lesson here is that one-factor-at-a time often fails when interaction is present.* This lesson appears to be at odds with a common principle in science, that *to study the effect of a factor, you need to hold everything else constant.* For observational studies, this can be a good cautionary principle, one that reminds us that there can be more than one reason why two response values differ. For experimental studies, however, you can often do better. The pig feed study is a good illustration of how a one-factor-at-a-time study can mislead. To visualize the general principle concretely and geometrically in your mind's eye, turn to the renowned statistician Mother Goose for a useful example.

Example 8.17: *Jack and Jill: What's the matter with one-at-a-time?*

Jack and Jill, two middle school geeks, have a hill to climb. (See the picture in Figure 8.11.) Their hill has response z (altitude) that depends on two quantitative explanatory variables x (east/west) and y (north/south). The ridge of this particular hill runs diagonally, which means that the "effect" of x depends on the value of y, and vice-versa: interaction is present.

Hopeful of becoming statisticians when they grow up, Jack and Jill decide to rely on science to plan their hill climb. Jack knows from his courses that "to study the effect of a variable, you should hold everything else constant." Proud of his knowledge and mistaking his chauvinism for chivalry, Jack decides it is his duty to choose and present a plan of action. He proposes a pair of one-way designs. "I'll execute the first one-way design. I'll keep my east/west position (x) constant at 100, and vary my south-to-north position (y), measuring altitude (z) every 100 yards. This will tell us the best (highest) north-south coordinate. Next, you'll do the second one-way design. You'll keep your north/south coordinate fixed at my best value, and go east-to-west, taking altitude measurements every 100 yards. Together, we'll find the top of the hill."

Poor Jack has already fallen down. Figure 8.12 shows a contour map of the hill, whose ridge runs diagonally. Use the map to follow Jack's plan. If he keeps his east/west coordinate fixed at 100, what will he decide is the best (highest) north/south coordinate? If Jill keeps her north/south coordinate fixed at Jack's "best" value, what east/west value will she decide is best? Where does Jack's one-at-a-time design locate the top of the hill?

394   CHAPTER 8. OVERVIEW OF EXPERIMENTAL DESIGN

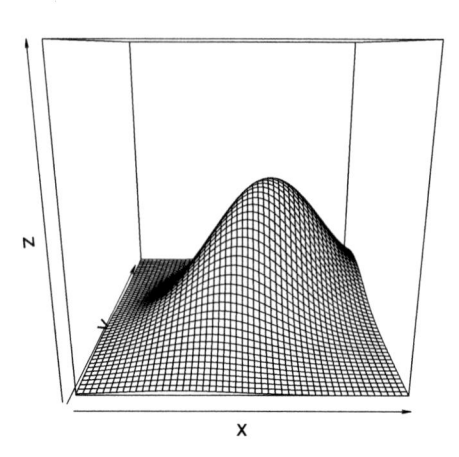

Figure 8.11: The hill. The diagonal ridge corresponds to interaction, which causes Jack's one-at-a-time method to fail.

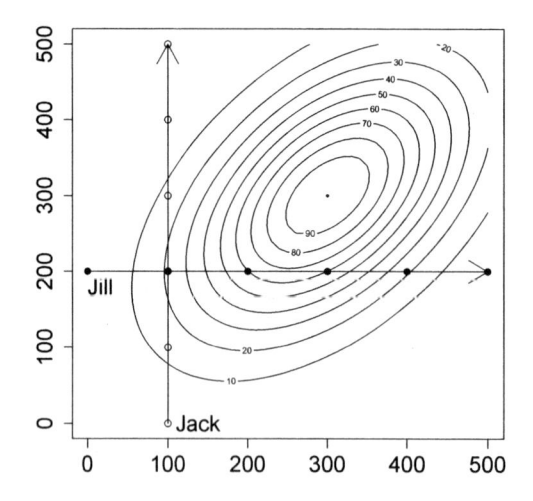

Figure 8.12: Jack's one-at-a-time plan for climbing the hill. The vertical line shows Jack's one-way design. The horizontal line shows Jack's one-way follow-up design for Jill. This leads to an estimate of the top of the hill being at (250,200). The top of the hill is at (300, 300).

8.4. DESIGN STRATEGY: FACTORIAL CROSSING

Fortunately, Jill is too smart to go tumbling after. She knows a two-way design that uses factorial crossing to choose combinations of the north/south and east/west factors will be able to detect interaction. Deploying her best diplomatic skills, she proposes the two-way design as just a small tweak to Jack's idea, and he goes along. Thanks to the strategy of factorial crossing, they reach the top of the hill together, and Jack's fragile crown survives intact. (Jill goes on to become founder and CEO of a successful consulting company, Downhill Research.)

◊

Some examples of design with crossed factors. The simplest way to extend the two-way design is by including a third factor, with all three factors completely crossed, as in the first of the four examples that follow. Some situations call for more than three crossed factors, as illustrated in the second example, which is a five-way design with all factors completely crossed. Both of these examples use completely randomized designs, with no blocks, but it is possible to use factorial crossing and blocking in the same design, as illustrated in the last two examples.

Example 8.18: *Plant nutrition: three crossed factors*

As any farmer or home gardener knows, there are three essential nutrients, *Nitrogen* (N), *Phosphorus* (P), and *Potassium* (K). Because different plants need different amounts of the three, it takes some study to find the right combination, but rather than rely on trial and error, plant scientists can find the right combination using a design with three crossed factors. Each factor has three levels, which for simplicity we'll call Low, Medium, and High. In all, there will be 27 treatment combinations.[12]

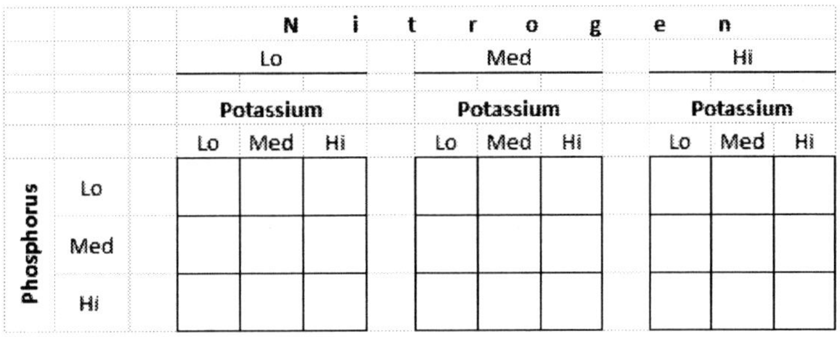

Figure 8.13:

Notice that three factors are completely crossed: Every possible combination of levels of all three factors appears in the design. To run this experiment as a completely randomized design, you

---
[12] For a three-factor experiment like this one, there will be four possible interactions: three 2-way interactions (NxP, NxK, PxK), and one 3-way interaction (NxPxK). Interpretation of 2-way interactions is similar to what you have seen in Chapter 6. Interpretation of 3-way interactions is a bit tricky, and beyond the scope of this chapter.

would need 27 units, for example, 27 styrofoam cups with vermiculite, which is nutrient-free. You would randomly assign a treatment combination to each cup, and add the appropriate levels of nutrients to the water.[13]

◇

A version of the last experiment was the starting point for an actual study of plant competition, which used a five-factor design.

Example 8.19: *Plant competition: Five crossed factors of interest*

Digitaria sanguinalis (D.s.) is a species of crabgrass that is larger than the related *Digitatria ischaemum (D.i.)*. In this experiment, the big plant and the little plant were forced to compete with each other for water and nutrients. One factor in the 5-way design was species mix, with 5 levels:

Level	A	B	C	D	E
D.s.	20	15	10	5	0
D.i.	0	5	10	15	20

The other four factors, nutrients N, P, K, and water, had two levels each, High and Low. As in the previous example, the experimental unit was a styrofoam cup of vermiculite. In all there were 80 = 5 x 2 x 2 x 2 x 2 treatment combinations. With two units per treatment combination, the experiment called for 160 cups of vermiculite.

◇

Both of the previous two examples used completely randomized designs. There were no blocks. The next examples reuse human subjects in a block design with two crossed factors of interest.

Example 8.20: *Remembering words: A two-way design in randomized complete blocks*

To get a feel for what it was like to be a subject in this experiment, read the list of words below, then turn away from the page, think about something else for the moment, and then try to recall the words. Which ones do you remember?

The words:

      dog, hangnail, fillip, love, manatee, diet, apostasy, tamborine, potato, kitten,
      sympathy, magazine, guile, beauty, cortex, intelligence, fauna, gestation, kale

In the actual experiment, there were 100 words, presented to each subject in a randomized order. The goal of the experiment was to see whether some kinds of words were easier to remember than

---

[13] If you use only 27 cups (units), you will be able to estimate all 27 cell means, but you will have 0 df for residuals, and no way to get separate estimates of 3-way interaction and residual error. There are ways to deal with this problem, but they are beyond the scope of this course.

8.4. DESIGN STRATEGY: FACTORIAL CROSSING

others, in particular, are common words like potato, love, diet, and magazine easier to remember than less common words like manatee, hangnail, fillip, and apostasy? Are concrete words like coffee, dog, kale, and tamborine easier than abstract words like beauty, sympathy, fauna, and guile? There were 25 words each of four kinds, obtained by crossing the two factors of interest, *Abstraction* (concrete or abstract) and *Frequency* (common or rare):

Think first about how you could run this experiment as a completely randomized two-way design. There are four treatment combinations, corresponding to the four kinds of words, so you would create four word lists, one for each combination of factor levels. Your experimental unit would be a subject, so you use chance to decide which subjects get List 1, which get List 2, and so on.

Although this design is workable, you can do much better at very little cost. The experimenters knew that different people have different capacities for remembering words, and that the main cost in using the subjects was getting them to show up. It would cost almost nothing extra to give them a list of 100 words, with all four kinds mixed in a random order. This reuse of subjects makes each subject into a block. Figure 8.14 shows a data set from an actual experiment. The response is the percentage of words of each kind that were recalled correctly.

	Abstract		Concrete		
Subject	Frequent	Infrequent	Frequent	Infrequent	Average
I	60	44	60	44	52
II	52	32	44	28	39
III	20	32	40	24	29
IV	36	32	44	36	37
V	40	32	52	44	42
VI	36	20	28	40	31
VII	36	40	32	28	34
VIII	44	64	44	64	54
IX	36	40	32	56	41
X	52	52	64	60	57
Average	41.2	38.8	44.0	42.4	41.6

Figure 8.14: Percentage of words recalled

Notice that all three factors – *Abstraction*, *Frequency*, and *Subjects* (= Blocks) – are completely crossed: all possible combinations of levels of the three factors are present. Notice also that there is only one observation per cell, so it is not possible to measure the three-way interaction. Nevertheless for each combination of the two factors of interest, that is, each combination of *Abstraction* and *Frequency*, there are multiple observations, one per subject, so it is possible to estimate and test for interaction between these two factors.[14]

---

[14] Remind yourself that the numbers in Figure 8.14 are percentages, and with that in mind, run your eyes over the observed values, ignoring the averages. Can you tell, just from the percentages, how many words were in each list? Hint: All percentages are divisible by what?

As a final example of a multifactor design, we consider a variation of the last design, created by including another factor.[15]

Example 8.21: *Remembering words: A split plot or repeated measures design.*

It is reasonable to think that your memory for English words might depend on whether or not you are a native speaker or learned English as a second language. This hypothesis suggests using a design with a fourth factor. To *Abstraction*, *Frequency*, and *Subject*, add the factor *Native Speaker* (yes/no). Here's one possible format for data from such a design:

Native Speaker?	Subject	Abstract Frequent	Abstract Infrequent	Concrete Frequent	Concrete Infrequent
Yes	I				
Yes	II				
Yes	V				
No	VI				
No	VII				
No	X				

Figure 8.15:

Several features of the situation deserved attention: (1) Two kinds of factors: *Abstraction*, *Frequency*, and *Native Speaker* are factors of interest; *Subject* is a nuisance factor. (2) Two kinds of factors: *Abstraction* and *Frequency* are experimental; *Native Speaker* is observational. (3) For this situation, complete factorial crossing is not possible. Think specifically about the two factors *Subject* and *Native Speaker*, and check that because it is not possible to use a coin toss to decide whether a subject will be a native speaker, it is not possible to cross the nuisance factor and the observational factor. Nevertheless, this is a sound design, one that is often used, and one that would be analyzed using ANOVA.

---

[15] There are fancier designs based on *partial* crossing, and *partial* replication. Like all designs, these involve a tradeoff between greater efficiency (fewer observations) and stronger conditions: the more you assume, the less you have to rely on data. BUT: The less you rely on data, the more vulnerable you are to flawed assumptions. (The dustbin of history is filled with faith-based analyses.)

8.5 Chapter Summary

In this chapter we introduce the important topic of **experimental design** and its role in the analysis of data. A careful study of this topic (which can easily be expanded to a full course of its own) involves lots of new terminology. For this reason, we summarize the main points of this chapter with a glossary of some of the important terms in this field.

A Glossary of Experimental Design

Experiment versus Observational Study
In an **experiment**, the condition(s) of interest are chosen and assigned by the investigator; in an **observational study**, they are not.

Treatment and Units
In an experiment, the conditions that get assigned are called **treatments**; the people, animals, or objects that receive the treatments are called **experimental units**, or just **units**.

Comparative Experiment, Controlled Experiment, Placebo
In a **comparative** experiment, there are two or more conditions of interest. In a controlled experiment, one of the conditions is a **control**, a condition that serves as a "non- treatment". Often the control condition is a **placebo**, a non-treatment designed to be as much like the treatment as possible, apart from the feature of interest.

Randomized Experiment
An experiment is **randomized** if the treatments are assigned to experimental units using random numbers or some other chance device. Randomization protects against bias, justifies inference about cause and effect, and provides a probability model for the data.

Randomization F-test
This test involves comparing the F statistic computed from the data as collected with a distribution of F statistics computed from randomly reassigning the observations to groups. The p-value is computed as the proportion of the distribution of F statistics that are as large as, or larger than, the F statistic computed from the original data. The only condition necessary to use the randomization F-test is that randomization is properly used in the data collection procedure (either in an experiment or when random sampling was used).

Balance
A comparative experiment or observational study is **balanced** if all groups to be compared have the same number of units.

Block Designs
A **block** is a group of similar experimental (or observational) units. "Similar" means similar with

respect to the response variable. Blocks can be created by matching units, by subdividing, or by reusing. In a **complete block design**, the number of units in a block equals the number of treatments, and each block of units gets all treatments, one to each unit.

Factors, Levels, and Crossing

In ANOVA, each categorical predictor is called a **factor**; the categories of a factor are called **levels**. Two factors are called **crossed** if every combination of factor levels appears in the design. Each combination of factor levels is called a **cell**. In order to get separate estimates of interaction and error, you must have more than one experimental unit per cell.

8.6 Exercises

Conceptual Exercises

8.1 List the three main reasons to randomize the assignment of treatments to units.

8.2 *North Carolina births.* In the study of North Carolina births (Exercise 5.29), the response was the birth weight of a newborn infant, and the factor of interest was the racial/ethnic group of the mother.

  a. Explain why random assignment is impossible, and how this impossibility limits the scope of inference: Why is inference about cause not possible?

  b. (b) Under what circumstances would inference from samples to populations be justified? Were the requirements for this kind of inference satisfied?

Exercises 8.3 — 8.8 For each of the following studies (a) name the factors. Then for each factor, tell (b) how many levels it has, (c) whether the factor is observational or experimental, and (d) what the experimental or observational units are.

8.3 *Rating Milgram.* Example 8.7.

8.4 *Pigs and vitamins.* Example 6.4.

8.5 *River iron.* Example 6.2.

8.6 *Finger tapping.* Example 6.1.

8.7 *Plant nutrition.* Example 8.18.

8.8 *Bee stings.* Exercise 8.13.

8.9 *Fenthion.* In the study of fenthion in olive oil (Exercise 5.26), the response was the concentration of the toxic chemical fenthion in samples of olive oil and the factor of interest was the time when the concentration was measured, with three levels: day 0, day 281, and day 365. Consider two versions of the study: In Version A, eighteen samples of olive oil are randomly divided into three groups of six samples each. Six samples are measured at time 1, six at time 2, and six at time 3. In Version B, the olive oil is divided into six larger samples. At each of Times 1 - 3, a subsample is taken from each, and the fenthion concentration is measured.

  a. One version is a one-way completely randomized design; the other is a complete block design. Which is which? Explain.

  b. What is the advantage of the block design?

8.10 Give one example each (from the examples in the chapter) of three kinds of block designs, one that creates blocks by reusing subjects, one that creates blocks by matching subjects, and one that creates blocks by sub-dividing experimental material. For each, tell the blocks and the experimental units.

8.11 Give the two most common reasons why creating blocks by reusing subjects may not make sense, and give an example from the chapter to illustrate each reason.

8.12 Recall the two versions of the finger tapping study in Example 6.1. Version 1, the actual study, used four subjects three times each, with subjects as blocks and time slots as units. Version 2, a completely randomized design, used twelve subjects as units, and had no blocks.

  a. List two advantages of the block design for this study.

  b. List two advantages of the completely randomized design for this study.

8.13 *Crossing.* Give two examples of one-way designs that can be improved by crossing the factor of interest with a second factor of interest. For each of your designs, tell why it makes sense to include the second factor.

8.14 True or false. In a complete block design, blocks and treatments are crossed. Explain.

8.15 Why is it that in a randomized complete block design, the factor of interest is nearly always experimental rather than observational?

8.16 Six subjects A, B, C, D, E, and F are available for a memory experiment. Three subjects, A, B, and C have good memories; the other three have poor memories. If you randomly divide the six subjects into two groups of three, there are 20 equally likely possibilities:

ABC/DEF	ACD/BEF	ADF/BCE	BCF/ADE	CDE/ABF
ABD/CEF	ACE/BDF	AEF/BCD	BDE/ACF	CDF/ABE
ABE/CDF	ACF/BDE	BCD/AEF	BDF/ACE	CEF/ABD
ABF/CDE	ADE/BCE	BCE/ADF	BEF/ACD	DEF/ABC

What is the chance that all three subjects with good memory end up in the same group?

8.17 Four subjects a, b, c, and d are available for a memory experiment. Two subjects, a, and b have good memories; the other two have poor memories. If you randomly divide the four subjects into two groups of two, what is the chance that both subjects with good memory end up in the same group?

8.6. EXERCISES

Guided Exercises

8.18 *Behavior therapy for stuttering.* Exercise 6.3 described a study[16] that compared two mild shock therapies for stuttering. Each of the 18 subjects, all of them stutterers, was given a total of three treatment sessions, with the order randomized separately for each subject. One treatment administered a mild shock during each moment of stuttering, another gave the shock after each stuttered word, and the third "treatment" was a control, with no shock. The response was a score that measured a subject's adaptation.

   a. Explain why the order of the treatments was randomized.

   b. Explain why this study was run as a block design with subjects as blocks of time slots instead of a completely randomized design with subjects as experimental units.

8.19 *Sweet smell of success.* Chicago's Smell and Taste Treatment and Research Foundation funded a study, "Odors and Learning," by A. R. Hirsch and L. H. Johnson. The purpose of the study was to see whether a pleasant odor could improve learning. Twenty subjects participated.[17] If you had been one of the subjects, you would have been timed while you completed a paper-and-pencil maze, two times: once under control conditions, and once in the presence of a "floral fragrance."

   a. Suppose that for all 20 subjects, the control attempt at the maze came first, the scented attempt came second, the results showed that average times for the second attempt were shorter, and that the difference in times was statistically significant. Explain why it would be wrong to conclude that the floral scent caused subjects to go through the maze more quickly.

   b. Consider a modified study. You have 20 subjects, as above, but you have two different mazes, and your subjects are willing to do both mazes. Your main goal is to see whether subjects solve the mazes more quickly in the presence of the floral fragrance or under control conditions, so the fragrance factor has only two levels, control and fragrance. Tell what design you would use. Be specific and detailed: What are the units? Nuisance factors? How many levels? What is the pattern for assigning treatments and levels of nuisance factors to units?

8.20 *Rainfall.* According to theory, if you release crystals of silver iodide into a cloud (from an airplane), water vapor in the cloud will condense and fall to the ground as rain. To test this theory, scientists randomly chose 26 of 52 clouds and seeded them with silver iodide. They measured the total rainfall, in acre-feet, from all 52 clouds.

Explain why it was not practical to run this experiment using a complete block design. (Your answer should address the three main ways to create blocks.)

---

[16] "Rate of stuttering adaptation under two electro-shock conditions," *Behavior Res. Therapy* 1967, pp. 49-54

[17] Actually, there were 21 subjects. (We've simplified reality for the sake of the exercise.)

8.21 *Hearing.* Audiologists use standard lists of 50 words to test hearing; the words are calibrated, using subjects with normal hearing, to make all 50 words on the list equally hard to hear. The goal of the study described here was to see how four such lists, denoted by L1-L4 in Table 8.1, compared when played at low volume with a noisy background. The response is the percentage of words identified correctly.

Sub	L1	L2	L3	L4	Avg
1	28	20	24	26	24.5
2	24	16	32	24	24.0
3	32	38	20	22	28.0
4	30	20	14	18	20.5
5	34	34	32	24	31.0
6	30	30	22	30	28.0
7	36	30	20	22	27.0
8	32	28	26	28	28.5
9	48	42	26	30	36.5
10	32	36	38	16	30.5
11	32	32	30	18	28.0
12	38	36	16	34	31.0
13	32	28	36	32	32.0
14	40	38	32	34	36.0
15	28	36	38	32	33.5
16	48	28	14	18	27.0
17	34	34	26	20	28.5
18	28	16	14	20	19.5
19	40	34	38	40	38.0
20	18	22	20	26	21.5
21	20	20	14	14	17.0
22	26	30	18	14	22.0
23	36	20	22	30	27.0
24	40	44	34	42	40.0
Avg	33	30	25	26	28.3

Table 8.1: Percentage of words identified for each of four lists

a. Is this an observational or experimental study? Give a reason for your answer. (Give the investigators the benefit of any doubts: if it was possible to randomize, assume they did.)

b. List any factors of interest, and any nuisance factors.

c. What are the experimental (or observational) units.

d. Are there blocks in this design? If so, identify them.

8.6. EXERCISES

8.22 *Burning calories.* (See Exercise 6.18 on page 291 for the original study.) The purpose of this study was to compare the times taken by men and women to burn 200 calories using two kinds of exercise machines: a treadmill and a rowing machine.[18] Suppose you have 20 subjects, 10 male and 10 female.

a. Tell how to run the experiment with each subject as a unit. How many factors are there? For each factor, tell whether it is experimental or observational.

b. Now tell how to run the experiment with each subject as a block of two time slots. How many factors are there this time?

c. Your design in (b) has units of two sizes, one size for each factor of interest. Subjects are the larger units. Each subject is a block of smaller units, the time slots. Which factor of interest goes with the larger units? Which with the smaller units?[19]

8.23 *Fiber in crackers.* This study uses a two-way complete block design, like Example 8.20 on page 396 (remembering words). Twelve female subjects were fed a controlled diet, with crackers before every meal. There were four different kinds of crackers: control, bran fiber, gum fiber, and a combination of both bran and gum fiber. Over the course of the study, each subject ate all four kinds of crackers, one kind at a time, for a stretch of several days. The order was randomized. The response is the number of digested calories, measured as the difference between calories eaten and calories passed through the system.

Subj	control	gum	combo	bran
1	2353.21	2026.91	2254.75	2047.42
2	2591.12	2331.19	2153.36	2547.77
3	1772.84	2012.36	1956.18	1752.63
4	2452.73	2558.61	2025.97	1669.12
5	1927.68	1944.48	2190.10	2207.37
6	1635.28	1871.95	1693.35	1707.34
7	2667.14	2245.03	2436.79	2766.86
8	2220.22	2002.73	1844.77	2279.82
9	1888.29	1804.27	2121.97	2293.27
10	2359.90	2433.46	2292.46	2357.40
11	1902.75	1681.86	2137.12	2003.16
12	2125.39	2166.77	2203.07	2287.52
Avg	2158.05	2089.97	2109.16	2159.97

a. What are the experimental units? What are the blocks?

---

[18] *Medicine and Science in Sports and Exercise* (August, 2001)

[19] Statisticians often call the design in (b) a split plot / repeated measures design. Psychologists call it a mixed design, "mixed" because there is both a between-subjects factor, and a within-subjects factor.

b. There are two crossed factors of interest. Show them by drawing and labeling a two-way table with rows as one factor and columns as the other factor.

c. Fill in the cell means.

d. Draw and label an interaction graph. Write a sentence or two telling what it shows.

8.24 *Noise and ADHD.* Exercise 6.4 on page 289 described a study to test the hypothesis that children with attention deficit and hyperactivity disorder tend to be particularly distracted by background noise.[20] The subjects were all second graders, some who had been diagnosed as hyperactive, and others who served as controls. All were given sets of math problems under two conditions, high noise and low noise. The response was their score. (Results showed an interaction effect: the controls did better with the higher noise level; the opposite was true for the hyperactive children.)

a. Describe how the study could have been done without blocks.

b. If the study had used your design in (a), would it still be possible to detect an interaction effect?

c. Explain why the interaction is easier to detect using the design with blocks.

8.25 *Heavy metal, lead foot?* Is there a relationship between the type of music teenagers like and their tendency to exceed the speed limit? A study[21] was designed to answer this question (among several). For this question, the response is the number of times a person reported driving over 80 mph in the last year. The study was done using random samples taken from four groups of students at a large high school: (1) those who described their favorite music as acoustic/pop; (2) those who preferred mainstream rock; (3) those who preferred hard rock; and (4) those who preferred heavy metal.

a. Results showed that on average, those who preferred heavy metal reported more frequent speeding. Give at least two different reasons why it would be wrong to conclude that listening to heavy metal causes teens to drive fast.

b. What kinds of generalizations from this study *are* justified?

8.26 *Happy face, sad design.* Researchers at Temple University[22] wanted to know whether drawing a happy face on the back of restaurant customers' checks would lead to higher tips. They enlisted

---
[20]S.S. Zentall and J.H. Shaw (1980), "Effects of classroom noise on performance and activity of second-grade hyperactive and control children," *J. Ed. Psych.*, 72, pp. 830-840.
[21]"The Soundtrack of Recklessness: Musical Preferences and Reckless Behavior Among Adolescents," *Journal of Adolescent Research*, 1992, pp. 313-331.
[22]B. Rind and P. Bordia (1996). "Effect on restaurant tipping of male and female servers drawing a happy face on the backs of customers' checks," *J. Soc. Psych*, 26, pp. 215-225.

8.6. EXERCISES

the cooperation of two servers at a Philadelphia restaurant, one male server, one female. Each server recorded tips for 50 consecutive tables. For 25 of the 50, following a predetermined randomization, they drew a happy face on the back of the check. The other 25 randomly chosen checks got no happy face. The response was the tip, expressed as a percentage of the total bill. See Exercise 6.20 on page 292 for the results of the study.

- a. Although the researchers who reported this study analyzed it as a two-way completely randomized design, with Sex of Server (Male/Female) and Happy Face (Yes/No) as crossed factors, and with tables serving as units, that analysis is deeply flawed. Explain how to tell from the description that the design is not completely randomized and why the two-way analysis is wrong. (Hint: Is a server a unit or a block? Both? Neither?)

- b. How many degrees of freedom are there for Male/Female? How many degrees of freedom for differences between servers? How many degrees of freedom for the interaction between those two factors?

- c. In what way is the design and published two-way analysis fatally flawed?

- d. Suppose you have six servers: three male, three female. Tell how to design a sound study using servers as blocks of time slots.

8.27 *Migraines.* People who suffer from migraine headaches know that once you get one, it's hard to make the pain go away. A comparatively new medication, Imatrex, is supposed to be more effective than older remedies. Suppose you want to compare Imatrex (I) with three other medications: Fiorinol (F), Acetaminophen (A), and Placebo (P).

- a. Suppose you have available four volunteers who suffer from frequent migraines. Tell how to use a randomized complete block design to test the four drugs: What are the units? How will you assign treatments (I, F, A, and P) to units?

- b. Notice that you have exactly the same number of subjects as treatments, and that it is possible to reuse subjects, which makes it possible to have the same number of time slots as treatments. Thus you have four subjects (I, II, III, and IV), four time slots (1, 2, 3, 4), and four treatments (A, F, I, P). Rather than randomize the order of the four treatments for each subject, which might by chance assign the placebo always to the first or second time slot, it is possible to balance the assignment of treatments to time slots, so that each treatment appears in each time slot exactly once. Such a design is called a **Latin square**. Create such a design yourself, by filling in the squares below with the treatment letters A, F, I, and P, using each letter four times, in such a way that each letter appears exactly once in each row and column.

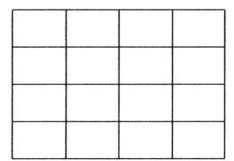

8.28 *Running dogs.* To test the common belief that racing dogs run faster if they are fed a diet containing vitamin C, scientists at the University of Florida used a randomized complete block design (*Science News*, July 20, 2002) with five dogs serving as blocks. See Exercise 6.5 on page 289 for more detail. Over the course of the study, each dog got three different diets, in an order that was randomized separately for each dog. Suppose the scientists had been concerned about carry over effects of the diets, and so had decided to use a completely randomized design with 15 dogs as experimental units.

a. Complete the following table to compare degrees of freedom for the two designs.

BLOCKS			NO BLOCKS	
Source	df		Source	df
Diets			Diets	
Blocks				
Error			Error	
Total	14		Total	14

b. Compare the advantages and disadvantages of the two designs.

8.29 *Fat rats.* Researchers investigating appetite control measured the effect of two hormone injections, leptin and insulin, on the amount eaten by rats (*Science News*, July 20, 2002). Male rats and female rats were randomly assigned to get one hormone shot or the other. See Exercise 6.6 on page 290 for more detail.

a. Tell how this study could have been run using each rat as a block of two time slots.

b. Why do you think the investigators decided not to use a design with blocks?

8.30 *Recovery times.* A study[23] of two surgical methods compared recovery times, in days, for two treatments, the standard and a new method. Four randomly chosen patients got the new treatment; the remaining three patients got the standard. Here are the results:

Recovery times, in days, for seven patients: Average:
New procedure: 19, 22, 25, 26   23
Standard: 23, 33, 40   32

There are 35 ways to choose three patients from seven. If you do this purely at random, the 35 ways are equally likely. What is the probability that a set of three randomly chosen from 19, 22, 23, 25, 26, 33, 40 will have an average of 32 or more?

---
[23]Ernst, M. (2004). Permutation Methods: A Basis for Exact Inference. Statistical Science, 19, 676-685.

8.6. EXERCISES

8.31 *Challenger: sometimes statistics IS rocket science.* Bad data analysis can have fatal consequences. After the Challenger exploded in 1987, killing all five astronauts aboard, an investigation uncovered the faulty data analysis that had led Mission Control to OK the launch despite cold weather. The fatal explosion was caused by the failure of an O-ring seal, which allowed liquid hydrogen and oxygen to mix and explode.

   a. For the faulty analysis, engineers looked at the seven previous launches with O-ring failures. The unit is a launch; the response is the number of failed O-rings. Here are the numbers:

   Number of failed O-rings for launches with failures:
   Above 65°:   1 1 2
   Below 65°:   1 1 1 3

   The value of F for the actual data is 0.0649. Consider a randomization F-test for whether the two groups are significantly different. We want to know how likely it is to get a value at least as large as 0.0649 purely by chance. Take the seven numbers 1, 1, 1, 1, 1, 2, 3 and randomly choose four to correspond to the four launches below 65 degrees. There are four possible samples, which occur with the percentages shown below. Use the table to find the p-value. What do you conclude about temperature and O-ring failure?

A =Above	B = Below 65°	% of samples	Means: A	B	MSE	F
3 2 1	1 1 1 1	14.3%	2.00	1.00	0.400	4.2857
3 1 1	2 1 1 1	28.6%	1.67	1.25	0.683	0.4355
2 1 1	3 1 1 1	28.6%	1.33	1.50	0.733	0.0649
1 1 1	3 2 1 1	28.6%	1.00	1.75	0.550	1.7532

   b. The flaw in the analysis done by the engineers is this: They ignored the launches with zero failures. There were 17 such zero-failure launches, and for all 17, the temperature was above 65 degrees. Here is the complete data set that the engineers should have chosen to analyze:

   Number of failed O-rings for all launches:
   Above 65°:   0 0 0 0 0 0 0 0 0 0 0 0 0 0 0 1 1 2
   Below 65°:   1 1 1 3

   The value of F for these two groups is 14.427. To carry out a randomization F-test, imagine putting the 24 numbers on cards and randomly choosing four to represent the four launches with temperatures below 65 degrees. The table in Figure 2 summarizes the results of repeating this process more than ten thousand times. Use the table to find the p-value, that is, the chance of an F greater than or equal to 14.427. What do you conclude about temperature and O-ring failure?

B = Below 65°				% of samples	Means: A	B	MSE	F
0	0	0	0	22.4%	0.50	0.00	0.619	1.346
1	0	0	0	32.0%	0.45	0.25	0.652	0.204
1	1	0	0	12.8%	0.40	0.50	0.657	0.051
2	0	0	0	6.4%	0.40	0.50	0.657	0.051
1	1	1	0	1.6%	0.35	0.75	0.633	0.842
2	1	0	0	6.4%	0.35	0.75	0.633	0.842
3	0	0	0	6.4%	0.35	0.75	0.633	0.842
1	1	1	1	0.0%	0.30	1.00	0.581	2.811
2	1	1	0	1.6%	0.30	1.00	0.581	2.811
3	1	0	0	6.4%	0.30	1.00	0.581	2.811
2	1	1	1	0.1%	0.25	1.25	0.500	6.667
3	1	1	0	1.6%	0.25	1.25	0.500	6.667
3	2	0	0	1.3%	0.25	1.25	0.500	6.667
3	1	1	1	0.1%	0.20	1.50	0.390	14.427
3	2	1	0	0.8%	0.20	1.50	0.390	14.427
3	2	1	1	0.1%	0.15	1.75	0.252	33.811

Figure 8.16: Randomization distribution of F-statistics for full Challenger data

Supplementary Exercises

8.32 *Social learning in monkeys.*[24] Some behaviors are "hardwired" — we engage in them without having to be taught. Other behaviors we learn from our parents. Still other behaviors we acquire through "social learning" by watching others and copying what they do. To study whether chimpanzees are capable of social learning, two scientists at the University of St. Andrews randomly divided seven chimpanzees into a treatment group of four and control group of three. Just as in Exercise 8.30 (recovery times), there are 35 equally likely ways to do this. During the "learning phase" each chimp in the treatment group was exposed to another "demo" chimp who knew how to use stones to crack hard nuts; each chimp in the control group was exposed to a "control" chimp who did not know how. During the test phase, each chimp was provided with stones and nuts, and the frequency of nut-cracking bouts was observed.

The results for the seven chimpanzees in the experiment:

Frequency of nut cracking bouts:	Average:
Treatment group: 5, 12, 22, 25	16
Control group: 2, 3, 4	3

Assume that the treatment has no effect, and that the only difference between the two groups is due to the randomization.

---
[24]Sarah Marshall-Pescini and Andrew Whiten (2008). "Social Learning of Nut-Cracking Behavior in East African Sanctuary-Living Chimpanzees (*Pan Troglodytes schweinfurthii*)," *J. Comp. Psych.*, v. 122, pp. 186-194.

8.6. EXERCISES

a. What is the probability that just by chance, the average for the treatment group would be greater than or equal to 16?

b. The F-value for the observed data is 5.65. Of the 35 equally likely random assignments of the observed values to a treatment group of 4 and control group of 3, one other assignment has an F-value as high or higher:

Control	Treatment	Group Means			
		Ctrl	Tr	MSE	F
12 22 25	2 3 4 5	19.67	3.50	19.53	22.94

What is the p-value from the randomization F-test?

c. Why is the probability in (a) different from the p-value in (b)?

d. For a comparison of just two groups, which randomization p-value, the one based on averages in (a), or the one based on the F-statistic in (b), is preferred? Explain.

e. "The correct answer in (d) is related to the difference between a one-tailed t-test to compare two groups, and the F-test for ANOVA." Explain.

Chapter 9

Logistic Regression

Is a single mom more likely to marry the birth father if she has a boy or a girl or does it matter? How much will additional studying help you pass a test? Is a new migraine medication more likely to offer relief than one currently on the market? Are campaign contributions associated with a senator's vote?

Like earlier chapters, each of these examples involves predicting a **response variable** using an **explanatory variable**: predicting the response of a *Mother's Marital Status* (married or not) using the explanatory variable *Baby's Sex* (girl or boy); predicting the response of the *Outcome of an Exam* (pass or fail) using the explanatory variable *Time Spent Studying*; predicting the response of *Headache Relief* (relief or no relief) using the explanatory variable *Brand of Medication*; or predicting a *Senator's Vote* (yes or no) using the explanatory variable *Campaign Contributions*. Unlike previous chapters, for each of these examples the response variable takes on one of only two possible values, such as, married or not, pass or fail, relief or no relief, yes or no vote. Variables which can take on only one of two possible values are referred to as **binary**. When the response is binary, we use **logistic regression** for modeling. It is tempting to simply replace the response Y with a 0 or 1 indicator variable—as we did with binary predictors in ordinary regression—but we'll see that linear regression models do not work well for modeling a binary response directly as a 0 or 1. With a binary response the key quantity to estimate is the probability that the response will be in a certain category. For example, if you study for seven hours for an exam, what is the probability that you pass it? Or if you use the experimental medication, what is the probability that you get relief from a migraine?

> **Notation for the probability of success**
>
> We use the letter π to denote the true probability of "success" for a particular population and $\hat{\pi}$ for an estimate of the probability based on a particular sample or fitted model. This is analogous to the way we used μ_Y as the population parameter of interest for a quantitative response in ordinary regression and $\hat{y}$ as the estimate based on a prediction equation from a fitted model.

As you will see in the next section, with logistic regression we do not model the response probability directly but rather we model a function of the probability known as a *logit* function. It is the logit function that is assumed to be linearly related to the predictor. The steps of choosing a model, fitting it to the data, assessing the fit, and using the model to address a question of interest are the same steps that we used in our earlier work in ordinary regression, but there will be some notable differences in carrying out these steps. As with ordinary regression, we begin by specifying and then fitting a model with a single predictor. We then discuss how to interpret coefficients for logistic regression and consider how to make inferences with logistic regression. In particular, we will identify and check conditions of logistic regression models before going on to test for associations or to construct confidence intervals.

9.1 The Logistic Regression Model

Example 9.1: *Medical school acceptance*

Every year in the U.S., over 120,000 undergraduates submit applications in hopes of realizing their dreams to become physicians. Medical school applicants invest endless hours studying to boost their GPAs. They also invest considerable time in studying for the medical school admission test or MCAT. Which effort, improving GPA or increasing MCAT scores, is more helpful in medical school admission? We investigate these questions using data gathered on 55 medical school applicants from a liberal arts college in the Midwest. For each applicant, medical school *Acceptance* status (accepted or denied), *GPA*, *MCAT* scores, and *Gender* were collected. The data are stored in **MedGPA**. The response variable *Acceptance* status is a binary response. When a response is binary, its categories can be coded as 0's and 1's, where 1 is often referred to as a *success* and 0 a *failure*. Lets begin by exploring the relationship between medical school acceptance and *GPA*.

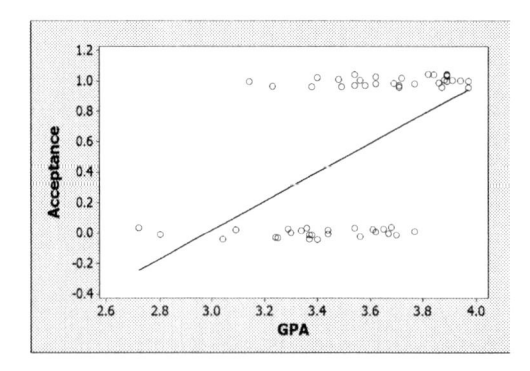

Figure 9.1: Acceptance Status by GPA: jittered plot with ordinary regression line

Proceeding as in previous chapters, we plot Y (*Acceptance*) by X (*GPA*), as shown in Figure 9.1. Because there are many instances where multiple data points have the same coordinates, we have "jittered" the points in the y-direction, to allow us to better see the individual points in the data

9.1. THE LOGISTIC REGRESSION MODEL

set. This plot looks markedly different from the simple linear regression scatterplots of Chapter 1, since the response variable only takes on two values. Ordinary linear regression models assume that the average response at each level of the predictor is linearly related to the predictor.

$$\mu_Y = \beta_0 + \beta_1 X$$

In the logistic regression setting, at each level of the predictor we have a binary 0 or 1 response so the "average" value is equivalent to the proportion of the responses that are successes; this is the sum (the number of ones) divided by the number of tries (successes + failures). That is, the average, or mean, of a binary response in a sample is $\hat{\pi}$, the sample proportion of successes; this estimates π, the probability of success in the population. The ordinary least squares regression line in Figure 9.1 assumes that this success probability increases at a linear rate as GPA increases. The prediction equation in this case is

$$\widehat{Acceptance} = -2.82 + 0.948 GPA$$

While the form of the scatterplot makes it difficult to judge whether this linear rate of increase in the acceptance probability is reasonable for these data, there are clear problems with such a model: If GPA = 2.9, then $\widehat{Acceptance}$ = -0.07. That is, the model predicts that students with GPAs below 3.0 have a negative probability of being admitted to medical school! Admittedly someone with a GPA less than 3.0 does not have a very good chance of being accepted, but that probability cannot be less than 0. In other situations a simple linear regression with a binary (0 or 1) response can produce predicted values above one, which are also impossible for a probability. ⋄

To get a model that captures these properties, we consider what is called a *logit* function of the success probability π.

Logistic Regression Model for a Single Predictor

The *logistic regression* model for the probability of success π of a binary response variable based on a single predictor X has either of two equivalent forms:

$$\text{Logit form:} \qquad \log\left(\frac{\pi}{1-\pi}\right) = \beta_0 + \beta_1 X$$

or

$$\text{Probability form:} \qquad \pi = \frac{e^{\beta_0 + \beta_1 X}}{1 + e^{\beta_0 + \beta_1 X}}$$

These two equivalent forms for the logistic regression model have different "response" variables (on the left-hand side). The *probability form* of the model looks a bit more traditional as a model for the quantity of interest (π), but uses a pretty complicated function of the predictor. Nevertheless, if we have values for the coefficients β_0 and β_1, we can use a calculator to compute a success probability for any value of the predictor, X. The *logit form* of the model has the familiar linear function of the predictor on the righthand side, but a more complicated function as the response. This logit function will look a bit less daunting when we introduce the concept of odds in Section 9.3. We can use some basic algebra to rewrite the logit form of the model into the probability form (see Exercise 11.2). Either form of the model allows for a relatively easy extension to include more than one predictor, which parallels the case of multiple regression. We explore this in more detail in Chapter 10. You will want to be comfortable working with either form of the logistic regression model since assessing and interpreting aspects of the model are often much easier in one form or the other.

For example, when looking at the probability form of the model we see that π could never be negative (since the exponential terms are always positive) and the denominator is larger than the numerator so the value can never be more than one. This avoids the embarrassment we encountered when using ordinary linear regression to produce a model that predicted impossible probabilities; the logistic model always keeps the success probability between 0 and 1. We also see, in either form of the model, that the success probabilities are *not* a linear function of the predictor. The probability form of the model shows that when the "slope," β_1, is positive the probability π starts near zero and grows smoothly (but not linearly) towards one as the predictor X increases.[1] When β_1 is negative we see the opposite effect, so that π approaches zero as X gets larger. The logistic model captures that S shape. (See Figure 9.5 for examples of positive and negative values of β_1.)

9.2 Fitting a Logistic Regression Model

Recall that we find the estimates of the intercept and slope for linear regression problems using the method of least squares. We saw in Chapter 1 that we could write a formula for the estimated intercept and a formula for the estimated slope.

When fitting a logistic regression model, we do not use the method of least squares, but rather a more general method called *maximum likelihood estimation*. With this method, the values that make our given data set "most likely" in a probabilistic sense are selected as the estimates of β_0 and β_1 and are referred to as *maximum likelihood estimates*. In general, maximum likelihood methods are computationally complex, requiring iterative methods that are implemented by computing software. We cannot write out explicit formulas for $\widehat{\beta_0}$ and $\widehat{\beta_1}$ in logistic regression, as we can in regular regression (when we use least squares with a single predictor).[2] Fortunately, the estimates are readily available in statistical software packages such as R and Minitab.

---

[1] We put quotes around the β_1 because it is not really a slope in the probability form of the model, only in the logit form.

[2] Interestingly, even though a completely different idea was used to obtain the least squares estimates for ordinary regression, it turns out that the least squares estimates that we saw in Chapter 1 are also maximum likelihood estimates in that setting.

9.2. FITTING A LOGISTIC REGRESSION MODEL

Example 9.2: *Medical school acceptance (continued)*

Below we give the Minitab output for fitting a simple logistic regression model to predict *Acceptance* status for medical school using GPA of the applicants in the **MedGPA** file.[3]

```
Logistic Regression Table
                                                   95% CI
Predictor      Coef    SE Coef      Z      P   Odds Ratio   Lower    Upper
Constant   -19.2065    5.62922  -3.41  0.001
GPA          5.45417   1.57931   3.45  0.001       233.73   10.58  5164.74
```

Based on the maximum likelihood criteria, we see that the estimates for the slope and intercept of the logit form of the model are $\hat{\beta}_0 = -19.21$ and $\hat{\beta}_1 = 5.454$, respectively. This gives a prediction equation for the logit of the success probability to be

$$\log\left(\frac{\hat{\pi}}{1-\hat{\pi}}\right) = -19.21 + 5.454 GPA$$

and the equivalent probability form of the prediction equation is

$$\hat{\pi} = \frac{e^{-19.21+5.45 GPA}}{1+e^{-19.21+5.454 GPA}}$$

For applicants with GPAs of 3.7 we see the linear part of the logit model is $-19.21 + 5.454 \cdot 3.7 = 0.9698$. Substituting this value in the exponent of the probability form of the model gives

$$\hat{\pi} = \frac{e^{0.9698}}{1+e^{0.9698}} = 0.725$$

So we would estimate that students with GPAs of 3.7 have about a 72.5% chance of being accepted into medical school.

However the chance of acceptance for applicants with GPA's of 3.25 is only 18.5%.

$$\hat{\pi} = \frac{e^{-19.21+5.454(3.25)}}{1+e^{-19.21+5.454(3.25)}} = 0.185$$

---

[3]Previously we have declined to identify output with specific software packages. We are altering that policy now because, for logistic regression, output from R and Minitab are more different.

Length of putt (in feet)	3	4	5	6	7
Number of successes	84	88	61	61	44
Number of failures	17	31	47	64	90
Total number of putts	101	119	108	125	134

Table 9.1: Putting success versus distance of the putt

We can do such a calculation for any value of GPA, Figure 9.2 shows a plot of the logistic prediction equation for the applicant success probability versus GPA. We can confirm the predicted probabilites for students with 3.25 and 3.7 GPA's on this graph. Note that the curve is not linear and approaches (but never reaches) zero as the GPA decreases and approaches (but never reaches) one as the GPA gets closer to a perfect 4.0.

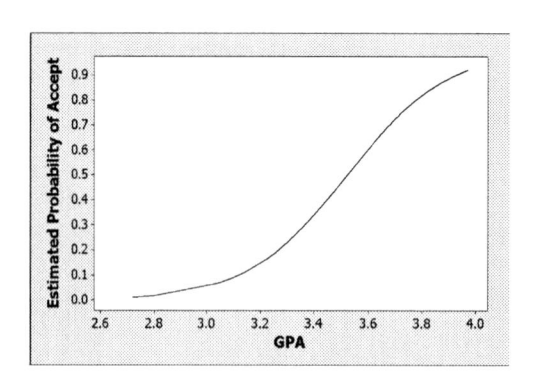

◇

Figure 9.2: Logistic regression fit for probability of medical school acceptance

Example 9.3: *Putting prowess*

A golfing statistician kept track of how often he made (or missed) putts of certain lengths. Part of his data are displayed in Table 9.1 and stored in the file **Putts1**.

In these data we have a large number of sample points with the same predictor value, so we can estimate a success proportion directly for each putting distance. For example, this golfer made 88/119=0.74 of his four-foot putts. Let's call this proportion $\hat{p}$—a simple observed proportion of successes in repeated trials—so as to not confuse it with an estimate given by a logistic regression model. From six feet his success rate was only 61/125=0.49. These proportions are summarized in the table below along with the logit function, $\log(\hat{p}/(1-\hat{p}))$, evaluated for each proportion.

Length of putt (in feet)	3	4	5	6	7
Proportion of successes, $\hat{p}$	0.832	0.739	0.565	0.488	0.328
Logit of proportion	1.60	1.04	0.26	-0.05	-0.72

9.2. FITTING A LOGISTIC REGRESSION MODEL

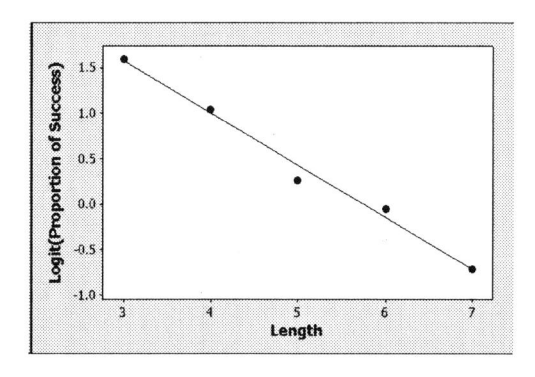

Figure 9.3: Logistic regression for putts made: Fitted model and empirical logits versus putt length.

We use R to fit a logistic model to these data. The R summary below shows the estimated intercept and slope.

```
Coefficients:
            Estimate Std. Error z value Pr(>|z|)
(Intercept)  3.25684    0.36893   8.828   <2e-16 ***
Length      -0.56614    0.06747  -8.391   <2e-16 ***
```

The two forms of the prediction equation for this fitted model are

$$\text{logit:} \quad \log\left(\frac{\hat{\pi}}{1-\hat{\pi}}\right) = 3.257 - 0.566 Length \qquad \text{probability:} \quad \hat{\pi} = \frac{e^{3.257-0.566 Length}}{1+e^{3.257-0.566 Length}}$$

◇

To construct Figure 9.3, we first compute the **odds** of a successful putt at each distance by calculating the proportion of successful putts to the proportion of unsuccessful putts. The **empirical logits** are computed by taking the log of the empirical odds for each distance. See the preceding table. The plot of the empirical logits of the five sample proportions versus the putt lengths exhibits a linear trend that is captured by the line with intercept $\hat{\beta}_0 = 3.257$ and slope $\hat{\beta}_1 = -0.566$.

We digress a moment to compare the medical school example (Example 9.1) to the putting example (Example 9.3): while their essential structures are the same, there is one important difference. In both examples, the response variable is binary: acceptance to medical school and success in making the putt. In both examples, the explanatory variable is quantitative: GPA in the medical school example and length of the putt in the putting example. Because of this structure, both are amenable to the fitting of a logistic regression model through the principle of maximum likelihood. The important difference is the concept of *replicate x-values*. The 587 points in the **Putts1** data set contain many points with the same value of *Length*. (Indeed, only 5 distinct values of *Length* are evident—3, 4, 5, 6, and 7—so that the 587 cases fall into 5 natural groups.) Because of these replicate points, we can compute the sample proportions and empirical logits described above. Having

(a) Empirical logits vs. GPA group (b) Empirical proportions vs. GPA group

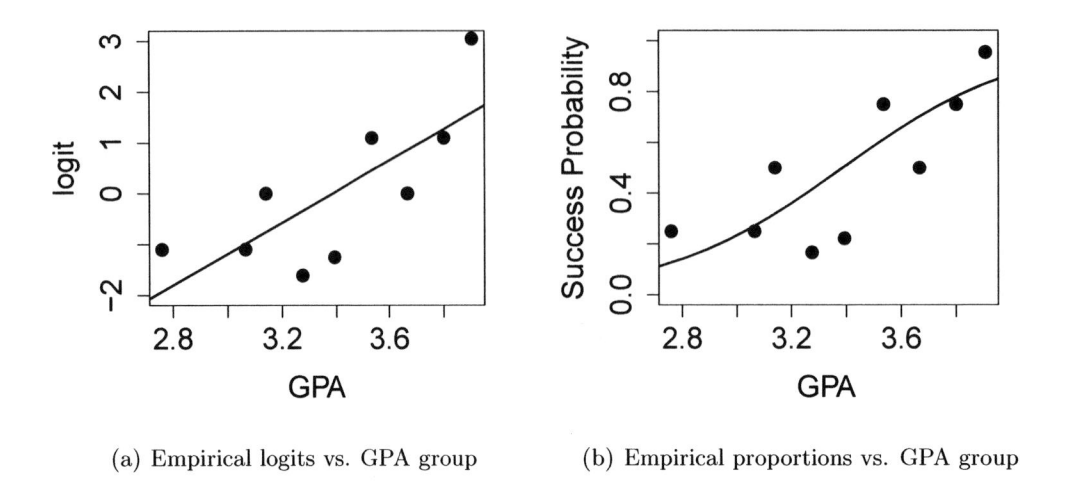

Figure 9.4: Empirical logits and proportions for medical school acceptance by GPA

empirical logits allows us to make the *empirical logit plot* in Figure 9.3, a plot which gives us a visual way to assess whether it seems reasonable to model success probability as a logistic function. If the model makes sense, the empirical logit plot should look straight. Assessing straightness is a visual task our brains are pretty good at, a fact we made use of in our diagnostic tools with regression analysis earlier in the book.

Having such replicate x-values, while nice, is not common. The medical school example is more typical: in most cases we will not have a set of empirical logits to plot. We can, however, obtain an approximation to the empirical logit plot by grouping the data according to the predictor variable.

For example, Figure 9.4 shows the results when we slice the GPA scale to create several groups of students with similar GPA's and find the sample proportions and logits within each group. Our choice of how to group observations is arbitrary, and we lose information from the data when we form groups, but creating the empirical logit plot can sometimes still provide a useful diagnostic about the nature of the relationship. In some cases (usually in the extremes of the predictor scale) a sample proportion may be 0 or 1, so that logits become undefined. In such cases, the logit scale graph is compromised—resulting in further information loss—but the probability scale graph (e.g., 9.4(b)) is not.[4] Figure 9.4 does give some reassurance that a logistic model would fit these data; panel (a) is roughly linear, and panel (b) exhibits the characterstic S-shape (at least, it does not deviate wildly from it).

In general though, diagnostic graphs will not be nearly as useful in logistic regression as we found

---

[4]To avoid this problem, some authors suggest adding or subtracting 0.5 to or from the number of successes for any group in which the proportion is 0 or 1.

9.3. LOGISTIC REGRESSION AND ODDS

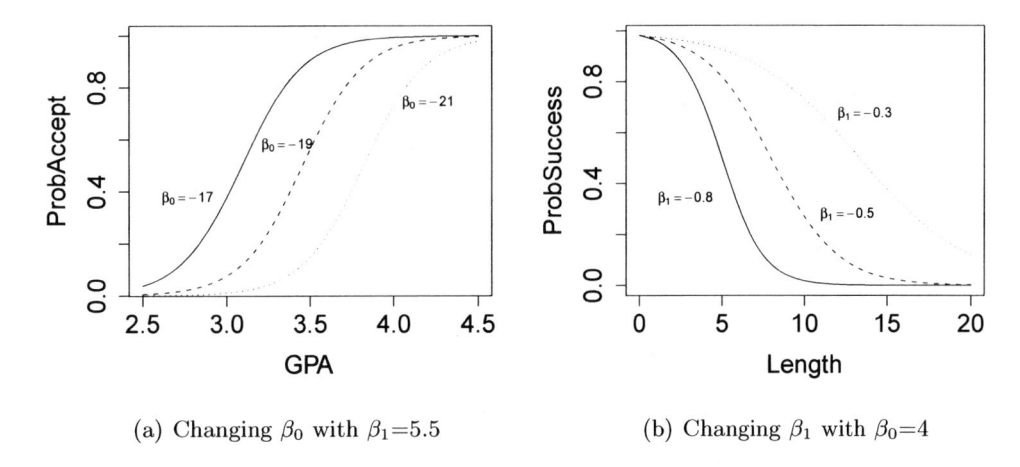

(a) Changing β_0 with β_1=5.5

(b) Changing β_1 with β_0=4

Figure 9.5: Effects of changing β_0 or β_1 in a logistic regression model

them to be in ordinary regression applications.

How parameter values affect shape

To help get a feel for how the slope and constant terms of the linear function in the logit form of the model affect the curve in the probability form of the model, Figure 9.5 shows several possible models for the medical school acceptance and putting data. In Figure 9.5(a) the slope is fixed at $\beta_1 = 5.5$, and the constant term takes on values $\beta_0 = -17$, $\beta_0 = -19$ and $\beta_0 = -21$. Figure 9.5(b) shows probability curves with the constant set to $\beta_0 = 4$ and the slope varying among $\beta_1 = -0.8$, $\beta_1 = -0.5$ and $\beta_1 = -0.3$.
We see that:

- Changing the constant term slides the probability curve horizontally to the right or left.

- Changing the slope makes the curve rise (or fall) more steeply for values farther from zero and more slowly for values closer to zero.

- Note that $\beta_1 = 0$ would give a horizontal line ($\hat{\pi} = 1/2$).

- The sign of β_1 determines whether the probabilities rise ($\beta_1 > 0$) or fall ($\beta_1 < 0$) as values of the predictor increase.

9.3 Logistic Regression and Odds

The term $\pi/(1-\pi)$ in the logit form of the model is the ratio of the probability of "success" to the probability of "failure." This quantity is called the *odds* of success.

> **Probability and Odds**
>
> If the probability of an event is denoted by π we define the *odds* of the event to be
>
> $$odds = \frac{\pi}{1-\pi}. \qquad \text{Some basic algebra verifies that} \qquad \pi = \frac{odds}{1+odds}.$$

In some situations odds are expressed with A:B notation. Odds of 3:2 mean 3 successes for every 2 failures, which is the same as odds of 1.5:1 or simply "odds=1.5;" in this case the probability is $1.5/(1+1.5) = 3/(2+3) = 0.6$.

From the logit form of the model we see that the log of the odds is assumed to be a linear function of the predictor: $\log(odds) = \beta_0 + \beta_1 X$. Thus, the odds in a logistic regression setting with a single predictor can be computed as $odds = e^{\beta_0 + \beta_1 X}$. To gain some intuition for what odds mean and how we use them to compare groups using an *odds ratio*, we consider some examples in which both the response and predictor are binary.

Odds and odds ratios

Example 9.4: *Zapping migraines*

A study investigated whether a handheld device that sends a magnetic pulse into a person's head might be an effective treatment for migraine headaches. Researchers recruited 200 subjects who suffered from migraines and randomly assigned them to receive either the TMS (transcranial magnetic stimulation) treatment or a sham (placebo) treatment from a device that did not deliver any stimulation. Subjects were instructed to apply the device at the onset of migraine symptoms and then assess how they felt two hours later.

The explanatory variable here is which treatment the subject received (a binary variable). The response variable is whether the subject was pain-free two hours later (also a binary variable). The results are summarized in the following 2×2 table:

	TMS	Placebo	Total
Pain-free two hours later	39	22	61
Not pain-free two hours later	61	78	139
Total	100	100	200

Notice that 39% of the TMS subjects were pain-free after two hours compared to 22% of the Placebo subjects, so TMS subjects were more likely to be pain free. Although comparing percentages (or

9.3. LOGISTIC REGRESSION AND ODDS

proportions) is a natural way to compare success rates, another measure—the one preferred for some situations, including the logistic regression setting—is through the use of odds. Although we defined the odds above as the probability of success divided by the probability of failure, we can calculate the odds for a sample by dividing the number of successes by the number of failures. Thus the odds of being pain-free for the TMS group is 39/61=0.639, and the odds of being pain-free for the placebo group is 22/78 = 0.282. Comparing odds gives the same basic message as does comparing probabilities: TMS increases the likelihood of success. The important statistic we use to summarize this comparison is called the *odds ratio* and is defined as the ratio of the two odds. In this case the odds ratio (OR) is given by:

$$OR = \frac{39/61}{22/78} = \frac{.639}{.282} = 2.27$$

We interpret the odds ratio by saying "the odds of being pain free were 2.27 times higher with TMS than with the placebo."

Suppose that we had focused not on being pain-free but on still having some pain. Let's calculate the odds ratio of still having pain between these two treatments. We calculate the odds of still having some pain as $61/39 = 1.564$ in the TMS group and $78/22 = 3.545$ in the placebo group. If we form the odds ratio of still having pain for placebo group compared to TMS group we get $3.545/1.564 = 2.27$, exactly the same as before. One could also take either ratio in reciprocal fashion. For example, $1.564/3.545 = 1/2.27 = .441$ tells us that the odds for still being in pain for the TMS group are .441 times the odds of being in pain for the placebo group. It is often more natural to interpret a statement with an OR greater than 1. ◇

Example 9.5: *Letrozole therapy*

The November 6, 2003 issue of the New England Journal of Medicine reported on a study of the effectiveness of letrozole in postmenopausal women with breast cancer who had completed five years of tamoxifen therapy. Over 5,000 women were enrolled in the study; they were randomly assigned to receive either letrozole or a placebo. The primary response variable of interest was disease-free survival. The article reports that 7.2% of the 2,575 women who received letrozole suffered death or disease, compared to 13.2% of the 2,582 women in the placebo group.

These may seem like very similar results, as 13.2% - 7.2% is a difference of only 6 percentage points. But the odds ratio is 1.97 (check this for yourself by first creating the 2×2 table; you should get the same answer as a ratio of the two proportions.). This indicates that the odds of experiencing death or disease were almost twice as high in the placebo group as in the group of women who received letrozole.

◇

These first two examples are both randomized experiments, but odds ratios also apply to observational studies, as the next example shows.

Example 9.6: *Transition to marriage*

Are single mothers more likely to get married depending on the sex of their child? Researchers investigated this question by examining data from the Panel Study of Income Dynamics. For mothers who gave birth before marriage, they considered the child's sex as the explanatory variable, *Baby's Sex*, and whether or not the mother eventually marries as the (binary) response, *Mother Married*. The data are summarized in the following table:

	Boy child	Girl child	Total
Mother eventually married	176	148	324
Mother did not marry	134	142	276
Total	310	290	600

We see that 176/310=0.568 of the mothers with a boy eventually married, compared with 148/290 =0.510 of mothers with a girl. The odds ratio of marrying between mothers with a boy vs. a girl is: $(176 \cdot 142)/(148 \cdot 134) = 1.26$. So, the odds of marrying are slightly higher (1.26 times higher) if the mother has a boy than if she has a girl.

This example differs from the previous one in two ways. First, the relationship between the variables is much weaker, as evidenced by the odds ratio being much closer to one (and the success proportions being closer to each other). Is that a statistically significant difference or could a difference of this magnitude be attributed to random chance alone? We will investigate this sort of inference question shortly. Second, these data come from an observational study rather than a randomized experiment. One implication of this is that even if the difference between the groups is determined to be statistically significant, we can only conclude that there is an association between the variables, not necessarily that there is a cause-and-effect relationship between them.

⋄

Example 9.7: *Putting prowess (continued)*

We return to the example of a statistician's putting ability in Example 9.3 to observe that the concept of an odds ratio can be applied in settings in which the explanatory variable is not binary. From the data in Table 9.1 we can calculate the empirical proportion of successful putts and the odds of making a putt from each of the lengths. The second and third rows of the table below give these empirical proportions and odds, respectively. The fourth and fifth rows of the table below give the estimated probabilities and estimated odds from a fitted logistic model.

Length of putt (in feet)	3	4	5	6	7
Empirical data:					
Proportion made $\hat{p}$	0.832	0.739	0.565	0.488	0.328
Odds of success	4.941	2.839	1.298	0.953	0.489
Fitted logistic model:					
$\hat{\pi}$	0.826	0.730	0.605	0.465	0.330
Odds of success	4.751	2.697	1.531	0.869	0.494

9.3. LOGISTIC REGRESSION AND ODDS

We can use the odds ratio to compare the odds for any two lengths of putts. For example, the odds ratio of making a four-foot putt to a three-foot putt is calculated as $2.84/4.94 = 0.57$. So, the odds of making a four-foot putt are about 57% of the odds of making a three-foot putt. Comparing five-foot putts to four-foot putts gives an odds ratio of 0.46, comparing six-foot to five-foot putts gives an odds ratio of 0.73, and comparing seven-foot to six-foot putts gives an odds ratio of 0.51. Each time we increase the putt length by one foot, the odds of making a putt are reduced by a factor somewhere between 0.46 and 0.73. These empirical odds ratios are the second row of the table below, the third row of which are the corresponding odds ratios computed from the fitted model.

Length of putt (in feet)	4 to 3	5 to 4	6 to 5	7 to 6
Empirical data: Odds ratio	0.575	0.457	0.734	0.513
Fitted logistic model: Odds ratio	0.568	0.568	0.568	0.568

◇

A consequence of the logistic regression model is that the model constrains these odds ratios (as we increase the predictor by one) to be constant. The third row of the table—the odds ratios from the fitted model—illustrate this principle by all being .568. In assessing whether a simple logistic model makes sense for our data, we should ask ourselves if the empirical odds ratios appear to be roughly constant. That is, are the empirical odds ratios very different from one another? In this example, the empirical odds ratios reflect some variability, but seemingly not an undue amount. This is analogous to the situation in ordinary simple linear regression where the predicted mean changes at a constant rate (the slope) for every increase of one in the predictor—even though the sample means at successive predictor values might not follow this pattern exactly.

Odds Ratios in Logistic Regression

Now that we have introduced how to calculate and interpret odds ratios, we return to developing a model for predicting a binary response variable from a quantitative explanatory variable.

Example 9.8: *Medical school admissions (continued)*

In our medical school acceptance example, we found that the estimated slope coefficient relating GPA to acceptance was 5.45.

```
                                                   95% CI
Predictor     Coef      SE        Z     P    Odds Ratio  Lower   Upper
Constant   -19.2065  5.62922   -3.41  0.001
Total GPA    5.45417  1.57931   3.45  0.001      233.73   10.58 5164.74
```

In ordinary simple linear regression we interpret the slope as the change in the mean response for every increase of one in the predictor. Since the logit form of the logistic regression model relates the log of the odds to a linear function of the predictor, we can interpret the sample "slope," $\hat{\beta}_1 = 5.45$, as the typical change in log(*odds*) for each one-unit increase in GPA.

However, log odds is not as easily interpretable as odds itself, so if we exponentiate both sides of the logit form of the model for a particular GPA, we have

$$\frac{\hat{\pi}_{GPA}}{1-\hat{\pi}_{GPA}} = odds_{GPA} = e^{-19.21+5.45GPA}$$

If we increase the GPA by 1 unit, we get

$$\frac{\hat{\pi}_{GPA+1}}{1-\hat{\pi}_{GPA+1}} = odds_{GPA+1} = e^{-19.21+5.45(GPA+1)}$$

So an increase of 1 GPA unit can be described in terms of the odds ratio

$$\frac{odds_{GPA+1}}{odds_{GPA}} = \frac{e^{-19.21+5.45(GPA+1)}}{e^{-19.21+5.45(GPA)}} = e^{5.45}$$

Therefore, a one-unit increase in GPA is associated with an $e^{5.45}$ or 233.7 fold increase in the odds of acceptance! We see here a fairly direct interpretation of the estimated slope, $\hat{\beta}_1$: Increasing the predictor by one unit gives an odds ratio of $e^{\hat{\beta}_1}$, that is, the odds of success is multiplied by $e^{\hat{\beta}_1}$. In addition to the slope coefficient, Minitab gives the estimated odds ratio (233.73) in its logistic regression output.

The magnitude of this increase appears to be extraordinary, but in fact it serves as a warning that the magnitude of the odds ratio is dependent upon the units we use for measuring the explanatory variable (just as the slope in ordinary regression depends on the units of the predictor). Increasing one's GPA from 3.0 to 4.0 is dramatic and one would certainly expect remarkable consequences. It might be more meaningful to think about a tenth of a unit change in grade point as opposed to an entire unit change. We can compute the odds ratio for a tenth of a unit increase by $e^{(5.454)(.1)} = 1.73$.

An alternative method would have been to redefine the X units into "tenths of GPA points." If we multiply the GPAs by 10 and refit the model we can see from the output below that the odds of acceptance nearly doubles (1.73) for each tenth unit increase in GPA, corroborating the result of the previous paragraph. The model assumes this is true no matter what your GPA is (e.g., increasing from 2.0 to 2.1 or from 3.8 to 3.9), your odds of acceptance go up by a factor of $e^{.5454} = 1.73$. Note that this refers to odds and not probability.

```
Logistic Regression Table
                                          Odds     95% CI
Predictor      Coef    SE Coef      Z     P  Ratio  Lower  Upper
Constant   -19.2065    5.62922  -3.41  0.001
GPA10      0.545417   0.157931   3.45  0.001   1.73   1.27   2.35
```

The summary for the logistic regression in R that is shown for the putting data in Example 9.3 does not include an odds ratio, but it is easy to compute it from $\hat{\beta}_1 = -0.566$ so that $e^{-0.566} = 0.568$ which is exactly what we observed to be the odds ratio as we increased each putt length by one foot in Example 9.7.

9.4 Assessing the Logistic Model

As with simple linear regression, there are conditions for the logistic regression model. Before using this model, it is a good idea to evaluate the conditions and assess how well the model fits. As part of that assessment, we should ask whether or not there even appears to be any association between the predictor and the binary response variable. Is an association that appears in a sample strong enough that we can attribute it to a relationship between the two variables rather than just random chance? Recall that when we assessed ordinary regression models we considered conditions of linearity, normality, independence, and constant variance. What are the equivalent conditions to consider for a logistic regression setting and how do we go about assessing them?

Conditions for Logistic Regression

Linearity. Our model specifies that the logit, the log of the odds (not the mean value of Y), is a linear function of the explanatory variable, X. If there are many replicates of Y at each level of X, we can use the plot of empirical logits against X to assess this condition. As we learned earlier however, for many logistic regression problems this is not feasible.

Independence. As in ordinary regression, we want there to be no pairing or clustering among the observations. Examining the study design and data collection methods are helpful in assessing independence.

Random. The data are obtained using a random process. Two common methods of data collection are (1) random sampling of a population of interest and (2) using randomization in an experiment.

The following two conditions were important in the ordinary regression setting but are specifically NOT applicable to logistic regression.

(No) Constant Variability. In ordinary simple linear regression we assume that the variability in the response Y around the regression line stays constant over all values of the predictor X. For a binary response the variability in the sample proportion of success or failures depends on the value of the proportion. When π is close to the extremes of 0 or 1 the variability is less than when π is near $1/2$, where there is less certainty. Thus, for more extreme values of X (either large or small) we expect more consistency (and less variability) in the response Y than we expect at values of X in the middle of the range.

(No) Normality. When the response values are only 0's and 1's we don't have "residuals" in the same sense as ordinary regression so we do not have a condition of normality in the response at each value of the predictor. [5]

Example 9.9: *Checking conditions for medical school admissions*

In Example 9.1 we considered a logistic regression model for medical school acceptance based on undergraduate GPA. The plot of the empirical logits when we grouped the similar GPAs (Figure 9.4(a)) does not exhibit any kind of extraordinary departure from linearity. However, the validity of the independence condition may be suspect. The pre-med students in the sample are all from the same small college. It is possible that students with similar GPAs study together and participate in similar extracurricular activities. This could affect admission in ways apart from GPA alone, so we need to keep this in mind when drawing our conclusions. This group cannot be considered a random sample of pre-med students from any population—not from their college nor any group of colleges. Thus, we will qualify conclusions with statements such as "if this sample could be considered representative of ..." ◇

Assessing Model Utility: Does the predictor help explain the response?

Recall that in simple linear regression we had several ways to assess whether the relationship between the predictor and response was stronger than one would expect to see by random chance alone and to assess the strength of the relationship. These included:

- A t-test to see if the slope coefficient of the predictor differs from zero.

- An ANOVA F-test to see if a significant amount of variability is explained by the linear fit.

---

[5] *Technical aside (for readers with some probability background):* A binary response coded as 0 and 1 follows what is known as a *Bernoulli* distribution with probability of "success" (or 1) equal to π. The mean value of a Bernoulli random variable is just its probability of success π and the variance is $\pi(1-\pi)$. This helps explain why the variability of Y is smaller when π is close to zero or one and largest at one-half. If we have m different data cases with the same value of the predictor, X, the total number of successes (assuming the independence condition) follows what's known as a *binomial* variable with m trials, probability of success π, and variance $m\pi(1-\pi)$. (We can distinguish this from ordinary regression, in which we assume a normal distribution at each value of the predictor.)

9.4. ASSESSING THE LOGISTIC MODEL

- Quantities such as the variability explained by the model ($SSModel$), the variability left unexplained (SSE), the percent of variability explained (R^2), or the estimated standard deviation of the error term ($\hat{\sigma}_\epsilon$), which can be used to compare the effectiveness of competing models.

Do we have analogous tests and measurements to help assess the effectiveness of a model in the logistic setting? The answer is "yes," although some of the analogous procedures will look more similar to the simple regression setting than others: the procedure to test to see if a slope is significantly different from zero will look very similar, but for many other procedures we need to discuss a substitute for the sums of squared deviations that forms the basis of most of the procedures in ordinary regression.

Here are the outputs for the two logistic models: first, to predict medical school acceptance with GPA, but using $GPA10$—which is GPA·10—as the predictor (Example 9.1) and, second, putting success based on the length of the putt (Example 9.3).

From Minitab output for medical school acceptance data:

```
                                          Odds     95% CI
Predictor      Coef    SE Coef     Z     P  Ratio  Lower  Upper
Constant   -19.2065    5.62922  -3.41  0.001
GPA10       0.545417   0.157931  3.45  0.001   1.73  1.27   2.35
```

From R output for putting data:

```
            Estimate Std. Error z value Pr(>|z|)
(Intercept)  3.25684    0.36893   8.828  <2e-16 ***
Length      -0.56614    0.06747  -8.391  <2e-16 ***
```

The output is interpreted much the same as the t-test in ordinary regression, although the p-value for logistic regression is computed using the normal distribution rather than the t-distribution. In both models, the small p-value shown for the test of the slope coefficient gives strong evidence that there is some relationship between the predictor and the binary response. Software typically supplies the estimated coefficient, $\hat{\beta}_1$, and its standard error, $SE_{\hat{\beta}_1}$, as well as the other details for the test.

> **Test and Confidence Interval for the Slope of a Simple Logistic Regression Model**
>
> significance tests for coefficients To test whether the slope in a model with a single predictor is significantly different from zero the hypotheses are
>
> $H_0 : \beta_1 = 0$
> $H_a : \beta_1 \neq 0$
>
> and the test statistic is
>
> $$z = \frac{\hat{\beta}_1}{SE_{\hat{\beta}_1}}.$$
>
> Assuming we have a reasonably large sample, the p-value is determined from a normal distribution.
>
> This z statistic is also called the *Wald statistic*.
>
> We can also compute a confidence interval for the slope using
>
> $$\hat{\beta}_1 \pm z^* \cdot SE_{\hat{\beta}_1}$$
>
> where z^* is found using the normal distribution and the desired level of confidence.

If we use the output for the medical school data using the $GPA10$ predictor (in tenths of GPA) from the previous section we can compute a 95% confidence interval for the slope of that model with
$$0.5454 \pm 1.96 \cdot 0.1579 = 0.5454 \pm .3095 = (0.2359, 0.8549)$$
However the slope, measuring the change in log(odds) for every unit change in the predictor, is often difficult to interpret in a practical sense. As we saw in the last section, we can convert the slope coefficient to an estimated odds ratio using $e^{0.5454} = 1.73$. Applying the same process to the confidence bounds for the slope produces a confidence interval for the odds ratio.
$$(e^{0.2359}, e^{0.8549}) = (1.27, 2.35)$$
Thus we can say with 95% confidence that every increase of 0.1 in GPA (which is a one-unit increase in $GPA10$) increases the odds of getting accepted to medical school by a factor of between 1.27 and 2.35. Note that the Minitab output includes this confidence interval for the odds ratio along with the estimated odds ratio.

The second important tool we used in assessing the fit of a simple linear regression model was the F-test based on the ANOVA table. The main principle in deriving that test was to see how much

9.4. ASSESSING THE LOGISTIC MODEL

improvement was gained by using the linear model instead of just a constant model. We measured "improvement" by how much the sum of squared residuals changed between the two models. Recall that estimates in the ordinary regression setting were obtained by minimizing that sum of squared residuals. In logistic regression we choose coefficients to maximize the likelihood of the observed data. Comparing this likelihood for models with and without the linear predictor serves as the basis for the corresponding test of how well the model fits.

We define the *likelihood* for each data case as the estimated probability (based on the fitted model) of getting that response. If the case is a "success," then the likelihood is $\hat{\pi}$ and if it's a "failure," then the likelihood is $(1 - \hat{\pi})$. The likelihood for the whole sample (assuming independence) is just the product of all of the individual likelihoods. For any reasonable sample size this product of many probabilities between zero and one is typically very close to zero, even for a very effective model. For example, for the medical school acceptance data and the logistic model based on GPA the likelihood of the sample is $L = 4.5 \times 10^{-13}$ and for the putting data the likelihood is $L = 4.7 \times 10^{-157}$. These are tiny numbers, but because we used maximum likelihood estimation, both are bigger than what we would get with any other choice of slope and intercept for those models.

To get somewhat more reasonable numbers to work with we generally use the natural logarithm of the likelihood. For the medical school data we find $\log(L) = -28.42$ and for the putting example we get $\log(L) = -359.95$. This log-likelihood is also displayed in the Minitab logistic regression output.

Logistic Regression Table

```
                                     Odds     95% CI
Predictor      Coef    SE Coef    Z      P  Ratio  Lower  Upper
Constant    -19.2065   5.62922  -3.41  0.001
GPA10        0.545417  0.157931  3.45  0.001  1.73  1.27   2.35

Log-Likelihood = -28.420
Test that all slopes are zero: G = 18.952, DF = 1, p-value = 0.000
```

You won't need to worry too much about the details of computing a likelihood or log-likelihood (although Exercise 9.9 has some additional details). The key point to remember is that it provides a measure of how likely our data are under one particular model, with larger values of the likelihood (or values closer to zero for the log-likelihood, which is always negative) indicating a better model.

To test $H_0 : \beta_1 = 0$ vs. $H_a : \beta_1 \neq 0$ we need to compare the models

$$\log\left(\frac{\pi}{1-\pi}\right) = constant \quad \text{vs.} \quad \log\left(\frac{\pi}{1-\pi}\right) = \beta_0 + \beta_1 X$$

The null model assumes the *same* probability of success for every data case. We shouldn't be surprised that the best estimate for this common probability is the proportion of successes in the entire sample; that is, $constant = \hat{\pi}_0 = \frac{\#successes}{\#trials}$. For the medical school acceptance data there were 30 students accepted to medical school out of 55 in the sample, so $\hat{\pi}_0 = 30/55 = 0.5454$ (and it's only a coincidence that these digits match the first four digits in the estimate of the logistic slope!). To compute the likelihood using the null model we have a product consisting of 30 values of 0.5454 times 25 more values of $(1 - 0.5454) = 0.4546$. Taking the log of this product shows the log-likelihood for the null (constant model) to be $\log(L_0) = -37.895$. As expected, this log-likelihood is farther from zero than the log-likelihood of the logistic model using GPA, which has $\log(L) = -28.42$; but is the difference so large that we can safely say there is some association between *GPA* and *Acceptance*?

It turns out we can use the log-likelihood to construct a formal significance test to ascertain the answer to the question just posed. This tests rests upon the fact that $-2\log(L)$ follows an approximate *chi-square* distribution with degrees of freedom equal to the sample size minus the number of coefficients estimated in the model. Let us lay out the calculation one would use for this formal test, after which we will give an explanation and then a summary of the test.

$$\begin{aligned} -2\log(L_0) &= 75.79 = \chi^2 \text{ value for constant model, } df = 55 - 1 = 54 \\ -2\log(L) &= 56.84 = \chi^2 \text{ value for logistic model, } df = 55 - 2 = 53 \\ 75.79 - 56.84 &= 18.95 = \text{difference in } \chi^2, df = 54 - 53 = 1 \end{aligned}$$

For the logistic model that predicts acceptance using GPA, we can interpret the value of $-2\log(L) = 56.84$ as a value from a chi-square distribution with 53 degrees of freedom. For the null or constant model, we can interpret the value of $-2\log(L_0) = 75.79$ as the value of a chi-square distribuion with 54 degrees of freedom. Informally, we can think of $-2\log(L)$ as playing the role SSE plays in ordinary regression. So the test rests on a difference in these $-2\log(L)$ values, the 18.95, which we view as the *improvement gained* by adding the GPA term to the model. This difference also follows a chi-square distribution, with degrees of freedom obtained by subtraction, and that difference is our test statistic.

The p-value in this example is, therefore, the area to the right of 18.95 in a chi-square distribution with 1 degree of freedom, which gives a p-value of .000013. We would surely conclude here that there is compelling evidence of a relationship between GPA and medical school acceptance.

Note that the value of our test statistic, the 18.95, is the same $G = 18.952$ in the Minitab output and the p-value of .000013 is essentially the 0.000 Minitab gives.

The test described above is called the "likelihood ratio test (LRT) for utility of a simple regression model." (You will also see it called the "G test" or the "drop-in-deviance test.") Later we will generalize this test to a "full-versus-reduced model test" analogous to what appeared earlier in our discussions of multiple regression using an F distribution. Although the p-value here is not *exactly*

9.4. ASSESSING THE LOGISTIC MODEL

the same as the p-value from the z-test for the slope, when there is only one predictor in the logistic regression model both are testing essentially the same thing. There are, however, instances where the z-test for slope and the likelihood ratio test will disagree more radically. In such instances, you should trust the likelihood ratio test to be the more accurate.

Likelihood ratio test for utility of a simple logistic regression model

To test the overall effectiveness of a logistic regression model with a single predictor

$H_0 : \beta_1 = 0$ vs. $H_a : \beta_1 \neq 0$

we use the test statistic $G = -2\log(L_0) - (-2\log(L))$

where L_0 is the likelihood for a constant model and L is the likelihood using the logistic model. We compare this improvement in -2loglikelihood to a chi-square distribution with one degree of freedom.

Example 9.10: *Putting prowess (continued)*

Let's check the utility of the logistic model to predict putting success based on the length of the putt. The complete summary of the logistic model in R is shown here.

```
Deviance Residuals:
    Min       1Q   Median       3Q      Max
-1.8705  -1.1186   0.6181   1.0026   1.4882

Coefficients:
            Estimate Std. Error z value Pr(>|z|)
(Intercept)  3.25684    0.36893   8.828   <2e-16 ***
Length      -0.56614    0.06747  -8.391   <2e-16 ***
---
Signif. codes:  0 *** 0.001 ** 0.01 * 0.05 . 0.1   1

(Dispersion parameter for binomial family taken to be 1)

    Null deviance: 800.21  on 586  degrees of freedom
Residual deviance: 719.89  on 585  degrees of freedom
AIC: 723.89
Number of Fisher Scoring iterations: 4
```

Note that R (unlike Minitab) does not explicitly calculate the statistic G, but it does give the $-2\log(L)$ values for both the null (constant) model and the logistic model based on *Length*, labeled as "Null deviance" and "Residual deviance" respectively. Thus the G statistic is

$$G = 800.21 - 719.89 = 80.32$$

When compared to a chi-square distribution with one degree of freedom (note 586-585=1) we find the p-value to be essentially zero. Thus we have very strong evidence that the success of making a putt depends on the length of the putt. Furthermore, we have quantified the effect of distance on the odds of making a putt.

◇

The previous example establishes that a model in which the log(odds) are assumed to be a linear function of the putt length is effective, but how do we know if something other than a linear function might give an even better model? We consider this question in Chapter 10.

9.5 Analyzing 2×2 Tables with Logistic Regression

What if both the explanatory and response variables are binary? We can still use logistic regression to model the relationship between the variables and test whether there is a statistically significant association between them.

Example 9.11: *Zapping migraines*

Recall the earlier example of a study that investigated whether a handheld device that sends a magnetic pulse into the head might be an effective treatment for migraine headaches. Both the explanatory and the response variable are binary. The results are summarized in the following 2×2 table:

	TMS	Placebo	Total
Pain-free two hours later	39	22	61
Not pain-free two hours later	61	78	139
Total	100	100	200

As we saw before, the success proportions are 0.39 for the TMS group and 0.22 for the placebo group, a difference of 0.17. The odds ratio turns out to be 2.27, indicating that the odds of being pain-free are more than twice as high with TMS than with the placebo. But how can we assess whether this is statistically significant? One answer is that we can continue to use logistic regression modeling.

The explanatory variable here is the *Treatment Received* (TMS or placebo), which is binary. We can represent this numerically with an indicator variable: 1 = TMS, 0 = placebo. It makes sense to use pain-free = 1 in coding the response. Fitting the logistic regression model produces the following Minitab output:

9.5. ANALYZING 2×2 TABLES WITH LOGISTIC REGRESSION

```
Logistic Regression Table
                                              Odds     95% CI
Predictor      Coef    SE Coef       Z      P Ratio  Lower  Upper
Constant   -1.26567   0.241402   -5.24  0.000
TMS?        0.818354  0.316717    2.58  0.010  2.27   1.22   4.22

Log-Likelihood = -119.566

Test that all slopes are zero: G = 6.885, DF = 1, p-value = 0.009
```

Note that the coefficient of the explanatory variable is positive (0.818), indicating that a subject is more likely to be pain-free with TMS than with the placebo. More specifically, the log of the odds ratio for comparing pain-free status between the TMS and placebo groups is 0.818. This means that the odds ratio is $e^{0.818} = 2.27$ (as the Minitab output also shows).

Logistic regression does more, though. It enables us to determine if the difference in success rates between the two groups is statistically significant. The p-value for the slope coefficient is 0.010, so we have strong evidence of an association between treatment and response. In fact, since this was a randomized experiment, the data provide strong evidence that the TMS treatment causes an increase in the odds of being pain-free. In other words, we have strong evidence that the TMS treatment is more effective than a placebo.

How effective is it? We can use a confidence interval for the odds ratio to address that question. The output above reveals that a 95% confidence interval for the population odds ratio is (1.22, 4.22). So, we can be 95% confident that TMS increases the odds of being pain-free, as compared to a placebo, by as little as a factor of 1.22 or as much as a factor of 4.22.

One last point on this example: We can use the logistic model to calculate predicted probabilities of success. In earlier examples, we predicted such probabilities for the chance of medical school acceptance at various values of GPA or of making putts of various lengths. But now we have a binary explanatory variable, so we only predict two success probabilities: one for the TMS treatment (obtained by plugging in X = 1) and one for the placebo (obtained by plugging in X = 0). This gives:

$$\hat{\pi}_{TMS} = \frac{e^{-1.266+0.818(1)}}{1+e^{-1.266+0.818(1)}} = 0.39 \quad \text{and} \quad \hat{\pi}_{Placebo} = \frac{e^{-1.266+0.818(0)}}{1+e^{-1.266+0.818(0)}} = 0.22$$

Notice that these are exactly the same as the observed success proportions for each group.

The fact that the probabilities given by the logistic model match the sample proportions in each group is no accident. With a binary explanatory variable coded as x = 0 and x = 1, the logistic regression model

$$\log\left(\frac{\pi}{1-\pi}\right) = \beta_0 + \beta_1 X$$

can be written as two equations:

$$\log\left(\frac{\pi}{1-\pi}\right) = \beta_0$$

for the placebo group and

$$\log\left(\frac{\pi}{1-\pi}\right) = \beta_0 + \beta_1$$

for the TMS group.

If $\hat{p}_0$ is the sample proportion for the placebo group, we can use it to find the log(odds) for the placebo group and thus estimate the constant term, $\hat{\beta}_0$, to match this case exactly.

$$\hat{\beta}_0 = \log\left(\frac{\hat{p}_0}{1-\hat{p}_0}\right)$$

Once we have found $\hat{\beta}_0$ using the proportion $\hat{p}_0$ from one group, we can use the other degree of freedom to compute a slope, $\hat{\beta}_1$, that estimates a probability to match the proportion $\hat{p}_1$ in the other group.

$$\hat{\beta}_1 = \log\left(\frac{\hat{p}_1}{1-\hat{p}_1}\right) - \hat{\beta}_0$$

We can check in the Minitab output that the estimated constant, -1.266, is the log of the odds for the placebo group, $\log(22/78)$, and the estimated slope is the change in the log odds when we move to the TMS group: $0.818 = \log(39/61) - \log(22/78)$. This is one of the rare cases in which we can compute the maximum likelihood estimates for the intercept and slope for the logistic regression without relying on a computer.

Two-sample z-test for proportions (review)

An alternative (and possibly more familiar) method for assessing whether the proportions differ between two groups is to use a two-sample test based on the normal distribution. This procedure is often covered in an introductory statistics course, so we only review it briefly here and compare the results to what we would see in binary logistic regresion.

9.5. ANALYZING 2×2 TABLES WITH LOGISTIC REGRESSION

> **Z-test for difference in two proportions**
>
> Assume we have two independent samples of size n_0 and n_1, respectively. Let X_0 and X_1 be the counts of successes in each group so that $\hat{\pi}_0 = X_0/n_0$ and $\hat{\pi}_1 = X_1/n_1$. To test
>
> $$H_0 : \pi_0 = \pi_1 \qquad \text{vs.} \qquad H_a : \pi_0 \neq \pi_1$$
>
> $$z = \frac{\hat{\pi}_1 - \hat{\pi}_0}{\sqrt{\hat{\pi}(1-\hat{\pi})\left(\frac{1}{n_1} + \frac{1}{n_0}\right)}}$$
>
> where $\hat{\pi} = \frac{X_1 + X_0}{n_1 + n_0}$ is the proportion obtained by pooling the two samples. Assuming the sample sizes are not small, the distribution of the test statistic z is approximately standard normal under H_0, so we use the normal distribution to compute a p-value.

For the TMS (1) vs. placebo (0) data we have the sample proportions:

$$\hat{\pi}_1 = \frac{39}{100} = 0.39 \qquad \hat{\pi}_0 = \frac{22}{100} = 0.22 \qquad \hat{\pi} = \frac{39 + 22}{100 + 100} = 0.305$$

and the test statistic is

$$z = \frac{0.39 - 0.22}{\sqrt{0.305(1 - 0.305)\left(\frac{1}{100} + \frac{1}{100}\right)}} = 2.61$$

The p-value=$2P(Z > 2.61) = 0.009$ where $Z \sim N(0,1)$ which again indicates that the proportion of patients who got relief from migraines using TMS was significantly different from the corresponding proportion for those using the placebo. Note that the test statistic and p-value for this two-sample z-test for a difference in proportions are similar to the test statistic (2.58) and p-value (0.010) from the Minitab output for testing whether the slope of the logistic regression model differs significantly from zero.

Chi-square test for a 2×2 table (review)

In an introductory statistics class you might also have studied a chi-square procedure for testing relationships between categorical variables using a two-way table. How does this test relate to the assessment of the binary logistic regression with a single binary predictor?

> **Chi-square test for a 2×2 table**
>
> Start with data of *observed* counts (O_{ij}) in a 2×2 table. Compute an *expected* count for each cell using $E_{ij} = \frac{(RowTotal)\cdot(ColumnTotal)}{n}$.
>
> | Group | Success | Failure |
> |---|---|---|
> | 0 | O_{00} (E_{00}) | O_{01} (E_{01}) |
> | 1 | O_{10} (E_{10}) | O_{01} (E_{11}) |
>
> The test statistic compares the *observed* and *expected* counts within each of the cells of the table and sums the results over the four cells.
>
> $$X^2 = \sum \frac{(Observed - Expected)^2}{Expected}$$
>
> If the sample sizes are not small (expected counts at least 5) the test statistic under H_0 follows a chi-square distribution with one degree of freedom.

Here are the observed and expected counts for the 2×2 table based on the TMS and placebo data.

Group	Success	Failure	Total
Placebo	22 (30.5)	78 (69.5)	100
TMS	39 (30.5)	61 (69.5)	100
Total	61	139	200

The chi-square test statistic is

$$t.s. = \frac{(22-30.5)^2}{30.5} + \frac{(78-69.5)^2}{69.5} + \frac{(39-30.5)^2}{30.5} + \frac{(61-69.5)^2}{69.5} = 6.82$$

This gives a p-value= $P(\chi^2_1 > 6.821) = 0.009$ which matches the two-sample z-test exactly (as it always must in the 2×2 case). While this chi-square statistic is not exactly the same as the chi-square test statistic based on $-2\log(L)$ in the logistic regression (6.885), the p-values are very similar and the conclusions are the same.

Example 9.12: *Transition to marriage (continued)*

Recall the study about whether single mothers are more likely to marry eventually depending on whether they have a boy child or a girl child. The 2×2 table is reproduced in Table 9.2

The explanatory variable here is the sex of the child (*boy child*) and the response is whether or not the mother eventually married (*MomMarried*). To conduct a logistic regression analysis on these

9.5. ANALYZING 2×2 TABLES WITH LOGISTIC REGRESSION

	Boy child	Girl child	Total
Mother eventually married	176	148	324
Mother did not marry	134	142	276
Total	310	290	600

Table 9.2: Mother's marital status by child's sex.

data, we'll code boy = 1, married = 1, and the other outcomes as 0. This leads to the following logistic regression output:

```
Logistic Regression Table
                                            Odds     95% CI
Predictor      Coef      SE Coef    Z     P  Ratio  Lower  Upper
Constant    0.0413852   0.117469  0.35  0.725
boy child   0.231259    0.164145  1.41  0.159  1.26   0.91   1.74
```

Notice that the odds ratio is 1.26, as we found earlier. But the p-value is .159, indicating that the difference in marriage rates between the two groups is not statistically significant. We can also see this by observing that the 95% confidence interval for the population odds ratio (0.91, 1.74) includes the value 1, again indicating that there's not much evidence of an association between the child's sex and whether or not the mother eventually marries.

◇

What have we learned in this section? The logistic regression model that we have used with a quantitative explanatory variable X can also be used with a binary explanatory variable, coded as X = 0 and X = 1. The odds ratio turns out to be $e^{\hat{\beta}_1}$, and the estimated probabilities turn out to be the observed success proportions. So why bother to use logistic regression here? Four reasons:

- Logistic regression allows for inferences to be drawn: the p-value assesses whether the association between the variables is significant, and a confidence interval estimates the magnitude of the population odds ratio.

- This approach provides consistency with the modeling approach that has been emphasized throughout this book.

- We can use the logistic regression modeling approach to analyze much more complicated contingency tables; the other two methods are far less flexible.

- Finally, we can extend this analysis to consider multiple explanatory variables in a logistic regression model including both categorical and continuous explanatory variables.

9.6 Summary

Our look at logistic regression reveals that a **binary response**, like a continuous normally distributed response, can be modeled as a function of an explanatory variable, X. We explored medical school admission (accepted or not) as a function of GPA and putting success (made or missed) as a function of distance. With a continuous normally distributed response such as Porsche price, our interest centers on the mean response (μ) which is modeled as a linear function of mileage, for example. With a binary response, our interest is in the probability of success (π) at any given level of X, but we model the log odds—$\log(\pi/(1-\pi))$—not simply π, as a linear function of the predictor. Logistic regression models produce estimates of odds ratios, such as the odds of admission for a tenth-unit increase in GPA, by exponentiating a coefficient. Logistic regression models also provide probability estimates, such as the probability of admission for a given GPA, which remain between 0 and 1.

This approach is consistent with the modeling approach that has been emphasized throughout this book. Logistic regression allows for inferences to be drawn: The p-value assesses whether the association between the binary response and explanatory variables is significant, and a confidence interval estimates the magnitude of the population odds ratio. Logistic regression can accommodate multiple explanatory variables including both categorical and continuous explanatory variables which is the subject of the next chapter.

9.7 Exercises

Conceptual Exercises

9.1 *Ordinary regression.* Why does simple linear regression used in previous chapters not work well when the response is binary?

9.2 *Equivalence of logit and probability forms of the logistic regression.* Use basic algebra to start with the logit form of the logistic regression model

$$\log\left(\frac{\pi}{1-\pi}\right) = \beta_0 + \beta_1 X$$

and derive the probability form of the model

$$\pi = \frac{e^{\beta_0 + \beta_1 X}}{1 + e^{\beta_0 + \beta_1 X}}$$

9.3 *Probability to odds.*

a. If the probability of an event occurring is 0.5, what are the odds?

b. If the probability of an event occurring is 0.9, what are the odds?

c. If the probability of an event occurring is 0.1, what are the odds?

9.4 *Odds to probabilities.*

a. If the odds of an event occurring are 2:1, what is the probability?

b. If the odds of an event occurring are 10:1, what is the probability?

c. If the odds of an event occurring are 1:4, what is the probability?

9.5 *Odds ratio for recovery.* If the probability of a recovery with treatment is 0.3 and without 0.1, what is the odds ratio for recovery when treated versus not treated?

9.6 *Odds ratio for birth defects.* If the probability of a birth defect with exposure to a potential teratogen is 0.6 and without exposure the probability is 0.01, what is the odds ratio for a birth defect when exposed versus not exposed?

9.7 *Effects of slope and intercept.* Suppose that we have a logistic model with intercept $\beta_0 = 5$ and slope $\beta_1 = 2$. Explain what happens to a plot of the probability form of the model in each of the following circumstances.

a. The slope β_1 decreases to 1.

b. The intercept β_0 increases to 8.

c. The slope changes sign to become $\beta_1 = -2$.

9.8 *Putting revisited.* Consider the putting data (Example 9.3). We established that there was no evidence of a problem with linearity. Comment on the conditions of independence and randomness in this setting.

9.9 *Looking at likelihoods.* You are trying to estimate a true proportion (π) using the results of several independent Bernoulli trials, each with the same probability of success (π). Open the Excel file labelled **likelihood-play.xls**. You will find a graph of a likelihood function for a series of Bernoulli trials. You can specify different values for the numbers of successes and failures and you will see a corresponding change in the likelihood function. What happens as the number of trials increases? How does the graph change as the proportion of successes changes? How does the corresponding graph of the log of the likelihood compare?

9.10 *Compute a likelihood.*
In Example 9.4 on zapping migraines we showed how to fit the logistic model, by hand, using sample proportions from the 2×2 table. One can also use these proportions to directly compute the G statistic given there in the Minitab output. Show this calculation and verify that your answer agrees with the Minitab output.

Guided Exercises

9.11 *Medical school acceptance.* The data file **MedGPA** used in Example 9.1 also contains information on the medical school admission test (MCAT) scores for the same sample of 55 students. Fit a logistic regression model to predict the *Acceptance* status using the *MCAT* scores.

a. Write down the estimated versions of both the logit and probability forms of this model.

b. What would the estimated model say about the chance that a student with $MCAT = 40$ gets accepted to medical school?

c. For approximately what *MCAT* score would a student have roughly a 50-50 chance of being accepted to medical school? Hint: You might look at a graph or solve one of the equations algebraically.

9.12 *Metastasizing cancer.* In a study of 31 patients with esophageal cancer it was found that in 18 of the patients the cancer had metastasized to the lymph nodes. Thus, an overall estimate of the probability of metastasis is 18/31=0.58. A predictor variable measured on each patient is *Size* of the tumor (in cm). A fitted logistic regression model is

9.7. EXERCISES

$$log\left(\frac{\hat{\pi}}{1-\hat{\pi}}\right) = -2.086 + 0.5117 \cdot Size$$

a. Use this model to estimate the odds of metastisis, $\pi/(1-\pi)$, if a patient's tumor size is 6 cm.

b. Use the model to predict the probability of metastasis if a patient's tumor size is 6 cm.

c. How much does the estimated odds change if the tumor size changes from 6 cm to 7 cm? Provide and interpret an odds ratio.

d. How much does the estimate of π change if the tumor size changes from 6 cm to 7 cm?

9.13 *Metastasizing cancer (continued).* Consider the fitted logit model from Exercise 9.12. How large does a tumor need to be for the estimated probability of metastasis to be at least 0.80?

9.14 *Sinking of the Titanic.* The *Titanic* was a British luxury ocean liner that sank famously in the icy North Atlantic Ocean on its maiden voyage in April of 1912. Of the approximately 2200 passengers on board, 1500 died. The high death rate was blamed largely on the inadequate supply of lifeboats, a result of the manufacturer's claim that the ship was "unsinkable." A partial data set of the passenger list was compiled by Philip Hinde in his *Encyclopedia Titanica* and is given in the datafile **Titanic**.

Two questions of interest are the relationship between survival and age and the relationship between survival and sex. The following variables will be useful for your work on the following questions.

Age   which gives the passenger's age in years
Sex   which gives the passenger's sex (male or female)
Survived   a binary variable where 1 indicates the passenger survived and 0 indicates death
SexCode   which numerically codes male as 0 and female as 1.

a. Use a plot to explore whether there is a relationship between survival and the passenger's age. What do you conclude from this graph alone?

b. Use software to fit a logistic model to the survival and age variables to decide whether there is a statistically significant relationship between age and survival, and if there is, what its direction and magnitude are. Write the estimated logistic model using the output and interpret the output in light of the question.

c. Use a two-way table to explore whether survival is related to the sex of the passenger. What do you conclude from this table alone? Write a summary statement that interprets *Sex* as the explanatory variable and *Survived* as the response variable and that uses simple comparisons of conditional proportions or percentages.

d. Use software to fit a logistic model to the survival and sex variables to decide whether there is a statistically significant relationship between sex and survival. If there is, what are the nature and magnitude of the relationship? Does the relationship found by the logistic model confirm the descriptive analysis?

e. Present a calculation that compares the estimated slope coefficient of the model with *Sex* as a predictor to the estimated odds ratio. Then give a sentence that interprets the odds ratio in the context of the Titanic episode.

f. Write a sentence that interprets a 95% CI for the odds ratio discussed in (e).

g. Present a calculation from the two-way table that leads to an estimated coefficient from the output for the model found in (d).

h. Use the model to estimate the probability that a female would have survived the sinking of the Titanic.

i. Confirm the results of the significance of the slope coefficient using a chi-square test on the two-way table.

j. Assess the model conditions for the model relating *Survival* to *Sex*. Write a short summary of the assessment.

k. Write a short paragraph that summarizes your analysis of the relationships between the *Sex* and *Age* of a passenger and the passenger's *Survival*.

9.15 *Leukemia treatments.*[6] A study involved 51 untreated adult patients with acute myeloblastic leukemia who were given a course of treatment, after which they were assessed as to their response. The variables recorded in the data set **Leukemia** are:

Age	at diagnosis in years
Smear	differential percentage of blasts
Infil	percentage of absolute marrow leukemia infiltrate
Index	percentage labeling index of the bone marrow leukemia cells
Blasts	absolute number of blasts, in thousands
Temp	highest temperature of the patient prior to treatment, in degrees Farenheit
Resp	1=responded to treatment and 0=failed to respond
Time	the survival time from diagnosis, in months
Status	0=dead and 1=alive

a. Fit a logistic model using *Resp* as the response variable and *Age* as the predictor variable. Interpret the results and state whether the relationship is statistically significant.

b. Form a two-way table that exhibits the nature of the relationship found in (a).

c. Redo parts (a) and (b), but using *Temp* as the single predictor variable.

---
[6]Data come from *Statistical Analysis Using S-Plus* (Brian S. Everitt; first edition 1994, Chapman & Hall).

9.7. EXERCISES

9.16 *First Year GPA.* In Example 4.2 we considered relationships between a number of variables in the **FirstYearGPA** dataset to help explain the variability of GPA for first year students. One of those variables was an indicator, *FirstGen*, for whether the student was a first generation college attendee (1 if so, 0 if not). In this exercise you will compare several different ways of seeing if there is some association between *GPA* and *FirstGen*.

- a. Use a two-sample t-test to see if there is a significant difference in the average GPA between students who are and are not first generation college attendees. Report the p-value of the test and the direction of the relationship, if significant.

- b. Use a simple linear regression to predict *GPA* using *FirstGen*. Report a p-value and indicate the nature of the relationship if a significant relationship is present.

- c. Do the comparison once more, this time using a logistic regression with *GPA* as the predictor and *FirstGen* as the response. Compare the conclusion (and p-value) you would draw from this model to the results from parts (a) and (b).

9.17 *Flight response of Pacific Brant.*[7] A 1994 study collected data on the effects of air traffic on the behavior of the Pacific Brant (a small migratory goose). Each fall nearly the entire population of 130,000 of this species uses the Izembek Lagoon in Alaska as a staging area, where it feeds and stores fat for its southerly migration. Because offshore drilling near this estuary had increased the necessity of air traffic, an impact study was timely. The data represent the flight response to helicopter "overflights" to see what the relationship between the proximity of a flight, both lateral and altitudinal, would be to the propensity of the Brant to flee the area. For this experiment, air traffic was restricted to helicopters because previous study had ascertained that helicopters created more radical flight response than other aircraft.

The data are in **FlightResponse**. Each case represents a flock of Brant that has been observed during one overflight in the study. Flocks were determined observationally as contiguous collections of Brants, flock sizes varying from 10 to 30,000 birds. For this study the variables we investigate are:

---

[7] Data come from the book *Statistical Case Studies: A Collaboration Between Academe and Industry*, Roxy Peck, Larry D. Haugh, and Arnold Goodman, editors; SIAM and ASA, 1998.

Altitude   The experimentally determined altitude of the overflight by the helicopter. Units are in 100m, with the variable range being 0.91 to 12.19, being recorded at 9 distinct values: 0.91, 1.52, 3.05, 4.57, 6.10, 6.71, 7.62, 9.14, and 12.19.

Lateral   The perpendicular or lateral distance (in 100m) between the aircraft and flock, as determined from studying area maps to the nearest 0.16 km.

AltCat   A categorical variable, derived from the Altitude variable. The range [0, 3) is category "low," the range [3, 6] is "mid," and the range (6, infinity) is "high."

LatCat   A categorical variable, derived from the *Lateral* variable. The range [0, 10) is category 1, [10, 20) is 2, [20, 30) is 3, and [30, infinity) is 4.

Flight   This is a binary variable in which 0 represents an outcome where fewer than 10% of a flock flies away during the overflight and 1 represents an outcome where more than 10% of the flock flies away. This is the response variable of interest in this study.

a. Create a two-way table of *AltCat* by *Flight*. For each level of *AltCat*, calculate the odds and log-odds for a flight response. Does there appear to be a relationship between altitude and flight, based solely on this table? Describe the nature and intensity of this response.

b. Calculate a logistic regression model using *Flight* as the response variable and *Altitude* as the explanatory variable. Does this model confirm your suspicion from part (a) about the existence and direction of a relationship between flight and altitude? Explain. Report model estimates and interpret the estimated slope coefficient.

c. Redo the analysis of parts (a) and (b), except to investigate the relationship between lateral distance and flight response. Use *LatCat* for the part (a) equivalent.

9.18 *Nikkei 225.* If we know how the New York stock market performs today, can we use that information to predict whether the stock market in Japan will go up or down tomorrow? Or is the movement of the Japanese stock market today better predicted by how the Japanese market did yesterday? The file **Markets** contains data from two stock markets for 56 days. The variables recorded are *Date*, *Nik*225*ch* (the one-day change in the Nikkei 225, a stock index in Japan), *DJIAch* (the one-day change in the New York-based Dow Jones Industrial Average from the previous day), *Up* (1 or 0 depending on whether or not the Nikkei 225 went up on a date), and *lagNik* (the one-day change in the Nikkei 225 from the previous day). Thus, if we want to predict whether the stock market in Japan will go up or down on a Tuesday, we might use the Monday result from Japan (*lagNik*) or the Monday result from New York (*DJIAch*)—remembering that when it is Monday evening in New York, it is Tuesday morning in Japan.

a. Fit a logistic model with *Up* as the response and *DJIAch* as the predictor. Is *DJIAch* a significant predictor of the direction the Nikkei 225 will go the next day? Explain the basis for your conclusion.

b. Fit a logistic model with *Up* as the response and *lagNik* as the predictor. Is *lagNik* a significant predictor of the direction the Nikkei 225 will go the next day? Explain the basis for your conclusion.

9.7. EXERCISES

c. Compare the models in part (a) and part(b). Which variable, $DJIAch$ or $lagNik$ is a more effective predictor of where the Nikkei 225 is going the next day? Explain the basis for your decision.

9.19 *Red states and Blue states in 2008.* Can we use state-level variables to predict whether a state votes for the Democratic vs. the Republican presidential nominee? The file **Election08** contains data from 50 states plus the District of Columbia. The variables recorded are

StateAbr   abbreviation for the state
Income   per capita income as of 2007
HS   percentage of adults with at least a high school education
BA   percentage of adults with at least a college education
Dem.Rep   %Democrat-%Republican in a state
       including those who lean toward either party according to a 2008 Gallup poll and
ObamaWin   1 or 0 indicating whether the Democratic candidate
       Barack Obama did or did not win a majority of votes in the state

a. Fit separate logistic regression models to predict *ObamaWin* using each of the predictors *Income*, *HS*, *BA* and *Dem.Rep*. Which of these variables does the most effective job of predicting this response? Which is the least effective? Explain the criteria you use to make these decisions.[8]

9.20 *Red states and Blue states in 2008 (continued).* Refer to the data in **Election08** that are described in Exercise 9.19. Run a logistic regression model to predict *ObamaWin* for each state using the per capita *Income* of the state.

a. Use the estimated slope from the logistic regression to compute an estimated odds ratio and write a sentence that interprets this value in the context of this problem.

b. Find a 95% confidence interval for the odds ratio in (a).

c. The units of the *Income* variable are dollars with values ranging from $28,845 (Mississippi) to $61,092 (District of Columbia). The odds ratio and interval in (a) and (b) are awkward to interpret since they deal with the change in the odds when state income changes by $1, a very trivial amount! To get an odds ratio that may be more meaningful, create a new variable (call it *IncomeTh*) using *Income*/1000 to express the state per capita incomes in $1,000s. Run the logistic regression using *IncomeTh* as the predictor of *ObamaWin*. How does the fitted prediction equation change? How (if at all) does the predicted probability of Obama winning a state change?

d. Repeat parts (a) and (b) using the slope and odds ratio from the model in which state incomes are in thousands of dollars.

---

[8]Note: We will consider models that include multiple predictors for logistic regression in the next chapter.

9.21 *THC vs. prochlorperazine.* An article in the *New England Journal of Medicine* described a study on the effectiveness of medications for combatting nausea in patients undergoing chemotherapy treatments for cancer. In the experiment, 157 patients were divided at random into two groups. One group of 78 patients were given a standard antinausea drug called prochlorperazine, while the other group of 79 patients received THC (the active ingredient in marijuana). Both medications were delivered orally and no patients were told which of the two drugs they were taking. The response measured was whether or not the patient experienced relief from nausea when undergoing chemotherapy. The results are summarized in the 2x2 table shown below.

Drug	Effective	Not Effective	Patients
THC	36	43	79
Prochlorperazine	16	62	78

a. Find the proportion of patients in each of the sample groups for which the treatment was effective.

b. Fit a binary logistic regression model to predict *Effectiveness* (yes or no) using type of *Drug* as a binary predictor. Give the logit form of the fitted model.

c. Use the model to predict the odds and the probability of effectiveness for each of the two drugs. Compare predicted proportions to the sample proportions in (a).

d. Find the odds ratio comparing the effectiveness of THC to prochlorperazine based on the binary logistic regression model. Write a sentence that interprets this value in the context of this problem and find a 95% confidence interval for the odds ratio.

e. The table shows that THC was effective in more cases, but is this difference statistically significant? Use information from the logistic regression output to do a formal test at a 1% level to see if there is evidence that THC is more effective than prochlorperazine in combatting nausea for chemotherapy patients.

f. Address the question in part (e) using a two-sample z-test to compare the proportions of effective cases between the THC and prochlorperazine groups.

g. Try one more test, this time using a chi-square test based on the observed and expected counts in the 2×2 table.

h. Are the results of the three tests in parts (e), (f), and (g) consistent with each other? Are any of them identical in the strength of evidence? Comment.

9.7. EXERCISES

Open-ended Exercises

9.22 *Backpack weights.* Is your back aching from the heavy backpack you are lugging around? A survey of students at California Polytechnic State University (San Luis Obispo) collected data to investigate this question. The data set **Backpack**[9] includes the following variables:
Back Problems (0=no, 1=yes)
Backpack Weight
Body Weight
Ratio

Use the data to explore the possibility of backpacks being responsible for back problems. What are the relative merits of using the backpack weight alone as a predictor compared to using the ratio of backpack to body weight? Which would you recommend and why?

9.23 *Risky youth behavior.*
What happens when someone who is underaged needs a ride home after a night of drinking? The concern of the media today focuses on those who drive home in this situation. However, what about those who ride with someone who has been drinking? We'll denote such an event as RDD. This choice can be as devastating as driving drunk, resulting in a large number of RDD traffic deaths. A recent study examines who accepts these rides and why.[10]

We'll examine RDD behavior with only a few variables from a very large dataset to examine questions concerning youth who ride with drivers who have been drinking (RDD) so as to identify where safety messages might best be targeted. We'll use a data set **YouthRisk2007** derived from the 2007 Youth Risk Behavior Surveillance System (YRBSS), which is an annual survey conducted by the Centers for Disease Control and Prevention (CDC) to monitor the prevalence of health-risk youth behaviors.[11] The following variables are included in **YouthRisk2007**:

ride.alc.driver	1=respondent rode with a drinking driver within past 30 days, 0=did not
female	1=female, 0=male
grade	A factor variable with levels "9", "10", "11", "12"
age4	14, 15, 16, 17, or 18 years of age
smoke	1= "yes" to "have ever tried cigarette smoking, even one or two puffs?", 0= "no"

Investigate the following hypotheses:

a. Some researchers proposed that the odds of riding with a driver who has been drinking are higher for young women compared to young men. Do these data provide evidence of this?

b. One study suggests that youth become more reckless about drinking and driving after they

---

[9] See "Oh, My Aching Back! A Statistical Analysis of Backpack Weights," by J. Mintz, J. Mintz, K. Moore, and K. Schuh, *Stats: The Magazine for Students of Statistics,* vol. 32, 2002, pp. 1719.

[10] The article "Which Young People Accept a Lift From a Drunk or Drugged Driver?" in *Accident Analysis and Prevention* (July 2009. pp. 703-9) provides more details.

[11] A more recent version of the full dataset is available at http://www.cdc.gov/brfss/technical_infodata/surveydata.htm. The full data set would be a rich source of information for those interested in updating the results of the study.

obtain their driver's license[12]. On the other hand, Poulin et al.[13] reported that a driver's license was found to be protective among Canadian youth. What does the evidence in this survey suggest about the effect of getting a driver's license in the U.S.?

c. Rybabc et al. (2007) have reported that smokers are more likely to RDD citing that injury-prone behaviors were also more common among the smokers than non-smokers. Is smoking associated with an increased risk of RDD?

9.24 *Good movies.* Statisticians enjoy movies as much as the next person. One statistician movie fan decided to use statistics to study the movie ratings in his favorite movie guide, *Movie and Video Guide* (1996), by Leonard Maltin. He was interested in discovering what features of Maltin's Guide might correlate to his view of the movie. Maltin rates movies on a one-star to four-star system, in increments of half-stars, with higher numbers being better. Our statistician has developed, over time, the intuition that movies rated 3 or higher are worth considering for view, but lower ratings can be ignored. He used a random number generator to select a simple random sample of 100 movies rated by the Guide. For each movie, he measured and recorded the following variables:

Title	The movie's title
Year	The year the movie was released (range is 1924 to 1995)
Time	The running time of the movie in minutes (range is 45 to 145)
Cast	The number of cast members listed in the guide (range is 3 to 13)
Rating	The movie's Maltin rating (range is 1 to 4, in steps of 0.5)
Description	the number of lines of text Maltin uses to describe the movie (range is 5 to 21)
Origin	The country where the movie was produced (0 = USA, 1 = Great Britain, 2 = France, 3 = Italy, 4 = Canada)

The data are in the file **Film**.

For the purposes of his study, he also defined a variable called Good?, where 1=a rating 3 stars or better and 0=any lower rating. He was curious about which variables might be good predictors of his personal definition of a good movie. Analyze the data to find out. Write a short report of your findings. (Note: restrict your explanatory variables to Year, Time, Cast, and Description.)

9.25 *Lost Letter.* In a 1965 article in *Public Opinion Quarterly*, psychologist Stanley Milgram reported on a study where he intentionally lost letters to see if the nature of the address on the letter would affect return rates when people found the letters lying on a public sidewalk. For example, he found that letters addressed to a subversive organization such as Friends of the Nazi Party were far less likely to be returned than those addressed to an individual or to a medical research organization.

---

[12] McCarthy and Brown, *J. Stud. Alcohol*, 65: 289-296, 2004
[13] See *Addiction in 2007*, Volume 102, Issue 1, pp. 1-172, January 2007.

9.7. EXERCISES

In 1999 Grinnell College students Laurelin Muir and Adam Gratch conducted an experiment similar to Milgram's for an introductory statistics class. They intentionally "lost 140 letters" in either the city of Des Moines, the town of Grinnell, or on the Grinnell College campus. Half of each sample were addressed to Friends of the Confederacy and the other half to Iowa Peaceworks. Do the return rates differ depending on address? Get the data in **LostLetter**. Use logistic regression and analyze the data to decide, location by location, if there is a difference. Summarize your findings.

9.26 *Political Activity of Students* Students Jennifer Wolfson and Meredith Goulet conducted a survey in the spring of 1992 of Grinnell College students to ascertain patterns of political behavior. They took a simple random sample of 60 students who were U.S. citizens and conducted phone interviews. Using several "call backs" they obtained 59 responses. The dataset for the study is **Political** and the variables are:

Year   The class-year of the student (1=first year, 2=second year, 3=junior, 4=senior)
Sex   The sex of the student (0=male, 1=female)
Vote   Voting pattern of student (0=not eligible, 1=eligible only,
      2=has registered, but not voted, 3=has voted)
Paper   "How often do you read newspapers or news magazines, not including comics?"
      (0=never, 1=less than once per week, 2=once a week, 3=2 or 3 times per week, 4=daily)
Edit   "Do you read the editorial page?" (0=no, 1=yes)
TV   "How often do you watch TV for news?" (0=never,
      1=less than once per week, 2=once a week, 3=2 or 3 times per week, 4=daily)
Ethics   "Should politics be ruled by ethical considerations or by practical power politics?"
      (scale of 1 to 5, with one being "purely ethical" and 5 being "purely practical")
Inform   "How informed do you consider yourself to be about national or international politics?"
      (scale of 1 to 5 with 1=generally uninformed, 5=very well informed)
Participate   a variable derived from Vote, with Vote=0 transformed to missing value,
      Vote=1 or 2 transformed to 0, and Vote=3 transformed to 1).

One question we can ask of these data is a classical concern in democracies: Do the better informed citizens tend to be the ones who vote? Use *Participate* as the response and use simple logistic regression to investigate this question. Summarize your findings.

Chapter 10

Multiple Logistic Regression

Is there evidence that congressional representatives' votes are influenced by campaign contributions? Does the data suggest that this occurs more with one party than another? Once you take into account a student's GPA, does the MCAT really matter for admission to medical school? Or can you ace the MCAT and not worry about your GPA's influence on admission? Like the previous chapter, both of these scenarios involve a binary response: favorable vote or not, admitted to med school or not. Unlike the previous chapter, these examples concern more than one explanatory variable: size of contribution and party, GPA and MCAT score.

In Chapter 1 we considered the simple linear regression model for a quantitative response based on a single quantitative predictor and in Chapter 3 we extended this model to include multiple predictors. In Chapter 5 we considered ANOVA models with a single categorical explanatory factor and then allowed more than one factor in Chapter 6. Having introduced logistic regression for a binary response with a single predictor in Chapter 9, we are now in a position to extend that model to allow for multiple predictors. That is the goal of this chapter. Although the specific techniques differ from the ordinary multiple regression setting, the issues that we need to address in order to choose, fit, assess, and use a binary logistic model with multiple predictors should seem familiar.

10.1 Multiple Logistic Regression Model

The logit form of the logistic regression model allows for an easy generalization to handle situations in which we are interested in the relationship between a binary response variable and several predictors or explanatory factors. Suppose π is the probability of success for some combination of k predictors $X_1, X_2, ..., X_k$. We extend the single predictor case to a model for the *log odds* of "success" as a linear function of those predictors:

$$log\left(\frac{\pi}{1-\pi}\right) = \beta_0 + \beta_1 X_1 + \cdots + \beta_k X_k$$

This gives an equivalent model for the probability of success:

$$\pi = \frac{e^{\beta_0+\beta_1 X_1+\cdots+\beta_k X_k}}{1+e^{\beta_0+\beta_1 X_1+\cdots+\beta_k X_k}}$$

As we saw in the multiple regression models of Chapter 3, the predictors can be in various forms including

- individual quantitative variables such as *GPA* or *Weight*,
- functions (such as squares or log transformations) of quantitative variables,
- categorical factors coded as indicator variables such as political party or species, and
- products of variables to investigate interaction.

As with ordinary multiple regression we need to decide which combination of predictors is most appropriate—how to model proportions for the binary response. We use statistical software to fit the model. The software is again using maximum likelihood to estimate coefficients, rather than the least-squares approach of ordinary regression. Once we have fit the model, we assess the contributions of individual terms in the model as well as judge the overall effectiveness of the model and check that the conditions for the model have been satisfied. Once we have determined a reasonably good model, we can use it to address questions of interest.

Example 10.1: *Campaign contributions and senate votes by party*

The *Corporate Average Fuel Economy* (CAFE) bill was proposed by Senators John McCain and John Kerry to improve the fuel economy of cars and light trucks sold in the United States. However a critical vote on an amendment in March of 2002 threatened to indefinitely postpone CAFE. The amendment charged the National Highway Traffic Safety Administration to develop a new standard, the effect being to put on indefinite hold the McCain-Kerry bill. It passed by a vote of 62-38.
A political question of interest is whether there is evidence of monetary influence on a senator's vote. Scott Preston, a professor of statistics at SUNY, Oswego, collected data on this vote which includes our response variable, the vote of each senator (Yes or No), and as an explanatory variable, monetary contributions that each of the 100 senators received over his or her lifetime from the car manufacturers. The data are in the **CAFE** data set. Figure 10.1 suggests that such influence is present: The higher the contribution, the more likely a "yes" vote. Anyone with an interest in U.S. politics might naturally be led to ask: What is the effect of party affiliation on the vote in the CAFE amendment context? How different are the votes of Republican and Democratic senators and to what extent is the effect of contribution dependent on party affiliation? We will now explore those questions.

10.1. MULTIPLE LOGISTIC REGRESSION MODEL

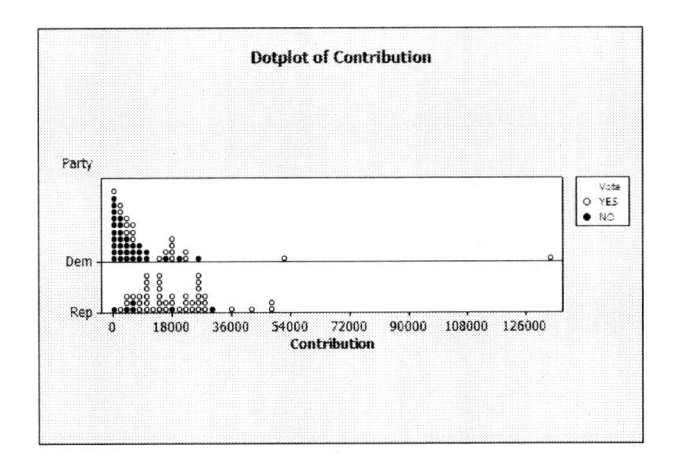

Figure 10.1: Contributions by Party

CHOOSE

Table 10.1 and Figure 10.1 show contribution comparisons (broken down by *Vote* and *Party*) and comparative dotplots. We observe that:

- For either party, "yes" voters tend to have been given higher contributions than "no" voters.

- The mean contribution difference between "yes" and "no" voters is somewhat larger for the Democrats than for the Republicans.

- Republicans are more likely to vote "yes" than Democrats.

- Republicans tend to get larger contributions (acknowledging a couple of high outliers among the Democrats).

The distribution of contributions is heavily skewed, so it is convenient to use the *base-10 log* of the contributions, *LogContr*. To indicate the party affiliation we use an indicator variable (*Dem*) which is 1 for Democrats and 0 otherwise. We can use a two-predictor logistic regression model to investigate these relationships.

$$log\left(\frac{\pi}{1-\pi}\right) = \beta_0 + \beta_1 LogContr + \beta_2 Dem$$

	Democrat	Republican	All
NO	4.7	10.5	5.6
YES	19.0	18.8	18.8
All	10.0	17.8	13.8

Table 10.1: Mean Contributions by Party (in $1,000s)

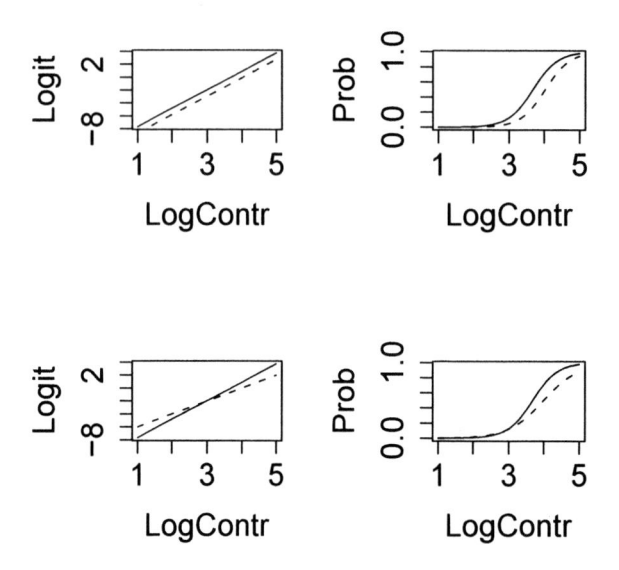

Figure 10.2: Left-side graphs: logit scale; Right-side graphs: probability scale—Top row: Only intercepts differ; Bottom row: Both intercept and slope differ

FIT

As we encountered in the multiple regression setting, with one continuous and one categorical predictor, we can illustrate the models we are considering. In particular, Figure 10.2 shows the two forms of the model, the logit form in the left column and the probability form in the right. As expected, the logit form is linear and the probability form consist of curves which begin near 0 and remain below 1 for increasingly large contributions. It is more difficult to discern the effect of equal or unequal slopes with the probability curves.

The procedure for fitting the model is similar to the single predictor case, in that the coefficients are chosen to maximize the likelihood of the observed sample. Minitab output follows.

```
Logistic Regression Table
                                              Odds      95% CI
Predictor        Coef    SE Coef      Z      P  Ratio  Lower  Upper
Constant     -6.84015    2.49032  -2.75  0.006
LogContr      2.16589   0.613083   3.53  0.000   8.72   2.62  29.01
Dem          -1.73280   0.580447  -2.99  0.003   0.18   0.06   0.55
Log-Likelihood = -43.668
Test that all slopes are zero: G = 45.477, DF = 2, P-Value = 0.000
```

10.1. MULTIPLE LOGISTIC REGRESSION MODEL

The coefficient estimates in the Minitab output produce a prediction equation for the logit form of the model.

$$\log\left(\frac{\hat{\pi}}{1-\hat{\pi}}\right) = -6.84 + 2.166 LogContr - 1.733 Dem \tag{10.1}$$

or in the probability form

$$\hat{\pi} = \frac{e^{-6.84+2.166 LogContr-1.733 Dem}}{1+e^{-6.84+2.166 LogContr-1.733 Dem}}$$

For example, a Democratic Senator with \$50,000 in lifetime contributions from the car industry ($LogContr = 4.699$)—remember, we are using base-10 logs—would be predicted to have log odds for voting "yes" on the CAFE amendment of

$$\log\left(\frac{\hat{\pi}}{1-\hat{\pi}}\right) = -6.84 + 2.166*4.699 - 1.733*1 = 1.605$$

which gives a probability of

$$\hat{\pi} = \frac{e^{1.605}}{1+e^{1.605}} = \frac{4.978}{5.978} = 0.83$$

A Republican Senator with the same contributions would have

$$\log\left(\frac{\hat{\pi}}{1-\hat{\pi}}\right) = -6.84+2.166*4.699-1.733*0 = 3.338 \quad \text{and} \quad \hat{\pi} = \frac{e^{3.338}}{1+e^{3.338}} = \frac{28.162}{29.162} = 0.97.$$

As in the case of a single predictor, coefficient estimates also translate to odds ratios. For example, exponentiating the logit form of the fitted model yields

$$\left(\frac{\hat{\pi}}{1-\hat{\pi}}\right) = e^{6.84+2.166 LogContr-1.733 Dem}$$

We can see that if two senators are in the same party, each additional "log dollar" contributed changes the odds of voting in favor of the bill by a factor of $e^{2.166}$. Equivalently a tenfold increase in the dollar contribution is associated with the odds of a favorable vote increasing by a factor of $e^{2.166}$ or 8.7, given the senators are in the same party. Note that, just as in ordinary multiple regression, the coefficient interpretations in models with more than a single explanatory variable are conditional on all other variable being equal.

ASSESS

The Minitab output provides tests for each of the coefficients in the model. These follow the same format that we saw in the simple logistic model and should be interpreted in a manner similar to the individual t-tests for predictors in a multiple regression model. In this example the tests suggest that the effects for both contribution (*LogContr*, P-value=0.000) and party (*Dem*, p-value=0.003) are highly significant. Democrats are less likely to vote "yes," as evidenced by the negative coefficient, -1.73280 which translates to an odds ratio of 0.18. Higher contributions (log base 10 in this case) correspond to higher probabilities of a "yes" vote, as evidenced by a positive coefficient of 2.16589 and an odds ratio of 8.72. As in ordinary multiple regression, the test for an individual coefficient assesses the contribution of that term to the model after all of the other terms are already in the model. We must take care when interpreting these results, especially when the predictors are related to each other. Concerns raised by issues such as multicollinearity are as valid in the multiple logistic regression setting as in the ordinary multiple linear regression setting.

In ordinary multiple regression we use the results of the ANOVA table and F-test to assess the overall effectiveness of the model as a whole. In logistic regression the analogous procedure uses the $-2\ln(Likelihood)$ value and a χ^2-distribution. In the current example, both predictors can be tested simultaneously where the hypotheses are

$H_0 : \beta_1 = \beta_2 = 0$

$H_a : \beta_1 \neq 0$ or $\beta_2 \neq 0$

The procedure is the same as for simple logistic regression. We determine whether it is worthwhile adding the two predictors of H_a as opposed to the null model (H_0)using the likelihood of our data as the criterion. The likelihood of our data for the larger model (H_a) uses the probabilities generated from this logistic fit. The likelihood for the null or constant model uses the same probability of success regardless of contributions or party (in this case the overall proportion of "yes" votes, $\hat{\pi}_0 = 62/100$, or "no", $1 - \hat{\pi}_0 = 38/100$) for finding the likelihood of every case, depending on the vote of that senator. Twice the difference in the log-likelihoods for the two models can be compared to a χ^2 distribution, thereby providing a way in which to assess the models. This G statistic helps to determine whether the likelihood for the larger model is significantly greater than the likelihood of our data with the smaller model. The degrees of freedom are the number of predictors in the multiple model (in this case df=2), and we use the upper tail of a chi-square distribution to find the p-value. For this two-predictor logistic model, the p-value is given as 0.000 so we have strong evidence that this model is useful for predicting votes on the CAFE amendment.

Table 10.1 suggests that the contribution effect might be larger for Democrats than Republicans. Thus we might want to add an interaction term to the model.

$$log\left(\frac{\pi}{1-\pi}\right) = \beta_0 + \beta_1 LogContr + \beta_2 Dem + \beta_3 Dem * LogContr$$

Here is some output for fitting this model.

10.1. MULTIPLE LOGISTIC REGRESSION MODEL

```
                                          Odds       95% CI
Predictor      Coef      SE Coef    Z       P    Ratio   Lower      Upper
Constant     -10.1636   5.40134   -1.88   0.060
LogContr       3.00151  1.35714    2.21   0.027  20.12    1.41     287.58
Dem            2.54364  5.97423    0.43   0.670  12.73    0.00  1549007.17
Dem*LogContr  -1.08829  1.51459   -0.72   0.472   0.34    0.02       6.56

Log-Likelihood = -43.391
Test that all slopes are zero: G = 46.031, DF = 3, p-value = 0.000
```

Despite the suggestion of an interaction from our descriptive analysis, the interaction term is not significant (p-value = 0.472). Note also that the extra predictor produces very little improvement in the log-likelihood of the sample (-43.391 compared to -43.668) and gives a slightly higher G statistic but "costs" one extra degree of freedom to achieve that improvement. Not surprisingly, the new interaction term is strongly related to both the original *LogContr* and *Dem* variables. This observation suggests we not put too much emphasis on the main effects of *Dem* and *logC* in this setting. Note that the p-value for *Dem* is also insignificant in this three-term model and the confidence interval for the odds ratio for *Dem* is remarkably noninformative! Given these results, we would choose to fit the simpler, more easily interpreted model with just the linear terms for *Party* and log of contribution. A confidence interval for a coefficient in the model can be constructed as before; likewise, a confidence interval for the odds ratio for a one-unit change in a predictor can be found in the same way as in Chapter 9.

Let us now summarize what we have learned about the workings of the United States Senate. Our current preferred model is additive—since the model contains no interaction term. The model tells us that contributions correlate positively with votes and that a tenfold increase in dollar contribution corresponds to an 8.7 increase in the odds of a "yes" vote. This answers Scott Preston's primary question and provides a comforting data-based confirmation of what many of us might have assumed to begin with. We are able to conclude that this result is about the same for Democrats and Republicans, and that we should attribute the Republicans' greater likelihood of a "yes" vote to their higher contribution amounts.

Finally, we take care to treat these conclusions as descriptive of patterns valid only for the 100 Senators making up our data set. Since they are not sampled from a larger population, we cannot infer their patterns to a larger population. Since campaign contributions are not randomly allocated, as they might be in a real experiment, we cannot infer causality to the observed patterns. We might, however, keep our patterns in mind as we or others study the effects of campaign contributions on other legislative bodies or on other issues. ◇

> **Tests for individual predictors and overall fit in multiple logistic regression**
>
> For a multiple binary logistic regression model with k predictors
>
> $$\log\left(\frac{\pi}{1-\pi}\right) = \beta_0 + \beta_1 X_1 + \beta_2 X_2 + \ldots + \beta_k X_k$$
>
> A test for the contribution of the predictor X_i, given the other predictors in the model, $H_0: \beta_i = 0$ vs. $H_a: \beta_i \neq 0$, uses test statistic
>
> $$z = \frac{\hat{\beta}_i}{SE_{\beta_i}}$$
>
> with p-value from a standard normal distribution.
>
> A test for the overall effectiveness of the model
> $H_0: \beta_1 = \beta_2 = \cdots = \beta_k = 0$ vs.
> $H_a:$ At least one $\beta_i \neq 0$
> uses test statistic
> $$G = -2\ln(L_0) - (-2\ln(L))$$
>
> where L_0 is the likelihood of the sample under a constant model and L is the likelihood under the larger model. The p-value comes from the upper tail of a χ^2-distribution with k degrees of freedom.
>
> A confidence interval for an individual coefficient is found by
>
> $$\hat{\beta}_i \pm z^* SE_{\beta_i}$$
>
> The estimated odds ratio for a one unit change in the predictor X_i is
>
> $e^{\hat{\beta}_i}$ with confidence interval $(e^{\hat{\beta}_i - z^* SE_{\beta_i}}, e^{\hat{\beta}_i + z^* SE_{\beta_i}})$.

Example 10.2: *Medical school acceptance with GPA and MCAT*

At the outset of Chapter 9 we posed the question of which strategy was most effective for medical school applicants: working to improve GPA or studying hard to do well on the MCAT? Using our simple model with the single explanatory variable GPA, we found (Example 9.7) that, indeed, increasing GPA is positively associated with probability of acceptance. In particular, with each one-tenth increase in GPA, the odds of acceptance increased by a factor of 1.73 (95% CI: 1.27, 2.35).

10.1. MULTIPLE LOGISTIC REGRESSION MODEL

We could pose a similar question for the MCAT. The Medical College Admission Test (MCAT) is a standardized, multiple-choice examination designed to assess the examinee's problem-solving ability, critical thinking ability, writing skills, and knowledge of science concepts and principles prerequisite to the study of medicine. Almost all U.S. medical schools require applicants to submit MCAT scores as part of the application process. The test has four sections: Physical Sciences, Verbal Reasoning, Writing Sample, and Biological Sciences, each with possible score scales ranging from 1 (low) to 15 (high).

Here is the output for fitting a simple logistic regression model to predict medical school acceptance based on $MCAT$ scores.

```
                                             Odds      95% CI
Predictor     Coef     SE Coef     Z      P  Ratio  Lower  Upper
Constant   -8.71245    3.23653  -2.69  0.007
MCAT        0.245964   0.0893806  2.75  0.006  1.28   1.07   1.52

Log-Likelihood = -32.349
Test that all slopes are zero: G = 11.094, DF = 1, p-value = 0.001
```

We have convincing evidence (p-value=0.006) that odds of acceptance to medical school will increase by an estimated factor of 1.28 for each single point increase on the MCAT (95% CI:1.07, 1.52). However the model based on $GPA10$ (GPA in tenths) was somewhat better (Log-likelihood= -28.42, $G = 18.95$ with 1 df).

Perhaps we can get an even better model for medical school acceptance if we use both $GPA10$ and $MCAT$ in the same model.

$$\log\left(\frac{\pi}{1-\pi}\right) = \beta_0 + \beta_1 MCAT + \beta_2 GPA10$$

Here is the Minitab output for this two-predictor model.

```
                                             Odds      95% CI
Predictor     Coef     SE Coef     Z      P  Ratio  Lower  Upper
Constant  -22.3727    6.45360  -3.47  0.001
MCAT        0.164501  0.103154   1.59  0.111  1.18   0.96   1.44
GPA10       0.467646  0.164155   2.85  0.004  1.60   1.16   2.20

Log-Likelihood = -27.007
Test that all slopes are zero: G = 21.777, DF = 2, p-value = 0.000
```

As in previous modeling, we interpret each coefficient assuming that the others are held constant. Thus a one-point increase on the MCAT does not have a statistically significant benefit for those with the same GPA (p-value=0.111). However for every one tenth increase in GPA the odds of acceptance to medical school will increase by an estimated factor of 1.60 for students with comparable MCAT scores. Note that $G = 21.8$ is now compared to a χ^2 with 2 df (two explanatory variables) and with p-value< .001 the model with $GPA10$ and $MCAT$ is significantly better than a model with neither. However, it is probably not much of an improvement over the simpler model using just $GPA10$.

Finally, there are a few caveats. First, the correlation between $MCAT$ and $GPA10$ is 0.541 (p-value< .001). It is a little tricky talking about increasing GPA without increasing MCAT and visa versa. Also, we should expect some multicollinearity issues when both predictors are in a model together (for example, $MCAT$ is significant by itself, but not in the presence of $GPA10$). Third, these are observational data that have not been collected via a random sample and, in fact, all come from students who attend a particular liberal arts college. The results would certainly be more generalizable if a random sample of medical school applicants was selected from a variety of schools.

⋄

A Nested LRT-test

Another technique we used when evaluating predictors in an ordinary multiple regression model was a nested F-test that allowed us to test subsets of predictors as a group. The basic idea of the test was to compare the *full* model with all predictors to a *reduced* model that omitted the predictors in the subset being tested. We were interested in whether the improvement gained through those extra predictors was more than we would expect to see by random chance alone. In ordinary regression we measured "improvement" by the change in the sum of squared errors (or equivalently the increase in the sum of squares explained by the model). In logistic regression we apply the same principle and measure improvement using $-2\ln(L)$.

Example 10.1 *Campaign contributions and Senate votes by party*
Although the interaction between log contributions and party was not particularly helpful to the model, we might wonder if the relationship between $LogContr$ and the log odds of a yes vote was actually quadratic rather than linear; this kind of a lack of fit is harder to detect with diagnostic plots in the logistic setting than it is in the OLS—ordinary least squares—setting. We might even wish to allow for different quadratic fits for Democrats than Republicans. To see if such a more complicated model is needed, we fit a full model that includes such terms and compare it to a model with those terms omitted.

Specifically, we fit the following (FULL) model:

$logit(\pi) = \beta_0 + \beta_1 Dem + \beta_2 LogContr + \beta_3 Dem * LogContr + \beta_4 LogContr^2 + \beta_5 Dem * LogContr^2$

10.1. MULTIPLE LOGISTIC REGRESSION MODEL

For Republicans ($Dem = 0$) the quadratic model is $logit(\pi) = \beta_0 + \beta_2 LogContr + \beta_4 LogContr^2$ and for Democrats we add β_1, β_3, and β_5 to the intercept, linear, and quadratic coefficients, respectively. Note that we don't add a Dem^2 term for the simple mathematical reason that $Dem^2 = Dem$ since $0^2=0$ and $1^2=1$.

To see if these extra terms are really helpful (and thus the simpler model including just $LogContr$ and Dem was inadequate) we test

$H_0 : \beta_3 = \beta_4 = \beta_5 = 0$ vs.
$H_a :$ Either $\beta_3 \neq 0$ or $\beta_4 \neq 0$ or $\beta_5 \neq 0$

by comparing the output for the full model above to the reduced model

$$logit(\pi) = \beta_0 + \beta_1 Dem + \beta_2 LogContr$$

Here are the results from Minitab:

```
FULL MODEL:
Predictor            Coef      SE Coef       Z       P     Odds Ratio
Constant          -37.0597     48.5112    -0.76    0.445
Dem                41.5836     50.8099     0.82    0.413    1.14692E+18
LogContr           16.9318     24.5197     0.69    0.490    22563623.84
Dem*LogContr      -21.7753     25.9324    -0.84    0.401           0.00
LogContrSq         -1.78859     3.08742   -0.58    0.562           0.17
DemxLogContrSq      2.71144     3.29990    0.82    0.411          15.05

Log-Likelihood = -42.847
Test that all slopes are zero: G = 47.118, DF = 5, p-value = 0.000

Goodness-of-Fit Tests
Method           Chi-Square   DF      P
Pearson            83.8195    75    0.227
Deviance           73.6960    75    0.521
Hosmer-Lemeshow     7.0752     8    0.529\footnote{David W. Hosmer}
------------------------------------------------------------------
REDUCED MODEL:
Predictor            Coef      SE Coef       Z       P    Ratio   Lower   Upper
Constant          -6.84015     2.49032    -2.75    0.006
Dem               -1.73280     0.580447   -2.99    0.003   0.18    0.06    0.55
LogContr           2.16589     0.613083    3.53    0.000   8.72    2.62   29.01

Log-Likelihood = -43.668
```

```
Test that all slopes are zero: G = 45.477, DF = 2, p-value = 0.000

Goodness-of-Fit Tests
Method           Chi-Square   DF     P
Pearson            81.0142    78   0.385
Deviance           75.3376    78   0.564
Hosmer-Lemeshow     9.9199     8   0.271
```

Immediately we notice that the Odds Ratios for the Dem and LogContr terms are extraordinarily large. When we see something like this in logistic regression, we are concerned with the computational integrity fitting this model. It is likely a signal of serious problems with the fitting algorithms, and one must be cautious about using these results. The improvement in going to the more complicated model is measured as the decrease in $-2\ln(L)$. Using the Log-likelihood values from the Minitab output for both models, we have

$$X^2 = -2\ln(L_{Reduced}) - (-2\ln(L_{Full})) = -2(-43.668) - (-2(-42.847)) = 87.336 - 85.694 = 1.642$$

We find the p-value for this test statistic using a χ^2-distribution with 3 degrees of freedom, since that is the number of extra predictors (coefficients) from the full model that we are testing. Thus the p-value is $P(\chi^2_3 \geq 1.642) = 0.65$. That, along with our concerns about the large odds ratio estimates, indicates that we lack evidence of any substantial improvement due to the three extra terms in the model. We conclude there is no gain in fitting a model beyond the simple additive and linear model based on *Dem* and *LogContr*.

$\diamond$

In practice there are several alternative methods for computing the improvement in $-2\ln(L)$ between a full and nested model.

(1) Note that the G statistic is $G = -2\ln(L_0) - (-2\ln(L_{Model}))$ so, since L_0 is the same for both the full and reduced model,

$$X^2 = G_{Full} - G_{Reduced} \text{ is the same as } X^2 = -2\ln(L_{Reduced}) - (-2\ln(L_{Full})).$$

Also, we can get the degrees of freedom for the nested test as the difference in the degrees of freedom for the G statistics.

From the output above we have $G_{Full} - G_{Reduced} = 47.118 - 45.477 = 1.641$ with $5 - 2 = 3$ d.f., matching (up to roundoff) the calculation for the nested test.

(2) Note that Minitab's output reports a "Deviance" Chi-square value for both the full and reduced models. The *deviance* of a model is a difference of the $-2\ln(L)$ quantity between the given model and a saturated model for the data set that fits a separate proportion for each value of the predictor. Each deviance has an associated degrees-of-freedom parameter. Because the deviance for

10.1. MULTIPLE LOGISTIC REGRESSION MODEL

the two models compares their $-2\ln(L)$ values to the same base value, a subtraction of deviances will result in the same value as the test statistic from (1). In this case, the computation gives $75.3376 - 73.6960 = 1.6416$ and $78 - 75 = 3$ d.f. For this reason the nested G test is also known as a "drop in deviance" test.

(3) Finally, if we consider the summary output provided by R for these same models (see below), the "Residual Deviance" is the value of $-2\ln(L)$ for the model so again we can subtract to find the improvement and degrees of freedom for the test.

FULL MODEL:

Deviance Residuals:
```
    Min      1Q  Median      3Q     Max
-2.4276 -0.6307  0.3401  0.4956  1.9876
```

Coefficients:
```
                  Estimate Std. Error z value Pr(>|z|)
(Intercept)        -37.060     48.511  -0.764    0.445
LogContr            16.932     24.520   0.691    0.490
Dem                 41.584     50.810   0.818    0.413
I(LogContr^2)       -1.789      3.087  -0.579    0.562
I(Dem * LogContr^2)  2.711      3.300   0.822    0.411
LogContr:Dem       -21.775     25.932  -0.840    0.401
```

(Dispersion parameter for binomial family taken to be 1)

```
    Null deviance: 132.813  on 99  degrees of freedom
Residual deviance:  85.694  on 94  degrees of freedom
AIC: 97.694
```

Number of Fisher Scoring iterations: 5

REDUCED MODEL:

Deviance Residuals:
```
    Min      1Q  Median      3Q     Max
-2.4117 -0.6570  0.3600  0.5622  2.3625
```

```
Coefficients:
            Estimate Std. Error z value Pr(>|z|)
(Intercept)  -6.8402     2.4903  -2.747 0.006020 **
LogContr      2.1659     0.6131   3.533 0.000411 ***
Dem          -1.7328     0.5804  -2.985 0.002833 **
---
Signif. codes:  0 *** 0.001 ** 0.01 * 0.05 . 0.1   1

(Dispersion parameter for binomial family taken to be 1)

    Null deviance: 132.813  on 99  degrees of freedom
Residual deviance:  87.336  on 97  degrees of freedom
AIC: 93.336

Number of Fisher Scoring iterations: 5
```

10.2 Case Study: Bird Nests

What factors influence nest type between species of birds?

Birds construct nests to protect their eggs and young from predators and from harsh environmental conditions. The variety of nest types varies widely. Many are familiar with the open, saucer-shaped nest of the American robin and the closed, pendant-shaped nest of the Baltimore oriole. A black-capped chickadee builds its nest in a tree hole or crevice, and other birds have other types. *The Birders Handbook*, by Ehrlich, et al. (1988), a reference book for birdwatchers, students, and ornithologists, identifies 7 categories of bird nest for birds belonging to the order Passeriformes: **saucer, cup, pendant, spherical, cavity, crevice,** and **burrow**. Passeriformes (also known as passerines) represent the order of so-called "perching birds." Included in the order are families of small birds such as warblers, sparrows, and finches, larger birds such as jays, blackbirds, and thrushes, and a few large species such as crows and (the largest North American passerines) ravens. Hawks, gulls, ducks, wading birds, owls, and so on, are not passerines. Passeriformes is the order of birds with the largest number of species.

Amy R. Moore, as a student at Grinnell College in 1999, wanted to study the relationship between species characteristics and the type of nest a bird builds, using data collected from available sources. For the study, she collected data by species for 84 separate species of North American passerines. While there are approximately 470 total species of North American passerines, she limited her study to the 84 species for which knowledge of the evolutionary relationships between species was known. Her analysis used this evolutionary relationship information in ways that we will not explore here, but her study also included the kind of analysis we present below, which is related to logistic regression. There was no apparent reason to suspect that the 84 species chosen for analysis represent a biased sample from the population of 470.

Moore's goal was not prediction, since we know through years of observation of all passerine species what type of nest each species builds. Rather, the goal of her study was to better understand the evolutionary process that led to birds building the nests they do.

We quote here from Moore's final paper:

> Ehrlich et al. (1988) described one hypothesis, which I will refer to as the body size hypothesis, which explains why some birds use open nests while others use closed nests. The body size hypothesis states that larger birds are more likely to have closed nests because they are able to out compete smaller birds for preferential nesting sites in cavities and crevices. The body size hypothesis assumes that all birds prefer closed nests because they provide better protection from predators than do open nests. Collias and Collins (1984) found that larger birds were more likely to have closed nests than were smaller birds, which seemed to support the body size hypothesis. I propose an alternative hypothesis, the nest use hypothesis, which asserts that the type of nest a

bird constructs is related to the amount of time the bird will spend using the nest. Specifically, the nest use hypothesis predicts that birds that construct closed nests will tend to use their nests for a longer period of time than birds that construct open nests. Like the body size hypothesis, the nest use hypothesis is based on the assumption that closed nests offer better protection from predators than do open nests.

The file **BirdNest** contains the data from Moore's study. For this discussion we focus on several of these variables:

Common   common name for the bird
Length   mean body length for a species (in cm.)
Incubate mean length of time the species incubates an egg in the nest (in days)
Nestling mean length of time the species cares for babies in the nest until fledged (in days)
TotCare mean total care time = $Incubate + Nestling$ (in days)
Closed   1=closed nest (pendant, spherical, cavity, crevice, burrow), 0=open nest (saucer, cup)

The response variable of interest here is *Closed*. We want to understand which variables help explain whether a species builds a closed nest. The research question revolves around whether size (*Length*) or time spent in the nest (*TotCare*) has a stronger association with the type of nest a species builds (*Closed*). For example, the first case in the data set is the Eastern Kingbird, a fly-catching species that has a mean adult length of 20.0 cm, incubates an egg on average 17 days, and takes an average of another 17 days to fledge its young, for a *TotCare* value of 34 days. The Eastern Kingbird builds an open cup-shaped nest (so *Closed* = 0).

CHOOSE

We plan to use logistic regression to model *Closed* based on *Length* and *TotCare*, but first we do some exploratory analysis on the data using boxplots. The boxplots on the left in Figure 10.3 give an initial surprise: Contrary to the body size hypothesis, closed nests tend to be associated with smaller species. On the other hand, the right side of Figure 10.3 supports the nest use hypothesis in that closed nests tend to be for birds with longer total care periods.

FIT

Next we fit a logistic regression model using *Closed* as the response variable and using two quantitative explanatory variables, *Length* and *TotCare*. Some output from Minitab follows:

```
Response Information
Variable  Value  Count
Closed      1       26   (Event)
            0       57
```

10.2. CASE STUDY: BIRD NESTS

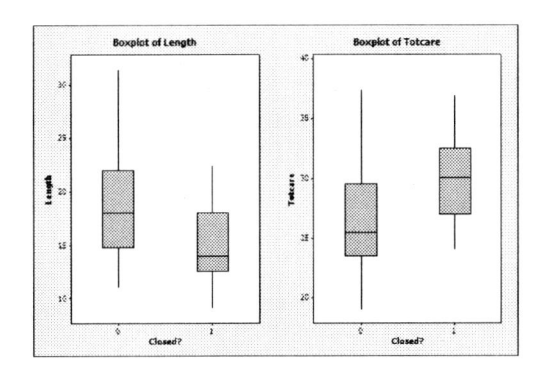

Figure 10.3: Bird length and total care time versus Closed or Open nests

```
           Total     83

* NOTE * 83 cases were used
* NOTE * 1 cases contained missing values

Logistic Regression Table
                                              Odds      95% CI
Predictor      Coef      SE Coef      Z       P    Ratio  Lower  Upper
Constant    -4.35134    1.93046    -2.25   0.024
Length      -0.303631   0.0875397  -3.47   0.001   0.74   0.62   0.88
TotCare      0.306070   0.0821818   3.72   0.000   1.36   1.16   1.60

Log-Likelihood = -36.571
Test that all slopes are zero: G = 30.056, DF = 2, p-value = 0.000

Goodness-of-Fit Tests
Method           Chi-Square   DF      P
Pearson           64.5690     76    0.822
Deviance          73.1427     76    0.572
Hosmer-Lemeshow   12.9317      8    0.114
```

ASSESS

Observe that both predictors are statistically significant as predictors for *Closed*. The p-values are 0.001 for *Length* and 0.000 for *TotCare*. The coefficient for *Length* is negative, confirming the counter-intuitive signal we saw in the boxplot above: longer species are less likely to build a closed nest. Controlling for total care, each increase of 1 cm in length decreases the odds of a closed nest by a factor of 0.74 ($e^{-0.304} = 0.738$). The coefficient for *TotCare* is positive, again affirming the message in the boxplot: Species that use the nest longer are more apt to build a closed nest.

The odds for a closed nest increase by a factor of 1.36 ($e^{0.306} = 1.36$). Both relationships are statistically significant. Also, the p-value for the G-test for overall fit (0.000) indicates that these two predictors provide an effective model for predicting open or closed nest types.

As a next step, we explore whether second-order terms improve the model. We fit a model adding an interaction term ($Length * Totcare$) and both quadratic terms ($Length^2$ and $TotCare^2$). We use a nested likelihood ratio (drop-in-deviance) test to see if this model is significantly better than the additive model above that includes just the linear $Length$ and $TotCare$ terms. The Minitab output follows:

```
Logistic Regression Table
                                                  Odds      95% CI
Predictor         Coef       SE Coef     Z      P  Ratio  Lower  Upper
Constant        -29.8414    15.8429    -1.88  0.060
Length           -0.0240951  0.693738  -0.03  0.972  0.98   0.25    3.80
TotCare           1.89831    1.03453    1.83  0.067  6.67   0.88   50.70
LengthxTotCare  -0.0027062  0.0238022  -0.11  0.909  1.00   0.95    1.04
LengthSq        -0.0044720  0.0149298  -0.30  0.765  1.00   0.97    1.03
TotCareSq       -0.0264259  0.0190296  -1.39  0.165  0.97   0.94    1.01

Log-Likelihood = -34.962
Test that all slopes are zero: G = 33.276, DF = 5, p-value = 0.000

Goodness-of-Fit Tests
Method            Chi-Square  DF     P
Pearson            58.3636    73   0.894
Deviance           69.9230    73   0.580
Hosmer-Lemeshow     8.1520     8   0.419
```

Interpretation: First we note that none of the added terms of second order are statistically significant, but what we are really interested most in testing is whether the quadratic model adds to the simpler additive model. We expect the second order terms to be strongly related to the linear terms so multicollinearity issues might be affecting the tests for individual coefficients. The drop-in-deviance test proceeds using the deviance measures from each model above and the corresponding degrees of freedom in a chi-square test. Here are the details.

The restricted model is the additive model with linear terms only. For this fitted model:

Deviance = 73.1427, with df = 76.

The full model is the model with interaction and quadratic terms. For this fitted model:

Deviance = 69.9230, with df = 73.

Therefore the "drop in deviance" is 73.1427 - 69.9230 = 3.2197. We compare this value to a chi-square distribution with 76 - 73 = 3 degrees of freedom. The probability that one would obtain a drop this great or greater by chance alone is obtained as $P(\chi^2_3 > 3.2197) = 1 - .641024 = .36$. Note that we could also have obtained this test statistic by comparing the values for the G-statistic of each model $33.276 - 30.056 = 3.22$ with $5 - 2 = 3$ df.

The p-value of 0.36 is well above any typical standard of statistical significance; we fail to reject the null hypothesis and we thus conclude that the simple additive model, with only linear terms in *Length* and *TotCare* is adequate.

Our analysis supports the nest-use hypothesis that species that need the nest for longer periods—to hatch and fledge offspring—tend to build nests in crevices or holes (closed nests). The data run counter to the body-size hypothesis: instead the conclusion appears to be that the smaller species are more likely to build closed nests than are larger species.

10.3 Summary

Our look at logistic regression reveals that a **binary response**, like a continuous normally distributed response, can be modeled as a function of one or more explanatory variables. Unlike a continuous normally distributed response, we model the logit—log(odds)—of a binary response as a linear function of explanatory variables. This approach provides consistency with the modeling approach that has been emphasized throughout this book. Where appropriate, logistic regression allows for inferences to be drawn: A *p*-value assesses whether the association between the binary response and explanatory variables is significant, and a confidence interval estimates the magnitude of the population odds ratio. Logistic regression can accommodate multiple explanatory variables including both categorical and continuous explanatory variables. When we have a categorical predictor with multiple categories, we can use several indicator variables to include information from that predictor in a logistic regression model. Logistic regression extends the set of circumstances where you can use a model and does so within a familiar modeling framework.

10.4 Exercises

Conceptual Exercises

10.1 *Risky youth behavior (continued).* In Exercise 9.23 we learned that Rybabc et al. (2007) have reported that smokers are more likely to ride with drivers who have been drinking (RDD):

> Injury-prone behaviors were more common among smokers than non-smokers: riding with drunk driver (38 percent versus 13 percent). In multiple logistic regression models adjusting for demographics, SES (socio-economic status), and substance abuse, smoking revealed significantly higher odds ratios (OR) for riding with drunk driver (OR = 2.2).

a. What is the unadjusted odds ratio for riding with a drinking driver for smokers versus non-smokers?

b. Explain why this odds ratio differs from the given odds ratio of 2.2.

10.2 *Sinking of the Titanic (continued).* In Exercise 9.14 we considered data on the passengers who survived and those who died when the ocean liner *Titanic* sank on its maiden voyage in 1912. The data in **Titanic** includes the following variables.

Age which gives the passenger's age in years
Sex which gives the passenger's sex (male or female)
Survived a binary variable where 1 indicates the passenger survived and 0 indicates death
SexCode which numerically codes male as 0 and female as 1

a. In Exercise 9.14 you fit separate logistic regression models for the binary response *Survived* using *Age* and then *SexCode*. Now fit a multiple logistic model using these two predictors. Write down both the logit and probability forms for the fitted model.

b. Comment on the effectiveness of each of the predictors in the two-predictor model.

c. According to the fitted model, estimate the probability and odds that an 18-year-old man would survive the *Titanic* sinking.

d. Repeat the calculations for an 18-year-old woman and find the odds ratio compared to a man of the same age.

e. Redo both (b) and (c) for a man and woman of age 50.

f. What happens to the odds ratio (female to male of the same age) when the age increases in the *Titanic* data? Will this always be the case?

10.3 *Sinking of the Titanic (continued).* Refer to the situation described in Exercise 10.2. Perhaps the linear relationship between log(odds) of survival and *Age* is much different for women than for men. Add an interaction term to the two-predictor model based on *Age* and *SexCode*.

10.4. EXERCISES

a. Explain how the coefficients in the model with *Age*, *SexCode* and *Age* ∗ *SexCode* relate to separate linear models for males and females to predict log(odds) of survival based on *Age*.

b. Is this model a significant improvement over one that uses just the *SexCode* variable (and not *Age*)? Justify your answer by showing the details of a nested likelihood ratio test for the two terms involving *Age*.

10.4 *Red states or Blue states in 2008 (continued).* In Exercises 9.19 and 9.20 you considered some state level variables for the data in **Election08** to model the probability that the Democratic candidate, Barack Obama, won a state (response=*ObamaWin*). Among the potential predictors were the per capita *Income*, percentages of adults with high school (*HS*) or college (*BS*) degrees, and a measure (*Dem.Rep*) of the %Democrat-%Republican leaning in the state.

a. Fit a logistic model with *ObamaWin* as the response and *Dem.Rep*, *HS*, *BA*, and *Income* as the predictors. Which predictor has the strongest relationship with the response in this model?

b. Consider the model from part (a). Which predictors (if any) are not significantly related to *ObamaWin* in that model?

c. Identify the state that has a very large positive deviance (residual) and the state that has a very large negative deviance (residual). Note: Deviance residuals are available in computer output. They can be treated similarly to residuals in the regression setting; we discuss them further in the next chapter.

d. Consider applying a backward elimination process, starting with the four-predictor model in (a). At each "step" find the least significant predictor, eliminate it, and refit with the smaller model—unless the worst predictor is significant, say at a 10% level, in which case you stop and call that your final model. Describe what happens at each step in this situation.

10.5 *Nikkei 225 (continued).* In Exercise 9.18 you considered models to predict whether the Japanese Nikkei 225 stock index would go up (*Up* = 1) or down (*Up* = 0) based on the previous day's change in either the New York-based Dow Jones Industrial Average (*DJIAch*) or previous Nikkei 225 change (*lagNik*). The data are in **Markets**.

a. Run a multiple logistic regression model to use both *DJIAch* and *lagNik* to predict *Up* for the Nikkei 225 the next day. Comment on the importance of each predictor in the model and assess the overall fit.

b. Suppose we had a remarkable occurrence and *both* the Nikkei 225 and Dow Jones Industrial average were unchanged one day. What would your fitted model say about the next day's Nikkei 225?

c. Based on the actual data, should we be worried about multicollinearity between the *DJIAch* and *lagNik* predictors? Hint: Consider a correlation coefficient and a plot.

10.6 *Leukemia treatments (continued).* Refer to Exercise 9.15 which describes data in **Leukemia** that arose from a study of 51 patients treated for a form of leukemia. The first six variables in that data set all measure pre-treatment variables: *Age*, *Smear*, *Infil*, *Index*, *Blasts*, and *Temp*. Fit a multiple logistic regression model using all six variables to predict *Resp*, which is 1 if a patient responded to treatment and 0 otherwise.

   a. Based upon values from a summary of your model, which of the 6 pre-treatment variables appear to add to the predictive power of the model, given that other variables are in the model?

   b. Specifically, interpret the relationship (if any) between *Age* and *Resp* and also between *Temp* and *Resp* indicated in the multiple model.

   c. If a predictor variable is nonsignificant in the fitted model here, might it still be possible that it should be included in a final model? Explain why or why not.

   d. Despite your answer above, sometimes one gets lucky, and a final model is, simply, the model that includes all "significant" variables from the full additive model output. Use a nested likelihood ratio (drop-in-deviance) test to see if the model that excludes precisely the nonsignificant variables seen in (a) is a reasonable choice for a final model. Also, comment on the stability of the estimated coefficients between the full model from (a) and the reduced model without the "nonsignificant" terms.

   e. Are the estimated coefficients for *Age* and *Temp* consistent with those found in the Exercise 9.15 using these data? Consider both statistical significance and the value of the estimated coefficients.

10.7 *Intensive care unit.* The data in **ICU**[1] show information for a sample of 200 patients who were part of a larger study conducted in a hospital's Intensive Care Unit (ICU). Since an ICU often deals with serious, life-threatening cases, a key variable to study is patient survival, which is coded in the *Survive* variable as 1 if the patient lived to be discharged and 0 if the patient died. Among the possible predictors of this binary survival response are the following.

$$\begin{aligned}
Age &\quad = \text{age (in years)} \\
AgeGroup &\quad = \text{1 if young (under 50), 2 if middle (50-69), 3 if old (70+)} \\
Sex &\quad = \text{1 for female, 0 for male} \\
Infection &\quad = \text{1 if infection is suspected, 0 if no infection} \\
SysBP &\quad = \text{systolic blood pressure (in mm of Hg)} \\
Pulse &\quad = \text{heart rate (beats per minute)} \\
Emergency &\quad = \text{1 if emergency admission, 0 if elective}
\end{aligned}$$

Consider a multiple logistic regression model for *Survive* using the three quantitative predictors in the data set *Age*, *SysBP*, and *Pulse*.

---

[1] Source: The Data and Story Library (DASL) *http://lib.stat.cmu.edu/DASL/Datafiles/ICU.html*

10.4. EXERCISES

a. After running the three-predictor model, does it appear as though any of the three quantitative variables are not very helpful in this model to predict survival rates in the ICU? If so, drop one or more of the predictors and re-fit the model before going on to the next part.

b. The first person in the data set (ID #4) is an 87-year-old man who had a systolic blood pressure of 80 and a heart rate of 96 beats per minute when he checked into the ICU. What does your final model from part (a) say about this person's chances of surviving his visit to the ICU?

c. The patient with ID #4 survived to be discharged from the ICU ($Survive = 1$). Based on your answer to (b) would you say that this result (this patient surviving) was very surprising, mildly surprising, reasonably likely, or very likely?

10.8 *Intensive care unit (continued).* Refer to the **ICU** data on survival in a hospital ICU as described in Exercise 10.7. The data set has three binary variables that could be predictors of ICU survival: *Sex*, *Infection*, and *Emergency*.

a. First consider each of the the three binary variables individually as predictors in separate simple logistic regression models for *Survive*. Comment on the effectiveness of each of *Sex*, *Infection*, and *Emergency* as predictors of ICU survival on their own.

b. A nice feature of the multiple linear model for the log odds is that we can easily use several binary predictors in the same model. Do this to fit the three-predictor model for *Survive* using *Sex*, *Infection*, and *Emergency*. How does the effectiveness of each predictor in the multiple model compare to what you found when your considered each individually?

c. The first person in the data set (ID #4) is an 87-year-old man who had an infection when he was admitted to the ICU on an emergency basis. What does the three-predictor model in part (b) say about this person's chances of surviving his visit to the ICU?

d. The patient with ID #4 survived to be discharged from the ICU ($Survive = 1$). Based on your answer to (c) would you say that this result (this patient surviving) was very surprising, mildly surprising, reasonably likely, or very likely?

e. Does the model based on the three binary predictors in this exercise do a better job of modeling survival rates than the model based on three quantitative predictors that you used at the start of Exercise 10.7? Give a justification for your answer.

10.9 *Intensive care unit (continued).* Refer to the **ICU** data on survival in a hospital ICU as described in Exercise 10.7. In addition to the quantitative *Age* variable, the data has a categorical variable that classifies patients into three broad groups based on age. We'll call patients "Young" if they are under 50 years old ($AgeGroup = 1$), "Middle" age if they are in their 50s or 60s ($ageGroup = 2$) and "Old" if they are 70 or older ($AgeGroup = 3$).

a. Produce a two-way table showing the number of patients who lived and died in each of the three age groups. Also compute the sample proportion who survived in each age group. What do those proportions tell us about a possible relationship (if any) between *AgeGroup* and *Survive*?

b. Suppose that we ignore the fact that *AgeGroup* is really a categorical variable and just treat it as a quantitative predictor of *Survive*. Fit that logistic regression model. Compute the estimated proportion surviving in each age group from this model.

c. Use the logit form of the model in (b) to find the log(odds) for survival in each of the age groups. Since the value coding the "Middle" group (*AgeGroup* = 2) is exactly between the coding for the other two groups (*AgeGroup* = 1 and *AgeGroup* = 3), what does that imply about estimated log(odds) for the "Middle" group? Given what you observed in the two-way table in part (a), is that reasonable for the ICU data and these age groups?

d. Create indicator variables for the age groups and use any two of them in a multiple logistic regression model to predict *Survive*. Explain what the coefficients of this model indicate about how the odds of ICU survival are related to the age groups.

e. Check that the survival proportions estimated for each age group using the model in (d) match the sample proportions you found part (a).

f. Perform a chi-square test on the two-way table from part (a). Compare the results to the chi-square test for effectiveness of the logistic model you fit using the indicator variables in part (d). Are the results consistent? What do they tell you about the relationship (or lack of a relationship) between the age groups and ICU survival?

g. Try a logistic model using the actual ages in the **ICU** data set, rather than the three age groups. Does this provide a substantial improvement? Give a justification for your answer.

10.10 *Sinking of the Titanic (continued).* The **Titanic** data considered in Exercises 10.2 and 10.3 also contains a variable identifying the travel class (1st, 2nd, or 3rd) for each of the passengers.

a. Create a 2 × 3 table of *Survived* (yes or no) by the three categories in the passenger class (*PClass*). Find the proportion surviving in each class. Make a conjecture in the context of this problem about why the proportions behave the way they do.

b. Use a chi-square test for the 2 × 3 table to see whether there is a significant relationship between *PClass* and *Survive*.

c. Create indicator variables (or use *PClass* as a factor) to run a logistic regression model to predict *Survived* based on the categories in *Pclass*. Interpret each of the estimated coefficients in the model.

d. Verify that the predicted probability of survival in each passenger class based on the logistic model matches the actual proportion of passengers in that class who survived.

10.4. EXERCISES

e. Compare the test statistic for the overall test of fit for the logistic regression model to the chi-square statistic from your analysis of the two-way table. Are the results (and conclusion) similar?

10.11 *Flight response of Pacific Brant (continued).* The data in Exercise 9.17 dealt with an experiment to study the effects of nearby helicopter flights on the flock behavior of Pacific Brant. The binary response variable is *Flight*, which is coded as 1 if the more than 10% of the flock flies away and 0 if most of the flock stays as the helicopter flies by. The predictor variables are the *Altitude* of the helicopter (in 100m) and the *Lateral* distance from the flock (also in 100m). The data are stored in **FlightResponse**.

a. Fit a two-predictor logistic model with *Flight* modeled on *Altitude* and *Lateral*. Does this model indicate that both variables are related to *Flight*, controlling for the other variable? Give the fitted model and then give two interpretive statements: one about the relationship of *Flight* to *Altitude* and the other about the relationship of *Flight* to *Lateral*. Incorporate the estimated slope coefficients in each statement.

b. The model fit in (a), the "additive model," assumes that the relationship between, say, flight response and lateral distance is the same for each level of altitude. That is, it assumes that the slope of the model regarding the Flight-versus-Lateral relationship does not depend on altitude. Split your data set into three subsets using the *AltCat* variable. For each subset, fit a logistic model of *Flight* on *Lateral*. Report your results. How, if at all, does the relationship between *Flight* and *Lateral* distance appear to depend on the altitude?

c. The lack of "independence" found in (b), suggests the existence of an interaction between *Altitude* and *Lateral* distance. Fit a model that includes this interaction term. Is this term significant?

d. Give an example of an overflight from the data for which using the model from (a) and the model from (c) would give clearly differing conclusions for the estimated probability of a flight response.

10.12 *March Madness.* Each year 64 college teams are selected for the NCAA Division I Men's Basketball tournament, with 16 teams placed in each of four regions. Within each region the teams are seeded from 1 to 16, with the (presumed) best team as the 1 seed and the (presumed) weakest team as the 16 seed; this practice of seeding teams began in 1979 for the NCAA tournament. Only one team from each region advances to the Final Four, with #1 seeds having advanced to the Final Four 54 times out of 128 possibilities during the years 1979 - 2010. Of the 128 #2 seeds, only 28 have made it into the Final Four; the lowest seed to ever make a Final Four was a #11 (this has happened twice). The file **FinalFourLong** contains data on *Year* (1979 to 2010), *Seed* (1 to 16), and *Final4* (1 or 0 according as a team did or did not make it into the Final Four). The file **FinalFourShort** contains the same information in more compact form, with *Year*, *Seed*, *In* (counting the number of teams that got into the Final Four at a given *Seed* level and *Year*), and

Out (counting the number of teams that did not get into the Final Four at a given *Seed* level and *Year*). We are interested in the relationship between *Seed* and *Final*4 - or equivalently, between *Seed* and *In/Out*.

  a. Fit a logistic regression model that uses *Seed* to predict whether or not a team makes it into the Final Four. Is there a strong relationship between these variables?

  b. Fit a logistic regression model that uses *Seed* and *Year* to predict whether or not a team makes it into the Final Four. In the presence of *Seed*, what is the effect of *Year*? Justify your answer.

10.13 *March Madness and Tom Izzo.* Consider the Final Four data from Exercise 10.12. A related dataset is **FinalFourIzzo**, which includes an indicator variable that is 1 for teams coached by Tom Izzo at Michigan State University and 0 for all other teams.

  a. Fit a logistic regression model to uses *Seed* and the *Izzo* indicator to predict whether or not a team makes it into the Final Four.

  b. Is the *Izzo* effect consistent with chance variation, or is there evidence that Izzo-coached teams do better than expected?

10.14 *Health care vote.* On 7 November 2009 the U.S. House of Representatives voted, by the narrow margin of 220-215, for a bill to enact health insurance reform. Most Democrats voted yes while almost all Republicans voted no. The file **InsuranceVote** contains data for each of the 435 representatives and records several variables: *Party* (D or R); congressional district; *InsVote* (1 for yes, 0 for no); *Rep* and *Dem* (indicator variables for Party affiliation); *Private*, *Public*, and *Uninsured* (the percentages of non-senior citizens who have private health insurance, public health insurance (e.g., through the Veterans Administration), or no health insurance); and *Obama* (1 or 0 according as the congressional district voted for Barack Obama or John McCain in November of 2008).

  a. Fit a logistic regression model that uses *Uninsured* to predict *InsVote*. As the percentage of uninsured residents increases, does the likelihood of a yes vote increase? Is this relationship stronger than would be expected by chance alone? (That is, is the result statistically significant?)

  b. Fit a logistic regression model that uses *Dem* and *Obama* to predict *InsVote*. Which of the two predictors (being a Democrat or representing a district that voted for Obama in 2008) has a stronger relationship with *InsVote*?

  c. (Optional) Make a graph of the Cook's Distance values for the fitted model from part (b). Which data point is highly unusual?

10.4. EXERCISES

Open-ended

10.15 *Risky youth behavior (continued).*
Use the Youth Risk Behavior Surveillance System (YRBSS) data (**YouthRisk2007**) to create a profile of those youth who are likely to ride with a driver who has been drinking. What additional covariates do you think would be helpful?

10.16 *Backpack weights.* Consider the **Backpack** data from Exercise 9.22. The data set contains the following more extensive set of variables than we introduced in that exercise:

Back Problems (0=no, 1=yes)
BackpackWeight
BodyWeight
Ratio (of backpack weight to body weight)
Major
Year
Sex
Status (undergraduate or graduate status)
Units (number of credits)

a. Use this more extensive data set to investigate which factors seem to figure into the presence of back pain for these students. Perform a thorough exploratory analysis. You may want to explore the relationships between potential predictors and consider interaction terms. In writing up your report, provide rationale for your results and comment on possible inferences that can be made.

b. One study that piqued the students' interest said that a person should not carry more weight in a backpack than 10% to 15% of body weight. Use these data to see which of the measured variables appears to relate to the likelihood that a student carries too much weight.

10.17 *Math placement: Does it work?* Students at a small liberal arts college took a placement exam prior to entry in order to provide them guidance when selecting their first math course. The data set **MathPlacement** contains the following variables:

ID
Gender
PSATM (PSAT math score)
SATM (SAT math score)
ACTM (ACT math score)
Rank (adjusted to size of high school class)
Size (of high school class)
GPAadj (adjusted gpa)
PlcmtScore (score on placement test)
Recommends (placement recommendation)
Course (student took the recommended course)
Grade (grade in the course)

Note: The *Recommends* variable comprises the following possible values:

- R0: Stat, 117 or 210
- R12: R0 course or 120
- R2: 120 or 122
- R4: 122
- R6: 126
- R8: 128

Use these data to decide how well the math placement process is working. Consider using only the placement score to place students into a course. If they take the recommended course how do they do? Define "success" as a grade of "B" or above. Does it improve the model to add other variables?

10.18 *More film.* Consider the data set **Film** giving information of a random sample of 100 films taken from the 19,000+ films in the Maltin film guide, which was fully introduced in Exercise 9.24. Build a multiple logistic regression model for the binary response variable *Good?* using possible explanatory variables *Year*, *Time*, *Cast*, or *Description*. Explain your results in a report.

10.19 *More on the lost letter.* Recall from Exercise 9.25 that in 1999 Grinnell College students Laurelin Muir and Adam Gratch conducted an experiment where they intentionally "lost 140 letters" in either the city of Des Moines, the town of Grinnell, or on the Grinnell College campus. Half of each sample were addressed to Friends of the Confederacy and the other half to Iowa Peaceworks. Would the return rates differ depending on address? Get the data in **LostLetter**. Use multiple logistic regression to ascertain the affect of both address and location on return rate. Report on your findings.

10.4. EXERCISES

10.20 *Basketball.*[2] Since 1991, David Arseneault, men's basketball coach of Grinnell College, has developed a unique, fast-paced style of basketball that he calls "the system." In 1997, Arseneault published a book entitled *The Running Game: A Formula for Success* in which he outlines the keys to winning basketball games utilizing his fast-paced strategy. In it, Arseneault argues that: (1) making opponents take 150 trips down the court, (2) taking 94 shots (50 percent of those from behind the three-point arc), (3) rebounding 33 percent of offensive missed shots, and (4) forcing 32 turnovers would ultimately result in a win. It is these four statistical goals that comprise Arseneault's "keys to success." The data set **Hoops** comes from the 147 games the team played within its athletics conference between the 1997-98 season through the 2005-06 season. Use these data to investigate the validity Coach Arseneault's "keys to success" for winning games in the system. Then, use the variables available to, if possible, improve on his keys. Write a report of your findings. The variables are:

Game	An ID number assigned to each game
Opp	The name of the opponent school for the game
Home	Dummy variable where 1 = home game and 0 = away game
OppAtt	The number of field goal attempts by the opposing team
GrAtt	The number of field goal attempts by Grinnell
Gr3Att	The number of three-point field goal attempts by Grinnell
GrFT	The number of free throw attempts by Grinnell
OppFT	The number of free throw attempts by the opponent
GrRB	The total number of Grinnell rebounds
GrOR	The number of Grinnell offensive rebounds
OppDR	The number of defensive rebounds the opposing team had
OppPoint	Total number of points scored in the game by the opponent
GrPoint	Total number of points scored in the game by Grinnell
GrAsst	The number of assists Grinnell had in the game
OppTO4	The number of turnovers the opposing team gave up
GrTO	The number of turnovers Grinnell gave up
GrBlocks	The number of blocks Grinnell had in the game
GrSteal	The number of steals Grinnell had in the game
40Point	A dummy variable that is 1 if some Grinnell player scored 40 or more points
BigGame	A dummy variable that is 1 if some Grinnell player score 30 or more
WinLoss	A dummy variable that is 1 if Grinnell wins the game
PtDiff	The point differential for the game (Grinnell score minus Opponents score)

---

[2] These data were collected by Grinnell College students Eric Ohrn and Ben Johannsen.

10.21 *Credit cards and Risk.*[3] Researchers conducted a survey of 450 undergraduates in large introductory courses at either Mississippi State University or the University of Mississippi. There were close to 150 questions on the survey, but only four of these variables are included in the dataset **Overdrawn**. (You can consult the paper to learn how the variables beyond these 4 affect the analysis.) These variables are as follows:

Age	The age of the student, in years
Sex	(0=male, 1 = female)
*DaysDrink*4	"During the past 30 days, on how many days did you drink alcohol?"
Overdrawn	"In the last year, I have overdrawn my checking account." (0=no, 1=yes)

The primary response variable of interest to the researchers was *Overdrawn*: What characteristics of a student predict his or her overdrawing their checking account? Explore this question and report your findings. (Note: There are several missing values, most notably for *Age*.)

---

[3] See "Sensation-Seeking, Risk-Taking, and Problematic Financial Behaviors of College Students" (Worthy S.L., Jonkman J.N., Blinn-Pike L., *Journal of Family and Economic Issues* 2010; 31: 161-170)

Chapter 11

Logistic Regression: Additional Topics

11.1 Assessing Logistic Regression Models

The previous two chapters have introduced what we consider to be the basics of logistic regression. They contain those topics you should know in order to understand someone's use of logistic regression in a report or journal or to perform a basic logistic analysis of your own. This chapter takes a deeper look at logistic regression and introduces concepts that may be unnecessary for a basic analysis, but that will strengthen many analyses that use logistic regression.

As we have done with OLS—ordinary least squares—models, we assess our fitted logistic regression model with an eye toward three general concerns.

  a. Is it worthwhile to even bother with the explanatory variable(s)? Does the probability of interest appear to be related to one or more of the explanatory variables?

  b. Are there any concerns about the model assumptions?

  c. How well does the model perform? Is it good at prediction?

Goodness of fit: Assessing the fit of the model

In Chapters 9 and 10, you investigated ways to obtain evidence of an association between a probability of interest and our explanatory variable(s). Like the OLS t-test, you found that in the case of simple logistic regression you could refer to a z-statistic (also known as a Wald statistic) associated with the explanatory variable coefficient. Alternatively, you could decide between a model with no explanatory variables and one with an explanatory variable by comparing likelihoods, or, more precisely, comparing the difference in $-2\ln(L)$ for the two models to a chi-square distribution.

In the earlier logistic regression chapters, we used the full-versus-reduced approach to decide on the overall effectiveness of the model (see Example 10.2). The question this test addresses is: Does the model under consideration indicate that there is an association between the explanatory variable(s) and the probability of interest? Is this model better at describing the probability of success than a model with no predictors, that is, a model which predicts a constant response value regardless of explanatory variable values? More formally we were comparing these two models:

$$H_0 : \beta_1 = \beta_2 = ... = 0 \quad \text{(reduced model)}$$

versus

$$H_a : logit(\pi) = \beta_0 + \beta_1 X_1 + \beta_2 X_2 + \ldots \ldots, \text{ at least one } \beta_i \neq 0$$
$$\text{(full model)}$$

The reduced model here is the *constant model* that predicts the same probability of success regardless of predictor values. The full model here is the model we are considering: *the linear logit model.*

We now change topics. Suppose we are satisfied that we have fit an effective model to the data, that is, we are satisfied that the linear logit model has some predictive power (beyond predicting a constant). Realizing that no model is perfect, we might ask, is there evidence of a problem with the fit of the model? That is, we are looking to see if there is evidence to reject the model we are considering because the probabilities of success do not follow very well the logistic function pattern. This question is another facet of assessing a model and we again attack it with the full-versus-reduced model approach, which will lead to what we call a *goodness-of-fit test*. In contrast to the previous test, with the goodness-of-fit test the model under consideration—the linear logit model—becomes the null or reduced model. Here are the full and reduced models in this context:

$$H_0 : logit(\pi) = \beta_0 + \beta_1 X_1 + \beta_2 X_2 + ... \quad \text{(reduced model)}$$

$$H_a : logit(\pi) = p_i, \quad\quad\quad\quad i = 1, 2, \ldots k \text{ (full model)}$$

In the full model, p_i represents a separate probability for each value of the independent variable (or each combination of the collection of independent variables); the notation assumes k distinct values of the explanatory variable (or k distinct combinations of the collection of explanatory variables). This full model is called the *saturated model* and does not assume the logits are linear in the explanatory variables.

The saturated model specifies a proportion for each level (or combination of levels) of the explanatory variables, and in doing so it provides the best fit possible.[1] This approach will only be feasible if there are enough replicates[2] at each level of the explanatory variable in the observed data. We'll

---
[1] "Saturated" suggests a model as full as can be, which makes sense here given that every possible combination of the predictors has its own, separately fitted probality estimate; a less saturated model would fit fewer parameters and would smooth out the probabilities across these combinations.

[2] Replicates at each level means multiple observations with the same value of the explanatory variable.

11.1. ASSESSING LOGISTIC REGRESSION MODELS

Length of putt (in feet)	3	4	5	6	7
Number of successes	84	88	61	61	44
Number of failures	17	31	47	64	90
Total number of putts	101	119	108	125	134

Table 11.1: Putting success vs. distance of the putt

Length of putt (in feet)	3	4	5	6	7
Saturated probability estimates $\hat{p}$	0.831	0.739	0.565	0.488	0.328
Logits of Saturated estimates	1.60	1.04	0.26	-0.05	-0.72
Probability estimates: linear logit model	.826	.730	.605	.465	.330
Logit estimates: linear logit model	1.558	.992	.426	-.140	-.706

Table 11.2: Goodness of Fit test: Saturated model is "full"; Linear logit model is "reduced."

see that this will work well for the putting example but not the MCAT data.

Example 11.1: *Putting*

The data for this application can be modeled as a set of binomial random variables. A **binomial** random variable counts the number of successes for a fixed number of independent trials, m. For example, we have $m = 101$ putts for the first distance of 3 feet. Our model states that these putts can be considered the result of 101 independent trials, each of which has a specified probability (call it p_3—we will index from 3 to 7, rather than 1 to 5) of success. For these data, the value of the binomial variable is 84. For the second distance of 4 feet, there was another set of 119 independent trials, with some fixed success probability (call it p_4); for these data the value the binomial variable takes on is 88. And so on for putt lengths of 5, 6, and 7 feet. To each of the 5 putt lengths we have a fixed number of independent trials with some fixed "success probability" that changes with each putt length.

This is the underlying model used when fitting the saturated model. It is important to keep in mind the assumptions of the binomial model:

- Each trial results in a success or failure. Here each putt results in making it or not.

- The probability of success is the same for each trial at a given level of the predictor. This implies in the putting example that the probability of making it is the same for every putt from the same distance.

- The trials are independent of one another at a given level of the predictor. When putting, this implies that what happens on one putt does not affect the probability of success or failure of other putts.

If any of these assumptions are suspect, the goodness of fit test is suspect. (For example, might you have reason to suspect that the set of all 3-foot putts are independent of one another? This is a question you can consult your golfing friends about.) If the assumptions do seem reasonable, we can fit the saturated model using maximum likelihood estimates, which turn out to be nothing more than the empirical proportions for each putt length. These proportions are the "saturated probability estimates" of Table 11.2.

The reduced model is the linear logit model that expresses the logit of π_d as a linear function of distance and from the computer output, we see that the fitted model has the form:

$$logit \frac{\widehat{\pi}}{1-\widehat{\pi}} = 3.26 - 0.566(distance)$$

or, equivalently

$$\widehat{\pi} = \frac{e^{3.26-.566(distance)}}{1+e^{3.26-.566(distance)}}$$

These give the "Linear logit probability estimates" of Table 11.2; their logits are also given there and it is these logits that fall exactly on a line as the model specifies.

The goodness-of-fit test compares the probability estimates from these two models: the closer they are to one another, the better the goodness of fit. As before, the comparison is made using a likelihood ratio test. Recall from the previous chapter that in the nested likelihood ratio test the test statistic is:

$$X^2 = G_{Full} - G_{Reduced} = -2\ln(L_{Reduced}) - (-2\ln(L_{Full})).$$

When the full model is the saturated model this becomes:

$$X^2 = -2\ln(L_{Reduced}) - (-2\ln(L_{Saturated})).$$

This last expression is called the **residual deviance** by R and is called the "Deviance" by Minitab . Small residual deviance values indicate a reduced model that fits well; large values indicate poorer fit. This is intuitive if we consider the deviance as representing variation unexplained by the model (although this is not in exactly the same form as we encountered in OLS.) We assess statistical significance using a chi-square distribution with degrees of freedom equal to the difference in the number of parameters estimated in the two models, in this case 5 - 2 = 3. (Note that this d.f. is given by Minitab and R output.) In this case the residual deviance is 1.0692 on 3 d.f., which gives a p-value of 0.784514. This result suggests that there is no compelling evidence that the fit of the model is problematic. So the null model is not rejected and we conclude that there are no

11.1. ASSESSING LOGISTIC REGRESSION MODELS

detectable problems with the fit of the linear logistic regression model.

Recall that in Chapter 10 we defined the test statistic for the nested likelihood ratio test for a full versus a reduced model via the equation:

$$X^2 = -2\ln(L_{Reduced}) - (-2\ln(L_{Full}))$$

which we can rewrite:

$$X^2 = [-2\ln(L_{Reduced}) - (-2\ln(L_{Saturated}))] - [(-2\ln(L_{Full})) - (-2\ln(L_{Saturated}))]$$

or more simply as:

$$X^2 = Deviance_{Reduced} - Deviance_{Full}$$

This derivation explains again why the nested likelihood ratio test can be called the "drop in deviance" test.

Before leaving the topic of residual deviance and the goodness-of-fit test, we consider one more way of understanding the logic of this test. The R output below fits two logit models. The first model—"model with linear distance"—fits the probability of success to a linear logistic model of distance. The second model—"saturated model"—fits separate estimates to the logit for each putt distance, treating distance as a factor with 5 unordered levels ("3", "4", ..., "7"). The residual deviance for the saturated model is 0 (i.e., -2.3315e-14) because there is no better fit of a model to these data: the fitted values of probability are precisely the empirical probabilities. Hence the drop in deviance statistic is simply 1.0692 - 0; the 1.0692 in the linear logit model being already a comparison to the saturated model.

Thus the **residual deviance** appearing on R output is a likelihood ratio statistic comparing that model to a saturated model. This applies when the data are entered as binomial counts (e.g., with the command `model1=glm(cbind(Makes,Misses)~ Lengths,family=binomial)` where Makes = Number of successes, Misses = Number of failures, and Lengths = Length or distance of putt from Table 11.2, and the data are entered as columns in a file.)

```
### Model with linear distance
glm(formula = cbind(n.made, n.missed) ~ distance, family = binomial)
            Estimate Std. Error z value Pr(>|z|)
(Intercept)  3.25684    0.36893   8.828   <2e-16 ***
distance    -0.56614    0.06747  -8.391   <2e-16 ***

    Null deviance: 81.3865  on 4  degrees of freedom
Residual deviance:  1.0692  on 3  degrees of freedom
```

AIC: 30.175

Saturated Model
# Model with Proportions for each Distance
glm(formula = cbind(n.made, n.missed) ~ factor(distance), family = binomial)

```
                  Estimate Std. Error z value Pr(>|z|)
(Intercept)         1.5976     0.2659   6.007 1.89e-09 ***
factor(distance)4  -0.5543     0.3382  -1.639    0.101
factor(distance)5  -1.3369     0.3292  -4.061 4.90e-05 ***
factor(distance)6  -1.6456     0.3205  -5.134 2.84e-07 ***
factor(distance)7  -2.3132     0.3234  -7.154 8.46e-13 ***

    Null deviance:       81.387     on 4  degrees of freedom
Residual deviance:  -2.3315e-14     on 0  degrees of freedom
AIC: 35.106
```

◇

GPA	Admitted	Denied	$\hat{p}$
3.54	2	1	0.67
3.56	1	1	0.50
3.58	1	0	1.00
3.61	0	1	0.00
3.62	2	1	0.67
3.65	0	1	0.00

Table 11.3: MCAT data

Example 11.2: *MCAT*

Table 11.3 illustrates the type of data found in the MCAT data set. It is apparent that the MCAT data are not in the form that the putting data are. That is, the MCAT data consists of many different possible predictor (GPA) levels and there are not replicates at each level. While we could think of the observation of the number of successful applicants at each GPA level as binomial variables, the number of trials (m) is small and often 1. (When $m = 1$, the variable is referred to as a binary observation or a Bernoulli trial.) In the table, we have $m = 2 + 1$ or 3, $m = 1 + 1$ or 2, $m = 1$, $m = 1$, etc. The residual deviance test is only appropriate when $m > 5$ for all values of the explanatory variable (or all combinations of the set of explanatory variables). The MCAT example is more typical than the putting example, that is, it is more common to have few, if any, replicate values or combinations of values of the explantory variables. Therefore, goodness-of-fit tests based

11.1. ASSESSING LOGISTIC REGRESSION MODELS

on the residual deviance are of limited value in logistic regression.

◇

There is a test proposed by Hosmer and Lemeshow[3] that groups data in order to compute a goodness-of-fit measure that can be useful in more general circumstances. The data cases are divided into several (up to 10) groups based on similar logistic predicted probabilities and the sample proportions in those groups are compared to the logistic fits—again with a chi-square procedure. This test performs better with large samples so that there is at least 5 in each group. In this case the d.f. will be the number of groups minus two. This option may be preferable when there are a large number of observations with little replication in the data. There are several other tests for goodness-of-fit whose results are provided routinely by software packages that we do not discuss here.

Reasons for lack of fit: While there was no problem detected with fit in the putting data, when a lack-of-fit is discovered, it is essential to investigate it. Here are three possible reasons for lack of fit, which we give in the order in which we suggest you consider them:

a. **Omitted covariates or transformation needed**

b. **Unusual observations**

c. **Overdispersion**

We consider each of these possibilities below.

Omitted explanatory variables or transformation needed

This is the classic challenge in modeling and it should be considered before any of the other reasons for lack-of-fit. Are there additional covariates you could include in your model? Do you need a quadratic or interaction term? Much of this book has provided ways in which to address just these kinds of concerns. Note that applications where there are many replicates also allow us to compute empirical logits and look at the assumption of linearity graphically. Without replicates, near-replicates, or large samples, graphical checks are not as easily interpreted with binary data. A common approach in logistic modeling is to compare models using a LRT to determine whether quadratic or interaction terms are useful (see chapter 10).

Unusual observations

In the following case, we are using the goodness-of-fit statistic as a guide to finding extreme or unusual observations. It is not being used to make formal inferences about fit. Nonetheless, it can be very helpful to construct such diagnostic plots.

---

[3]*Applied Logistic Regression*, by David W. Hosmer and Stanley Lemeshow

In addition to looking at summary statistics for goodness-of-fit, it is often useful to examine each individual observation's contribution to those statistics. Even if the summary statistic suggests there are no problems with fit, we recommend that individual contributions be explored graphically. Each observation is removed one at a time from the data set and the summary goodness-of-fit statistic, X^2, is recalculated. The change (delta) in X^2 provides an idea of how each particular observation affects the X^2. A large goodness-of-fit X^2 suggests problems with the fit of the model. An observation contributing a proportionately large amount to the goodness-of-fit X^2 test statistic requires investigation.

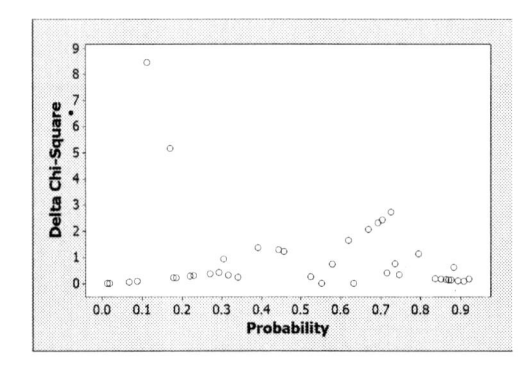

Figure 11.1: Contributions to X^2 using Delta Chi-square criterion

Figure 11.1 shows the contribution of each point to the goodness-of-fit statistic for the medical school acceptances based on GPA. It is quite clear that there are two observations strongly affecting goodness-of-fit. These two students have relatively low GPAs (3.14, 3.23) yet both were accepted to medical school. This does not fit the pattern for the rest of the data.

Overdispersion

It is easier to understand overdisperson in the context of binomial data, for example, with $m > 5$ for most combinations of explanatory variables, although it is possible for overdipersion to occur in other circumstances as well. As established above, we may model such data using a **binomial** variable with m trials and with probability of success π for that combination. Intuitively, we expect the number of successes to be $m\pi$, on average, and from theory of binomial variables, the variance is known to be $m\pi(1-\pi)$.

When R fits a logistic model, part of the standard output is the sentence (`Dispersion parameter for binomial family taken to be 1`) which implies that the variance is assumed to be $m\pi(1-\pi)$. In practice, it could happen that *extra-binomial dispersion* is present, meaning that the variance is larger than expected for levels of the predictor. This could happen for a number of different reasons and there are a number of different ways to take this into account with your model. One approach that is often helpful is to fit a model identical to the usual linear-logistic model, but with one extra parameter, ϕ, that inflates the variance for each combination of the predictors

11.1. ASSESSING LOGISTIC REGRESSION MODELS

Length of putt (in feet)	3	4	5	6	7
Number of successes	79	94	60	65	40
Number of failures	22	25	48	60	94
Total number of putts	101	119	108	125	134
Sample proportion ($\hat{p}$)	0.782	0.790	0.556	0.520	0.299
Logistic model $\hat{\pi}$	0.826	0.730	0.605	0.465	0.330

Table 11.4: Another putting data set to illustrate overdisperson

by a factor of ϕ, so that the model variance is $\phi m \pi (1 - \pi)$ rather than $m \pi (1 - \pi)$, for some ϕ greater than 1. When we inflate the variance in this way we are no longer working with a true likelihood but rather we use a *quasi-likelihood*, the details of which are left to our computer software.

In the case of binomial data, the deviance goodness-of-fit test, discussed earlier in this section, provides a check on dispersion. If the residual deviance is much greater than the degrees of freedom, then overdispersion may be present. For example, the putting data produce a residual deviance of 1.0692 on 3 degrees of freedom, so there is no evidence of overdispersion here. But consider the following similar example.

Example 11.3: *Another putting data set*

Table 11.4 gives another data set that leads to a similar fit, but with an overdispersion problem. Fitting the logistic regression of successes on length of putt in R gives the same fit as for the original data but with a much larger residual deviance, due to the fact that the sample variances for the different lengths of putts do not match very well with a binomial distribution:

```
Coefficients:
            Estimate Std. Error z value Pr(>|z|)
(Intercept)  3.25684    0.36893   8.828   <2e-16 ***
Lengths     -0.56614    0.06747  -8.391   <2e-16 ***

(Dispersion parameter for binomial family taken to be 1)

    Null deviance: 87.1429  on 4  degrees of freedom
Residual deviance:  6.8257  on 3  degrees of freedom
```

Overdispersion might be evident, for example, when our golfer has good or bad days. In those cases, data from the same type of day may be more similar or correlated so that the counts of successes are more spread out than we would expect if every putt was made when the probability of success was the same. By using an overdispersion parameter, we take into account unmeasured variables such as good and bad days.

Comparing the two sets of R output illustrates the problems of fitting a regular linear logistic model when an overdispersed model is more appropriate. The slope and intercept estimates are identical, which will always occur. The differences lie in the standard errors. Not taking overdispersion into account leads to standard error estimates that are too small and, thereby inflates the chance of statistical significance in the z-tests for coefficients. Notice here that with the usual model, the coefficient for Length appears highly statistically significant (p-value $\approx$ 0), whereas with the overdispersion model fit the standard errors are, appropriately, larger and in this case, the p-value increases to .011, a less statistically significant result. In some cases, the change in p-value could be even more pronounced.

In general, if we do not take the larger variance into account, the standard errors of the fitted coefficients in the model will tend to be too small, so the z-tests for terms in the model are significant too often.

Note: To fit the overdispersion model in R change the `family=binomial` part of the glm command to `family=quasibinomial`. Here is output from using `family=quasibinomial` in R for the data in the previous table:

```
             Estimate Std. Error t value Pr(>|t|)
(Intercept)    3.2568     0.5552   5.866  0.00988 **
Lengtho       -0.5661     0.1015  -5.576  0.01139 *

(Dispersion parameter for quasibinomial family taken to be 2.264714)

    Null deviance: 87.1429  on 4  degrees of freedom
Residual deviance:  6.8257  on 3  degrees of freedom
```

◇

11.2 Residual Diagnostics: Assessing the Conditions

Are there any concerns about the model assumptions? We began this chapter assessing the fit of a logistic regression model for binomial data with $m > 5$ using the residual deviance, which is equivalent to a likelihood ratio test comparing the model to a saturated model. If we have concerns about the fit, we check out potential reasons for lack-of-fit. We assess the assumption that the model includes the correct covariates in forms that are not in need of a transformation. If we have binomial data with $m > 5$, we can construct empirical logits and examine this question graphically. If there are no repeat observations (the more usual case), we can compare increasingly complex models, models with quadratic terms or interactions, for example, using nested LRTs. If we are satisfied with the covariates in the model, we can use the delta chi-square plot, as in Figure 11.1, to identify unusual observations. We can also use residuals. There are a number of different kinds of residuals used in logistic regression including Pearson residuals (which you may have encountered in chi-square analysis) and deviance residuals (whose sum of squares gives the residual deviance).

Beyond simply identifying unusual observations, many other diagnostic methods have been proposed for assessing the conditions of a logistic model. Recall that when we studied ordinary least-squares regression, we examined plots of residuals, leverages, and Cook's Distance[4] values to determine which points in a dataset deviate from what the model predicts and which points have high influence on the fitted model. Analogous ideas apply to logistic regression.

Example 11.4: *Campaign contributions and votes*

For example, consider the CAFE data from Example 10.1. We considered there campaign contributions, *LogContr* (the base-10 log of contributions, in thousand of dollars) and political party (1 for Democrats and 0 for Republicans and independents) as predictors of a senator's vote on the CAFE amendment. The logistic model that uses *LogContr* and *Dem* provides useful predictions, but there are a couple of unusual points present. In many circumstances we try to eliminate the effect of unusual points, but here it is the extreme observations that are interesting. The more money that senators had received from car manufacturers, the more likely they were to vote "yes," but Senator McCain voted "no" despite having received a lot of money. At the other extreme, Senator Kohl was a Democrat who had received no money from car manufacturers but, unlike 5 other Democrats who had received no money and all voted "no," Senator Kohl voted "yes."

A plot of Pearson residuals shows that the McCain residual is a bit less than -4 and the Kohl residual is a bit greater than 4. (In R the command `plot(residuals(mymodel,type="pearson"))` produces the desired plot and the command
    `identify(residuals(mymodel,type="pearson"),label=Senator)`
helps us find the senator who corresponds to a point on the plot.) Since Pearson residuals have, roughly, mean zero and standard deviation one, values as far from zero as -4 or 4 are quite extreme.

---

[4]See section 4.3.

A plot of Cook's Distance values shows that Senator Kohl is extreme on this measure. (Here the R command is `plot(cooks.distance(mymodel))`.) A plot of leverages shows that the two senators with the most leverage in the model fitting process are Senators Chafee and Enzi—two Republicans who had received little money from car manufacturers, whereas most Republicans had received quite a bit of money. (Here the R command is `plot(influence(mymodel)$hat)`.)

⋄

Assessing Prediction

In our previous example, our objective was to describe a relationship, not to predict a new outcome. If your purpose in fitting a logistic regression model is prediction, you would like to know more than whether it seems to fit well or not. It would be helpful to know how well the model performs at prediction.

Many of the ways to assess prediction are intuitively appealing but not all are particularly effective at judging the predictive capabilities of a model. A natural and easily understood measure is the percentage of points that are correctly classified. This statistic counts up the number of successes predicted (based on the number of observations for which the model estimates $\hat{\pi} > .5$) out of the actual number of successes and the number of failures predicted out of the actual failures in the data. Some packages allow the user to specify the cut-off predicted probability to be different from 0.5. Consequently this percentage can differ markedly even using the same data. Users should be aware of this sensitivity to small changes in the cut-off. As an example, the Minitab output below shows the model of med school acceptance on MCAT and GPA10. We used this model to construct Table 11.5, a contingency table of whether a student is denied or accepted to med school (row headings) versus whether the model predicts denied or accepted based upon a .5 cutoff of estimated acceptance probability from the fitted model. We see that of the 55, cases 41 (= 18+23) are correct predictions from the model based upon a .5 cutoff, a success rate of 74.5%.

		Predicted $\widehat{Deny}$	$\widehat{Accept}$
Actual	Deny	18	7
	Accept	7	23

Table 11.5: Rows are the actual admissions outcome: Deny or Accept; Columns are predicted by the model: Deny or Accept; based upon 0.5 cut-off

Minitab reports several measures to indicate how well a model is performing. One commonly used measure of prediction is the percentage of *concordant pairs*. The percentage of concordant pairs is often referred to as a **c-statistic** or **c-index**. To calculate this, consider each possible pair of a success and a failure in the dataset. Since there are 30 successes and 25 failures, there are $30 \times 25 = 750$ total pairs. Those pairs for which the model-predicted probability ($\hat{\pi}$) is greater for the success observation than it is for the failure observation are pairs considered to be concordant

11.2. RESIDUAL DIAGNOSTICS: ASSESSING THE CONDITIONS

with the fitted model. From the Minitab output, we see that in this example 625 of the 750 total pairs were concordant, giving a c-statistic of $625/750 = .833 = 83.3\%$. When the c-statistic is 0.5, the model is not helping at all and we can do just as well by guessing. A c-statistic of 1.0 indicates that the model perfectly discriminates between successes and failures. Our 83.3% indicates a good degree of accuracy in the model predictions.

If your goal is prediction, you will want to see good performance as indicated by the c-statistic or other similar measures. Other packages report one or more of the many logistic regression versions of R^2. Unfortunately, these **Pseudo-R^2s** are controversial and each suffers inadequacies. In general, this group of "measures of association" is not recommended nor discussed further here.

```
Response Information

Variable     Value   Count

Acceptance   1       30      (Event)

             0       25

             Total   55


Logistic Regression Table

                                              Odds       95% CI

Predictor      Coef       SE Coef    Z      P    Ratio  Lower  Upper

Constant     -22.3727    6.45360   -3.47  0.001
MCAT           0.164501  0.103154   1.59  0.111   1.18   0.96   1.44
GPA10          0.467646  0.164155   2.85  0.004   1.60   1.16   2.20

Log-Likelihood = -27.007

Test that all slopes are zero: G = 21.777, DF = 2, P-Value = 0.000

Goodness-of-Fit Tests
Method              Chi-Square   DF      P
Pearson               47.5644    52    0.649
Deviance              54.0142    52    0.397
```

496   CHAPTER 11. LOGISTIC REGRESSION: ADDITIONAL TOPICS

```
Hosmer-Lemeshow          8.6006   8   0.377
Brown:
General Alternative      2.1237   2   0.346
Symmetric Alternative    0.9513   1   0.329
```

Table of Observed and Expected Frequencies:

(See Hosmer-Lemeshow Test for the Pearson Chi-Square Statistic)
                                Group

Value	1	2	3	4	5	6	7	8	9	10	Total
1											
Obs	0	2	2	3	0	4	4	4	5	6	30
Exp	0.3	1.1	1.4	2.4	2.5	3.8	3.6	4.8	4.5	5.7	
0											
Obs	5	4	3	3	5	2	1	2	0	0	25
Exp	4.7	4.9	3.6	3.6	2.5	2.2	1.4	1.2	0.5	0.3	
Total	5	6	5	6	5	6	5	6	5	6	55

Measures of Association:
(Between the Response Variable and Predicted Probabilities)

Pairs	Number	Percent	Summary Measures	
Concordant	625	83.3	Somers' D	0.67
Discordant	124	16.5	Goodman-Kruskal Gamma	0.67
Ties	1	0.1	Kendall's Tau-a	0.34
Total	750	100.0		

Issues

We discuss here three issues that are important to logistic regression analysis, the first of which is unique to the logistic setting, the other two of which are analogous to OLS (ordinary least squares regression).

   a. **Perfect separation.** Strangely enough, when you include a covariate in the model which can perfectly discriminate between successes and failures you will run into computational difficulties. This situation can be easily anticipated in a thorough exploratory data analysis where tables of each covariate by the response are constructed. Even so, it is still possible to have a combination of covariates completely separate successes from failures. A clue that this may be happening is the presence of extraordinarily large SE for the coefficients. In general, if you spot very large SEs you may be dealing with some kind of computational issue. For example, the algorithm that searches for a maximum likelihood may not have converged for any number of reasons. Models with these large SEs should not be used. A number of sophisticated methods have been proposed for modeling in this situation, but none are included here. A simpler approach, that of a randomization (or permutation) test is described in the next section.

b. **Multicollinearity.** Like OLS, we need to remain vigilant about the possibility of multicollinearity in logistic regression. As before, highly correlated predictors will probably not be a big problem for prediction, but one must be very cautious in interpreting coefficients when explanatory variables are highly correlated. Again for categorical covariates, tables can be constructed and examined.

c. **Overfitting.** Overfitting—that is, closely fitting the idiosyncrasies of a particular data set at the cost of producing generalizable results—can also be a problem with logistic regression. As with OLS, there are several ways in which to avoid overfitting, including cross-validation and bootstrapping.

498 CHAPTER 11. LOGISTIC REGRESSION: ADDITIONAL TOPICS

11.3 Randomization Tests

In logistic regression modeling, randomization tests can be used for testing how well a model fits and whether an explantory variable is associated with the response. Why would we bother with randomization tests? The tests and confidence intervals we have constructed using logistic regression results rely on approximations to distributions that require relatively large samples. For example, when we compare models by examining the difference in the deviances we assume that when the null hypothesis is true, the difference would look as if it is coming from a chi-square distribution. When the samples are small, this approximation may not be correct and it may be better to perform a randomization test. Here we will treat the simple case of a single binary explanatory variable with a binary response. With a little ingenuity you can take these methods and extend them to more complex situations.

Example 11.5: *Archery*

In 2002, Heather Tollerud, a Saint Olaf College student, undertook a study of the archery scores of students at the college who were enrolled in an archery course. Students taking the course record a score for each day they attend class from the first until the last day. Hopefully the instruction they receive helps them to improve their game. The data in **ArcheryData** contains student id, average score, first and last day scores, number of days attended, and gender. Let's begin by determining whether males and females exhibit similar improvements over the course of the semester.

Both variables we are considering, improvement, the response, and sex, a possible explanatory variable, are binary variables. Hence, the information we have can be summarized in a 2×2 table. See Table 11.6.[5]

	improved	did not improve
males	9	1
females	6	2

Table 11.6: Archery improvement by sex.

```
glm(formula = improve ~ sex, family = binomial)

Coefficients:
            Estimate Std. Error z value Pr(>|z|)
(Intercept)   1.0986     0.8165   1.346    0.178
```

[5]The `improve` variable is binary, with value 1 when the last day score exceeds the first day score and value 0 otherwise.

11.3. RANDOMIZATION TESTS

```
sexm           1.0986     1.3333   0.824    0.410
```

(Dispersion parameter for binomial family taken to be 1)

```
    Null deviance: 16.220  on 17  degrees of freedom
Residual deviance: 15.499  on 16  degrees of freedom
  (2 observations deleted due to missingness)
AIC: 19.499
```

Using R to fit a logistic regression to this data, we estimate that males improve $e^{1.0986} = 3$ times what females improve on average. However, we have a very small number of students so it might be better to use a randomization test here as opposed to plain old logistic regression. Our approach will be to randomly assign the 15 improvements and 3 which are not improvements to the 8 women and 10 men. After each random assignment, we will use the "new data," fit a logistic regression and take note of the Odds Ratio and residual deviance. These ORs are a measure of the extent of the differences between males and females produced by chance. Repeating many times—1000 or 5000 or 10,000 times—we'll look at the distribution of ORs and determine which proportion are greater than or equal to our observed odds ratio of 3. With 10,000 samples, we found 1964 greater than the OR of 3 that we observed. This is consistent with the logistic regression results. Nearly 20% of the samples generated by chance yielded Odds Ratios larger than what we observed. It is difficult to argue that our observed result is unusually large and deserving of "statistically significant" status. Note that this result is consistent with what we found using logistic regression.

One of the reasons logistic regression is so useful is that it takes us beyond analyzing table counts and allows us to incorporate continuous variables. For example, we could find out about how attendance affected the students' improvement.

```
glm(formula = improve ~ attendance, family = binomial)

Coefficients:
            Estimate Std. Error z value Pr(>|z|)
(Intercept)   5.5792    11.0819   0.503    0.615
attendance   -0.1904     0.5278  -0.361    0.718
```

(Dispersion parameter for binomial family taken to be 1)

```
    Null deviance: 16.220  on 17  degrees of freedom
Residual deviance: 16.071  on 16  degrees of freedom
  (2 observations deleted due to missingness)
```

AIC: 20.071

The coefficient of -0.1904 is surprising: Does it really pay to skip class? A closer look at the output reveals that this negative association in not statistically significant. We might have considered a randomization (or permutation) test for this problem as well. And there's no reason not to proceed in a manner similar to the previous analysis. Randomly assign an attendance to each person and record attendance and improvement status for each person. Once again repeat this many times and determine the proportion of the samples with an OR smaller than what we observed, $e^{-.1904} = 0.83$.

Despite the fact that the point estimate suggests that each additional class period reduces your improvement by a factor of 17%, the p-values from the logistic regression (p-value=0.718; two-sided) and the permutation test(p=0.5150) do not provide convincing evidence that the observed reduction in improvement is significant, keeping in mind that this is an observational study. Nonetheless, you might be able to come up with reasons for why those who do not attend as much still show improvement. Possibly they are excellent archers at the start of the course and know they will not benefit from more instruction, which in turn suggests that it is the poorer archers attending class and struggling with improvement.

◇

```
# A Simple Randomization Test for Logistic Regression

#Obs Data
y=improve
#x.obs=sex
x.obs=attendance

fit.obs=glm(y~x.obs,family=binomial(link="logit"))
summary(fit.obs)

or.obs = exp(coef(fit.obs)[2])

# Getting a distribution of estimated Odds Ratios
# from randomly generated data sets
# We'll leave y as is and randomly assign each explanatory value
n.samples=10000

OR=rep(NA,n.samples)

for (i in 1:n.samples){
```

11.3. RANDOMIZATION TESTS

```
x.random=sample(x.obs,length(x.obs),replace=F)
fit=glm(y~x.random,family=binomial(link="logit"))
OR[i]= exp(coef(fit)[2])
}

# What proportion of the random samples have
# as small or smaller than the OR we observed?

OR.small = OR<= or.obs
p.OR= sum(OR.small)/n.samples
p.OR
```

11.4 Applying Logistic Regression to Larger Tables

In Section 9.5 we saw how logistic regression can be used with a binary response and binary predictor to analyze the information in a 2×2 table. Now that we have a model for logistic regression with multiple predictors, we can extend those ideas to work with larger $2 \times k$ tables. While we still assume a binary categorical response, the categorical predictor may now have any number of categories. The key to this analysis is to create 0-1 indicator variables for each of the categories in the predictor. If the predictor has k categories, we build a multiple logistic regression model using any $k-1$ of the indicator variables. The next example illustrates why we need to omit one of the indicators and how to interpret the coefficients to learn about the probability of the response in each of the k predictor categories.

Example 11.6: *Joking for a tip*

Can telling a joke affect whether or not a waiter in a coffee bar receives a tip from a customer? A study investigated this question at a coffee bar at a famous resort on the west coast of France[6]. The waiter randomly assigned coffee-ordering customers to one of three groups: When receiving the bill one group also received a card telling a joke, another group received a card containing an advertisement for a local restaurant, and a third group received no card at all. Results are stored in **TipJoke** and are summarized in the following 2×3 table:

	Joke Card	Advertisement Card	No Card	Total
Left a Tip	30	14	16	60
Did Not Leave a Tip	42	60	49	151
Total	72	74	65	211

The explanatory variable here is the type of card (if any) given to the customer. This variable is categorical but not binary, because it has three categories. The response variable is whether or not the customer left a tip. We will regard leaving a tip as a success, coded with a 1.

We can calculate the observed success (tipping) rates to be $30/72 = 0.417$ for the joke group, $14/74 = 0.189$ for the advertisement group, and $16/65 = 0.246$ for the no card group. Figure 11.2 shows a segmented bar graph of the proportion of tips within each card category. It appears that perhaps the joke card does produce a higher success rate, but we need to investigate whether the differences are statistically significant.

To use the categorical information of the card type in a logistic model, we construct new indicator variables for each of the card groups.

---

[6]Gueguen, 2002

11.4. APPLYING LOGISTIC REGRESSION TO LARGER TABLES

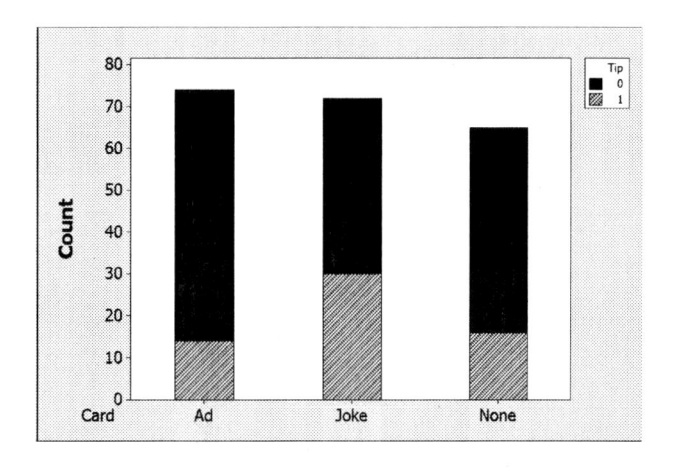

Figure 11.2: Proportion of tips within each card group

$$Joke = \begin{cases} 1 & \text{if Card = Joke} \\ 0 & \text{otherwise} \end{cases} \qquad Ad = \begin{cases} 1 & \text{if Card = Ad} \\ 0 & \text{otherwise} \end{cases} \qquad None = \begin{cases} 1 & \text{if Card = None} \\ 0 & \text{otherwise} \end{cases}$$

Note that if we know the value of any two of these indicator variables we can find the value of the third, since

$$Joke + Ad + None = 1$$

Thus we should not put all three indicator variables in the same regression model, since any one is exactly collinear with the other two. Besides the information for three indicators being redundant, mathematically having all three in a single model will cause computational problems. Specifically, it is not possible to find a unique solution, so most software packages will either produce an error message or automatically drop one of the predictors if we try to include them all in a model.

In general, when we want to use indicator variables to code the information for a predictor with k categories in a regression model, we should include all but one of the indicators, that is, use any $k-1$ indicator predictors. It doesn't really matter which indicator is omitted although you may have a preference depending on the context. As we shall see, we can recover information about the omitted category from the constant term in the model; the other coefficients show how each of the other categories differs from the one that was left out. For this reason, we often call the group that was omitted the *reference* category. In the tipping example, we can choose to omit the *None* indicator if we are interested in assessing how the other two treatments relate to doing nothing. While it may seem odd at first to leave out one category, note that we have been doing that quite naturally with binary variables; for example, we did this when coding gender with a single indicator(1=Female, 0=Male) for the two groups.

We are now in a position to fit a logistic regression model to predict tip success based on the type of card. We omit the indicator for *None* and fit a multiple logistic model to predict *Tip* (1=Yes,

0=No) using the *Joke* and *Ad* indicators.

$$\log(\text{tip odds}) = \beta_0 + \beta_1 Joke + \beta_2 Ad$$

Computer output includes the following fit and assessment of the model:

Logistic Regression Results for the 2×3 table

Predictor	Coef	SE Coef	Z	P	Odds Ratio	Lower CI	Upper CI
Constant	-1.12	0.29	-3.89	0.000			
Joke	0.78	0.37	2.09	0.036	2.19	1.05	4.56
Ad	-0.34	0.41	-0.81	0.416	0.71	0.32	1.61

Thus, the fitted model for the tip data in these groups is

$$\log\left(\frac{\hat{\pi}}{1-\hat{\pi}}\right) = -1.12 + 0.78 \cdot Joke - 0.34 \cdot Ad$$

$$\left(\frac{\hat{\pi}}{1-\hat{\pi}}\right) = e^{-1.12} \cdot e^{0.78 \cdot Joke} \cdot e^{-0.34 \cdot Ad}$$

$$\hat{\pi} = \frac{e^{-1.12+0.78 \cdot Joke-0.34 \cdot Ad}}{1+e^{-1.12+0.78 \cdot Joke-0.34 \cdot Ad}}$$

For the *None* group (the indicator that was omitted from the model) the logit form of the model with $Joke = Ad = 0$ gives the estimated log odds of $\hat{\beta}_0 = -1.12$. Exponentiating yields the estimated odds of getting a tip with no card as $e^{-1.12} = 0.326$ and from the probability form of the fitted model (11.1), the estimated probability is $\hat{\pi} = 0.326/1.326 = 0.246$.

For the *Joke* group the fitted equation indicates that the log odds for the *Joke* group can be found by increasing the log odds for *None* by $\hat{\beta}_1 = 0.78$. Exponentiating to get to the odds scale implies that we multiply the odds for *None* $e^{-1.12} = 0.326$ by $e^{0.78} = 2.181$ to get the estimated odds for *Joke*, that is $0.326 * 2.181 = 0.711$. Nonetheless, it is the odds ratio that is often of interest with logistic regression and it was easily determined by exponentiating the coefficient. Here, the odds of getting a tip with a joke is $e^{0.78} = 2.181$ times the odds of getting a tip with no card. In addition, we can use the probability form (11.1) of the fitted equation on the right to obtain the predicted probability of a tip with a joke to be $\hat{\pi} = 0.711/1.711 = 0.416$.

The fitted equation also provides an estimate of the odds ratio for *Ad* ($e^{-0.34} = 0.712$). To estimate the probability of a tip with an *Ad*, first compute $\log\left(\frac{\hat{\pi}}{1-\hat{\pi}}\right) = -1.12 - .34 = -1.46$. Exponentiating we get $\left(\frac{\hat{\pi}}{1-\hat{\pi}}\right) = e^{-1.46} = .232$. Solving we get $\hat{\pi} = 0.232/1.232 = 0.188$. Note that, as we saw earlier with a 2×2 table, the estimated probabilities from the logistic model with indicator predictors match (up to round off) the sample proportions of tips in each group from the 2×3 table.

11.4. APPLYING LOGISTIC REGRESSION TO LARGER TABLES

From the computer output, we see that the indicator variable for the joke card has a small P-value (.036) and a positive coefficient (0.783), so we can conclude that the joke card does have a significantly higher success rate than the reference (no card) group. But the p-value for advertisement card is not small (0.416), so these data provide insufficient evidence that an advertisement card leads to a higher rate of tipping than no card at all. From the estimated odds ratio for the *Joke* indicator, we can conclude that the odds of receiving a tip are roughly doubled when a joke is presented with the bill. Because this study was a randomized experiment, we can conclude that the joke card caused an increase in the tipping rate.

◇

Chi-square test for a $2 \times k$ table (review)

We saw in Section 9.5 that a binary logistic regression with a single binary predictor was closely related to a chi-square test for a 2×2 table (as well as to a z-test for the difference in two proportions). You should not be surprised that this parallel extends to a chi-square test for a $2 \times k$ table by using logistic regression with $k - 1$ indicator variables. First we review the basics of the chi-square test for these larger tables.

Chi-square test for a $2 \times k$ table

Start with data of observed counts (O_{ij}) in a $2 \times k$ table. Compute an *expected* count for each cell using $E_{ij} = \frac{(RowTotal) \cdot (ColumnTotal)}{n}$.

Response	Group #1	Group #2	...	Group # k	Total
0=No	O_{01} (E_{01})	O_{02} (E_{02})	...	O_{0k} (E_{0k})	# No
1=Yes	O_{11} (E_{11})	O_{12} (E_{12})	...	O_{1k} (E_{1k})	# Yes

The test statistic compares the *observed* and *expected* counts within each of the cells of the table and sums the results over the $2k$ cells.

$$X^2 = \sum \frac{(Observed - Expected)^2}{Expected}$$

If the sample sizes are not small (expected counts at least 5) the test statistic under H_0 follows a chi-square distribution with $k - 1$ degrees of freedom.

Note: This procedure works the same way for a $k \times 2$ table where the binary response determines the columns.

The expected counts are just the frequencies we would see if the counts in each cell exactly matched the overall proportion of success and failures for the entire sample. This is essentially the "constant" model that we use as a base for comparison when computing the G statistic to test the overall fit of a logistic regression model. Indeed, while not identical, the results of the chi-square test for a $2 \times k$ table are very similar to the overall G test for a model with $k-1$ indicator variables as predictors.

Example 11.6: *Joking for a tip*

The Minitab output for a chi-square test on the table of results from the tipping experiment is shown below, along with the G-test of $H_0: \beta_1 = \beta_2 = 0$ for the model that uses *Joke* and *Ad* indicators in predicting *Tips*. Note that setting both of those coefficients equal to zero in the model is equivalent to saying that the tip proportion should be the same for each of the three groups (in this case estimated as $60/211 = 0.284$)—which is the same assumption that generates the expected counts in the chi-square test for the table.

Minitab output for a chi-square test of the 2x3 table:

```
Rows: Tip    Columns: Card

         Ad    Joke    None     All
0        60      42      49     151
      52.96   51.53   46.52  151.00

1        14      30      16      60
      21.04   20.47   18.48   60.00

All      74      72      65     211
      74.00   72.00   65.00  211.00

Cell Contents:         Count
                       Expected count

Pearson Chi-Square = 9.953, DF = 2, P-Value = 0.007
```

Minitab output for logistic regression to predict *Tips* based on *Joke* and *Ad*:

```
                                              Odds      95% CI
Predictor       Coef   SE Coef       Z      P Ratio  Lower  Upper
Constant    -1.11923  0.287938   -3.89  0.000
Joke         0.782759  0.374234    2.09  0.036  2.19   1.05   4.56
Ad          -0.336056  0.413526   -0.81  0.416  0.71   0.32   1.61
```

11.4. APPLYING LOGISTIC REGRESSION TO LARGER TABLES

```
Log-Likelihood = -121.070
Test that all slopes are zero: G = 9.805, DF = 2, P-Value = 0.007
```

We see that the chi-square statistics for the two procedures are quite similar (9.953 and 9.805) with the same degrees of freedom (2) and essentially the same P-value (0.007). So the tests are consistent in showing that the tip rate differs depending on the waiter's action (*Joke*, *Ad*, or *None*). To see the nature of this difference in the chi-square test we see that the observed number of tips in the joke group (30) is quite a bit higher than expected (20.47), while the observed counts for both the advertising and no card groups are lower than expected. This is consistent with the coefficients we observed in the logistic model, which showed that the odds of a tip were significantly better for the joke group, but not much different between the advertising and no card groups. So if you happen to find yourself waiting on tables at a coffee bar in western France, you probably want to learn some good jokes!

◇

Logistic versus Chi-square: The consistency between the logistic and chi-square results is pleasing and you might wonder whether it is worth the trouble to use logistic regression when a chi-square test is so easy to understand and perform. There are a couple of reasons in favor of logistic regression. First, as you saw in earlier examples, logistic regression can handle categorical and continuous predictors. The $2 \times k$ tables are based only on categorical data. Second, you have have been presented several different tests to compare models. In particular, you've seen the Pearson chi-square test statistic for the $2 \times k$ tables, the Wald test for comparing models which differ by a single coefficient, and the likelihood ratio test (LRT, the comparison of likelihoods for two models in logistic regression sometimes referred to as G). Each of these depends upon an approximation and it is generally accepted that the LRT is preferred. Finally, the logistic approach to $2 \times k$ tables allows you to assess the statistical significance for values of the categorical variable in a way the chi-square approach does not. For example, in the tipping example, the logistic analysis gave us the power to say that including a joke, or not, had a statistically significant effect on tipping, whereas including an ad, or not, did not. The chi-square analysis only establishes a global statistical significance of an association between type of card and tipping behavior, with further significance requiring some kind of inferential follow-up procedures.

Factors: Many statistical software packages will automatically handle the creation of indicator variables to use in a regression model. For example, in Minitab's binary logistic regression procedure we can list a variable as a "factor" as well as a term in the model. When this occurs Minitab creates the requisite indicator variables for each category in the variable, chooses one to leave out and puts the rest in the model. This also allows us to have a predictor with values coded alphabetically (such as 'M' and 'F' rather than a more cryptic 0 and 1 for gender). In R this is accomplished with the function `as.factor( )`. For example, if X is a variable that can have any of the value 1, 2, 3 or 4, a model with Y ∼ X would treat X as a quantitative predictor, while Y ∼ as.factor(X) would include indicator variables for three of the four categories.

11.5 Summary

This chapter introduces a variety of topics in logistic regression beyond the basics of fitting and interpreting a model for your own data analysis or appreciating the use of a logistic regression that you might encounter in a scientific report.

Probably the most important idea to take away from this chapter is that there are methods of assessing the quality of a logistic regression model or choosing a good set of predictors, just as there are in regular OLS regression. The logistic assessment methods are somewhat less driven by diagnostic plots and more driven by comparing models using the full-and-reduced significance test. In logistic regression, this test uses a chi-square distribution and a change in log-likelihood, rather than an F distribution and a change in error sums of squares.

We introduced in this chapter: (1) the role of regression diagnostics in assessing model assumptions, (2) how logistic regression gives an alternative to the chi-square test for analyzing a $2 \times k$ contingency tables, (3) randomization tests in the logistic setting, (4) reasons for lack of fit, including the topic of overdispersion and how to adjust the model fitting to it, and (5) summary statistics for the quality of the model's ability to predict outcomes.

11.6 Exercises

Conceptual Exercises

11.1 *Two kinds of deviance.* In the archery example (Example 11.5), we gave logistic output from R that models improvement on attendance.

  a. Write down the full and reduced models corresponding to the null deviance, using notation related to the context of these data.

  b. Write down the full and reduced models corresponding to the residual deviance, using notation related to the context of these data.

11.2 *Related significance tests.* Again looking at the archery example, consider the model of improvement on attendance, the R output for which we included in the text.

  a. Describe a test based on one of the deviances that tests the same hypothesis as the one tested by the z statistic of -0.361.

  b. Do the two tests in part (a) necessarily lead to the same p-values? Explain.

  c. Explain why, in this example, you might not want to trust either of the tests referred to in (b).

  d. What does the text suggest for the remedy to the situation described in (c)?

11.3 *Tipping study.* Consider Example 11.6, the tipping study, where we created three indicator variables for the type of card. Answer the following questions using the results given in the text, but NOT using any computer software.

  a. Write down the linear logit equation that includes *Joke* and *None* as the two explanatory variables.

  b. Write down the linear logit equation that includes *Ad* and *None* as the two explanatory variables.

11.4 *The "other" putting data.* Consider Example 11.3, the summary table, and the R output given there.

  a. Which rows of the summary table are compared by the residual deviance?

  b. The null deviance compares the row of logistic model $\hat{\pi}$'s to something. What is that thing?

  c. Compute a p-value related to the null deviance and explain what other piece of the output tests the same hypotheses.

11.6. EXERCISES

d. Do the two p-values discussed in (c) lead to the same conclusion? Are they guaranteed to always do so?

e. What piece of the R output does the text suggest should lead to our considering an overdispersion, quasi-likelihood fit?

Guided Exercises

11.5 *Lost letters again.* Consider the data file **LostLetter** from Exercise 9.25, which described a study of intentionally losing letters to ascertain how different addresses affect return rates.

a. Do a chi-square analysis to decide if there is a relationship between *Location* and *Returned*. Report a p-value and state a conclusion. If there is a significant relationship, explore the tables of observed and expected counts and describe the nature of the relationship in simple terms.

b. Do a logistic model of Returned on Location. Describe the statistical significance of the coefficients and interpret the results in the context of the data set.

c. Use the residual deviance for the logistic model to test the goodness of fit of the model fit in (b). State a conclusion, making clear what two models you are comparing.

d. Redo the chi-square analysis of part (a) separately for both addresses: Iowa Peaceworks and Friends of the Confederacy. Summarize your findings.

e. Now, fit a logistic model that models Returned on Location and Address. Summarize your findings.

f. What advantages or disadvantages does the logistic analysis of part (e) have over the chi-square analysis of part (d)?

g. Use the residual deviance to test for goodness of fit for the logistic model.

h. Write down explicitly the empirical proportions for the saturated model being considered by the goodness of fit test and the corresponding linear logit estimated probabilities that these saturated logits are being compared to.

i. Does the output suggest the need to fit an overdispersion model here? Explain.

11.6 *Bird nests revisited.* In the "Case Study: Bird Nests" from Chapter 10, we landed on a simple additive model with linear terms for the length of nest use and the size of the bird species. Assess the predictive power of this model using the "percentage correctly classified" approach of this chapter. Try it two different ways:

a. using 0.5 as the cut-off point.

b. using $0.313 = 26/83$ as the cut-off point—this being the overall sample proportion of closed nests.

c. Then discuss the affectiveness of this model to predict a closed nest and discuss the relative merits of the two choices of cut-off.

Open-ended Exercises

11.7 *Field goals in the NFL.*[7] Data in **FGByDistance** summarize all field goals attempted by place kickers in the National Football League during regular season games for the 2000 through the 2008 seasons. We derived this data set from a larger data set where each case represents one attempted kick, and includes information on the game, date, and kicker, team of the kicker, and whether the kick was made, missed, or the kick was blocked. **FGByDistance** summarizes the 8520 attempted kicks by distance. Distances of kicks ranged from 18 yards to 76 yards. The minimum possible kick distance is 18 yards—because the shortest possible kick is from the opponent's 1-yard line, when the kicker would stand at the 8-yard line, and then another 10 yards would be added to the distance of the kick because of the 10-yard depth of the end zone.

The variables for FGByDistance are:

Dist	The distance of the attempted kick from the goal posts
N	The number of attempted kicks, of the 8520, from that distance
Makes	The number of those attempted kicks that were successful
PropMakes	The proportion of successes
Blocked	The number of those attempted kicks that were blocked
PropBlocked	The proportion of attempted kicks that were blocked

The first 10 cases of the data set are printed here:

```
   Dist  N  Makes PropMakes Blocked PropBlocked
1    18  15    15 1.0000000       0 0.000000000
2    19 107   105 0.9813084       1 0.009345794
3    20 204   201 0.9852941       1 0.004901961
4    21 197   194 0.9847716       0 0.000000000
5    22 244   239 0.9795082       2 0.008196721
6    23 298   291 0.9765101       2 0.006711409
7    24 241   233 0.9668050       1 0.004149378
8    25 211   202 0.9573460       5 0.023696682
9    26 237   226 0.9535865       1 0.004219409
10   27 238   226 0.9495798       1 0.004201681
```

[7] We thank Sean Forman and Doug Drinen of Sports Reference LLC for providing us with the NFL field goal data set.

11.6. EXERCISES

a. Create a model to describe the relationship between the probability of a make as the response variable and the distance of the attempted kick as the explanatory variable. Report your findings, including a summary of your model, a description of your model-building process, and relevant computer output and graphics.

b. Repeat the previous exercise, except using probability of a kick being blocked as the response variable.

11.8 *Small sample of NFL field goals* We extracted 30 cases at random from the 8520 cases in the NFL field goal data described in Exercise11.7. The data are in **SampleFG**. Suppose we had only this much data and wanted to model the probabilty of a make as a function of distance.

a. Create such a model using the maximum likelihood approach to fit a linear logistic model with distance as the one predictor. Report your results.

b. Repeat part (a), except using a randomization test.

c. Compare the approaches in (a) and (b), especially with respect to the significance of the relationship of probability of make to distance. At the common significance levels such as 5 or 1 percent, would conclusions be the same using either approach? Do you have any reason to prefer one approach to the other?

The data are printed here:

ID	Yards	Results2
5255	35	1
1802	51	1
7941	42	1
5836	44	1
7168	28	1
3763	41	1
5256	22	1
7445	48	0
2992	43	0
6523	19	1
8285	33	1
7612	27	1
6659	48	0
1505	40	1
2190	26	0
5372	36	1
226	51	0
4332	51	1

2196	36	1
2387	33	1
5319	49	0
5133	22	1
5919	45	0
4486	44	1
5402	52	0
4134	48	1
1198	20	1
1212	24	1
8484	26	1
6127	43	0

General Index

added variable plot, 145
adjusted R^2, 89
ANCOVA
    conditions, 341
    model, 340
ANOVA
    SSE, oneway, 205
    Bonferroni adjustment, 307
    comparisons, one-way, 227, 314
    conditions, 202
    contrasts, one-way, 314
    equal variance condition, 211
    error rates, 227, 306
    F-ratio, 209
    F-test, one-way, 210
    Fisher's LSD, one-way, 229, 307
    Fisher's LSD, two-way, 262
    independence condition, 212
    inference about cause, 219
    inference about populations, 219
    interaction graph, 265
    interaction, two-way, 263
    mean zero condition, 212
    model, one-way, 201
    model, two-way additive, 251
    model, two-way with interaction, 272
    MSE, oneway, 205
    multiple regression, 85
    normality condition, 210
    one-way as regression, 327
    parameter estimates, 203
    partition variability, 205
    randomization, 218
    residual, 203
    scope of inference, one-way, 218
    scope of inference, two-way, 257
    simple linear regression, 62, 63
    SSE, 207
    SSGroups, 207
    SSTotal, 207
    table, one-way, 209
    table, two-way additive, 256
    table, two-way with interaction, 277
    transformations, 214
    Tukey's HSD, 308
    two-way as regression, 330

blocking in experiments, 383
    blocks by grouping, 386
    blocks by subdividing, 384
boostrap
    distribution, 182
bootstrap
    confidence interval, 185
    regression slope, 182
Box, George, 2

c-statistic, 494
coefficient of determination, R^2, 63, 87
compare two regression lines, 90
    intercepts, 94
    slopes, 97
comparisons, 227, 314
    standard error, 317
    t-test, 318
complete block design, 385
complete factorial design, 276
complete second order model, 109
contrasts, 314
    standard error, 317
    t-test, 318
Cook's distance, 166
correlated predictors, 111
correlation
    t-test, 64
covariate, 6, 338
cross validation, 189

deviance, 464
    drop in deviance test, 487
    residual deviance, 486

empirical logit plot, 420
empirical logits, 419
error rates, 227
    family-wise error rate, 228, 306
    individual error rate, 228
experiment, 6
    comparisons, 362
    experimental units, 369
    factorial crossing, 390
    randomization, 363
    randomized comparative design, 369
    repeated measures block design, 384
    subjects, 6
    treatment, 369
    units, 6
explanatory variable, 413

F-distribution, 62
four-step process, 7

goodness of fit, 484
    logistic regression, 483

indicator variable, 92
    multiple categories, 169
interaction
    ANOVA, two-way, 263
    multiple regression, 101
interaction graph, 265

least squares line, 27
Levene's test, 301
leverage, 50, 160
likelihood ratio test, 432, 433, 462, 470
logistic regression, 430
    lack of fit, 489
    model assessment, 428, 493, 494
    model conditions, 427
    multiple logistic regression model, 453, 454
    randomization tests, 498
    replicate x-values, 419
    significance tests for coefficients, 429
    two-way tables, 434, 502
logistic regression model, 415
    logit form, 415
    probability form, 415
logit, 415
logit function, 414

maximum likelihood, 416, 431
mean square error, MSE, 62
mean square model, 62
multicollinearity, 113
multiple linear regression, 79
    adjusted R^2, 89
    ANOVA, 85
    choosing predictors, 149
    coefficient of determination, R^2, 87
    confidence interval for coefficients, 85
    confidence interval for mean response, 89
    correlated predictors, 111
    fit, 82
    leverage, 160
    model, 81
    nested F-test, 119
    prediction interval, 89
    standard error of regression, 83
    t-test for coefficients, 84

nested F-test, 119
normal plot, 33
    ANOVA, 210
    probability plot, 33
    quantile plot, 33
normal plot:probability plot, 10

observational study, 6
observational units, 3
odds, 419, 421
    relationship to probability, 422
odds ratio, 422
overdispersion, 490

parameter, 6
partition variability
    ANOVA, one-way, 205
    simple linear regression, 60
polynomial regression, 104, 109

GENERAL INDEX

population, 6
predictor selection
    backward elimination, 154
    best subsets, 151
    cautions, 159
    stepwise, 157

quadratic regression, 105
quasi-likelihood, 491

randomization distribution, 179, 374
randomization test
    correlation, 178
    one-way ANOVA, 373, 380
randomized comparative experiment, 369
regression models
    compare two lines, 90
    complete second order, 109
    indicators for multiple categories, 170
    interaction, 101
    polynomial, 104, 109
    quadratic, 105
residual, 26
    ANOVA, one-way, 203
    deleted-t, 46, 165
    standardized, 46, 165
    studentized, 46, 165
residual versus fits plot, 31
    ANOVA, 211
response variable, 413

sample, 6
saturated model, 484
simple linear regression, 23
    ANOVA, 62, 63
    coefficient of determination, r^2, 63
    conditions, 28
    confidence interval for individual response, 66
    confidence interval for mean response, 66
    confidence interval for slope, 60
    influential points, 47
    least squares line, 27
    model, 25
    MSE, 62
    MSModel, 62
    outliers, 44
    partition variability, 60
    regression standard error, 30
    residual, 26
    SSE, 27
    standard error of slope, 58
    t-test for slope, 58
    transformations, 37
SSE, 27
    ANOVA, one-way, 205
statistic, 6
statistical model, 1, 3
Studentized range distribution, 309

t-distribution, 59
tests for coefficients
    multiple logistic regression, 460
transformations, 37
    ANOVA, one-way, 214
    log, 43
    square root, 40
two-sample t-test, 10
    pooled, 324

variables, 3
    binary, 4
    categorical, 4
    explanatory, 5
    indicator, 92
    quantitative, 4
    response, 5
variance inflation factor, VIF, 116

Wald statistic, 430

Dataset Index

Alfalfa, 294
ArcheryData, 498
AutoPollution, 247, 354, 356

Backpack, 449, 479
BaseballTimes, 55, 74, 189
BirdNest, 468
Blood1, 193, 245, 351, 354
BritishUnions, 137

CAFE, 454, 462, 493
CancerSurvival, 220, 355
Cereal, 72, 190
ChildSpeaks, 72
Clothing, 122
CO2, 144

Day1Survey, 19
Diamonds, 104, 117, 139, 214
Diamonds2, 214

Election08, 447, 473

FantasyBaseball, 242, 352
FGByDistance, 512
Film, 450, 480
FinalFourIzzo, 478
FinalFourLong, 477
FinalFourShort, 477
FirstYearGPA, 141, 190, 445
FishEggs, 135
FlightResponse, 445, 477
FruitFlies, 199, 324, 328, 353

Grocery, 338, 358

Hawks, 224, 245, 247
HighPeaks, 188, 191, 194
Hoops, 481
HorsePrices, 75, 358
Houses, 114, 145

ICU, 474
InfantMortality, 75
InsuranceVote, 478

Jurors, 91, 136

Kids198, 97

Leukemia, 444, 474
LongJumpOlympics, 44
LostLetter, 450, 480, 511

Marathon, 19
Markets, 446, 473
MathPlacement, 479
MedGPA, 414, 417, 425, 428, 442, 460, 488
MetroHealth83, 4, 37, 142
MLB2007Standings, 134, 143, 189

NCBirths, 192
NCbirths, 245, 351, 352
NFLStandings2007, 79

Olives, 244, 355
Overdrawn, 481

PalmBeach, 47, 161
Perch, 102, 111, 162, 166, 195
PigFeed, 264, 274, 276, 330, 332
Political, 451
Pollster08, 141
Popcorn, 295
PorscheJaguar, 143
PorschePrice, 23, 58, 182, 195
Pulse, 344
Putts1, 418, 424, 433

ReligionGDP, 191
RiverElements, 259

SampleFG, 513
SandwichAnts, 334

DATASET INDEX

SATGPA, 193, 195
SeaSlugs, 247, 352, 353, 356
SpeciesArea, 42
Speed, 137, 143
Swahili, 296, 351

TextPrices, 54
ThreeCars, 169, 358
Titanic, 443, 472, 476
TomlinsonRush, 73

Volts, 54

WeightLossIncentive, 7, 18, 357

YouthRisk2007, 449, 479